VOLUME ONE HUNDRED AND NINETEEN

VITAMINS AND HORMONES

Vitamin B12

VOLUME ONE HUNDRED AND NINETEEN

VITAMINS AND HORMONES

Vitamin B12

Series Editor

GERALD LITWACK, PhD
Toluca Lake, North Hollywood, California

ELSEVIER

ACADEMIC PRESS
An imprint of Elsevier

Academic Press is an imprint of Elsevier
50 Hampshire Street, 5th Floor, Cambridge, MA 02139, United States
525 B Street, Suite 1650, San Diego, CA 92101, United States
The Boulevard, Langford Lane, Kidlington, Oxford OX5 1GB, United Kingdom
125 London Wall, London, EC2Y 5AS, United Kingdom

First edition 2022

ISBN: 978-0-323-99223-7
ISSN: 0083-6729

For information on all Academic Press publications
visit our website at https://www.elsevier.com/books-and-journals

Publisher: Zoe Kruze
Acquisitions Editor: Leticia M. Lima
Developmental Editor: Federico Paulo S. Mendoza
Production Project Manager: Vijayaraj Purushothaman
Cover Designer: Greg Harris

Typeset by STRAIVE, India

Contents

Contributors

David H. Alpers
Department of Internal Medicine, Division of Gastroenterology, Washington University School of Medicine, St Louis, MO, United States

Joris Beld
Department of Microbiology and Immunology, Center for Advanced Microbial Processing, Institute for Molecular Medicine and Infectious Disease, Drexel University College of Medicine, Philadelphia, PA, United States

T. Bito
Division of Applied Bioresource Chemistry, United Graduate School of Agricultural Sciences, Tottori University; Department of Agricultural, Life and Environmental Sciences, Faculty of Agriculture, Tottori University, Tottori, Japan

Olivier Cases
Centre de Recherche des Cordeliers, Sorbonne Université, INSERM, Université de Paris, Paris, France

Santosh K. Chaturvedi
Leicestershire Partnership NHS Trust, Leicester, United Kingdom

Montserrat Elías-Arnanz
Departamento de Genética y Microbiología, Área de Genética (Unidad Asociada al IQFR-CSIC), Facultad de Biología, Universidad de Murcia, Murcia, Spain

Arghya Pratim Ghosh
Department of Chemistry, University of Louisville, Louisville, KY, United States

Hadi Goubran
College of Medicine, University of Saskatchewan & Saskatoon Cancer Centre, SK, Canada

Ralph Green
Department of Pathology and Laboratory Medicine, University of California Davis Medical Center, Sacramento, CA, United States

Jean-Louis Guéant
University of Lorraine, INSERM UMR_S 1256, Nutrition, Genetics, and Environmental Risk Exposure (NGERE), Faculty of Medicine of Nancy; Department of Biochemistry, Molecular Biology, Nutrition, and Metabolism; Department of Hepato-Gastroenterology, University Hospital of Nancy; Université de Lorraine, Inserm, UMRS 1256, NGERE—Nutrition, Genetics, and Environmental Risk Exposure; CHRU-Nancy, National Center of Inborn Errors of Metabolism, Nancy, France

Rosa-Maria Guéant-Rodriguez
University of Lorraine, INSERM UMR_S 1256, Nutrition, Genetics, and Environmental Risk Exposure (NGERE), Faculty of Medicine of Nancy; Department of Biochemistry, Molecular Biology, Nutrition, and Metabolism, University Hospital of Nancy, Nancy, France

Luciana Hannibal
Laboratory of Clinical Biochemistry and Metabolism, Department of General Pediatrics, Adolescent Medicine and Neonatology, Faculty of Medicine, Medical Center—University of Freiburg, Freiburg, Germany

Donald W. Jacobsen
Department of Cardiovascular and Metabolic Sciences, Lerner Research Institute, Cleveland Clinic, Cleveland, OH, United States

Daniel S. Kantner
Department of Microbiology and Immunology, Center for Advanced Microbial Processing, Institute for Molecular Medicine and Infectious Disease, Drexel University College of Medicine, Philadelphia, PA, United States

K. Koseki
Division of Applied Bioresource Chemistry, United Graduate School of Agricultural Sciences, Tottori University, Tottori, Japan

Pawel M. Kozlowski
Department of Chemistry, University of Louisville, Louisville, KY, United States

Renata Kozyraki
Centre de Recherche des Cordeliers, Sorbonne Université, INSERM, Université de Paris, Paris, France

Bernhard Kräutler
Institute of Organic Chemistry and Centre of Molecular Biosciences, University of Innsbruck, Innsbruck, Austria

B.S. Lickey
Department of Chemistry, University of Vermont, Burlington, VT, United States

M.D. Liptak
Department of Chemistry, University of Vermont, Burlington, VT, United States

Wen-Sui Lo
Max Planck Institute for Biology Tübingen, Tübingen, Germany

Amy T. Ma
Department of Microbiology and Immunology, Center for Advanced Microbial Processing, Institute for Molecular Medicine and Infectious Disease, Drexel University College of Medicine, Philadelphia, PA, United States

Jose M. Martínez Felices
University of Groningen, Faculty of Science and Engineering, Groningen Biomolecular Sciences and Biotechnology, Membrane Enzymology Group, Groningen, Netherlands

Joshua W. Miller
Department of Nutritional Sciences, Rutgers University, New Brunswick, NJ, United States

J.A. Morris
Department of Chemistry, University of Vermont, Burlington, VT, United States

Mark Nijland
University of Groningen, Faculty of Science and Engineering, Groningen Biomolecular Sciences and Biotechnology, Membrane Enzymology Group, Groningen, Netherlands

Lucía Osete-Alcaraz
Departamento de Genética y Microbiología, Área de Genética (Unidad Asociada al IQFR-CSIC), Facultad de Biología, Universidad de Murcia, Murcia, Spain

S. Padmanabhan
Instituto de Química Física "Rocasolano", Consejo Superior de Investigaciones Científicas, Madrid, Spain

Ricardo Pérez-Castaño
Departamento de Genética y Microbiología, Área de Genética (Unidad Asociada al IQFR-CSIC), Facultad de Biología, Universidad de Murcia, Murcia, Spain

María Carmen Polanco
Departamento de Genética y Microbiología, Área de Genética (Unidad Asociada al IQFR-CSIC), Facultad de Biología, Universidad de Murcia, Murcia, Spain

Grégory Pourié
Université de Lorraine, Inserm, UMRS 1256, NGERE—Nutrition, Genetics, and Environmental Risk Exposure, Nancy, France

Guruvaiah Praveen
Department of Biochemistry, ICMR-National Institute of Nutrition, Hyderabad, India

Edward V. Quadros
Department of Medicine, SUNY Downstate Medical Center, Brooklyn, NY, United States

Gaafar Ragab
Internal Medicine Department, Rheumatology and Clinical Immunology Unit, Faculty of Medicine, Cairo University, Egypt

G. Bhanuprakash Reddy
Department of Biochemistry, ICMR-National Institute of Nutrition, Hyderabad, India

David S. Rosenblatt
Department of Human Genetics, McGill University, The Research Institute of the McGill University Health Centre, Montreal, QC, Canada

Waleed Sabry
College of Medicine, University of Saskatchewan & Saskatoon Cancer Centre, SK, Canada

Prashant Sahu
Prospect Park Hospital, Berkshire Healthcare NHS Foundation Trust, Reading, Berkshire, United Kingdom

Hisashi Shimakoshi
Department of Chemistry and Biochemistry, Graduate School of Engineering, Kyushu University, Fukuoka, Japan

Mudili Sivaprasad
Department of Biochemistry, ICMR-National Institute of Nutrition, Hyderabad, India

Dirk J. Slotboom
University of Groningen, Faculty of Science and Engineering, Groningen Biomolecular Sciences and Biotechnology, Membrane Enzymology Group, Groningen, Netherlands

Ralf J. Sommer
Max Planck Institute for Biology Tübingen, Tübingen, Germany

Chancievan Thangaratnarajah
University of Groningen, Faculty of Science and Engineering, Groningen Biomolecular Sciences and Biotechnology, Membrane Enzymology Group, Groningen, Netherlands

Harish Thippeswamy
National Institute of Mental Health & Neurosciences, Bangalore, India

Megan J. Toda
Department of Chemistry, University of Louisville, Louisville, KY, United States

Pierre Verroust
Centre de Recherche des Cordeliers, Sorbonne Université, INSERM, Université de Paris, Paris, France

Angela Yee-Moon Wang
Department of Medicine, Queen Mary Hospital, The University of Hong Kong, Pokfulam, Hong Kong, China

F. Watanabe
Division of Applied Bioresource Chemistry, United Graduate School of Agricultural Sciences, Tottori University; Department of Agricultural, Life and Environmental Sciences, Faculty of Agriculture, Tottori University, Tottori, Japan

David Watkins
Department of Human Genetics, McGill University, The Research Institute of the McGill University Health Centre, Montreal, QC, Canada

Henry H.L. Wu
Department of Renal Medicine, Lancashire Teaching Hospitals NHS Foundation Trust, Preston; Faculty of Medical and Human Sciences, University of Manchester, Manchester, United Kingdom

Preface

The naturally occurring form of vitamin B12 is methylcobalamin. The structure of this vitamin, synthetized by bacteria (but not by plants or humans), is the most complex of the B vitamins. Bacteria that are utilized for commercial production of the vitamin are *Pseudomonas denitrificans*, *Propionibacterium freudenreichii*, and *Bacillus megaterium*. There are bacteria in the colonic flora that can synthesize vitamin B12, but the colon is located far below the small intestine where most of the absorption of the vitamin occurs and, consequently, it is not considered to be a utilizable source in humans. Methylcobalamin is a coenzyme of methionine synthase (most of the vitamins function generally in the body as coenzymes). Methionine synthase is needed for methylation of DNA and proteins. Methylcobalamin binds to the intrinsic factor, a glycoprotein synthesized in parietal cells of the human stomach. The vitamin B12-intrinsic factor complex then binds to its receptor, known as cubilin. Cubilin is a 460 kDa peripheral membrane protein that, in complex with vitamin B12-intrinsic factor in the ileal mucosa, supports its endocytosis into cells. Proteins named transcobalamin I or II, particularly the latter, serve to protect vitamin B12 from acid degradation in the stomach once it has been absorbed by enterocytes of the terminal ileum and the vitamin B12-intrinsic factor complex has been degraded. Cubilin lines the cells of the small intestine primarily (but is also located in kidney cells). Cubilin is anchored to the outer membrane of cells of the small intestine by its attachment to a protein called amnionless.

Gram-negative anaerobic bacteria bind the cobalamin-intrinsic factor complex, and therefore bacterial overgrowth in the small intestine will suppress cobalamin absorption. When the body does not have enough healthy red blood cells due to a deficiency of vitamin B12, it poses a problem in carrying oxygen to the tissues because vitamin B12 is required to make red blood cells.

The recommended dietary allowance for men and women aged 14 years and older is 2.4 μg/day. The requirement increases during pregnancy to 2.6–2.8 μg/day. There is no established toxic level for vitamin B12 owing to its water solubility (a characteristic of the B vitamins) and excretion in the urine. Lack of vitamin B12 or intrinsic factor can cause megaloblastic anemia or pernicious anemia.

Food sources of vitamin B12 are fish, shellfish, liver, red meat, eggs, poultry, dairy products, nutritional yeast, fortified breakfast cereals, and enriched soy or rice milk.

The first part of this volume comprises chapters that deal with the basic aspects of vitamin B12, while the second part deals with studies that are of clinical relevance. This book should appeal to basic scientists interested in vitamin B12 as well as to clinicians.

The cover reflects a new design. The background that fills the cover is based on a portion of a molecule of an engineered enzyme modified by a process called "Foldit." The designed loop in pink stabilizes the ligand in green, thus increasing the catalytic activity of the enzyme. Future covers in this series may utilize other models of proteins, especially those involved in some way with vitamins or hormones. This wonderful series of surface structures of molecules has been developed by David S. Goodsell and is incorporated into the Protein Data Bank (PDB) as "Molecule of the Month." Structures can be obtained in great enough resolution for reproduction by becoming a licensee, and one can find the list of molecules by setting a browser at "pdb Molecule of the Month" and selecting "by title." I am indebted to Dr. Goodsell for making available these amazing surface structures of molecules.

In conclusion, I thank Paulo Mendoza and Leticia Lima of Elsevier for their help with the processing of manuscripts as well as the excellent production group in Chennai, India.

Gerald Litwack
Toluca Lake, North Hollywood, California
January 31, 2022

CHAPTER ONE

Insertion of cobalt into tetrapyrroles

J.A. Morris, B.S. Lickey, and M.D. Liptak*
Department of Chemistry, University of Vermont, Burlington, VT, United States
*Corresponding author: e-mail address: matthew.liptak@uvm.edu

Contents

Abstract

Vitamin B_{12} is one of the most complex cofactors known, and this chapter will discuss current understanding with regards to the cobalt insertion step of its syntheses. Two total syntheses of vitamin B_{12} were reported in the 1970s, which remain two of the most exceptional achievements of natural product synthesis. In subsequent years, two distinct biosynthetic pathways were identified in aerobic and anaerobic organisms. For these biosynthetic pathways, selectivity for Co(II) over other divalent metal ions with similar ionic radii and coordination chemistry remains an open question with three competing hypotheses proposed: metal affinity, tetrapyrrole distortion, and product inhibition. A 20 step biosynthetic route to convert 5-aminolevulinic acid (ALA) to vitamin B_{12} was elucidated in aerobic organisms in the 1990s, where cobalt is inserted relatively late in the pathway by the CobNST multi-protein complex. This chapter includes a mechanistic proposal for this reaction, but the majority of the proposal is based upon analogy to the ChlDHI magnesium chelatase complex as critical data for the cobalt chelatase is lacking. Later, in the 2010s, a distinct 21 step pathway from ALA to vitamin B_{12} was reported in anaerobic organisms, where cobalt is inserted early in the pathway

Vitamins and Hormones, Volume 119
ISSN 0083-6729
https://doi.org/10.1016/bs.vh.2022.01.013

by the enzyme CbiK. A recent study strongly suggests that the cobalt affinity of CbiK is the origin of cobalt selectivity for CbiK, but several important mechanistic questions remain unanswered. In general, it is expected that significant insight into the cobalt insertion mechanisms of CobNST and CbiK could be derived from additional structural, spectroscopic, and computational data.

1. Introduction

1.1 Vitamin B_{12} synthesis

The chemical synthesis of cobalamin was a 12year long *tour de force* by the Eschenmoser and Woodward groups, yet bacteria biosynthesize this complex vitamin. These two groups collaborated to complete two distinct total syntheses of cobyric acid in 1972, which both represent extraordinary achievements in natural product synthesis. One of these syntheses was relatively traditional for the field; synthesis of the "left" side of the corrin ring by the Woodward group, synthesis of the "right" side of the ring by the Eschenmoser group, and a collaborative effort to combine both moieties (Woodward, 1973). The second synthesis of cobyric acid incorporated a novel photocyclization reaction to combine the "top" and "bottom" parts of the corrin ring (Eschenmoser & Wintner, 1977). With regards to these total syntheses, the Nobel laureate Robert Woodward noted "cobaltation of these substances turned out to represent a major problem" (Woodward, 1973). In both cases, the authors noted the importance of using a cobalt halide salt as other forms of cobalt led to destruction of the tetrapyrrole ring through unwanted side reactions. Nevertheless, both of these total syntheses of vitamin B_{12} require approximately 70 synthetic steps, and modern industrial production of vitamin B_{12} instead relies upon biosynthesis.

The first biosynthetic pathway for vitamin B_{12} biosynthesis to be elucidated was an oxygen-dependent pathway where cobalt insertion into the tetrapyrrole occurs relatively late. The complete pathway for aerobic biosynthesis of vitamin B_{12} by *Pseudomonas denitrificans* was reported in 1994 (Battersby, 1994), based upon decades of research by groups from around the world. The aerobic pathway employs 13 distinct enzymatic reactions to convert the first committed precursor to metal tetrapyrroles, ALA, to hydrogenobyrinic acid *a,c*-diamide (HBAD). Next, Co(II) is inserted into HBAD in an ATP-dependent manner by the CobNST complex, yielding cobyrinic acid *a,c*-diamide (CBAD) (Debussche et al., 1992) (Fig. 1). Finally, six additional enzymatic reactions are required to convert CBAD to coenzyme B_{12}, and several of these reactions rely upon organic substrates that

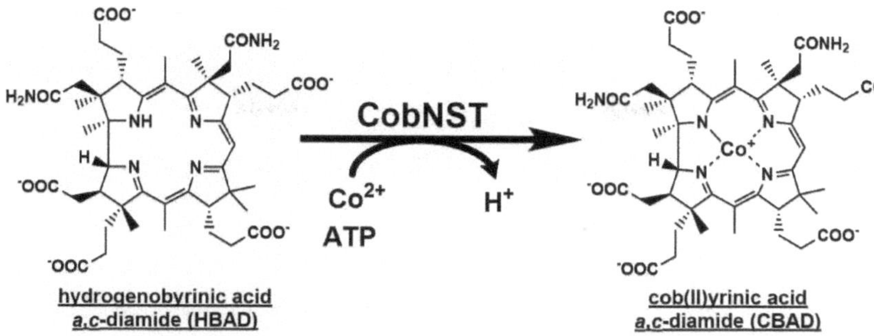

Fig. 1 Cobalt is inserted by the CobNST multi-enzyme complex during aerobic B_{12} biosynthesis. CobNST converts hydrogenobyrinic acid *a,c*-diamide (HBAD) to cobyrinic acid *a,c*-diamide (CBAD) in an ATP-dependent reaction.

themselves are biosynthesized by multi-enzyme pathways (Blanche et al., 1995)! This 20 step biosynthetic pathway from ALA to coenzyme B_{12} is employed by the biotechnology sector to produce tons of vitamin B_{12} on an annual basis *via* bacterial fermentation (Martens, Barg, Warren, & Jahn, 2002). Nevertheless, despite the full elucidation of this pathway in the 1990s and its current use in the biotechnology sector, it is not the only vitamin B_{12} biosynthesis pathway in nature.

Anaerobic organisms employ an oxygen-independent biosynthesis of vitamin B_{12} where cobalt is inserted relatively early in the pathway. The complete pathway for anaerobic B_{12} biosynthesis was elucidated much more recently than the aerobic pathway, in 2013, based upon characterization of *Bacillus megaterium* enzymes overexpressed in *Escherichia coli* (Moore et al., 2013). Prior to Co(II) insertion, just five enzymes of this pathway convert ALA to sirohydrochlorin (SHC). An ATP-independent chelatase, most commonly annotated as CbiK, is responsible for Co(II) insertion into SHC (Raux, Thermes, Heathcote, Rambach, & Warren, 1997) (Fig. 2). Following cobalt insertion, 15 additional enzymatic reactions are required to convert cobalt SHC (Co–SHC) to coenzyme B_{12} *via* the adenosylcobyric acid (ACA) intermediate that is common to both the aerobic and anaerobic pathways. In total, the aerobic and anaerobic pathways of vitamin B_{12} biosynthesis share six common steps, three at the beginning to convert ALA to uroporphyrinogen III and three at the end to convert ACA to coenzyme B_{12}. A comprehensive review of the two biosynthetic pathways was published recently (Bryant, Hunter, & Warren, 2020); this chapter will place particular emphasis on the Co(II) insertion reactions catalyzed by CobNST and CbiK.

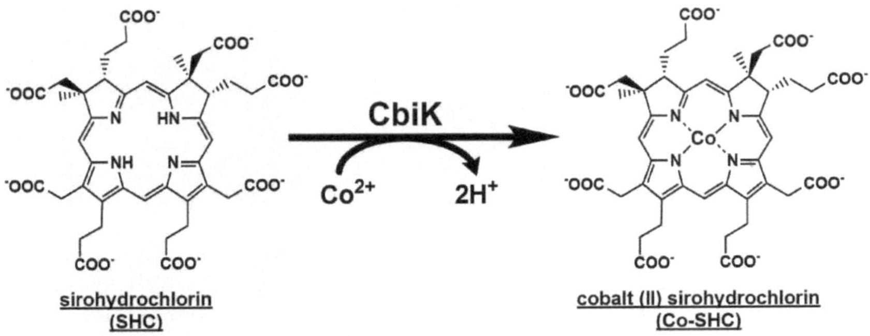

Fig. 2 Cobalt is inserted by CbiK during anaerobic B_{12} biosynthesis. CbiK converts sirohydrochlorin (SHC) to cobalt SHC (Co–SHC) in an ATP-independent reaction.

In both the aerobic and anaerobic pathways, cobalt is inserted into a tetrapyrrole ligand by a chelatase enzyme. In general, chelatase enzymes catalyze metal insertion into an organic tetrapyrrole during chlorophyll, cofactor F_{430}, heme, siroheme, and vitamin B_{12} biosynthesis (Bryant et al., 2020). These enzymes have been divided into three classes based upon their substrate requirements (Brindley, Raux, Leech, Schubert, & Warren, 2003). The CobN enzyme utilized by the aerobic B_{12} biosynthesis pathway is a class I chelatase due to its reliance upon ATP (Debussche et al., 1992), similar to the ChlH magnesium chelatase employed during chlorophyll biosynthesis (Gibson, Willows, Kannangara, von Wettstein, & Hunter, 1995). In contrast, the CbiK chelatase utilized by anaerobic B_{12} biosynthesis is a class II chelatase due to its ATP independence (Raux et al., 1997). Other class II chelatases are utilized for iron or nickel insertion during the biosyntheses of cofactor F_{430} (Zheng, Ngo, Owens, Yang, & Mansoorabadi, 2016), heme (Dailey, 1977), and siroheme (Leech, Raux-Deery, Heathcote, & Warren, 2002). As CobN and CbiK are both members of chelatase classes that utilize more than one metal ion substrate, a major mechanistic challenge for these enzymes is cobalt selectivity.

1.2 Cobalt selectivity

Considering the similarities of the metal ions present in the cell, the selectivity of chelatase enzymes is remarkable. From a classical standpoint where a metal ion is considered a charged sphere, the native metal substrates of chelatases (Mg, Fe, Co, and Ni) are incredibly similar with an ionic charges of 2+ and an ionic radii of 88 ± 4 pm (Shannon, 1976). Even from a viewpoint that considers covalent bonding, three (Fe, Co, and Ni) of the four metal ions have similar coordination chemistry due to their categorization as 3*d* transition metals. The affinity of metalloproteins for divalent transition metals is primarily governed by the Irving-Williams series (Irving & Williams, 1948), which would favor Cu(II) binding by all chelatases in the absence of other factors. Nevertheless, the class I chelatases ChlH and CobN selectively bind Mg(II) and Co(II), respectively (Debussche et al., 1992; Gibson et al., 1995). The class II chelatases CbiK (Raux et al., 1997), CfbA (Zheng et al., 2016), HemH (Dailey, 1977), and SirB (Leech et al., 2002) must distinguish between Fe(II), Co(II), and Ni(II) using similar enzyme active sites (Romão et al., 2011) (Fig. 3). Currently, there are three competing hypotheses in the literature for chelatase metal specificity: the metal affinity of the enzyme active site (Osman et al., 2019), enzyme-induced out-of-plane distortion of the tetrapyrrole substrate (McIntyre, Franco, Shelnutt, & Ferreira, 2011), and product inhibition by mismetallated products (Medlock, Carter, Dailey, Dailey, & Lanzilotta, 2009).

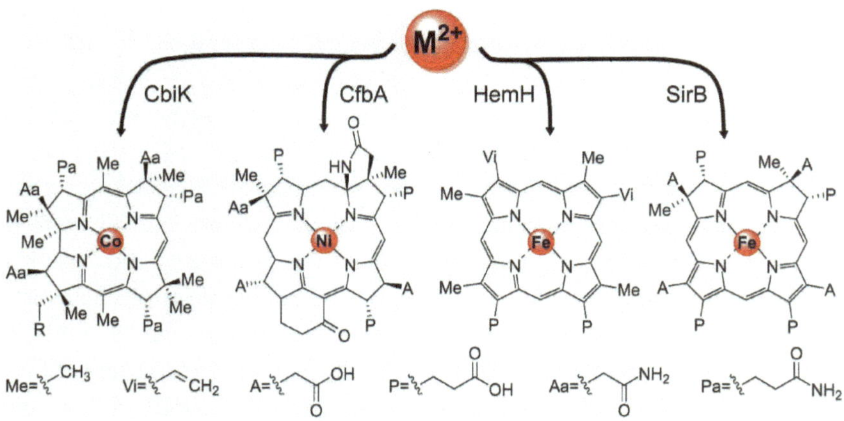

Fig. 3 Class II chelatases catalyze metal insertion within the biosynthetic pathways of heme (top left), siroheme (top right), vitamin B_{12} (bottom left), and cofactor F_{430} (bottom right). HemH and SirB catalyze iron insertion during heme and siroheme biosynthesis, respectively. CbiK catalyzes cobalt insertion into a precursor of vitamin B_{12} and CfbA catalyzes nickel insertion during cofactor F_{430} biosynthesis.

One hypothesis for the remarkable metal selectivity of chelatases is that it is driven by thermodynamics, specifically the metal affinities of the chelatase enzymes. In this proposal, selective metal chelation by the enzyme, prior to catalytic metal insertion into the tetrapyrrole substrate, is responsible for the overall metal selectivity of chelatases. This hypothesis was supported by a recently published article where the metal affinities of all known metal-sensing transcription regulators in *Salmonella enterica* and *S. enterica* CbiK were measured, which revealed that CbiK metal affinity followed the Irving-Williams series (Irving & Williams, 1948; Osman et al., 2019). However, Co(II) was the only metal where CbiK binding was competitive with sensor binding. These data provide a compelling explanation for the cobalt specificity of *S. enterica* CbiK, but it remains an open question whether this thermodynamic argument is generalizable to all chelatases or whether kinetics also have an influence on metal selectivity. If enzyme-bound metal is kinetically inert, then a metal trafficking protein could deliver a metal to a site where a kinetic barrier to dissociation prevents metal exchange even in the presence of a thermodynamic driving force. Thus, more data is needed to evaluate the generality of metal affinity as a driving force for chelatase metal selectivity and to provide satisfactory explanations for why older hypotheses regarding the metal specificity of chelatases are incorrect.

A second hypothesis for the metal selectivity of chelatases is that it is derived from the mode of tetrapyrrole substrate distortion by each chelatase enzyme. An opinion article was written outlining this hypothesis, which suggested that the out-of-plane distortion of the tetrapyrrole substrate by a chelatase enzyme determines which metal ion is inserted into the ring (Al-Karadaghi et al., 2006). Tetrapyrrole out-of-plane distortions have been categorized based upon symmetry into irreducible representations of the D_{4h} point group (Jentzen, Song, & Shelnutt, 1997) (Fig. 4), and a variety of these distortions have been observed in biological systems (Shelnutt et al., 1998). At least as far back as 1989, researchers proposed a role for tetrapyrrole

Fig. 4 There are six out-of-plane deformations of porphyrin from D_{4h} symmetry: propellering (a_{1u}), doming (a_{2u}), ruffling (b_{1u}), saddling (b_{2u}), and waving (e_g). Saddling and ruffling have been observed in chelatases.

distortion in the enzymatic mechanism of iron chelatases where saddling was proposed to drive the pyrrole nitrogen lone pairs out of the tetrapyrrole plane and minimize the kinetic barrier to iron–nitrogen bond formation (Dailey, Jones, & Karr, 1989). Subsequently, resonance Raman evidence was acquired for a saddling deformation of the tetrapyrrole substrate in eukaryotic iron chelatases (Lu et al., 2002), which was later corroborated by X-ray crystal structure data (Medlock, Swartz, Dailey, Dailey, & Lanzilotta, 2007). The observation that the CfbA chelatase induces a ruffling deformation of its tetrapyrrole substrates, instead of saddling, seems to further support the proposal that chelatase metal selectivity is derived from tetrapyrrole distortion (Schuelke-Sanchez, Stone, & Liptak, 2020). However, a global model for how ruffling or saddling would promote selective iron, cobalt, or nickel chelation has not yet been proposed. Thus, the hypothesis that chelatase metal specificity is derived from tetrapyrrole substrate deformation is still under investigation.

A third hypothesis for the metal selectivity of chelatases is that mismetallated products inhibit enzymatic turnover and metal selectivity emerges from removal of unwanted side-products from the labile tetrapyrrole pool. This model yields net selective insertion of the correct metal into the tetrapyrrole substrate by a chelatase enzyme because production of the correct metal–tetrapyrrole product is catalytic, due to the lability of the enzyme-product complex, whereas generation of incorrect metal–tetrapyrrole products are stoichiometric, due to the formation of an inert enzyme-product complex. This story starts in 2007 with the discovery that the rate-limiting step of human ferrochelatase is product release based upon steady state and pre-steady state kinetic measurements (Hoggins, Dailey, Hunter, & Reid, 2007). Later, X-ray crystal structures were acquired for several mismetallated ferrochelatase-product complexes, and based upon the data researchers proposed that chelatase metal specificity is determined by product release (Medlock et al., 2009). However, the cytosol is a more reducing environment than ambient conditions on the benchtop of a laboratory (López-Mirabal & Winther, 2008), and *in vivo* conditions may favor a different product redox state with different dissociation kinetics. In addition, chelatases are part of multi-protein biosynthetic pathways (Bryant et al., 2020), and product release *in vivo* may be driven by product affinity for the next protein in the pathway. Consequently, additional spectroscopic and kinetic data are needed in order to evaluate whether product inhibition determines chelatase metal specificity.

2. Aerobic B_{12} biosynthesis

2.1 CobNST structure

As noted above, Co(II) is inserted into HBAD to yield CBAD during aerobic vitamin B_{12} biosynthesis. The CobNST complex was first isolated from *P. denitrificans* in 1992 (Debussche et al., 1992), but structural information was not available until much later. Electron microscopy data for the CobS and CobT subunits was first published in 2009 (Lundqvist et al., 2009), while the X-ray crystal structure of the CobN subunit was not available until 2020 (Zhang et al., 2020). CobNST is a heterotrimeric, ATP-dependent complex that consists of two functionally distinct subunits: CobN, which is responsible for cobalt chelation, and CobST, which drives CobN activity *via* ATP hydrolysis. The overall architecture bears striking similarities with the magnesium chelatase complex ChlDHI used in the biosynthesis of chlorophyll, as well as the homologous BchDHI complex utilized in bacteriochlorophyll biosynthesis; sequence and structural analyses support an evolutionary linkage between the chelatases (Adams et al., 2020; Antonov, 2020; Reid & Hunter, 2002; Rodionov, Vitreschak, Mironov, & Gelfand, 2003; Walker & Willows, 1997). These ATPase-coupled heterotrimeric chelatases are often referred to as the class I chelatases (Brindley et al., 2003). As the magnesium chelatases have been the subject of a greater volume of research, this relationship remains a crucial point of reference for ongoing studies of CobNST and its chemistry. As will be evident throughout the remainder of this section, the lesser volume of work for CobNST means that a significant portion of what is currently understood about the complex must be extrapolated from studies of magnesium chelatases.

The CobST subunit consists of two ring-shaped hexamers, CobS and CobT, which stack to form a two-tiered barrel-like structure. Both CobS and CobT are trimers of dimers, which are capable of assembling into the active and isolable CobST complex in solution (Lundqvist et al., 2009). Many of the key structural components associated with the ATPase activity of CobST have now been identified. Most notable of these is the αβα core domain structure of CobS containing a phosphate-binding Walker A motif, as well as the catalytically-essential Walker B motif. These features are highly conserved among the AAA+ protein family (Snider, Thibault, & Houry, 2008), and had previously led to similar classifications for the analogous ChlI and BchI subunits of the magnesium chelatases. CobT shares similarities in its C-terminal domain with ChlD and BchD by virtue of a

Mg(II)-binding metal ion-dependent adhesion site (MIDAS). The MIDAS motifs of these subunits are in turn embedded as part of a larger von Willebrand factor type A (vWA) domain, a metal-coordinating domain typically associated with mediating interactions with other vWA-containing proteins. Current understanding of these complexes affixes a predominantly structural role to the larger subunit CobT, while the smaller CobS is understood to be the ATP-hydrolase of the CobNST complex.

The recent publication of the *Mycobacterium tuberculosis* CobN crystal structure has provided valuable insight into this subunit. The N-terminal domains of CobN have been dubbed the Head and Neck domains by analogy to the magnesium chelatase ChlH (Zhang et al., 2020). While the functions of these domains remain unknown, significant conformational differences have been found between these N-terminal domains of CobN and ChlH. The four C-terminal domains constitute two large structures referred to as the Body domains, which contain a number of residues that appear to be of catalytic significance, as well as a loop structure thought to be involved in substrate binding. In ChlH, the two Body domains form a cage-like structure at their interface to yield a central cavity proposed to be the active site (Adams et al., 2020), and the same role is hypothesized for these domains of CobN. Several key residues are conserved within the central cavity, including a Glu residue that has recently been proposed to be the primary catalytic residue involved in both proton abstraction and metal insertion by ChlH. In the CobN structure, this corresponds to Glu561, while two other residues, His522 and Asp1037, are also proposed to be of catalytic significance (Fig. 5). These structural data for the CobNST complex have laid the foundation for detailed studies of the cobalt insertion mechanism.

2.2 CobNST function

The motor subunit CobST is understood to operate similarly to many of its relatives within the AAA+ superfamily, in that it induces conformational changes to the bound CobN subunit through changes in its own conformation caused by hydrolysis of ATP. As is the case with most discourse surrounding the CobNST complex, this understanding of the CobST mechanism remains largely speculative, and is primarily a reflection of current understanding for the magnesium chelatases. Based upon the observation that certain prokaryotes can still synthesize cobalamin without the genes encoding for CobST by utilizing a chimeric system that pairs CobN with the

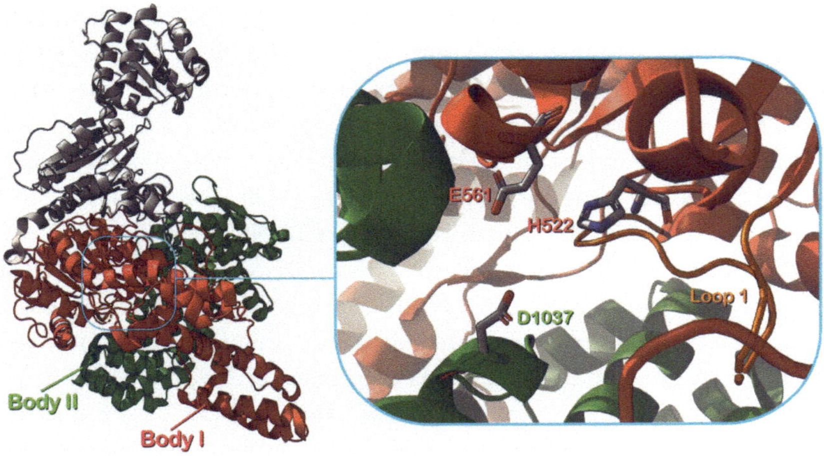

Fig. 5 X-ray crystal structure of *Mycobacterium tuberculosis* CobN featuring the Body I and II domains that enclose the putative active site cavity. The upper, uncolored region consists of the Head and Neck domains. Inset: Close-up view of the proposed catalytic triad, as well as the Loop 1 structure that has been hypothesized to interact with and secure HBAD.

motor units of magnesium chelatases (Antonov, 2020; Rodionov et al., 2003), the catalytically essential aspects of the motor units can most likely be attributed to those highly conserved features between CobST, ChlDI, and BchDI. Notably, a C-terminal MIDAS motif is conserved in the vWA domains of CobT, ChlD, and BchD, which is currently understood to be the primary structural linkage that mediates the chemomechanical coupling between the ATPase and chelatase subunits (Farmer et al., 2019; Lundqvist et al., 2010). Thus, the structural changes effected in the CobS subunit upon ATP hydrolysis are likely to alter the structure of CobT due to their nested geometry, which could in turn transmit a torque onto the bridging MIDAS motif to ultimately cause catalytically-relevant changes to the conformation of CobN. These conformational changes are thought to drive the delivery of the Co(II) ion into the active site cavity, upon which the CobN subunit can catalyze its insertion into the HBAD ligand.

Two recent publications provide synergistic insight into the function of the CobN active site. One of these key publications is the X-ray crystal structure of CobN (Zhang et al., 2020); the other is a detailed structural and mechanistic study of several ChlH variants (Adams et al., 2020). The structural and mechanistic studies of ChlH identified a catalytically-essential Glu residue, Glu660 in *Synechosystis sp.* PCC6830 ChlH, by virtue of the observation that all variants lacking this residue experienced a total loss of enzymatic activity. (For clarity, CobN and ChlH residue numbers will be based upon the genes from *M. tuberculosis* and *S. sp.* PCC6830, respectively, throughout this chapter). Based upon the X-ray crystal structure of CobN, this enzyme contains an analogous Glu residue, Glu561, within the central cavity that is also expected to have a critical functional role (Fig. 5). In ChlH, a conserved His residue, His1174, is located on the opposite side of the protoporphyrin IX substrate relative to Glu660. CobN does not have an analogous His residue; instead, Asp1037 is located in an equivalent position. The Glu to Asp substitution has been proposed to play a role in the Co(II) specificity of CobN, but the means by which these residues select for Co(II) or Mg(II) remain unclear.

Based upon data now available in the primary literature, it is possible to propose a mechanism for CobN-catalyzed cobalt insertion. As significantly more is known about magnesium insertion by ChlH, most of the mechanism proposed below for CobN is based upon current understanding of ChlH (Adams et al., 2020). Starting with CobN in an open conformation, the substrate HBAD is proposed to bind to the inner cavity of CobN, where it is anchored in place by electrostatic interactions between CobN and the propionate side-chains of HBAD (Fig. 6). Next, the active site cavity could be enclosed by the Loop 1 structure of the Body I domain. At that point, Glu561 would be in proximity to the pyrrole hydrogens of HBAD, where it could abstract a proton from the substrate and shuttle it out of the active site *via* a nearby solvent channel. Following this, a conformational rearrangement driven by CobST-mediated ATP hydrolysis could push a Co(II) ion into the active site where Glu561 could deliver it to deprotonated HBAD yielding CBAD. Finally, relaxation of the CobN subunit back to its open configuration would permit release of CBAD. As noted above, this mechanism is highly speculative, and is primarily based upon structural comparisons between CobN and ChlH.

2.3 CobNST selectivity

The origin of metal selectivity remains enigmatic not only within the study of CobNST, but across the field of chelatase research as a whole. Divalent first-row transition metals all have similar ionic radii and coordination chemistry, which makes metal selectivity a challenge for all metalloproteins (Irving & Williams, 1948). As noted above, there is a His to Asp substitution in CobN as compared to ChlH, which could be the origin of metal

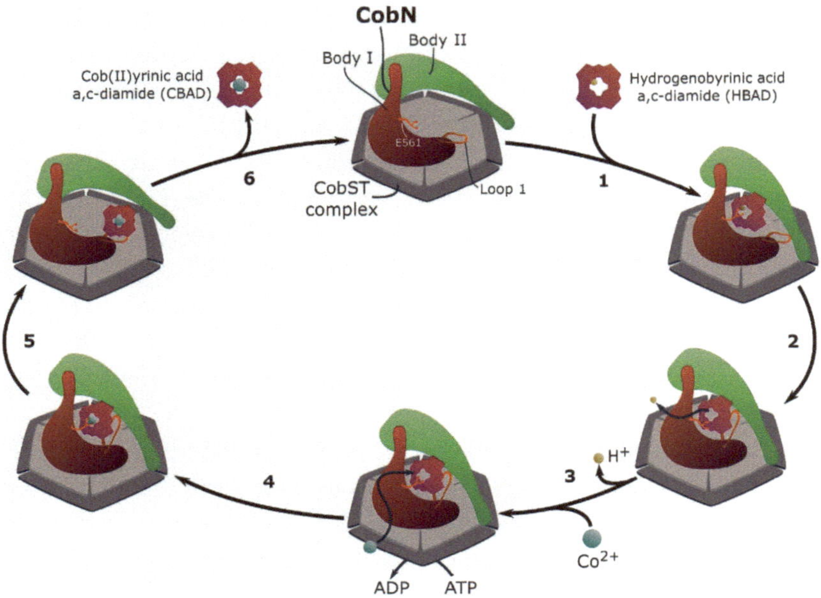

Fig. 6 Proposed catalytic mechanism for the CobNST system. The CobST motor complex is depicted as two stacked, hexagonal platforms underneath CobN. The Head and Neck domains of CobN have been omitted to simplify the representation. (**1**) With CobN in its open configuration, HBAD binds to the inner cavity *via* an induced-fit mechanism. (**2**) Loop 1 binds to HBAD, fully enclosing the active site. (**3**) The pyrrole hydrogen is abstracted by E561 and is shuttled out *via* a nearby solvent channel. A Co(II) ion is depicted as subsequently being introduced, though it is also possible that this ion has already been bound to and 'stored' by the complex. (**4**) A structural rearrangement caused by the hydrolysis of ATP by CobS drives the movement of a cobalt ion through a solvent channel, delivering it to the active site where is captured by Glu561 and inserted into the central cavity of HBAD to form CBAD. (**5**) CobN reverts to its open configuration. (**6**) CBAD is released, completing the catalytic cycle.

selectivity (Zhang et al., 2020). However, there is no simple explanation for why a His to Asp substitution would favor Co(II) binding over Mg(II) binding. Mg(II) is a harder acid than Co(II), and Asp is a harder ligand than His, which would suggest that CobN preferentially binds Mg(II) over Co(II) (Pearson, 1963). Clearly this is incorrect, but hard and soft acids and bases theory does not explicitly consider the contribution of coordination geometry. The different coordination geometries imposed by the CobN and ChlH active sites could certainly be responsible for the metal selectivities of these enzymes. The metal affinities of CobN merit further investigation.

More recently, data has been reported that suggests the cobalt selectivity of CobN may be determined by the Cob pathway, rather than by CobN itself. The protein CobW is known to play a role in aerobic B_{12} biosynthesis

(Crouzet et al., 1991), but it is not directly implicated in any of the chemical transformations of the pathway (Battersby, 1994). CobW is a GTPase, which has been proposed to function as a Co(II) chaperone to CobN (Heldt et al., 2005). Upon binding GTP and Mg(II), CobW forms a metal binding site for Fe(II), Co(II), Ni(II), Cu(I), or Zn(II) (Young et al., 2021). Although CobW binds Cu(I) more tightly than the other four metals, the Co(II) occupancy of this protein is expected to be the largest *in vivo* when intracellular metal ion concentrations are considered. As the Co(II) affinity of CobW decreases by three orders of magnitude following GTP hydrolysis, one can envision a mechanism whereby GTP hydrolysis facilitates Co(II) delivery to CobN. Although this relationship between CobN selectivity and CobW is still in its infancy, these findings highlight the potential for considering metal selectivity not only through the lens of chelatase-specific metal ion affinities, but also through the relative affinities of chaperones for these ions.

3. Anaerobic B_{12} biosynthesis

3.1 Historical background

Prior to the elucidation of the aerobic vitamin B_{12} biosynthesis pathway in *P. denitirificans* (Battersby, 1994), pulse labelling studies had already clued researchers into a separate pathway that inserts cobalt earlier. *Propionibacterium shermanii* inserts cobalt at the precorrin-2 stage (Müller et al., 1991), which is incompatible with the known sequence of events in *P. denitirificans*. Furthermore, *S. enterica* had previously been shown to biosynthesize vitamin B_{12} under anaerobic growth conditions (Jeter, Olivera, & Roth, 1984), which is incompatible with the oxygen-dependent pathway of *P. denitirificans*. Shortly after the groundbreaking publication of the complete aerobic pathway, the function of genes in the anaerobic *S. enterica cbi* operon were probed by a series of articles in the mid-1990s (Raux et al., 1996; Raux et al., 1997). The latter of these articles identified the role of the *cbi* operon gene *cbiK* as encoding an anaerobic cobalt chelatase. Interestingly, this work also demonstrated that the CbiK functions analogously to the siroheme synthase enzyme CysG in that it will insert Fe(II) into SHC, although this function is turned off in the presence of exogenous cobalt. Over the next several years the roles for the remainder of the genes involved in anaerobic vitamin B_{12} biosynthesis were elucidated, which indicated that cobalt insertion by CbiK, following oxidation of precorrin-2 to SHC by SirC (Raux et al., 2003), is the first committed step of the anaerobic pathway (Moore et al., 2013).

3.2 CbiK structure

The X-ray crystal structure of CbiK was solved in 1999 shortly after the elucidation of its physiological role yielding insight into its function *via* comparison to the PPIX iron chelatase, HemH. CbiK is a bilobal protein with two α/β domains that have pseudo two-fold symmetry suggesting a gene duplication event (Schubert, Raux, Wilson, & Warren, 1999). Each domain contains four parallel β strands, a characteristic shared with HemH (Al-Karadaghi, Hansson, Nikonov, Jönsson, & Hederstedt, 1997) (Fig. 7). CbiK has an additional helix in its N-terminal domain that breaks the two-fold symmetry, whereas HemH has two additional helices in its N-terminal domain. As a result, the two domains of CbiK are arranged in a more linear manner than for HemH; the later exhibits an additional 17° tilt between domains. Nonetheless, a structure-based sequence alignment revealed 14 conserved residues between CbiK and HemH; nine are involved in maintaining the protein structure and five are implicated in the catalytic mechanism of HemH. Based upon these alignments the active site of CbiK is located at the domain interface, and it is more open than the HemH active site.

A second anaerobic cobalt chelatase has been discovered, annotated $CbiX^L$, which appears to derive from a common ancestor of CbiK. $CbiX^L$ and CbiK are both class II chelatases that catalyze cobalt insertion into

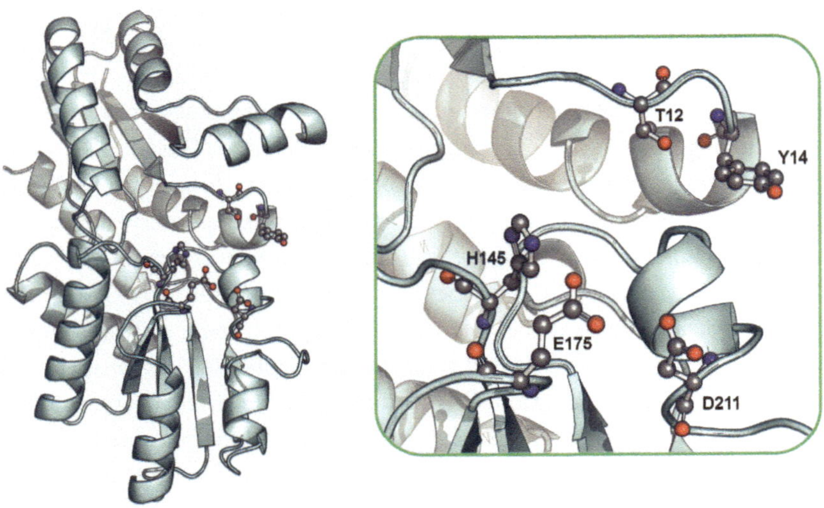

Fig. 7 X-ray crystal structure of *Salmonella enterica* CbiK displaying the five residues conserved between CbiK and HemH. His145 and Glu175 contribute to the first coordination sphere of the Co(II) substrate. His145 has also been proposed to contribute to SHC deprotonation by CbiK.

SHC yielding Co–SHC (Raux et al., 1997; Raux et al., 2003). However, the gene encoding for $CbiX^L$ first discovered in *B. megaterium* has minimal sequence similarity with the gene encoding CbiK in *S. enterica* (Raux, Lanois, Warren, Rambach, & Thermes, 1998). Nevertheless, partial alignment of the N-terminal domain of $CbiX^L$ with the C-terminal domain of CbiK reveals that the critical His residues in the active site are conserved, suggesting evolution from a common ancestor. Indeed, the nickel chelatase CfbA, originally annotated $CbiX^S$, has been identified as the putative ancestor of all class II chelatases (Brindley et al., 2003). It has been proposed that $CbiX^L$ and the iron chelatase SirB arose from duplication of the *cfbA* gene followed by evolution of a metal binding site in the N-terminal domain, whereas CbiK and HemH arose from *cfbA* duplication followed by evolution of the C-terminal domain (Romão et al., 2011). These observations suggest that class II chelatases are a versatile enzyme template that can evolve to insert a variety of divalent metal ions into a range of tetrapyrrole ligands.

3.3 CbiK function

CbiK and $CbiX^L$ must promote several bond breaking and forming steps in order to catalyze cobalt insertion into SHC, and one of these is desolvation of a Co(II) ion. Prior to chelation, an aqueous Co(II) ion is in an octahedral coordination environment with six water ligands. To facilitate metal insertion, CbiK and $CbiX^L$ need to: desolvate Co(II), deprotonate SHC, and catalyze formation of four new Co—N bonds between Co(II) and SHC (Raux et al., 1997). Based upon the X-ray crystal structure of *Desulfovibrio vulgaris* CbiK (Fig. 8), the Co(II) ion is coordinated by His145, Glu175, and His207, plus additional water and peroxide ligands (Lobo et al., 2017). (For clarity, the CbiK residue numbering in this chapter will refer to the gene from *S. enterica* throughout). Of these ligands, site-directed mutagenesis in combination with enzyme activity assays suggested that His145 is the most essential ligand based upon the dramatic reduction of cobyric acid synthesis by the H145A variant (Schubert et al., 1999). Co(II) is a d^7 transition metal ion, which means that an octahedral complex would be labile regardless of spin state. As noted previously within the context of Ni(II) chelation by CfbA (Schuelke-Sanchez et al., 2020), this lability means that there is a rapid equilibrium between five- and six-coordinate forms of cobalt-bound CbiK, and there is no kinetic barrier to Co(II) desolvation by CbiK. Consequently, CbiK-bound Co(II) will rapidly exchange ligands to achieve its most thermodynamically-stable coordination environment.

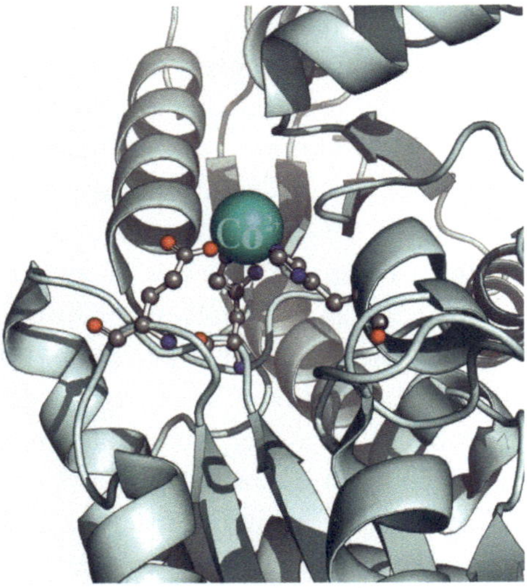

Fig. 8 X-ray crystal structure of *D. vulgaris* CbiK with Co(II) ligated by His145, Glu175, and His207. The Co(II) coordination sphere is completed by additional water and peroxide ligands. His145 has also been proposed to participate in SHC deprotonation.

X-ray crystallography has also provided insight into SHC binding by CbiK. Based upon the X-ray crystal structure of CbiK with bound Co–SHC product (Fig. 9), the enzyme active site induces a ruffling distortion of the tetrapyrrole (Graves, Graves, & Liptak, 2016; Jentzen et al., 1997; Romão et al., 2011). This is markedly different from the saddling deformation of PPIX induced by HemH (Lu et al., 2002; Medlock et al., 2007), which facilitates deprotonation of the tetrapyrrole substrate (Al-Karadaghi et al., 2006). Ruffling of SHC by CbiK would not push a pyrrole proton out of plane to favor deprotonation, but an interaction between Phe10 and SHC has been proposed to encourage pyrrole deprotonation. The Co–SHC product is held in place within the CbiK active site by hydrogen bonds between the backbone amides of four Asn residues and the propionate side-chains of SHC rings C and D. In addition, the ring C acetate forms a hydrogen bond with His207. Although there is X-ray crystallographic evidence for His207 to form a hydrogen bond with both the Co(II) substrate and the Co–SHC product (Lobo et al., 2017), it clearly cannot bind to both the Co(II) and SHC substrates simultaneously. Furthermore, the observation of a ruffled Co–SHC product does not necessarily mean that the SHC substrate is also ruffled. These observations highlight the need for additional structural data for SHC-bound CbiK, since the structural role of His207 and the out-of-plane distortion of SHC have important mechanistic implications.

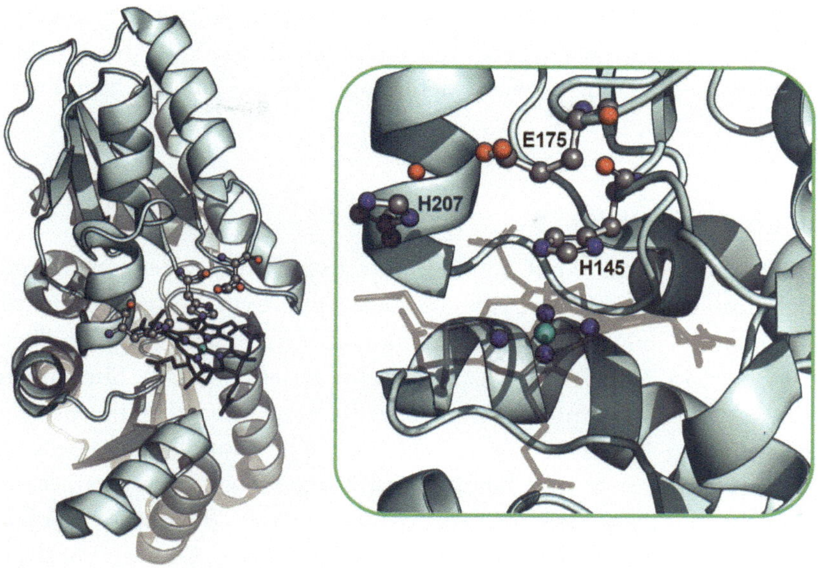

Fig. 9 X-ray crystal structure of *Salmonella enterica* CbiK with the Co–SHC product bound. The Co(II) binding triad is shown. Based upon currently available structural data, it is unclear whether His207 binds to Co(II), SHC, or both substrates.

The insights into cobalt and SHC binding by CbiK described above lay the foundation for a mechanistic proposal. As noted previously, CbiK is thought to bind a labile Co(II) ion and a ruffled SHC substrate (Graves et al., 2016; Jentzen et al., 1997; Lobo et al., 2017; Romão et al., 2011). These substrate binding modes suggest that CbiK-catalyzed cobalt insertion into SHC may follow a mechanism similar to that recently suggested for CfbA-catalyzed nickel insertion into SHC (Schuelke-Sanchez et al., 2020). Octahedral Co(II) coordination complexes are labile, and a rapid equilibrium between five- and six-coordination Co(II) in the CbiK active site will open a vacant Co(II) coordination site for dissociative substitution. Deprotonation of SHC by CbiK has been proposed to be facilitated by His145, as suggested by an unusual binding geometry that leaves a lone pair on His145 free to abstract a proton from a pyrrole nitrogen. However, SHC deprotonation prior to Co(II) ligation may be unnecessary due to the smaller ionic radius of Co(II) compared to high-spin Fe(II) (Shannon, 1976). Regardless, theoretical methods suggest that formation of the first Co–N_{SHC} bond is the rate-limiting step that will be rapidly followed by breakage of the five remaining CbiK–Co bonds, deprotonation of SHC, and formation of three additional Co–N_{SHC} bonds (Shen & Ryde, 2005). There remain several unclear features of the CbiK mechanism that would be elucidated by further characterization of cobalt-bound CbiK, to determine the Co(II) spin state and approximate ionic radius, and SHC-bound CbiK, to determine the out-of-plane deformation mode of this ligand upon binding to CbiK.

3.4 CbiK selectivity

A comparison of the CbiK and HemH active sites provides some limited insight into the metal selectivities of these two enzymes. Five residues surrounding the CbiK active site are conserved with HemH (Fig. 7): Thr12, Tyr14, His145, Glu175, and Asp211 (Schubert et al., 1999). Site-directed mutagenesis and enzyme activity assays have revealed that the principal cobalt binding ligand in CbiK is His145. The homologous residue in HemH, His183, is an iron binding ligand suggesting similarities between the metal binding sites of the two chelatases (Hansson, Karlberg, Rahardja, Al-Karadaghi, & Hansson, 2007; Kohno, Okuda, Furukawa, Tokunaga, & Taketani, 1994). Similar experiments demonstrated that a second His, His207, also has a critical role in CbiK-catalyzed cobalt insertion into SHC. The equivalent residue of HemH is Glu264, which also participates in iron ligation by HemH (Gora, Grzybowska, Rytka, & Labbe-Bois, 1996). It has been clearly demonstrated that both His145 and His207 are necessary for the cobalt selectivity of CbiK, which suggests that the Glu to His substitution plays an important role in the metal selectivities of CbiK and HemH. However, the means by which substitution of His for Glu would select for Co(II) over Fe(II) remains unclear. Thus, there are structural similarities between the class II chelatases CbiK and HemH, but small differences result in enzymes with different substrate specificities.

A recent study has provided futher insight into the cobalt selectivity of CbiK. Competitive binding assays were employed to determine the affinity of *S. enterica* CbiK for seven metal ions (Osman et al., 2019). Of these, Cu(I) binding by CbiK had the most exergonic standard free energy in accord with the Irving-Williams series (Irving & Williams, 1948). However, since the actual free energy of a chemical reaction depends upon both the standard free energy and the reagent concentrations, the authors also measured the affinities of the metal-sensing transcription regulators for *Salmonella*. A comparison of the free energies derived from the metal affinities of the transcription regulators and CbiK correctly predicts the metallation of CbiK *in vivo*. To be specific, the Co(II) occupancy of CbiK is expected to be the highest *in vivo* for the seven metal ions examined at 15.4%. In fact, the only other metal with non-zero occupancy was Fe(II), which is consistent with the observation that CbiK can function as an iron chelatase in a *cysG* deletion strain of *E. coli* (Raux et al., 1997). Finally, it is worth noting that when the Co(II) concentration is buffered to the level expected *in vivo* based upon the cobalt affinity of RcnR, cobalt insertion into SHC requires CbiK. Thus, despite the uncertainties with respect to the enzymatic mechanism, CbiK clearly catalyzes selective insertion of Co(II) into SHC *in vivo*.

4. Conclusions and future directions

In summary, at least four distinct syntheses of vitamin B_{12} are known, two abiological and two biological, and each of these syntheses have had to overcome the challenge of selectively inserting a Co(II) ion into a tetrapyrrole substrate. Cobalt insertion was a significant synthetic challenge for both abiological syntheses of vitamin B_{12}, and this challenge was overcome by using a cobalt halide salt (Woodward, 1973). In biological systems, the aerobic biosynthetic pathway employs the CobNST multi-protein complex to insert cobalt late in the pathway (Debussche et al., 1992), whereas the anaerobic pathway uses a single enzyme to insert cobalt as the first committed step of the pathway (Raux et al., 1997; Raux et al., 2003). A question at the forefront of the field is that of cobalt selectivity: these biological systems are challenged to select for Co(II) among all naturally-available divalent metals. In these authors' opinion, the hypotheses that cobalt selectivity in the aerobic and anaerobic pathways is driven by the metal affinities of CobW and CbiK, respectively, is currently the most compelling (Osman et al., 2019; Young et al., 2021). However, metal selectivity could also be derived from tetrapyrrole distortion (McIntyre et al., 2011), or mismetallated product inhibition (Medlock et al., 2009). Finally, although the overall reactions catalyzed by CobNST and CbiK have been determined, several details regarding the enzymatic mechanisms remain ill-defined. As such, the enzymology of cobalt chelatases remains a field with significant open questions to be addressed by collaborative efforts between microbiologists, biochemists, and chemists.

References

Adams, N. B. P., Bisson, C., Brindley, A. A., Farmer, D. A., Davison, P. A., Reid, J. D., et al. (2020). The active site of magnesium chelatase. *Nature Plants*, *6*(12), 1491–1502.

Al-Karadaghi, S., Franco, R., Hansson, M., Shelnutt, J. A., Isaya, G., & Ferreira, G. C. (2006). Chelatases: Distort to select? *Trends in Biochemical Sciences*, *31*(3), 135–142.

Al-Karadaghi, S., Hansson, M., Nikonov, S., Jönsson, B., & Hederstedt, L. (1997). Crystal structure of ferrochelatase: The terminal enzyme in heme biosynthesis. *Structure*, *5*(11), 1501–1510.

Antonov, I. V. (2020). Two cobalt chelatase subunits can be generated from a single chlD gene via programmed frameshifting. *Molecular Biology and Evolution*, *37*(8), 2268–2278.

Battersby, A. R. (1994). How nature builds the pigments of life: The conquest of vitamin B_{12}. *Science*, *264*(5165), 1551–1557.

Blanche, F., Cameron, B., Crouzet, J., Debussche, L., Thibaut, D., Vuilhorgne, M., et al. (1995). Vitamin B_{12}: How the problem of its biosynthesis was solved. *Angewandte Chemie International Edition*, *34*, 383–411.

Brindley, A. A., Raux, E., Leech, H. K., Schubert, H. L., & Warren, M. J. (2003). A story of chelatase evolution. *Journal of Biological Chemistry*, *278*(25), 22388–22395.

Bryant, D. A., Hunter, C. N., & Warren, M. J. (2020). Biosynthesis of the modified tetrapyrroles-the pigments of life. *Journal of Biological Chemistry*, *295*(20), 6888–6925.

Crouzet, J., Levy-Schil, S., Cameron, B., Cauchois, L., Rigault, S., Rouyez, M. C., et al. (1991). *Pseudomonas dentirificans* DNA fragment containing five *cob* genes and identification of structural genes encoding cob(I)alamin adenosyltransferase, cobyric acid synthase, and bifunctional cobinamide kinase-cobinamide phosphate guanylyltransferase. *Journal of Bacteriology*, *173*(19), 6074–6087.

Dailey, H. A. (1977). Purification and characterization of the membrane-bound ferrochelatase from *Spirillum itersonii*. *Journal of Bacteriology*, *132*(1), 302–307.

Dailey, H. A., Jones, C. S., & Karr, S. W. (1989). Interaction of free porphyrins and metalloporphyrins with mouse ferrochelatase. A model for the active site of ferrochelatase. *Biochimica et Biophysica Acta*, *999*, 7–11.

Debussche, L., Couder, M., Thibaut, D., Cameron, B., Crouzet, J., & Blanche, F. (1992). Assay, purification, and characterization of cobaltochelatase, a unique complex enzyme catalyzing cobalt insertion in hydrogenobyrinic acid *a,c*-diamide during coenzyme B_{12} biosynthesis in *Pseudomonas denitrificans*. *Journal of Bacteriology*, *174*(22), 7445–7451.

Eschenmoser, A., & Wintner, C. E. (1977). Natural product synthesis and vitamin B_{12}: Total synthesis of vitamin B_{12} provided a framework for exploration of several areas of organic chemistry. *Science*, *196*(4297), 1410–1420.

Farmer, D. A., Brindley, A. A., Hitchcock, A., Jackson, P. J., Johnson, B., Dickman, M. J., et al. (2019). The ChlD subunit links the motor and porphyrin binding subunits of magnesium chelatase. *Biochemical Journal*, *476*(13), 1875–1887.

Gibson, L. C. D., Willows, R. D., Kannangara, C. G., von Wettstein, D., & Hunter, C. N. (1995). Magnesium-protoporphyrin chelatase of *Rhodobacter sphaeroides*: Reconstitution of activity by combining the products of the *bchH, -I, and -D* genes expressed in *Escherichia coli*. *Proceedings of the National Academy of Sciences*, *92*, 1941–1944.

Gora, M., Grzybowska, E., Rytka, J., & Labbe-Bois, R. (1996). Probing the active-site residues in *Saccharomyces cerevisiae* ferrochelatase by directed mutagenesis. *Journal of Biological Chemistry*, *271*(20), 11810–11816.

Graves, A. B., Graves, M. T., & Liptak, M. D. (2016). Measurement of heme ruffling changes in MhuD using UV/Vis spectroscopy. *Journal of Physical Chemistry B*, *120*, 3844–3853.

Hansson, M. D., Karlberg, T., Rahardja, M. A., Al-Karadaghi, S., & Hansson, M. (2007). Amino acid residues His183 and Glu264 in *Bacillus subtilis* ferrochelatase direct and facilitate the insertion of metal ion into protoporphyrin IX. *Biochemistry*, *46*(1), 87–94.

Heldt, D., Lawrence, A. D., Lindenmeyer, M., Deery, E., Heathcote, P., Rigby, S. E. J., et al. (2005). Aerobic synthesis of vitamin B_{12}: Ring contraction and cobalt chelation. *Biochemical Society Transactions*, *33*(4), 815–819.

Hoggins, M., Dailey, H. A., Hunter, C. N., & Reid, J. D. (2007). Direct measurement of metal ion chelation in the active site of human ferrochelatase. *Biochemistry*, *46*(27), 8121–8127.

Irving, H., & Williams, R. J. P. (1948). Order of stability of metal complexes. *Nature*, *162*, 746–747.
Jentzen, W., Song, X.-Z., & Shelnutt, J. A. (1997). Structural characterization of synthetic and protein-bound porphyrins in terms of the lowest-frequency Normal coordinates of the macrocycle. *Journal of Physical Chemistry B*, *101*(9), 1684–1699.
Jeter, R. M., Olivera, B. M., & Roth, J. R. (1984). *Salmonella typhimurium* synthesizes cobalamin (vitamin B_{12}) *de novo* under anaerobic growth conditions. *Journal of Bacteriology*, *159*(1), 206–213.
Kohno, H., Okuda, M., Furukawa, T., Tokunaga, R., & Taketani, S. (1994). Site-directed mutagenesis of human ferrochelatase: Identification of histidine-263 as a binding site for metal ions. *Biochimica et Biophysica Acta*, *1209*, 95–100.
Leech, H. K., Raux-Deery, E., Heathcote, P., & Warren, M. J. (2002). Production of cobalamin and sirohaem in *Bacillus megaterium*: An investigation into the role of the branchpoint chelatases sirohydrochlorin ferrochelatase (SirB) and sirohydrochlorin cobalt chelatase (CbiX). *Biochemical Society Transactions*, *30*(4), 610–613.
Lobo, S. A. L., Videira, M. A. M., Pacheco, I., Wass, M. N., Warren, M. J., Teixeira, M., et al. (2017). *Desulfovibrio vulgaris* $CbiK^P$ cobaltochelatase: Evolution of a haem binding protein orchestrated by the incorporation of two histidine residues. *Environmental Microbiology*, *19*(1), 106–118.
López-Mirabal, H. R., & Winther, J. R. (2008). Redox characteristics of the eukaryotic cytosol. *Biochimica et Biophysica Acta*, *1783*, 629–640.
Lu, Y., Sousa, A., Franco, R., Mangravita, A., Ferreira, G. C., Moura, I., et al. (2002). Binding of protoporphyrin IX and metal derivatives to the active site of wild-type mouse ferrochelatase at low porphyrin-to-protein ratios. *Biochemistry*, *41*(26), 8253–8262.
Lundqvist, J., Elmlund, D., Heldt, D., Deery, E., Söderberg, C. A. G., Hansson, M., et al. (2009). The AAA+ motor complex of subunits CobS and CobT of cobaltochelatase visualized by single particle electron microscopy. *Journal of Structural Biology*, *167*(3), 227–234.
Lundqvist, J., Elmlund, H., Wulff, R. P., Berglund, L., Elmlund, D., Emanuelsson, C., et al. (2010). ATP-induced conformational dynamics in the AAA+ motor unit of magnesium chelatase. *Structure*, *18*, 354–365.
Martens, J.-H., Barg, H., Warren, M. J., & Jahn, D. (2002). Microbial production of vitamin B_{12}. *Applied Microbiology and Biotechnology*, *58*, 275–285.
McIntyre, N. R., Franco, R., Shelnutt, J. A., & Ferreira, G. C. (2011). Nickel(II) chelatase variants directly evolved from murine ferrochelatase: Porphyrin distortion and kinetic mechanism. *Biochemistry*, *50*(9), 1535–1544.
Medlock, A. E., Carter, M., Dailey, T. A., Dailey, H. A., & Lanzilotta, W. N. (2009). Product release rather than chelation determines metal specificity for ferrochelatase. *Journal of Molecular Biology*, *393*, 308–319.
Medlock, A. E., Swartz, L., Dailey, T. A., Dailey, H. A., & Lanzilotta, W. N. (2007). Substrate interactions with human ferrochelatase. *Proceedings of the National Academy of Sciences of the United States of America*, *104*(6), 1789–1793.
Moore, S. J., Lawrence, A. D., Biedendieck, R., Deery, E., Frank, S., Howard, M. J., et al. (2013). Elucidation of the anaerobic pathway for the corrin component of cobalamin (vitamin B_{12}). *Proceedings of the National Academy of Sciences*, *110*(37), 14906–14911.
Müller, G., Zipfel, F., Hlineny, K., Savvidis, E., Hertle, R., Traub-Eberhard, U., et al. (1991). Timing of cobalt insertion in vitamin B_{12} biosynthesis. *Journal of the American Chemical Society*, *113*, 9893–9895.
Osman, D., Martini, M. A., Foster, A. W., Chen, J., Scott, A. J. P., Morton, R. J., et al. (2019). Bacterial sensors define intracellular free energies for correct enzyme metalation. *Nature Chemical Biology*, *15*, 241–249.

Pearson, R. G. (1963). Hard and soft acids and bases. *Journal of the American Chemical Society, 85*(22), 3533–3539.

Raux, E., Lanois, A., Levillayer, F., Warren, M. J., Brody, E., Rambach, A., et al. (1996). *Salmonella typhimurium* cobalamin (vitamin B_{12}) biosynthetic genes: Functional studies in *S. typhimurium* and *Escherichia coli*. *Journal of Bacteriology, 178*, 753–767.

Raux, E., Lanois, A., Warren, M. J., Rambach, A., & Thermes, C. (1998). Cobalamin (vitamin B_{12}) biosynthesis: Identification and characterization of a *bacillus megaterium cobI* operon. *Biochemical Journal, 335*, 159–166.

Raux, E., Leech, H. K., Beck, R., Schubert, H. L., Santander, P. J., Roessner, C. A., et al. (2003). Identification and functional analysis of enzymes required for precorrin-2 dehydrogenation and metal ion insertion in the biosynthesis of sirhaem and cobalamin in *bacillus megaterium*. *Biochemical Journal, 370*, 505–516.

Raux, E., Thermes, C., Heathcote, P., Rambach, A., & Warren, M. J. (1997). A role for *Salmonella typhimurium cbiK* in cobalamin (vitamin B_{12}) and siroheme biosynthesis. *Journal of Bacteriology, 179*(10), 3202–3212.

Reid, J. D., & Hunter, C. N. (2002). Current understanding of the function of magnesium chelatase. *Biochemical Society Transactions, 30*(4), 643–645.

Rodionov, D. A., Vitreschak, A. G., Mironov, A. A., & Gelfand, M. S. (2003). Comparative genomics of the vitamin B_{12} metabolism and regulation in prokaryotes. *Journal of Biological Chemistry, 278*(42), 41148–41159.

Romão, C. V., Ladakis, D., Lobo, S. A. L., Carrondo, M. A., Brindley, A. A., Deery, E., et al. (2011). Evolution in a family of chelatases facilitated by the introduction of active site asymmetry and protein oligomerization. *Proceedings of the National Academy of Sciences of the United States of America, 108*(1), 97–102.

Schubert, H. L., Raux, E., Wilson, K. S., & Warren, M. J. (1999). Common chelatase design in the branched tetrapyrrole pathways of heme and anaerobic cobalamin synthesis. *Biochemistry, 38*(33), 10660–10669.

Schuelke-Sanchez, A. E., Stone, A. A., & Liptak, M. D. (2020). CfbA promotes insertion of cobalt and nickel into ruffled tetrapyrroles *in vitro*. *Dalton Transactions, 49*, 1065–1076.

Shannon, R. D. (1976). Revised effective ionic radii and systematic studies of interatomic distances in halides and chalcogenides. *Acta Crystallographica, A32*(5), 751–767.

Shelnutt, J. A., Song, X.-Z., Ma, J.-G., Jia, S.-L., Jentzen, W., & Medforth, C. J. (1998). Nonplanar porphyrins and their significance in proteins. *Chemical Society Reviews, 27*(1), 31–41.

Shen, Y., & Ryde, U. (2005). Reaction mechanism of porphyrin metallation studied by theoretical methods. *Chemistry - A European Journal, 11*(5), 1549–1564.

Snider, J., Thibault, G., & Houry, W. A. (2008). The AAA+ superfamily of functionally diverse proteins. *Genome Biology, 9*(4), 216.

Walker, C. J., & Willows, R. D. (1997). Mechanism and regulation of mg-chelatase. *Biochemical Journal, 327*(2), 321–333.

Woodward, R. B. (1973). The total synthesis of vitamin B_{12}. *Pure and Applied Chemistry, 33*, 145–178.

Young, T. R., Martini, M. A., Foster, A. W., Glasfeld, A., Osman, D., Morton, R. J., et al. (2021). Calculating metalation in cells reveals CobW acquires Co^{II} for vitamin B_{12} biosynthesis while related proteins prefer Zn^{II}. *Nature Communications, 12*, 1195.

Zhang, J., Yuan, H., Wang, X., Dai, H., Zhang, M., & Liu, L. (2020). Crystal structure of the large subunit of cobaltochelatase from *Mycobacterium tuberculosis*. *Proteins, 89*(4), 462–467.

Zheng, K., Ngo, P. D., Owens, V. L., Yang, X.-P., & Mansoorabadi, S. O. (2016). The biosynthetic pathway of coenzyme F430 in methanogenic and methanotrophic archaea. *Science, 354*(6310), 339–342.

CHAPTER TWO

Application of bioorganometallic B_{12} in green organic synthesis

Hisashi Shimakoshi*
Department of Chemistry and Biochemistry, Graduate School of Engineering, Kyushu University, Fukuoka, Japan
*Corresponding author: e-mail address: shimakoshi@mail.cstm.kyushu-u.ac.jp

Contents

Abstract

Bioorganometallic structure found in coenzyme B_{12} is a key component in B_{12}-dependent enzymatic reactions in natural enzymes. Cleavage of a cobalt-carbon bond in organometallic B_{12} compound provide reactive intermediate for molecular transformations. Application of the bioorganometallic B_{12} in organic synthesis have been developed using natural vitamin B_{12} as well as synthetic vitamin B_{12} derivatives as a bioinspired catalyst in organic solvent. Vitamin B_{12} derivatives composed of corrinoid structure should form stable organometallic compound having a cobalt-carbon bond. Using the unique property of the organometallic vitamin B_{12} derivatives, various catalytic reactions have been developed in synthetic organic chemistry. The dual catalytic system of vitamin B_{12} derivatives and photocatalyst, such as Ru(II) polypyridyl complex or titanium oxide, could construct light-driven molecular transformations. The B_{12}-dependent enzymes mimic reactions, such as the dechlorination of organic halides and the radical mediated isomerization reactions, catalytically proceed in the

Vitamins and Hormones, Volume 119
ISSN 0083-6729
https://doi.org/10.1016/bs.vh.2022.01.015

dual catalyst system. Electroorganic syntheses mediated by the vitamin B_{12} derivatives have been developed as green molecular transformations. The redox active vitamin B_{12} derivatives shows a unique catalysis in the electroorganic synthesis, such as alkene and alkyne reductions.

1. Vitamin B_{12}

Bioorganometallic compounds emerged in biological systems are unique metal complexes having a metal–organic σ-bond (Hillard & Jaouen, 2011). The multidisciplinary applications of these compounds in bio-imaging, bio-analysis, bio-drug, and bio-inspired catalysis are caused by unique chemical properties of the organometallic complexes (Patra, Gasser, & Metzler-Nolte, 2012; Zhang & Sadler, 2017). Coenzyme B_{12}, adenosylcobalamin (Fig. 1A), is a well-known example of bioorganometallic compound present in various B_{12}-dependent mutases, such as methylmalonyl-CoA mutase and lysine 5,6-aminomutase (Fig. 1B) (Banerjee, 2003; Bridwell-Rabb & Drennan, 2017; Brown, 2005; Buckel & Golding, 2006). The reduction of ribonucleotides to 2′-deoxyribonucleotides during DNA biosynthesis is also catalyzed by adenosylcobalamin in a

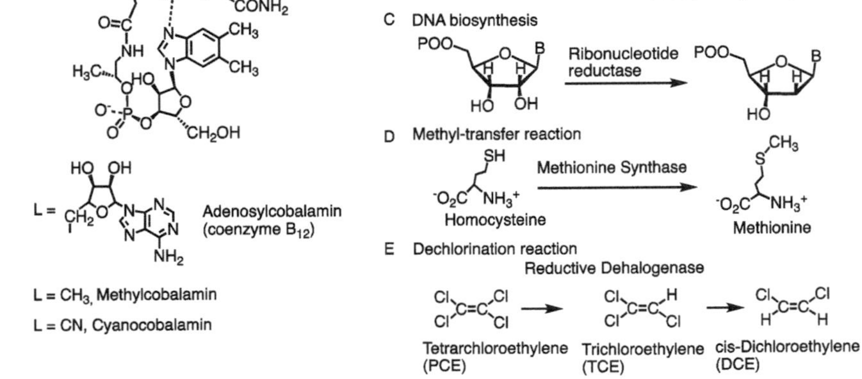

Fig. 1 (A) Structure of B_{12} derivatives. (B–E) Examples of B_{12}-dependent enzyme reactions.

ribonucleotide reductase (Fig. 1C) (Licht, Gerfen, & Stubbe, 1996). Methylcobalamin found in methyltransferases include methionine synthase transfer a methyl group to nucleophilic substrates (Fig. 1D) (Matthews, Koutmos, & Datta, 2008). Other classes of B_{12}-dependent enzymes that do not appear to use an alkyl ligand were also discovered. Reductive dehalogenase isolated from anaerobic bacteria contains the vitamin B_{12} derivative as a cofactor lacking an upper ligand for dehalorespiration by reducing 1,1,2,2-tetrarchloroethene (PCE) to *cis*-dichloroethene (DCE) via trichloroethene (TCE) (Fig. 1E) (Bommer et al., 2014).

By focusing on the organometallic compound, adenosylcobalamin or methyl cobalamin are alkylated cobalt complex having a cobalt-carbon σ-bond in which the adenosyl or methyl groups are bound to the cobalt ion as an axial ligand, respectively. One of the unique properties of these bioorganometallic compounds is the facile formation of carbon-centered radical species by homolysis of the cobalt-carbon bond. It is also noted that cycle between the Co(III) and Co(I) states with methyl cation transfer to substrate is caused by heterolytic cobalt-carbon bond cleavage in methylcobalamin. The bioorganometallic compounds in the B_{12}-dependent enzymes have been utilized for development for B_{12}-inspired catalytic reactions using the unique property of the alkylated cobalt complexes (Giedyk, Goliszewska, & Gryko, 2015; Shimakoshi & Hisaeda, 2017). A construction of bioinspired catalyst system for organic synthesis is described in this chapter.

1.1 Design for vitamin B_{12} inspired reactions

The vitamin B_{12} derivative consists of a cobalt ion coordinated with the tetrapyrrole corrin ring and linked to a heterocyclic base, 5,6-dimethylbenzimidazole (DMB), as an axial ligand (Fig. 1A). The axial base is replaced with adenine or 2-methyl-adenine in the pseudo vitamin B_{12} (Gruber, Puffer, & Kräutler, 2011). The corrin ring resembles porphyrin, which is found in heme proteins, chlorophyll in photosynthesis, or cytochromes, but different from porphyrin. The porphyrin is a dianionic ligand with a complete conjugated structure, on the other hand, the corrin ring is a monoanionic ligand with partially saturated structure. The monoanionic corrin ring can stabilize the low oxidation state of the central cobalt ion. The alkylated cobalt complex is generally formed by the reaction of the Co(I) state of the complex with electrophiles such as an organic halide in vitro. Therefore, the reductive formation of the

Co(I) species is important for the construction of the B_{12} catalytic reaction system. Chemical reductants, such as sodium borohydride, zinc, amalgam, dithiothreitol, and titanium (III) citrate, were used for the reduction of the cobalt complex to form nucleophilic Co(I) species. Recently, various kinds of photosensitizers have been utilized for the reductive activation of a metal complex (Skubi, Blum, & Yoon, 2016). The significant advantages of the photoredox system are reducing chemical waste using a clean and abundant light source and a simple and facilitated reaction system. The Co(I) species should be produced by the photoredox system.

Another approach for reductive formation of Co(I) species is electrochemical reduction of the cobalt complex. Electrolysis is recognized as a green process in synthetic organic chemistry since the redox process of a substrate occurs due to an electric current without any chemical reagent and waste after the reaction (Francke & Little, 2014; Yan, Kawamata, & Baran, 2017; Yoshida, Kataoka, Horcajada, & Nagaki, 2008). Electroorganic synthesis mediated by the B_{12} complexes have also been developed based on these backgrounds (Shimakoshi & Hisaeda, 2018). Various kinds of B_{12}-inspired reactions with a bioorganometallic character were developed both by a photochemical method or an electrochemical method.

1.2 Vitamin B_{12} derivatives

As the natural vitamin B_{12}, cyanocobalamin (Fig. 1A), has a low solubility in almost organic solvents, the amphiphilic vitamin B_{12} derivative with a high solubility in an organic solvent was developed for the application of the vitamin B_{12} derivative in organic synthesis. Among the amphiphilic vitamin B_{12} derivatives, heptamethyl cobyrinate perchlorate with peripheral methyl ester groups instead of natural amide groups has most simple structure and an excellent solubility in various organic solvents (Fig. 2) (Murakami, Hisaeda, & Kajihara, 1983). Furthermore, it well reproduces the redox property of cobalamin. As the monoanionic corrin ring can stabilize the low oxidation state of the cobalt atom in the cobalamin, the Co(II)/Co(I) redox couple of heptamethyl cobyrinate appeared at a relatively positive potential around $-0.4\,V \sim -0.6\,V$ vs Ag/AgCl in a non-aqueous solvent, such as DMF and CH_3OH, compared to those of the synthetic cobalt complexes such as cobalt porphyrins and cobalt phthalocyanines with dianionic ligands.

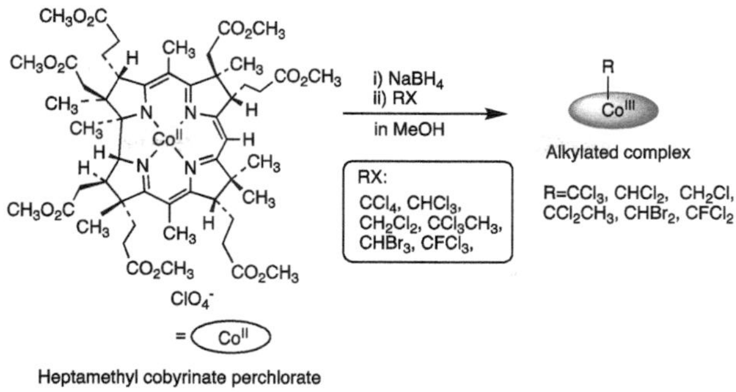

Fig. 2 Reactivity of heptamethyl cobyrinate toward organic halides, CCl_4, $CHCl_3$, CH_2Cl_2, CCl_3CH_3, $CHBr_3$, and $CHBr_3$.

To evaluate the reactivity of heptamethyl cobyrinate toward organic halides, reaction between Co(I) species of heptamethyl cobyrinate and CCl_4, $CHCl_3$, CH_2Cl_2, CCl_3CH_3, $CHBr_3$, and $CHBr_3$ were investigated (Fig. 2) (Shimakoshi et al., 2005). The formations of corresponding alkylated complexes with dehalogenation were confirmed by UV–vis and ESI-mass spectroscopic analysis. The potent application of the heptamethyl cobyrinate toward molecular transformation of organic halides via alkylated complex was expected from these results. Based on these backgrounds, various B_{12}-mimetic and -inspired reactions were developed using the heptamethyl cobyrinate as catalyst combined use of several reductive activation method of cobalt center.

2. B_{12} inspired reactions

2.1 Photo-driven B_{12} catalyzed dechlorination reaction

A photoluminescent compound, such as organic dyes and transition metal complexes are used as photocatalyst. These compounds support the light-induced molecular transformation by electron transfer process using its abundant light absorbance and redox property (Arias-Rotondo & McCusker, 2016). The combined use of photocatalyst with vitamin B_{12} derivatives could develop a light-driven B_{12} catalytic reactions. Among the photoluminescent compound, the Ru(II) trisbipyridine complex, $[Ru(II)(bpy)_3]Cl_2$ (where bpy is 2,2′-bipyridine), has been utilized as a photocatalyst for the reductive activation of the B_{12} derivative to form Co(I) species. In the early stage of these studies, pioneering works for photo-driven vitamin B_{12} (cyanocobalamin) catalytic reaction were reported with carbon dioxide and water reductions using the [Ru(II)(bpy)$_3$]Cl$_2$ by visible light irradiation (Ziessel, Hawecker, & Lehn, 1986). As reductive dechlorination of organic chlorides by the catalysis of cyanocobalamin with chemical reductants, such as NaBH$_4$, Zinc, Ti(III) citrate have been reported (Kim & Carraway, 2002; Krone, Thauer, & Hogenkamp, 1989; Schrauzer & Katz, 1978),photo-driven dechlorination of organic chloride was developed using the [Ru(II)(bpy)$_3$]Cl$_2$ in the place of chemical reductants. Dechlorination of 1,1,1-trichloro-2,2-bis(4-chlorophenyl)ethane (DDT) by the heptamethyl cobyrinate with the [Ru(II)(bpy)$_3$]Cl$_2$ to form 1,1-dichloro-2,2-bis(4-chlorophenyl)ethane (DDD) (Fig. 3) (Shimakoshi, Kudo, & Hisaeda, 2005; Shimakoshi, Tokunaga, Baba, & Hisaeda, 2004) under visible light irradiation. Since DDT is characterized by a pronounced insecticidal property and has been used worldwide for the past decades despite its known hazardous effects on human health and wildlife (Kiflom, Wandiga, & Kamau, 2001), the efficient and effective dechlorination method of DDT is required. In the advance of the photocatalytic reaction, both the photocatalyst and B$_{12}$ derivative were incorporated into the same polymer unit to enhance the photocatalytic reactivity (Fig. 4) (Shimakoshi, Nishi, Tanaka, Chikama, & Hisaeda, 2011; Zhang, Shimakoshi, Houfuku, Song, & Hisaeda, 2014). The hybrid polymer was synthesized by the radical polymerization of a B$_{12}$ derivative and a Ru(II) complex having styrene moieties in each of the peripheral positions. The hybrid polymer showed an over 10 times higher reactivity for the catalytic DDT dechlorination reaction than that of the non-polymer system. The fast UV–vis spectral change for the Co(I) formation of the B$_{12}$ complex, $\lambda_{max} = 390$ nm, under visible light irradiation showed that the intra-polymer electron transfer from the photosensitizer to the B$_{12}$ complex should be prior to that of the non-polymer system.

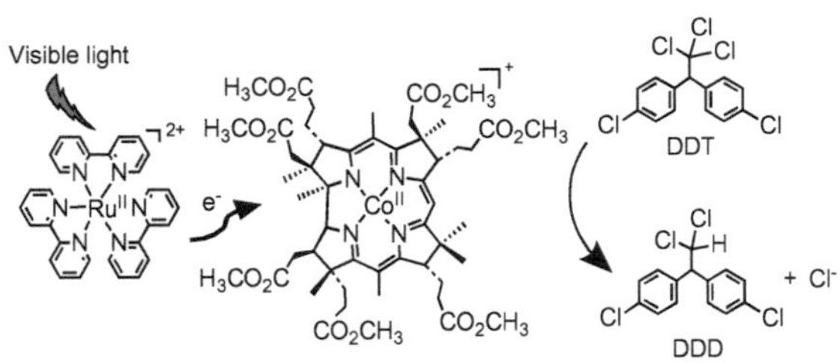

Fig. 3 Photocatalytic dichlorination of DDT catalyzed by heptamethyl cobyrinate with the photosensitizer, [Ru(II)(bpy)$_3$]Cl$_2$.

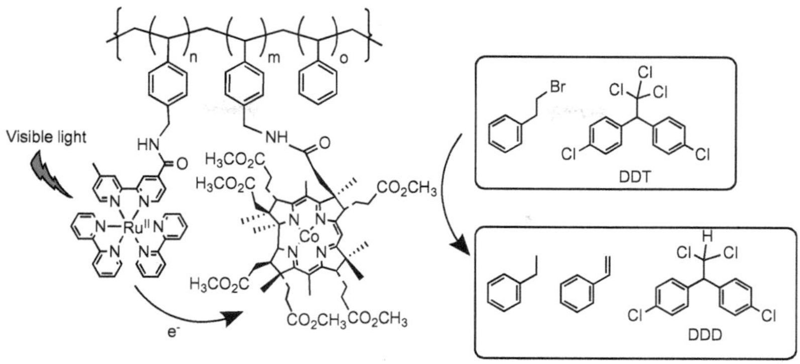

Fig. 4 Photocatalytic reaction by the polymer-supported vitamin B_{12} derivatives and photosensitizer.

Another approach in photo-driven B_{12} catalytic reaction is a heterogeneous system. The metal–organic frameworks (MOFs) have rapidly evolved in recent years and they are a class of crystalline materials, which are constructed from well-defined molecular building blocks and metal or metal-cluster connecting nodes. The cationic $[Ru(II)(bpy)_3]^{2+}$ and heptamethyl cobyrinate, are immobilized to the MOF, B_{12}-Ru$(bpy)_3$ @MOF, by a cation exchange reaction (Fig. 5) (Xu, Shimakoshi, & Hisaeda, 2015). The B_{12}-Ru$(bpy)_3$@MOF hybrid compound worked as a heterogeneous catalyst for the DDT dichlorination by visible light irradiation. It should be noted that B_{12}-Ru@MOF can be recovered by filtration and subsequently reused several times and maintain its photocatalytic activity.

Recently, cyclometalated iridium(III) complexes have been rapidly developing due to their superior photophysical and photochemical properties (Singh et al., 2015). The combined use of the cyclometalated iridium(III) complex and heptamethyl cobyrinate achieved a significantly high efficiency for the DDT dechlorination reaction, and the turnover numbers for the catalytic reaction were over 10,000 and 80,000 based on the heptamethyl cobyrinate and iridium(III) complex, respectively (Fig. 6) (Tian et al., 2018) The robust organometallic structure in the cyclometalated iridium(III) complex should affected the high catalytic performance.

2.2 Photo-driven B_{12} catalyzed organic synthesis

Photo-driven B_{12} catalyzed organic synthesis has developed using organic halides as substrate. The cross-coupling reactions of organic halides with

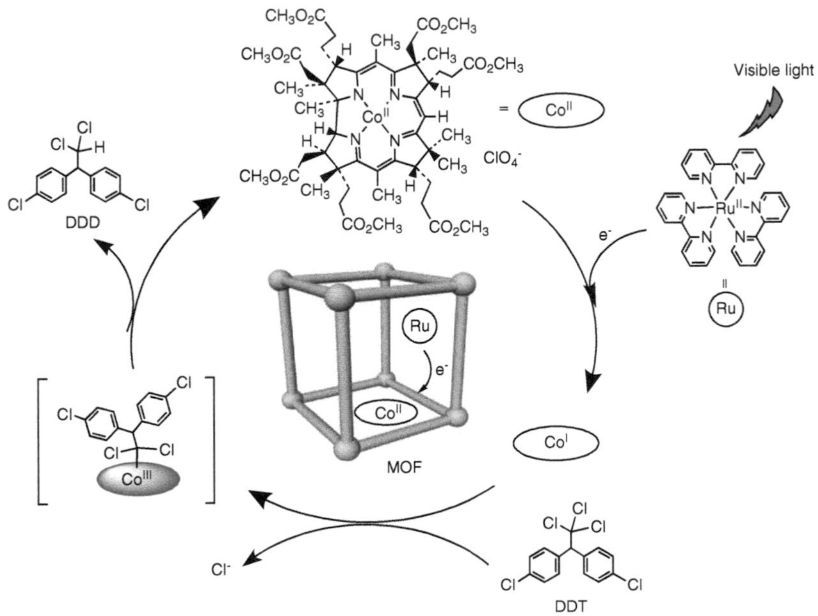

Fig. 5 Photocatalytic system of heptamethyl cobyrinate with the photosensitizer [Ru(II)(bpy)$_3$]Cl$_2$ in the MOF.

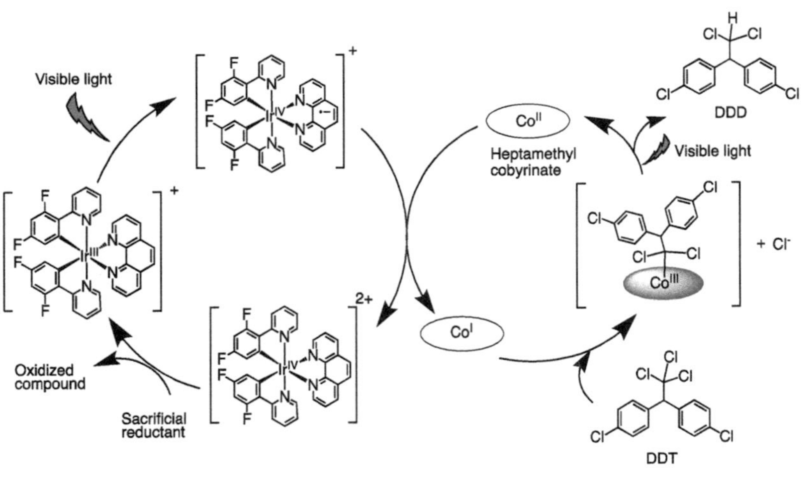

Fig. 6 Photocatalytic system of heptamethyl cobyrinate with cyclometalated iridium(III) complex.

terminal alkenes are useful methods of C(sp^3)-C(sp^2) bond formation for substituted alkenes in organic synthesis. A visible light driven coupling reaction of alkyl halides with styrene and its derivatives catalyzed by the heptamethyl cobyrinate with the [Ru(bpy)$_3$]Cl$_2$ has been developed (Fig. 7A)

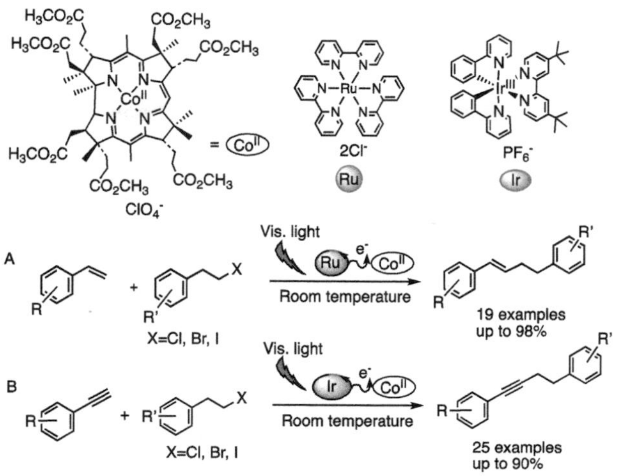

Fig. 7 Photocatalytic system of visible light-driven coupling reaction. (A) Mizoroki-Heck-type reaction of phenethyl bromide with styrene derivatives by heptamethyl cobyrinate and $[Ru(bpy)_3]Cl_2$, and (B) $C(sp^3)$-C(sp) bond formation catalyzed by heptamethyl cobyrinate and $[Ir(dtbbpy)(ppy)_2]PF_6$.

(Chen, Hisaeda, & Shimakoshi, 2019). Organometallic B_{12} complex was formed as intermediate and subsequent cleavage of cobalt carbon bond to form alkyl radical species. The alkyl radical was trapped by alkene to form the desired substituted alkenes as the product. The catalytic efficiencies of the B_{12} catalyst were compared to that of other B_{12} model complexes, such as cobaloxime. Mild reaction conditions at room temperature by visible light irradiation using an environmentally benign vitamin B_{12} derivative provided a practical protocol for the synthetic organic chemistry of the B_{12} catalysis.

Furthermore, visible light-driven cross-coupling reactions of alkyl halides with alkyne, such as phenylacetylene and its derivatives, for the $C(sp^3)$-C(sp) bond formation are reported using the heptamethyl cobyrinate and cyclometalated iridium(III) complex (Fig. 7B) (Chen et al., 2019). The robust B_{12} catalyst and iridium photocatalyst having a strong organometallic structure provided high turnover numbers of over 33,000 for the reaction.

Unique molecular transformation of organic halides was reported in B_{12} catalysis. Generally, the B_{12} mediated reactions are conducted under anaerobic conditions to inhibit the auto-oxidation of the Co(I) species of B_{12} derivatives. Oxygen-incorporated products were obtained from the trichlorinated organic compounds including DDT by the B_{12} catalyst and the cyclometalated iridium(III) complex in aerobic condition

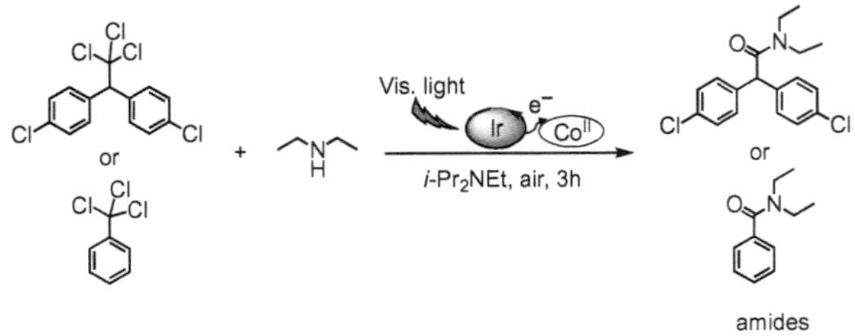

Scheme 1 Amide formation from DDT and benzotrichloride catalyzed by heptamethyl cobyrinate with cyclometalated iridium(III) complex.

(Tian, Shimakoshi, Ono, & Hisaeda, 2019). As the carbon-centered radical formed by cleavage of the cobalt-carbon bond of the alkylated complex easily reacts with oxygen, oxygen-incorporated amides were formed from benzotrichloride or DDT in the presence of amines (Scheme 1). Acyl chloride was considered as an intermediate for the aerobic dichlorination of trichloromethylated organic compounds.

2.3 B_{12}-titania hybrid catalyst system

Heterogeneous B_{12} hybrid catalyst was developed using the nanocrystal semiconductor. A semiconductor is a functional material, which is activated by light irradiation corresponding to its band gap energy. Among the semiconductors, the application of titanium oxide (TiO_2) has increased in various areas, such as solar cell devices, antibiotic and environmental applications, surface self-cleaning, and organic synthesis due to its chemical and physical stabilities and inexpensive price as well as its semiconductor property (Fujishima, Rao, & Tryk, 2000; Ohtani, 2010). The TiO_2 generates electron–hole pairs during band gap excitation (3.0 ~ 3.2 eV based on their crystal forms) by UV light irradiation. The excited electron existed in a conduction band (e_{CB}-) for TiO_2 has an E_{red} of ca. −0.5 V vs NHE (in pH 7 aqueous solution) in the anatase crystal form of TiO_2. This is negative enough to reduce the B_{12} derivatives to form the catalytical active Co(I) species. Thus, the light driven B_{12}-TiO_2 hybrid catalyst system in which B_{12} derivative is immobilized on TiO_2 surface was developed.

The B_{12}-TiO_2 hybrid catalyst was synthesized by surface modification of TiO_2 with cobyrinic acid having seven carboxylic acid group. Formation of Co(I) species of cobyrinic acid on TiO_2 surface during UV light irradiation was confirmed by appearance of typical UV–vis

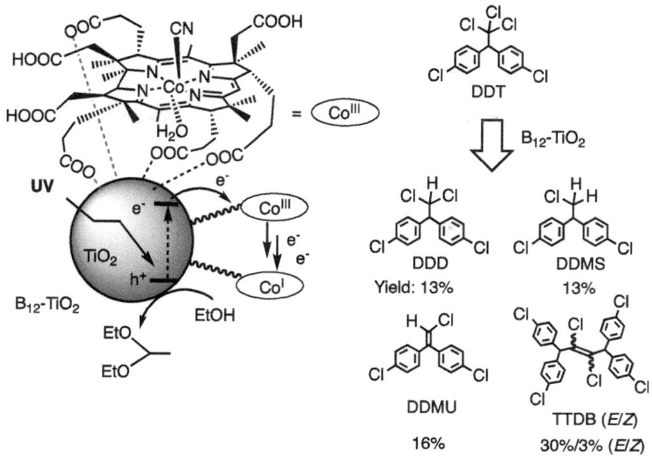

Fig. 8 Structure of the B_{12}-TiO_2 hybrid catalyst and photocatalytic dichlorination of DDT.

absorption of Co(I) species of B_{12} derivative around 390 nm. By the catalysis of B_{12}-TiO_2, the dechlorination of DDT or DDD occurred by UV light irradiation under nirogen to form various dechlorinated compounds in ethanol (Fig. 8) (Shimakoshi, Abiru, Izumi, & Hisaeda, 2009; Shimakoshi, Sakumori, Kaneko, & Hisaeda, 2009).

Applications of the B_{12}-TiO_2 hybrid catalyst for the radical-mediated organic synthesis were developed (Shimakoshi, Abiru, et al., 2009; Shimakoshi, Sakumori, et al., 2009). Various isomerization with 1,2-migration of functional groups were reported as mimic of B_{12}-devendent enzyme reaction. For example, UV light irradiation of a diethyl 2-bromomethyl-2-phenylmalonate produced the phenyl migrated product, diethyl benzylmalonate, by the catalysis of B_{12}-TiO_2 (Scheme 2). The B_{12}-TiO_2 catalyst was also used for the ring-expansion reactions of the alicyclic ketones (Scheme 3) (Izumi, Shimakoshi, Abe, & Hisaeda, 2010). This is a kind of isomerization with 1,2-migration of the acyl group.

To increase the stability of B_{12}-TiO_2, B_{12} derivative covalently bounded hybrid catalyst was prepared using a B_{12} derivative having two trimethoxysilyl groups at peripheral position. Surface sol–gel reaction of this B_{12} derivative with TiO_2 nano particle provided stable hybrid catalyst. The aerobic dechlorinations of benzotrichloride, DDT, 1,1,1-trichloroethane, and chloroform were achieved by the robust B_{12}-TiO_2 to form the corresponding ester or amide under air (Fig. 9) (Shimakoshi & Hisaeda, 2015).

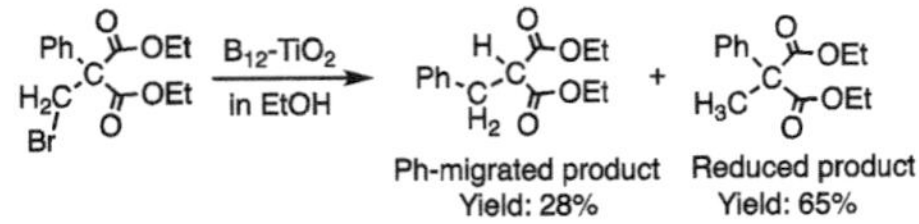

Scheme 2 1,2-Migration of a phenyl group catalyzed by the B_{12}-TiO_2.

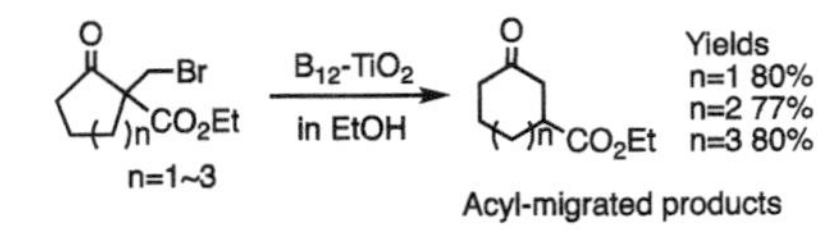

Scheme 3 Ring expansion reaction catalyzed by the B_{12}-TiO_2.

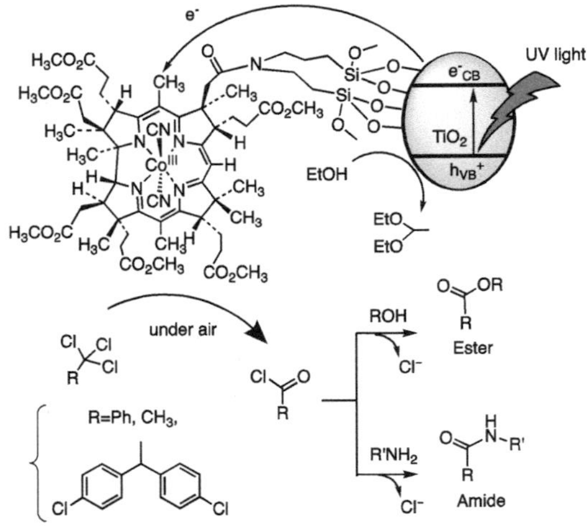

Fig. 9 Structure of covalently bound B_{12}-TiO_2 and aerobic dechlorination of trichloromethylated organic compounds by the B_{12}-TiO_2.

3. Electrochemical reactions catalyzed by vitamin B_{12} derivative

The electrochemical behaviors of vitamin B_{12} and its derivatives have been studied for a long time (Dereven'kov, Salnikov, Silaghi-Dumitrescu, Makarov, & Koifman, 2016; Lexa & Savéant, 1983). The cobalt complex of the B_{12} derivatives is electrochemically active, and the Co(I) species should be formed by an electrode process and its applications for electroorganic synthesis have been developed. Since indirect electrolysis using a mediator facilitates the redox process of the substrate to lower the applied potential resulting in a reduced energy cost and achieving an efficient molecular transformation, vitamin B_{12} and its derivatives were utilized as the mediator for the electroorganic reactions.

3.1 B_{12} catalyzed electrochemical dehalogenation of organic halides pollutants

The Co(II)/Co(I) redox couple of B_{12} derivative appeared at a relatively positive potential around $-0.4\,V \sim -0.6\,V$ vs Ag/AgCl in a nonaqueous solvent as the monoanionic corrin ring can stabilize the low oxidation state of the cobalt atom (Murakami, Hisaeda, Kajihara, & Ohno, 1984). The electrochemically generated Co(I) species of the vitamin B_{12} derivative reacts with organic halides by dehalogenation via an alkylated complex. The electrolysis of DDT catalyzed by heptamethyl cobyrinate was conducted at $-1.4\,V$ vs Ag/AgCl in DMF under anaerobic conditions (Shimakoshi, Tokunaga, & Hisaeda, 2004). Monodechlorinated products, such as DDD and 1,1-dichloro-2,2-bis(4-chlorophenyl)ethylene (DDE), and didechlorinated products, such as 1,1-bis(4-chlorophenyl)-2-chloroethylene (DDMU) and 1,1,4,4-tetrakis(4-chlorophenyl)-2,3-dichloro-2-butene (TTDB) (*E/Z*), were produced (Fig. 10). The electrolysis of DDT was also conducted in a conductive ionic liquid, 1-butyl-3-methylimidazolium tetrafluoroborate, without the use of the DMF electrolyte solution. Interestingly, further dechlorinated products, such as 1-chloro-2,2-bis (4-chlorophenyl)ethane (DDMS), DDNU, and 1,1-bis(4-chlorophenyl)

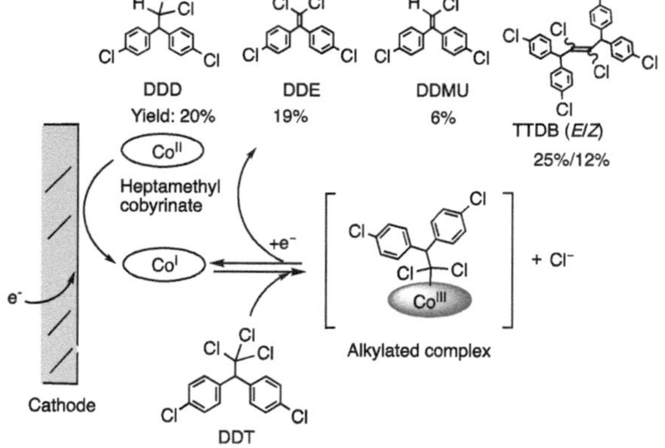

Fig. 10 Electrochemical dechlorination of DDT catalyzed by heptamethyl cobyrinate.

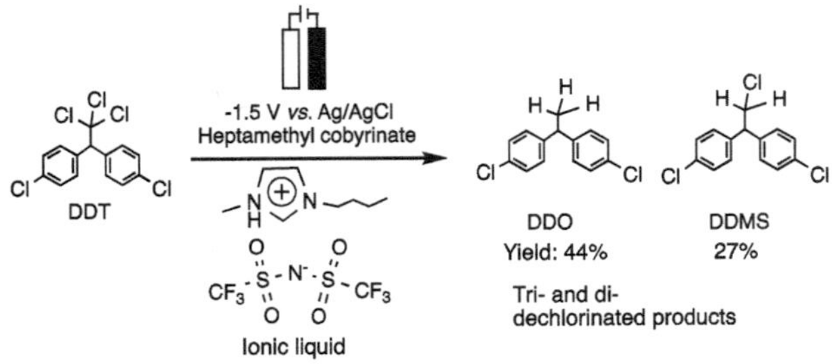

Scheme 4 Electrochemical dechlorination of DDT in ionic liquid catalyzed by heptamethyl cobyrinate.

ethane (DDO), were obtained in an ionic liquid (Scheme 4) (Jabbar, Shimakoshi, & Hisaeda, 2007). The enhancement of the B_{12} derivative reactivity in ionic liquid over the that of in DMF could be explained by the Hughes-Ingold predictions of the solvent polarity effects on the reaction rates (Cooper et al., 1948). Furthermore, the B_{12} derivative solution of an ionic liquid was recovered for the second electrolysis of DDT without any loss of catalysis efficiency due to the high stability and solubility of the catalyst in the ionic liquid. The advantages of an ionic liquid, such as high conductivity, very low volatility, and immiscible in an organic solvent combined with such a solvent effect, could provide a greener methodology for B_{12} derivative catalyzed electroorganic synthesis.

3.2 Electroorganic synthesis catalyzed by vitamin B_{12} derivative

The electrolysis of an organic halide under aerobic condition catalyzed by vitamin B_{12} derivative was also reported (Shimakoshi, Luo, Inaba, & Hisaeda, 2016). Though the labile Co(I) species of heptamethyl cobyrinate should be consumed during the electrolysis by auto-oxidation in air, the aerobic electrolysis of trichloromethylated organic compounds by the B_{12} catalyst afforded oxygen-incorporated dechlorinated products. The electrolysis of benzotrichloride under nitrogen afforded the 1,2-dichlorostilibenes (*E* and *Z* forms) and a small amount of the 1,1,2,2-tetrachloro-1,2-diphenylethane. On the other hand, the electrolysis of benzotrichloride in ethanol under aerobic conditions provided a 56% yield of ethylbenzoate with a 92% selectivity (Scheme 5A). The oxygen-controlled electrolysis of trichlorinated organic compounds was applied for the hazardous environmental pollutant, 1,1,1-trichloro-2,2,2-trifluoroethane (CFC-113a). The CFC-113a has been used as a cleaning agent that is responsible for

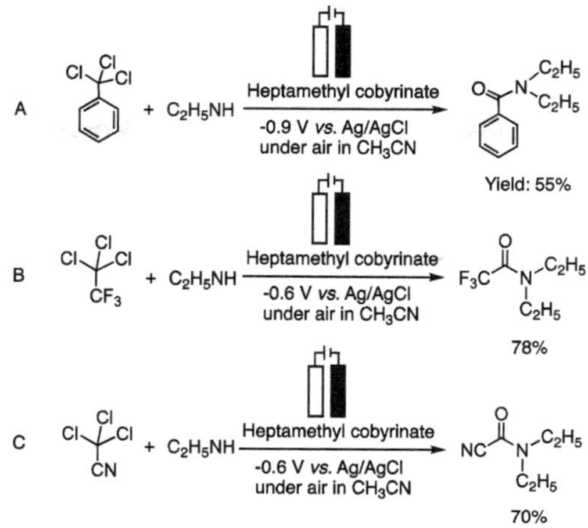

Scheme 5 Electrochemical transformation of trichloromethylated compounds to amides catalyzed by heptamethyl cobyrinate.

the ozone layer depletion (Adcock et al., 2018). From the CFC-113a, *N,N*-dialkyltrifluoroacetamides was obtained in moderate to good yields by the aerobic electrolysis with diamines (Scheme 5B).

The reductive electrolysis of an organic halide under aerobic conditions catalyzed by heptamethyl cobyrinate was applied for cynoformamides synthesis from trichloroacetonitrile (Moniruzzaman et al., 2021). The electrolysis of trichloroacetonitrile at −0.6 V vs Ag/AgCl in the presence of diamines in air provided cynoformamides in moderate to good yield by heptamethyl cobyrinate (Scheme 5C). The substrate radical should react with oxygen to form the oxyl-radical, then changes to cyanoformyl chloride by disproportionation. Thus, cyanoformyl chloride was deduced as an intermediate for the amide formation.

3.3 Co—H complex of vitamin B_{12} derivative for alkene and alkyne reduction

Vitamin B_{12} derivative catalyzed the C—C double and triple bonds reduction of alkene and alkyne, respectively. Alkene reductions catalyzed by the cyanocobalamin were reported by the Fischli and Donk groups using Zn and Ti(III) citrate as the reducing reagent, respectively (Scheme 6A and B) (Fischli & Suss, 1979; Shey et al., 2002). The electrochemically produced Co(I) species of the vitamin B_{12} derivative reacts with a proton to form the putative intermediate, cobalt-hydrogen complex (Co(III)-H complex), which catalyze the alkene and alkyne reductions. The direct reduction of

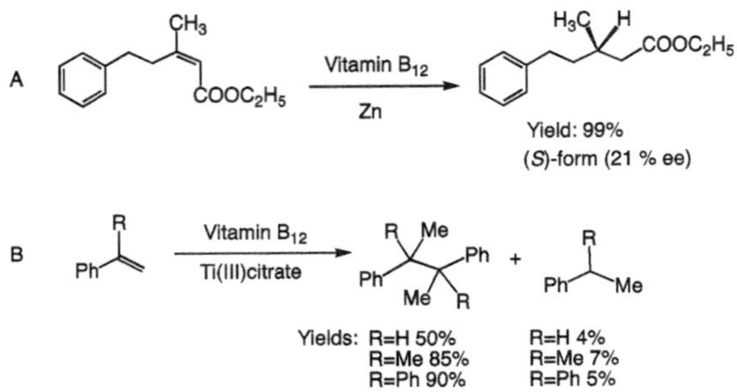

Scheme 6 Alkenes reduction catalyzed by vitamin B_{12}.

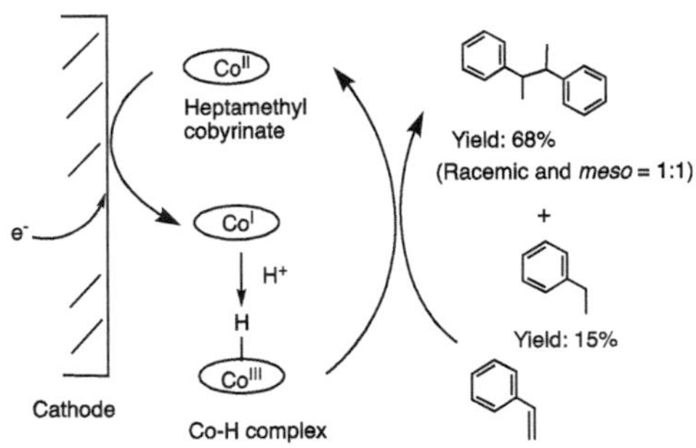

Fig. 11 Electrolysis of styrene catalyzed by heptamethyl cobyrinate in the presence of acid.

alkenes requires an extremely negative potential over −2 V vs Ag/AgCl in an organic solvent, while the alkene reduction may occur at a more positive potential in the presence of the vitamin B_{12} derivative since the Co(I) species can be formed at the potential and reacts with a proton to form the Co(III)-H complex as a reactive intermediate for the alkene reduction. The electrochemical reductive coupling and hydrogenation of alkenes catalyzed by heptamethyl cobyrinate were conducted in the presence of an acid (Fig. 11) (Shimakoshi, Luo, Tomita, & Hisaeda, 2017). The electrolysis was applied to a variety of conjugated alkenes to form moderate yields of the products. The electrolysis was also applied to alkynes, such as phenylacetylene and diphenylacetylene (Scheme 7). The spin-trapping experiment using 5,5-dimethylpyrroline *N*-oxide (DMPO) provided the DMPO-H ESR signal during the electrolysis. These results imply that the Co(III)-H complex has a hydrogen radical character and the attack of the hydrogen radical on the C=C bond of the alkene should be proceed. In this way, the bioinspired reduction of the alkene and alkyne by the

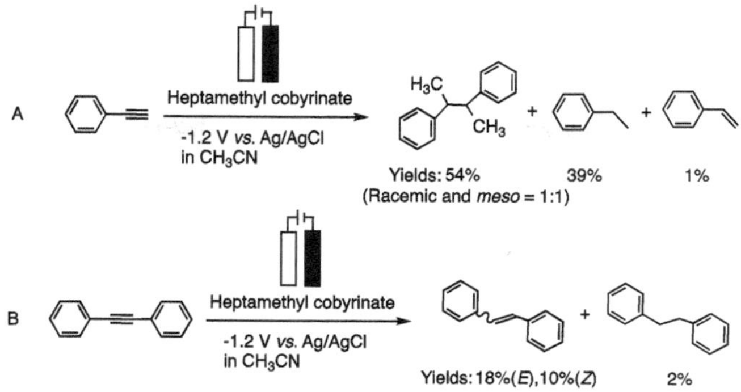

Scheme 7 Electrochemical reduction of alkynes catalyzed by heptamethyl cobyrinate.

vitamin B_{12} derivative proceeds under mild conditions, at room temperature and normal pressure without use of the hydrogen gas and a precious metal catalyst, such as platinum or ruthenium.

4. Conclusion and future directions

Bioorganometallic B_{12} acts as useful catalyst in various organic synthesis as described in this chapter. Photochemical transformations of organic halides catalyzed by vitamin B_{12} derivative were achieved in various fields of molecular transformations not only for organic synthesis but also reductive dichlorination of environmental pollutant. Applications of the bioinspired intermediate, the Co—H complex, as well as biomimetic intermediate, the alkylated cobalt complex, have been developed for electrochemical catalytic systems. In the electrochemical system, the combination of the bioorganometallic B_{12} with recent techniques in electrochemistry, such as a flow microreactor and light assisted electrolysis, will offer a new area for electroorganic synthesis. Furthermore, the asymmetric reaction by the vitamin B_{12} derivative should be developed in future as several chiral centers are existed in the corrin ring (Chen & Zhang, 2004). The progress of bioorganometallic B_{12} in molecular transformation will expand to many fields.

Acknowledgments

Parts of this study were supported by JSPS KAKENHI grant No. JP19H02735 from MEXT, Takahashi Industrial and Economic Research Foundation grant No. 11-003-359, and the Cooperative Research Program of "Network Joint Research Center for Materials and Devices: Dynamic Alliance for Open Innovation Bridging Human, Environment and Materials" grant No. 20214031.

References

Adcock, K. E., Reeves, C. E., Gooch, L. J., Elvidge, E. C. L., Ashfold, M. J., Brenninkmeijer, C. A. M., et al. (2018). Continued increase of CFC-113a (CCl_3CF_3) mixing ratios in the global atmosphere: Emissions, occurrence and potential sources. *Atmospheric Chemistry and Physics*, *18*, 4737–4751.

Arias-Rotondo, D. M., & McCusker, J. K. (2016). The photophysics of photoredox catalysis: A roadmap for catalyst design. *Chemical Society Reviews*, *45*, 5803–5820.

Banerjee, R. (2003). Radical carbon skeleton rearrangements: Catalysis by coenzyme B_{12}-dependent mutases. *Chemical Reviews*, *103*, 2083–2094.

Bommer, M., Kunze, C., Fesseler, J., Schubert, T., Diekert, G., & Dobbek, H. (2014). Structural basis for organohalide respiration. *Science*, *346*, 455–458.

Bridwell-Rabb, J., & Drennan, C. L. (2017). Vitamin B_{12} in the spotlight again. *Current Opinion in Chemical Biology*, *37*, 63–70.

Brown, K. L. (2005). Chemistry and enzymology of vitamin B_{12}. *Chemical Reviews*, *105*, 2075–2149.

Buckel, W., & Golding, B. T. (2006). Radicals enzymes in anaerobes. *Annual Review of Microbiology*, *60*, 27–49.

Chen, L., Hisaeda, Y., & Shimakoshi, H. (2019). Visible light-driven, room temperature heck-type reaction of alkyl halides with styrene derivatives catalyzed by B_{12} complex. *Advanced Synthesis and Catalysis*, *361*, 2877–2884.

Chen, L., Kametani, Y., Imamura, K., Abe, T., Shiota, Y., Yoshizawa, K., et al. (2019). Visible light-driven cross-coupling reactions of alkyl halides with phenylacetylene derivatives for C(sp^3)-C(sp) bond formation catalyzed by a B_{12} complex. *Chemical Communications*, *55*, 13070–13073.

Chen, Y., & Zhang, X. P. (2004). Vitamin B_{12} derivatives as natural asymmetric catalysts: Enantioselective cyclopropanation of alkenes. *The Journal of Organic Chemistry*, *69*, 2431–2435.

Cooper, K. A., Dhar, M. L., Hughes, E. D., Ingold, C. K., MacNulty, B. J., & Woolf, L. I. (1948). *Journal of the Chemical Society*, 2043–2049.

Dereven'kov, I. A., Salnikov, D. S., Silaghi-Dumitrescu, R., Makarov, S. V., & Koifman, O. I. (2016). Redox chemistry of cobalamin and its derivatives. *Coordination Chemistry Reviews*, *309*, 68–83.

Fischli, A., & Suss, D. (1979). Cob(I)alamin as catalyst. 5. Communication [1]. Enantioselective reduction of α,β-unsaturated carbonyl derivatives. *Helvetica Chimica Acta*, *62*, 2361–2373.

Francke, R., & Little, R. D. (2014). Redox catalysis in organic electrosynthesis: Basic principles and recent developments. *Chemical Society Reviews*, *43*, 2492–2521.

Fujishima, A., Rao, T. N., & Tryk, D. A. (2000). Titanium dioxide photocatalysis. *Journal of Photochemistry and Photobiology C*, *1*, 1–21.

Giedyk, M., Goliszewska, K., & Gryko, D. (2015). Vitamin B_{12} catalyzed reactions. *Chemical Society Reviews*, *44*, 3391–3404.

Gruber, K., Puffer, B., & Kräutler, B. (2011). Vitamin B_{12}-derivatives-enzyme cofactors and ligands of proteins and nucletic acids. *Chemical Society Reviews*, *40*, 4346–4363.

Hillard, E. A., & Jaouen, G. (2011). Bioorganometallics: Future trends in drug discovery. *Organometallics*, *30*, 20–27.

Izumi, S., Shimakoshi, H., Abe, M., & Hisaeda, Y. (2010). Photo-induced ring-expansion reactions mediated by B_{12}-TiO_2 hybrid catalyst. *Dalton Transactions*, *39*, 3302–3307.

Jabbar, M. A., Shimakoshi, H., & Hisaeda, Y. (2007). Enhanced reactivity of hydrophobic vitamin B_{12} towards the dechlorination of DDT in ionic liquid. *Chemical Communications*, 1653–1655.

Kiflom, W. G., Wandiga, S. O., & Kamau, G. N. (2001). Status of organochlorine (DDT) pollutants and steps toward electrocatalytic reductions. *Pure and Applied Chemistry*, *73*, 1907–1916.

Kim, Y.-H., & Carraway, E. (2002). R. Reductive dichlorination of PCE and TCE by vitamin B_{12} and ZVMs. *Environmental Technology*, *23*, 1135–1145.

Krone, U. E., Thauer, R. K., & Hogenkamp, H. P. C. (1989). Reductive dehalogenation of chlorinated C_1-hydrocarbons mediated by corrinoids. *Biochemistry*, *28*, 4908–4914.

Lexa, D., & Savéant, J. (1983). M. the electrochemistry of vitamin B_{12}. *Accounts of Chemical Research*, *16*, 235–243.

Licht, S., Gerfen, G. J., & Stubbe, J. (1996). Thiyl radicals in ribonucleotide reductases. *Science*, *271*, 477–481.

Matthews, R. G., Koutmos, M., & Datta, S. (2008). Cobalamin-dependent and cobamide-dependent methyltransferases. *Current Opinion in Structural Biology*, *18*, 658–666.

Moniruzzaman, M., Yano, Y., Ono, T., Imamura, K., Shiota, Y., Yoshizawa, K., et al. (2021). Electrochemical synthesis of cyanoformamides from trichloroacetonitrile and secondary amines mediated by the B_{12} derivatives. *The Journal of Organic Chemistry*, *86*, 16134–16143.

Murakami, Y., Hisaeda, Y., & Kajihara, A. (1983). Hydrophobic vitamin B_{12}. I. Preparation and axial ligatation behavior of hydrophobic vitamin B_{12r}. *Bulletin of the Chemical Society of Japan*, *56*, 3642–3646.

Murakami, Y., Hisaeda, Y., Kajihara, A., & Ohno, T. (1984). Hydrophobic vitamin B_{12}. II. Coordination geometry and redox behavior of heptamethyl cobyrinate in nonaqueous media. *Bulletin of the Chemical Society of Japan*, *57*, 405–411.

Ohtani, B. (2010). Photocatalysis a to Z-what we know and what we do not know in a scientific sense. *Journal of Photochemistry and Photobiology C Photochemistry Reviews*, *11*, 157–178.

Patra, M., Gasser, G., & Metzler-Nolte, N. (2012). Small organometallic compounds as antibacterial agents. *Dalton Transactions*, *41*, 6350–6358.

Schrauzer, G. N., & Katz, R. N. (1978). Reductive dichlorination and degradation of mirex and kepone with vitamin B_{12}. *Bioorganic Chemistry*, *9*, 123–143.

Shey, J., McGinley, C. M., McCauley, K. M., Dearth, A. S., Young, B. T., & van der Donk, W. A. (2002). Mechanistic investigation of a novel vitamin B_{12}-catalyzed carbon-carbon bond forming reaction, the reductive dimerization of arylalkenes. *The Journal of Organic Chemistry*, *67*, 837–846.

Shimakoshi, H., Abiru, M., Izumi, S., & Hisaeda, Y. (2009). Green molecular transformation by a B_{12}-TiO_2 hybrid catalyst as an alternative to tributyltin hydride. *Chemical Communications*, 6427–6429.

Shimakoshi, H., & Hisaeda, Y. (2015). Oxygen-controlled catalysis by vitamin B_{12}-TiO_2: Formation of esters and amides from trichlorinated organic compounds by photoirradiation. *Angewandte Chemie, International Edition*, *54*, 15439–15443.

Shimakoshi, H., & Hisaeda, Y. (2017). A hybrid catalyst for light-driven green molecular transformations. *ChemPlusChem*, *82*, 18–29.

Shimakoshi, H., & Hisaeda, Y. (2018). Electrochemistry and catalytic properties of vitamin B_{12} derivatives in nonaqueous media. *Current Opinion in Electrochemistry*, *8*, 24–30.

Shimakoshi, H., Kudo, S., & Hisaeda, Y. (2005). Recycled catalysis of a hydrophobic vitamin B_{12} in a ionic liquid. *Chemistry Letters*, *34*, 1096–1097.

Shimakoshi, H., Luo, Z., Inaba, T., & Hisaeda, Y. (2016). Electrolysis of trichloromethylated organic compounds under aerobic conditions catalyzed by B_{12} model complex for ester and qmide formations. *Dalton Transactions*, *45*, 10173–10180.

Shimakoshi, H., Luo, Z., Tomita, K., & Hisaeda, Y. (2017). Cathodic reductive coupling and hydrogenation of alkenes and alkynes catalyzed by the B_{12} model complex. *Journal of Organometallic Chemistry*, *839*, 71–77.

Shimakoshi, H., Maeyama, H., Kaieda, T., Matsuo, T., Matsui, E., Naruta, Y., et al. (2005). Hydrophobic vitamin B_{12}. Part 20: Supernucleophilicity of co(I) heptamethyl cobyrinate toward various organic halides. *Bulletin of the Chemical Society of Japan*, *78*, 859–863.

Shimakoshi, H., Nishi, M., Tanaka, A., Chikama, K., & Hisaeda, Y. (2011). Photocatalytic function of a polymer-supported B_{12} complex with a ruthenium trisbipyridine photosensitizer. *Chemical Communications*, *47*, 6548–6550.

Shimakoshi, H., Sakumori, E., Kaneko, K., & Hisaeda, Y. (2009). B_{12}-TiO_2 hybrid catalyst for dehalogenation of organic halides. *Chemistry Letters*, *38*, 468–469.

Shimakoshi, H., Tokunaga, M., Baba, T., & Hisaeda, Y. (2004). Photochemical dechlorination of DDT catalyzed by a hydrophobic vitamin B_{12} and a photosensitizer under irradiation with visible light. *Chemical Communications*, 1806–1807.

Shimakoshi, H., Tokunaga, M., & Hisaeda, Y. (2004). Hydrophobic vitamin B_{12}. Part 19: Electroorganic reaction of DDT mediated by hydrophobic vitamin B_{12}. *Dalton Transactions*, 878–882.

Singh, A., Teegardin, K., Kelly, M., Prasad, K. S., Krishnan, S., & Weaver, J. D. (2015). Facile synthesis and complete characterization of homoleptic and heteroleptic cyclometalated iridium(III) complexes for photocatalysis. *Journal of Organometallic Chemistry*, *776*, 51–59.

Skubi, K. L., Blum, T. R., & Yoon, T. P. (2016). Dual catalysis strategies in photochemical synthesis. *Chemical Reviews*, *116*, 10035–10074.

Tian, H., Shimakoshi, H., Ono, T., & Hisaeda, Y. (2019). Visible light-driven, one-pot amide synthesis catalyzed by the B_{12} model complex under aerobic conditions. *ChemPlusChem*, *84*, 237–240.

Tian, H., Shimakoshi, H., Park, G., Kim, S., You, Y., & Hisaeda, Y. (2018). Photocatalytic function of B_{12} complex with cyclometalated iridium(III) complex as photosensitizer under visible light irradiation. *Dalton Transactions*, *47*, 625–982.

Xu, J., Shimakoshi, H., & Hisaeda, Y. (2015). Development of metal-organic framework (MOF)-B_{12} system as new bio-inspired heterogeneous catalyst. *Journal of Organometallic Chemistry*, *782*, 89–95.

Yan, M., Kawamata, Y., & Baran, P. S. (2017). Synthetic organic electrochemical methods since 2000: On the verge of a renaissance. *Chemical Reviews*, *117*, 13230–13319.

Yoshida, J., Kataoka, K., Horcajada, R., & Nagaki, A. (2008). Modern strategies in electroorganic synthesis. *Chemical Reviews*, *108*, 2265–2299.

Zhang, P., & Sadler, P. J. (2017). Advances in the design of organometallic anticancer complexes. *Journal of Organometallic Chemistry*, *839*, 5–14.

Zhang, W., Shimakoshi, H., Houfuku, N., Song, X., & Hisaeda, Y. (2014). A polymerized ionic liquid-supported B_{12} catalyst with a ruthenium trisbipyridine photosensitizer for photocatalytic dechlorination in ionic liquid. *Dalton Transactions*, *43*, 13972–13978.

Ziessel, R., Hawecker, J., & Lehn, J.-M. (1986). Photogeneration of carbon monoxide and of hydrogen *via* simultaneous photochemical reduction of carbon dioxide and water by visible-light irradiation of organic solutions containing tris(2,2′-bipyridine)ruthenium(II) and cobalt(II) species as homogeneous catalysts. *Helvetica Chimica Acta*, *69*, 1065–1084.

CHAPTER THREE

Cobamide remodeling

Amy T. Ma*, Daniel S. Kantner, and Joris Beld
Department of Microbiology and Immunology, Center for Advanced Microbial Processing, Institute for Molecular Medicine and Infectious Disease, Drexel University College of Medicine, Philadelphia, PA, United States
*Corresponding author: e-mail address: atm84@drexel.edu

Contents

Abstract

Cobamides are a family of structurally-diverse cofactors which includes vitamin B_{12} and over a dozen natural analogs. Within the nucleotide loop structure, cobamide analogs have variable lower ligands that fall into three categories: benzimidazoles, purines, and phenols. The range of cobamide analogs that can be utilized by an organism is dependent on the specificity of its cobamide-dependent enzymes, and most bacteria are able to utilize multiple analogs but not all. Some bacteria have pathways for cobamide remodeling, a process in which imported cobamides are converted into compatible analogs. Here we discuss cobamide analog diversity and three pathways for cobamide remodeling, mediated by amidohydrolase CbiZ, phosphodiesterase CbiR, and some homologs of cobamide synthase CobS. Remodeling proteins exhibit varying degrees of specificity for cobamide substrates, reflecting different strategies to ensure that imported cobamides can be utilized.

1. Introduction

Cobamides are required micronutrients utilized by organisms in all three domains of life, but they are produced *de novo* by only some

Vitamins and Hormones, Volume 119
ISSN 0083-6729
https://doi.org/10.1016/bs.vh.2022.01.004

prokaryotes (Martens, Barg, Warren, & Jahn, 2002). Most bacteria encode cobamide-dependent enzymes and transporters for cobamides and their intermediates (Shelton et al., 2019). However, because only some bacteria have the ability to produce cobamides, they are thought of as a shared resource in microbial communities (Sokolovskaya, Shelton, & Taga, 2020). Approximately 40% of bacteria encode the genes required for catalyzing the ~30 enzymatic steps required for *de novo* biosynthesis, and almost 20% of bacteria encode the same last steps as in *de novo* biosynthesis (Shelton et al., 2019), which are associated with partial biosynthesis pathways called salvaging and remodeling. In salvaging, biosynthesis begins with imported intermediates from which the cobamide is built. During remodeling, imported cobamides are converted into compatible analogs *via* cleavage into biosynthetic intermediates or through a direct pathway. Notably, two algal species were shown to remodel cobamides (Baum, Menezes, Svatos, & Schubert, 2020; Helliwell et al., 2016) but the mechanism is unknown. Cobamide remodeling is reflective of cobamide-dependent enzymes having specificity for certain analogs, which ranges widely from utilization of all three classes of cobamides to utilization of only one analog, as recently reviewed (Sokolovskaya et al., 2020). In this chapter, we cover the diversity and distribution of cobamide analogs and three pathways for remodeling them.

2. Cobamide diversity and distribution

2.1 Cobamide analogs

Cobamides consist of a contracted tetrapyrrole ring coordinating a central cobalt ion, a convertible upper ligand (R_1 in Fig. 1), and an aminopropanol arm linking a nucleotide loop containing the variable moiety (R_2 in Fig. 1). This variable position defines the three major classes of cobamides: benzimidazolyl-, purinyl-, and phenolylcobamides. Here we refer to all constituents at the variable position as lower ligands, although only benzimidazolyl and purinyl lower ligands coordinate the central cobalt ion while phenolyl ones do not. Another class of cobamide analogs are the norcobamides, which lack a methyl group from the aminopropanol arm. Norpseudocobalamin is produced by *Sulfurospirillum multivorans* and supports reductive dehalogenation (Kräutler et al., 2003).

The most well-known cobamide is vitamin B_{12} ([DMB]Cba, also referred to as cobalamin), which contains 5,6-dimethylbenzimidazole (DMB) at the lower ligand position. It was discovered as a vital nutrient

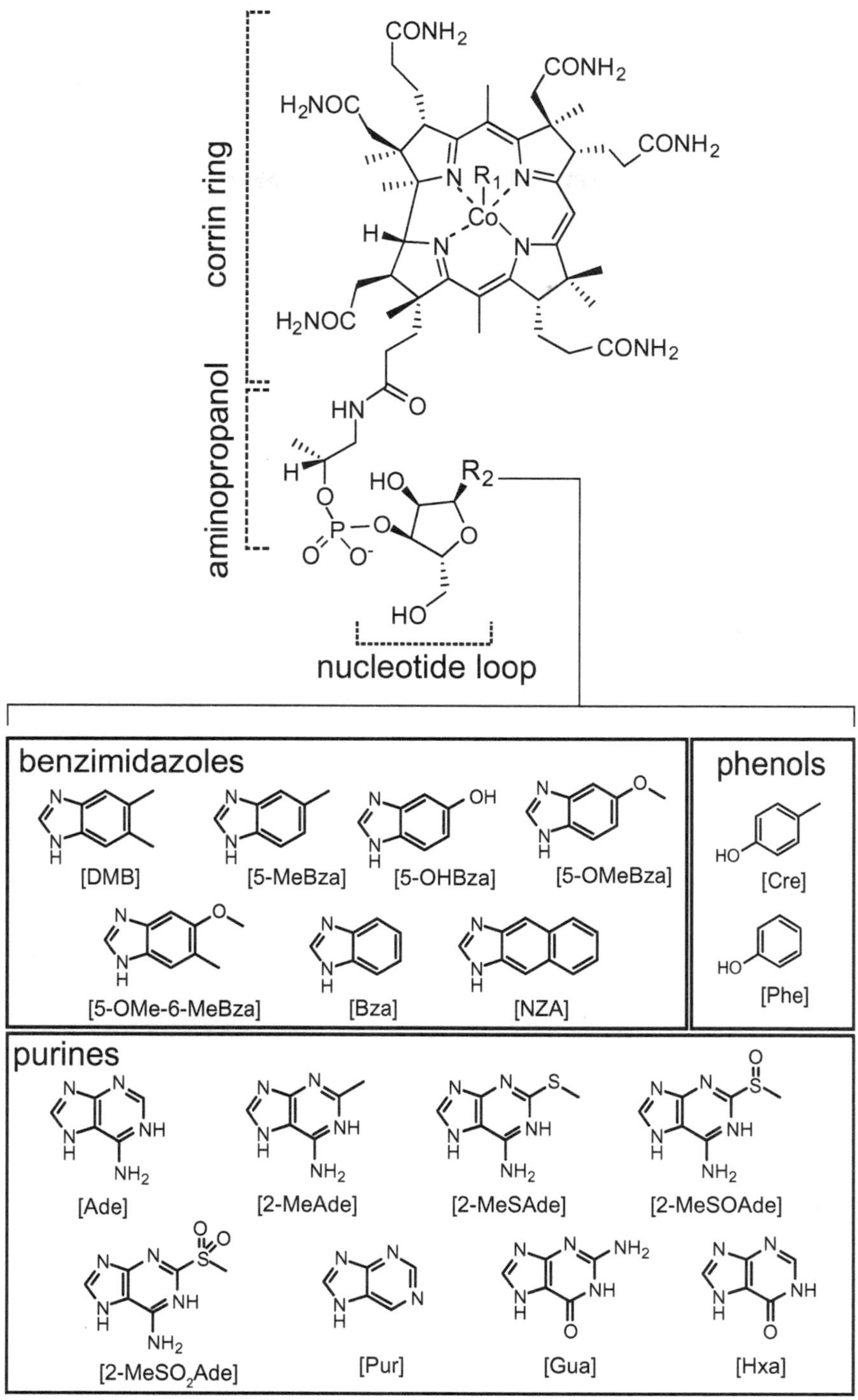

Fig. 1 Structural diversity of cobamides. Cobamides are composed of a contracted tetrapyrrole surrounding a central cobalt ion (referred to as a corrin ring), an aminopropanol linker, and a nucleotide loop containing the lower ligand (R_2). Upper ligands (R_1) include -CH_3, —Ado, -OH and -CN. Lower ligands (R_2) can be grouped into three classes: benzimidazoles, purines, and phenols (see Table 1).

in liver extracts that cured pernicious anemia (Bunn, 2014). Bioassays with *Lactobacillus lactis* were used for purification of cobalamin from liver extracts and the structure of cobalamin was elucidated by X-ray crystallography in the 1950s (Hodgkin et al., 1956). Around this time, it was discovered that several Actinomycetes, *Bacillus megaterium*, *Mycobacterium smegmatis*, and species of *Pseudomonas* and *Propionibacterium* produce cobalamin *de novo* (Table 1) (Leviton & Hargrove, 1952). Currently, *Pseudomonas denitrificans, Propionibacterium shermanii,* and *Propionibacterium freudenreichii* are used for industrial production of cobalamin (Balabanova, Averianova, Marchenok, Son, & Tekutyeva, 2021). Other known *de novo* producers of cobalamin include species of *Eubacterium*, Rhodobacteraceae, and Rhizobia (see Table 1). Producers of other benzimidazole-containing cobamides include *Methanobacillus omelianskii* and other methanogens ([5-OHBza]Cba) (Kräutler et al., 1988; Lezius & Barker, 1965), *Methanothrix soehngenii* and other sulfate reducers ([5-MeBza]Cba) (Kohler, 1988; Kräutler et al., 1988), *Clostridium thermoaceticum* ([5-OMeBza]Cba) (Irion & Ljungdahl, 1965), and *Clostridium formicoaceticum* ([5-OMe-6-MeBza]Cba) (Stupperich, Eisinger, & Kräutler, 1988). In sewage sludge, [Bza]Cba (Friedrich & Bernhauer, 1958) and [NZA]Cba have been found (Renz, 1999).

Pseudocobalamin ([Ade]Cba) is an abundant purinylcobamide that contains adenine at the lower ligand position and was discovered as the cobamide analog produced by *Clostridium tetanomorphum* (Barker et al., 1958). Other known *de novo* producers of pseudocobalamin include *Clostridium cochlearium*, cyanobacteria*, Lactobacillus reuteri*, *Salmonella enterica*, and *Eubacterium hallii* (Table 1). Producers of other purinyl-containing cobamides include *C. cochlearium* ([2-MeAde]Cba) (Hoffmann et al., 2000), *Desulfovibrio vulgaris* ([Hxa]Cba and [Gua]Cba) (Guimarães et al., 1994) and *Desulfitobacterium hafniense* ([Pur]Cba) (Yan et al., 2018). In sewage sludge, 2-methylthioadenine cobamide ([2-MeSAde]Cba) (Friedrich & Bernhauer, 1957) and its oxidized forms were discovered (Renz et al., 1987).

The third class of cobamides contain phenol groups at the lower ligand position. *De novo* producers of *para*-cresolylcobamide ([Cre]Cba) include *Sporomusa ovata* (Stupperich & Eisinger, 1989), *Veillonella parvula* (Crofts et al., 2013), and *Pelosinus fermentans* (Men et al., 2014), and phenolylcobamide ([Phe]Cba) has also been isolated from *S. ovata* (Stupperich et al., 1989).

2.2 Distribution of cobamide analogs

Cobamide profiles in mammalian gastrointestinal tracts have been characterized through analysis of rumen and fecal samples. Using bioassays paired with

Table 1 Cobamide diversity: natural producers and other sources.

Abbrev (name)	Lower ligand	Source(s)	References
benzimidazoles			
[DMB]Cba (cobalamin)	5,6-dimethylbenzimidazole	*Streptomyces griseus*	Rickes, Brink, Koniuszy, Wood, and Folkers (1948)
		other Actinomycetes	Saunders, Otto, and Sylvester (1952)
		Bacillus megaterium[a]	Garibaldi, Ijichi, Snell, & Lewis, 1953
		Mycobacterium smegmatis	Hendlin and Ruger (1950)
		Pseudomonas species	Cameron, Briggs, Pridmore, Brefort, and Crouzet (1989)(Ma, Beld and Brahamsha, 2017)
		Propionibacterium species	Leviton and Hargrove (1952)
		Eubacterium species	Höllriegl, Lamm, Rowold, Hörig, and Renz (1982); Schulze, Vogler, and Renz (1998)
		Rhodobacteraceae	Gray and Escalante-Semerena (2009b); McGoldrick, Deery, Warren, and Heathcote (2002)
		Sinorhizobium meliloti	Campbell et al. (2006)
		Rhizobium cobalaminogenum	Martens, Barg, Warren, & Jahn, 2002
		Salmonella species	Anderson et al. (2008); Keck and Renz (2000)
		other bacteria	Martens, Barg, Warren, & Jahn, 2002; Shelton et al. (2019)

Continued

Table 1 Cobamide diversity: natural producers and other sources.—cont'd

Abbrev (name)	Lower ligand	Source(s)	References
[5-MeBza]Cba	5-methylbenzimidazole	*Methanothrix soehngenii*	Kohler (1988)
		sulfate reducing bacteria	Kräutler, Kohler, and Stupperich (1988)
[5-OHBza]Cba	5-hydroxybenzimidazole	*Methanobacillus omelianskii*	Lezius & Barker, 1965
		archaea and methanogens	Kräutler et al. (1988)
		Geobacter sulfurreducens	Hazra et al. (2015)
[5-OMeBza]Cba	5-OMe-benzimidazole	*Clostridium thermoaceticum*[b]	Irion and Ljungdahl (1965)
[5-OMe-6-MeBza]Cba	5-OMe-6-Me-benzimidazole	*Clostridium formicoaceticum*	Stupperich and Eisinger (1989); Stupperich, Eisinger and Krautler (1988)
[Bza]Cba	benzimidazole	sewage sludge	Friedrich and Bernhauer (1958)
[NZA]Cba	napthimidazole	sewage sludge	Friedrich and Bernhauer (1959)
purines			
[Ade]Cba (pseudocobalamin)	adenine	*Clostridium tetanomorphum*	Barker, Weissbach, and Smyth (1958)
		other *Clostridium* species	Hoffmann et al., 2000; Stadtman (1960)
		methanogens	Stupperich and Kräutler (1988)
		Propionibacterium species	Hoffmann et al. (2000)
		Lactobacillus species	Santos et al. (2007); Torres et al. (2016)

		Eubacterium hallii	Belzer et al. (2017)
		Salmonella species	Anderson et al., 2008); Keck and Renz (2000)
		cyanobacteria	Heal et al. (2017); Helliwell et al. (2016); Ma, Beld, and Brahamsha (2017); Miyamoto et al. (2006); Tanioka et al. (2009)
[2-MeAde]Cba	2-methyladenine	*Clostridium cochlearium*	Hoffmann et al. (2000)
		Salmonella enterica, serovar Typhimurium	Keck and Renz (2000)
[Hxa]Cba	hypoxanthine	*Desulfovibrio vulgaris*	Guimarães, Weber, Klaiber, Vogler, and Renz (1994)
[Gua]Cba	guanine	*Desulfovibrio vulgaris*	Guimarães et al. (1994)
[Pur]Cba	purine	*Desulfitobacterium hafniense*	Yan et al. (2018)
[2-MeSAde]Cba	2-(methylthio)adenine	sewage sludge	Friedrich and Bernhauer (1957)
[2MeSOAde]Cba	(2-methylsulfinyl)adenine	sewage sludge	Renz, Blickle, and Friedrich (1987)
[2MeSO$_2$Ade]Cba	(2-methylsulfonyl)adenine	sewage sludge	Renz et al. (1987)
phenols			
[Cre]Cba	*para*-cresol	*Sporomusa ovata*	Stupperich and Eisinger (1989)
		Veillonella parvula	Crofts, Seth, Hazra, and Taga (2013)
		Pelosinus fermentans	Men et al. (2014)
[Phe]Cba	phenol	*Sporomusa ovata*	Stupperich, Eisinger, and Kräutler (1989)

[a]Renamed *Priestia megaterium*.
[b]Renamed *Moorella thermoacetical*.

paper electrophoresis and chromatography in comparison to isolated and synthesized standards, it was determined that cow rumen contains predominantly [2-MeAde]Cba, significant amounts of pseudocobalamin, [5-OHBza]Cba, and the intermediate cobinamide, and some cobalamin (Dryden & Hartman, 1971; Ford, Holdsworth, Kon, & Porter, 1953). These results were replicated more recently using mass spectrometry and [2-MeSAde]Cba and [Cre]Cba were also detected (Girard, Berthiaume, Stabler, & Allen, 2009). Similar analysis of human feces detected seven cobamides (cobalamin, pseudocobalamin, [2-MeAde]Cba, [5-OHBza]Cba, [2-MeSAde]Cba, [Phe]Cba and [Cre]Cba) and the cobinamide intermediate, and that [2-MeAde]Cba and [Cre]Cba were the dominant analogs (Allen & Stabler, 2008). Cobalamin supplementation was also assessed in this study, resulting in higher overall cobamide levels, including not only cobalamin but other analogs as well, providing early evidence of cobamide remodeling. A more recent study examining cobamides in murine fecal samples detected the cobinamide intermediate and four cobamide analogs: cobalamin, pseudocobalamin, [2-MeAde]Cba, and [2-MeSAde]Cba (Kelly et al., 2019).

Mammalian gastrointestinal tracts are a rich source of diverse cobamide analogs, but examination of environmental samples showed much more limited cobamide profiles. Analysis of marine samples showed that surface waters contained pseudocobalamin and cobalamin, with cyanobacteria and α-proteobacteria as the likely producers, respectively, while deeper waters contained cobalamin likely produced by archaea (Heal et al., 2017). Analysis of soil samples showed the presence of cobalamin in most samples but pseudocobalamin in only one (Lu, Heal, Ingalls, Doxey, & Neufeld, 2020). However, other cobamide analogs are likely present in the environment because known producers of cobamide analogs are environmental bacteria and archaea. For example, microbial enrichments of dechlorinating communities from soil or ground water contained seven different cobamides (cobalamin, pseudocobalamin, [Cre]Cba, [5-OHBza]Cba, [5-MeBza]Cba, [2-MeSAde]Cba, and [2-MeAde]Cba) and cobinamide (Men et al., 2015).

2.3 Biosynthesis of cobamide analogs

The lower ligand moiety is the distinguishing element of different cobamide analogs and lower ligands can be produced by some bacteria. There are two pathways for DMB biosynthesis: the oxygen-dependent BluB pathway which strips DMB from flavin mononucleotide (Taga, Larsen, Howard-Jones, Walsh, & Walker, 2007) and the anaerobic BzaABCDE pathway

(Hazra et al., 2015). Biosynthesis *via* the latter pathway derives DMB from 5-aminoimidazole ribotide, with sequential activity of BzaA/B producing [5-OHBza], BzaC producing [5-OMeBza], BzaD producing [5-OMe-6-MeBza] and finally BzaE producing DMB. Some bacteria encode partial Bza pathways and produce DMB intermediates which are incorporated as lower ligands. Sources of other benzimidazolyl lower ligands, [5-MeBza], [Bza] and [NZA], are unknown. Adenine and purine are common cellular metabolites, but it is unknown how other purinyl lower ligands are produced. *Para*-cresol and phenol are byproducts of tyrosine fermentation (Saito, Sato, Nomoto, & Tsuji, 2018). Lower ligands can also be imported from the environment, and some Firmicutes import activated lower ligands. CblT mediates import of ribosylated DMB with subsequent phosphorylation by CblS (Mattes & Escalante-Semerena, 2017), producing the DMB-R5P intermediate required for cobalamin biosynthesis (see below).

The penultimate step for cobamide biosynthesis is catalyzed by cobamide synthase CobS, which acts upon substrates Ado-cobinamide-GDP and DMB ribose-5′-phosphate (DMB-R5P, also referred to α-ribozole-5′-phosphate or α-RP) for the production of cobalamin (Cameron et al., 1991; Maggio-Hall & Escalante-Semerena, 1999). The DMB lower ligand is activated by the protein CobT (E.C. 2.4.2.21) which catalyzes the addition of R5P to DMB (Trzebiatowski & Escalante-Semerena, 1997). Production of purinyl- and phenolylcobamides requires the same activation of lower ligands before incorporation into cobamides. Some CobT proteins can also activate purinyl lower ligands, but a distinct group of distant CobT homologs were found in cyanobacteria, which largely produce the pseudocobalamin analog. While close homologs of bacterial CobT proteins were not found in cyanobacteria, it was discovered that they encode ones with more similarity to archaeal ones (Jeter, Mattes, Beattie, & Escalante-Semerena, 2019). Phenolyl lower ligands are activated by ArsAB, specialized CobT homologs which function as heterodimers. *arsAB* genes are encoded in tandem and are found exclusively in Negativicutes bacteria, some of which were shown to produce phenolylcobamides (Chan & Escalante-Semerena, 2011). Analysis of a diverse range of CobT proteins showed that they can accept different lower ligand substrates, with many demonstrating preferential activation of DMB (Crofts et al., 2013; Hazra, Tran, Crofts, & Taga, 2013). Supplementation of bacteria with exogenous lower ligands can shift their cobamide profile, and this process is called "guided biosynthesis" (Bernhauer & Friedrich, 1954; Mok & Taga, 2013). This process exploits the promiscuity of CobT and CobS enzymes to incorporate exogenous lower ligands and can be used for producing unnatural cobamides in the laboratory. Guided biosynthesis is reflective of natural interactions between bacteria and their environment, which can provide access to biosynthetic intermediates and different lower ligands for the production of cobamides.

3. Cobamide remodeling proteins

Import of cobamides is common in bacteria, including ones that can produce them. Some bacteria require specific analogs to support function their cobamide-dependent enzymes. Cobamide remodeling is a way for bacteria to import available cobamides and ensure that they are converted to compatible analogs. CbiZ and CbiR remodel cobamides by cleaving them into biosynthetic intermediates, while some homologs of cobamide biosynthesis protein CobS can directly replace the nucleotide loop.

3.1 Amidohydrolase CbiZ

CbiZ (E.C. 3.5.1.90) cleaves the amide bond between the corrin ring and the aminopropanol arm to produce cobyric acid (Woodson & Escalante-Semerena, 2004). CbiZ is encoded in 9% of bacterial genomes, in organisms with partial biosynthesis pathways and also in ones with *de novo* pathways (Shelton et al., 2019). *cbiZ* genes are sometimes encoded as multiple copies within the same genome, are frequently located within gene clusters related to corrinoid transport or biosynthesis, and are sometimes encoded as fusions to transport proteins. CbiZ homologs are encoded by many archaea, Bacillaceae, Actinomycetales, and Negativicutes bacteria, among others. CbiZ activity was first characterized in the cobinamide salvaging pathway of archaea and was later found to also cleave cobamides for remodeling in bacteria.

The *cbiZ* gene was identified as a part of a four gene operon containing cobamide biosynthesis genes in archaeon *Halobacterium* species strain NRC-1 and was shown to be required for cobinamide salvaging (Woodson & Escalante-Semerena, 2004). Archaea lack homologs of the bacterial bi-functional CobU protein (CobP in *Pseudomonas* nomenclature), which is required for *de novo* biosynthesis and plays an additional role in cobinamide salvaging. CobU phosphorylates cobinamide after import, making the same cobinamide-phosphate intermediate produced by CbiB during *de novo* biosynthesis (Fig. 2). Cobinamide-phosphate is then converted to cobinamide-GDP through the addition of GMP by CobU in bacteria or CobY in archaea. Although archaea lack the cobinamide kinase activity of CobU, they were

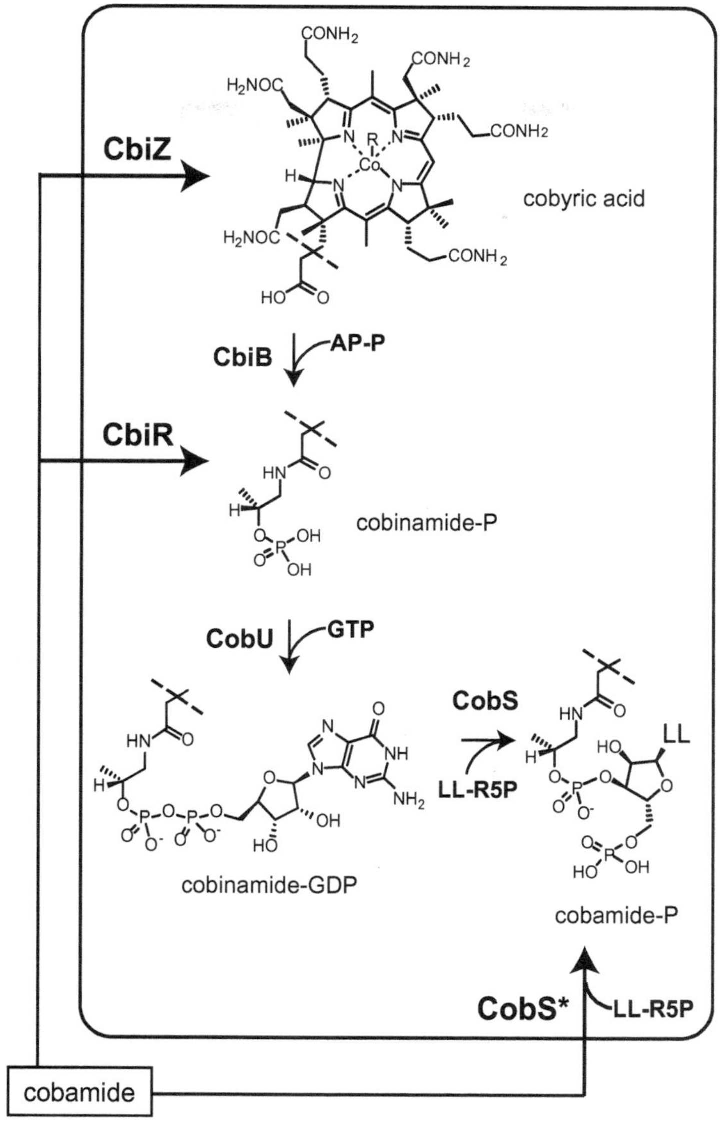

Fig. 2 Cobamide remodeling pathways. The last biosynthetic steps of *de novo* biosynthesis of cobamides, all or some of which are shared by salvaging and remodeling pathways. Cleavage by CbiZ or CbiR produces cobyric acid or cobinamide-phosphate intermediates, respectively. Some homologs of CobS, denoted by CobS*, perform direct remodeling in which the LL-R5P structure is directly replaced. Initial cobamide adenosylation by CobA and final dephosphorylation by CobC are not shown. R, adenosine upper ligand; AP-P, aminopropanol-phosphate; LL, lower ligand; R5P, ribose-5′-phosphate.

found to salvage cobinamide by converting it to an earlier intermediate *via* amidohydrolase CbiZ. The activity of CbiZ was characterized through *in vitro* reactions with purified protein, demonstrating that it cleaved the aminopropanol arm from cobinamide to produce the cobyric acid intermediate (Woodson & Escalante-Semerena, 2004). Other bacterial and archaeal homologs of CbiZ were shown to be functional in cobinamide salvaging assays performed heterologously in *Salmonella* (Gray, Tavares, & Escalante-Semerena, 2008). Additionally, the CbiZ-domain of bifunctional protein CbiS from *Methanopyrus kandleri* was shown to cleave cobinamide as well as cobalamin (Woodson & Escalante-Semerena, 2006).

The role of CbiZ in cobamide remodeling was first demonstrated in the photosynthetic α-proteobacterium *Rhodobacter sphaeroides* (Gray & Escalante-Semerena, 2009a). *R. sphaeroides* encodes a complete pathway for *de novo* biosynthesis of cobamides (including CobU), BluB for biosynthesis of lower ligand DMB, and the amidohydrolase CbiZ (Gray & Escalante-Semerena, 2009b). *R. sphaeroides* CobU was demonstrated to have guanylyltransferase activity associated with *de novo* cobamide biosynthesis and also cobinamide kinase activity associated with salvaging (Gray et al., 2008). Although it was seemingly redundant for *R. sphaeroides* to encode both CobU and CbiZ pathways for cobinamide salvaging, the discovery of cobamide substrates for CbiZ clarified its role in *R. sphaeroides* (Gray & Escalante-Semerena, 2009a). Acetate catabolism in *R. sphaeroides* is cobamide-dependent due to two enzymes, ethylmalonyl-CoA and methylmalonyl-CoA mutases. An *R. sphaeroides* mutant deficient for corrin ring biosynthesis was grown on acetate as a carbon source, and growth was supported by addition of the cobalamin analog. Addition of cobinamide and pseudocobalamin also supported growth, but both conditions required the activity of BluB or supplementation with DMB. CbiZ was also required for growth with pseudocobalamin but not cobinamide due to the kinase activity of CobU. HPLC analysis of *R. sphaeroides* extracts showed that supplementation with pseudocobalamin resulted in conversion to cobalamin, but that pseudocobalamin accumulated in a *cbiZ* mutant. Purified *R. sphaeroides* CbiZ was assayed for activity and was most active against pseudocobalamin, somewhat active against cobinamide, and minimally active against cobalamin. Together, this work showed that cobalamin is the required analog for *R. sphaeroides* growth on acetate and that growth was also supported by cobalamin production *via* cobinamide salvaging and CbiZ-dependent remodeling of pseudocobalamin. This was the first demonstration of cobamide remodeling, the process by which one analog is converted to another.

Cobamide remodeling was also demonstrated in *Dehalococcoides mccartyi*, a bacterium used for bioremediation due to its ability to dechlorinate groundwater contaminants (Yi et al., 2012). *D. mccartyi* requires cobamides for growth but only encodes a partial biosynthesis pathway and multiple copies of *cbiZ*. *D. mccartyi* grew when supplemented with cobalamin or [5-MeBza]Cba, but not with cobamides containing purinyl, phenolyl, or other benzimidazolyl ([Bza] and [5-OHBza]) lower ligands. Cobamides produced through salvaging and remodeling also supported growth. *D. mccartyi* grew when supplemented with cobinamide and some benzimidazolyl lower ligands (including [5-MeBza]); further, cobamide analog production was analyzed by mass spectrometry and showed preferential incorporation of the three lower ligands that best supported growth. Additionally, *D. mccartyi* could grow when supplemented with any cobamide when also supplied with DMB. Interestingly, some remodeling occurred when *D. mccartyi* was supplemented with cobalamin and a different benzimidazolyl lower ligand, suggesting that *D. mccartyi* also cleaves analogs that it can utilize. Cobamide remodeling by *D. mccartyi* was demonstrated in soil and ground water enrichments (Men et al., 2015) and in a defined tripartite consortium that also contained Veillonella species *Pelosinus fermentans* and *Desulfovibrio vulgaris* (Men et al., 2014). Supplementation with DMB supported growth of the consortium and *D. mccartyi* remodeled [Cre] Cba produced by *P. fermentans* to cobalamin.

3.2 Phosphodiesterase CbiR

Analysis of cobamide specificity in gut microbe *Akkermansia muciniphila* led to the recent discovery of a second cobamide remodeling protein, phosphodiesterase CbiR (Mok et al., 2020). *A. muciniphila* is known for its ability to degrade mucin and is associated with health benefits. Almost all *A. muciniphila* strains encode four cobamide-dependent enzymes: methionine synthase MetH, methyl-malonyl CoA mutase MCM, ribonucleotide reductase NrdJ, and epoxyqueuosine reductase QueG (Kirmiz et al., 2020). The impact of cobamides on *A. muciniphila* metabolism was demonstrated *via* coculture with pseudocobalamin-producing *Eubacterium hallii* which resulted in higher production of propionate by *A. muciniphila* (Belzer et al., 2017). *A. muciniphila* does not encode MetE, a cobamide-independent methionine synthase, and therefore relies on cobamide-dependent MetH for methionine production. Supplementation of *A. muciniphila* with a range of cobamides supported MetH-dependent growth, including representatives from each of the three cobamide groups (Mok et al., 2020). Further analysis showed that *A. muciniphila* converted supplemented cobamides into pseudocobalamin, suggesting that cobamide remodeling occurred despite the lack of *cbiZ* in its genome.

A. muciniphila encodes a partial biosynthesis pathway that supported salvaging of the intermediate cobinamide to produce pseudocobalamin (Mok et al., 2020). The *cbiR* gene is encoded within the cluster of cobamide biosynthesis genes and was found to have remodeling activity. CbiR is an enzyme in the apurinic/apyrimidinic endonuclease 2 superfamily which hydrolyzes cobamides at the phosphoribosyl bond of the nucleotide loop, forming the cobinamide-phosphate intermediate. Purified CbiR was tested using *in vitro* reactions and cleaved seven different cobamides from all three cobamide families, including pseudocobalamin, to form the cobinamide-phosphate intermediate. Interestingly, CbiR reactions were inhibited upon addition of metal chelator EDTA, suggesting that a metal cofactor is required for function. Heterologous expression of CbiR in *E. coli* confirmed the cleavage of cobamides and remodeling to produce another analog. Most CbiR homologs are encoded by species within the phyla Chlorobi, Cloroflexi, and Proteobacteria, but few homologs are found in bacteria more closely related *Akkermansia*. However, CbiR is highly conserved in *Akkermansia* species and its activity could greatly influence cobamide profiles in the gut environment.

3.3 CobS-mediated direct remodeling

A third pathway for cobamide remodeling was recently identified in *Vibrio cholerae*. Cobamide biosynthesis genes are conserved in *Vibrio* species (Ma, Tyrell, & Beld, 2020), with nearly all encoding partial biosynthesis pathways associated with salvaging but lacking remodeling enzymes CbiZ and CbiR. Examination of cobamide specificity in *Vibrio* species has led to the discovery of the direct cobamide remodeling pathway (Ma & Beld, 2021). In contrast to CbiZ and CbiR-mediated remodeling, in which cobamides are cleaved into biosynthetic intermediates, CobS-mediated remodeling occurs through direct replacement of the nucleotide loop by the biosynthesis protein cobamide synthase CobS (EC 2.7.8.26) (CobV in *Pseudomonas* nomenclature).

The first indication of cobamide remodeling in *Vibrio* species arose during characterization of an amoeba isolated as a cyanobacterial grazer (Ma et al., 2017). Amoebal strain LPG3 was isolated and maintained with

Synechococcus elongatus but was later found to harbor a *Pseudomonas* co-isolate which was required for amoebal growth. In place of *Pseudomonas*, a defined transposon library of *Vibrio cholerae* was used in a genetic screen which identified cobamide biosynthesis and transporter genes. Although *Synechococcus* produces pseudocobalamin, amoebal growth was only supported when also cultured with cobalamin-producing *Pseudomonas* and not a biosynthetic mutant. *V. cholerae* encodes only a partial biosynthesis pathway, suggesting that it remodeled the *Synechococcus*-derived pseudocobalamin to cobalamin, thereby supplying the required cobamide analog for amoebal growth.

Because *V. cholerae* does not encode CbiZ or CbiR, its remodeling pathway occurs through a distinct mechanism. In studies examining cobamide specificity of *V. cholerae*, remodeling of pseudocobalamin to cobalamin was confirmed and the link to CobS was established through heterologous assays (Ma et al., 2020). It was observed that wild-type *V. cholerae* did not accumulate pseudocobalamin in the absence of lower ligand DMB, but that a Δ*cobS* strain did. These results were replicated upon heterologous expression of *V. cholerae* CobS (VcCobS); pseudocobalamin did not accumulate in the absence of DMB, and pseudocobalamin was remodeled to cobalamin in the presence of DMB. In contrast, expression of *E. coli* CobS (EcCobS) resulted in pseudocobalamin accumulation, both in the presence and absence of DMB, matching the results observed in wild-type *E. coli*. These results indicated that only some CobS homologs are capable of CbiZ- and CbiR-independent cobamide remodeling.

Further characterization of the *V. cholerae* remodeling pathway showed that VcCobS acts directly on pseudocobalamin, and the term "direct remodeling" was introduced (Ma & Beld, 2021). Genetic characterization of remodeling in *V. cholerae* suggested that biosynthetic intermediates were not produced. During *de novo* biosynthesis and salvaging, CobU produces the typical cobinamide-GDP substrate for CobS (Fig. 2). CobU was required for cobinamide salvaging by *V. cholerae* but not for remodeling. *In vitro* remodeling reactions were performed with pseudocobalamin as a substrate, showing that VcCobS contained in membrane fractions supported remodeling, but that EcCobS did not. Additionally, VcCobS incorporated a thio analog of the activated lower ligand in both remodeling and salvaging reactions, showing that the ribosyl moiety in the nucleotide loop is derived from the lower ligand substrate and not retained from the pseudocobalamin substrate. These results support a catalytic model in which VcCobS attaches the DMB-R5P substrate to cobinamide-phosphate derived from pseudocobalamin. Additionally, other distantly-related *Vibrio* species were examined for direct remodeling (Ma & Beld, 2021); emerging pathogen *V. fluvialis*, abundant environmental species *V. splendidus*, and their CobS proteins were also able to remodel pseudocobalamin. This suggests that CobS-mediated direct remodeling could be widespread among diverse *Vibrio* species. Direct remodeling has not been examined in other organisms, but protein networks show that some CobS proteins cluster with VcCobS independent of phylogeny (Ma & Beld, 2021), suggesting that they might support direct remodeling. *In vitro* examination of CobS is difficult because it is a polytopic integral membrane protein, but recently reported detergent solubilization methods and liposome insertion approaches will facilitate characterization of CobS activity (Jeter & Escalante-Semerena, 2021).

4. Conclusions and outlook

Cobamide remodeling pathways are present in organisms that have specificity for certain cobamides, ensuring that compatible cobamides are present within the cell. Examination of cobamide remodeling substrate specificity reveals that two strategies are taken: selective and nonselective remodeling. In selective remodeling, only cobamides that cannot be utilized are remodeled. Both *R. sphaeroides* CbiZ and *V. cholerae* CobS act upon pseudocobalamin, converting it into cobalamin for utilization in acetate catabolism or methionine biosynthesis, respectively. In nonselective remodeling, all imported cobamides are cleaved irrespective of their lower ligand. *D. mccartyi* CbiZs and *A. muciniphila* CbiR cleave imported cobamides for cobalamin and pseudocobalamin production, respectively. Both bacteria cleaved all cobamides tested, including representatives of the three cobamide families and also the preferred analog produced by the bacteria. While nonselective remodeling could produce a futile cycle, in which cobamides produced by the bacteria are cleaved and rebuilt again, factors such as sequestration by cobamide-dependent enzymes and localization of remodeling proteins could protect the synthesized cobamides from being remodeled. Indeed, CbiZ is often encoded as a fusion to cobamide transporter machinery, suggesting that cleavage occurs as the cobamides are being imported.

Cobamides play a critical role in metabolic interactions within microbial communities because most bacteria can utilize cobamides but not all can produce them. Cobamide-dependent enzymes are involved in a range of metabolic pathways, including primary metabolism, catabolic pathways, and production of secondary metabolites. Specificity for different cobamide analogs adds another layer of complexity to vitamin cross-feeding and complicates metabolic reconstructions of microbial communities. Such modeling efforts will be improved by further examination of cobamides, such as defining cobamide profiles in communities and characterizing cobamide specificity in additional groups of bacteria, including specificity for cobamide utilization, remodeling, and production. The multitude of cobamide remodeling pathways suggests that remodeling plays a major role in cobamide flux within microbial communities and potentially impacts microbiome structure. CbiR and CobS-mediated direct remodeling were only very recently identified, and further examination of cobamides in other bacteria could reveal additional pathways for remodeling that are yet to be discovered.

References

Allen, R. H., & Stabler, S. P. (2008). Identification and quantitation of cobalamin and cobalamin analogues in human feces. *The American Journal of Clinical Nutrition*, *87*(5), 1324–1335.

Anderson, P. J., Lango, J., Carkeet, C., Britten, A., Kräutler, B., Hammock, B. D., et al. (2008). One pathway can incorporate either adenine or dimethylbenzimidazole as an α-axial ligand of B_{12} cofactors in *Salmonella enterica*. *Journal of Bacteriology*, *190*(4), 1160–1171.

Balabanova, L., Averianova, L., Marchenok, M., Son, O., & Tekutyeva, L. (2021). Microbial and genetic resources for cobalamin (vitamin B_{12}) biosynthesis: From ecosystems to industrial biotechnology. *International Journal of Molecular Sciences*, *22*(9), 4522.

Barker, H., Weissbach, H., & Smyth, R. (1958). A coenzyme containing pseudovitamin B_{12}. *Proceedings of the National Academy of Sciences*, *44*(11), 1093–1097.

Baum, C., Menezes, R. C., Svatos, A., & Schubert, T. (2020). Cobamide remodeling in the freshwater microalga *Chlamydomonas reinhardtii*. *FEMS Microbiology Letters*, *367*(20), fnaa171.

Belzer, C., Chia, L. W., Aalvink, S., Chamlagain, B., Piironen, V., Knol, J., et al. (2017). Microbial metabolic networks at the mucus layer lead to diet-independent butyrate and vitamin B_{12} production by intestinal symbionts. *mBio*, *8*(5). e00770-17.

Bernhauer, K., & Friedrich, W. (1954). Über die Vitamine der B_{12}-Gruppe. *Angewandte Chemie*, *66*(24), 776–780.

Bunn, H. F. (2014). Vitamin B_{12} and pernicious anemia—The dawn of molecular medicine. *New England Journal of Medicine*, *370*(8), 773–776.

Cameron, B., Blanche, F., Rouyez, M.-C., Bisch, D., Famechon, A., Couder, M., et al. (1991). Genetic analysis, nucleotide sequence, and products of two *Pseudomonas denitrificans cob* genes encoding nicotinate-nucleotide: Dimethylbenzimidazole phosphoribosyltransferase and cobalamin (5′-phosphate) synthase. *Journal of Bacteriology*, *173*(19), 6066–6073.

Cameron, B., Briggs, K., Pridmore, S., Brefort, G., & Crouzet, J. (1989). Cloning and analysis of genes involved in coenzyme B_{12} biosynthesis in *Pseudomonas denitrificans*. *Journal of Bacteriology*, *171*(1), 547–557.

Campbell, G. R., Taga, M. E., Mistry, K., Lloret, J., Anderson, P. J., Roth, J. R., et al. (2006). *Sinorhizobium meliloti bluB* is necessary for production of 5, 6-dimethylbenzimidazole, the lower ligand of B_{12}. *Proceedings of the National Academy of Sciences*, *103*(12), 4634–4639.

Chan, C. H., & Escalante-Semerena, J. C. (2011). ArsAB, a novel enzyme from *Sporomusa ovata* activates phenolic bases for adenosylcobamide biosynthesis. *Molecular Microbiology*, *81*(4), 952–967.

Crofts, T. S., Seth, E. C., Hazra, A. B., & Taga, M. E. (2013). Cobamide structure depends on both lower ligand availability and CobT substrate specificity. *Chemistry & Biology*, *20*(10), 1265–1274.

Dryden, L. P., & Hartman, A. M. (1971). Variations in the amount and relative distribution of vitamin B_{12} and its analog in the bovine rumen. *Journal of Dairy Science*, *54*(2), 235–246.

Ford, J. E., Holdsworth, E. S., Kon, S. K., & Porter, J. W. G. (1953). Occurrence of the various vitamin B_{12} active compounds. *Nature*, *171*, 150–151.

Friedrich, W., & Bernhauer, K. (1957). Zur Chemie und Biochemie der "Cobalamine", VII. 2-Methylmercapto-Adenin-Cobalamin-Analogon, ein Neuer B_{12}-Faktor des Faulschlammes. *Chemische Berichte*, *90*(9), 1966–1973.

Friedrich, W., & Bernhauer, K. (1958). Zur Chemie und Biochemie der "Cobalamine", IX. Die Isolierung der Benzimidazol-und 5-Methyl-benzimidazol-cobalamin-Analoga aus Faulschlamm. *Chemische Berichte*, *91*(10), 2061–2065.

Friedrich, W., & Bernhauer, K. (1959). Vitamin B_{12}-Analoga des Belebtschlammes. *Angewandte Chemie*, *71*(8), 284.

Garibaldi, J. A., Ijichi, K., Snell, N. S., & Lewis, J. C. (1953). *Bacillus megatherium* for biosynthesis of cobalamin. *Industrial & Engineering Chemistry*, *45*(4), 838–846.

Girard, C. L., Berthiaume, R., Stabler, S. P., & Allen, R. H. (2009). Identification of cobalamin and cobalamin analogues along the gastrointestinal tract of dairy cows. *Archives of Animal Nutrition*, *63*(5), 379–388.

Gray, M. J., & Escalante-Semerena, J. C. (2009a). The cobinamide amidohydrolase (cobyric acid-forming) CbiZ enzyme: a critical activity of the cobamide remodelling system of *Rhodobacter sphaeroides*. *Molecular Microbiology*, *74*(5), 1198–1210.

Gray, M. J., & Escalante-Semerena, J. C. (2009b). In vivo analysis of cobinamide salvaging in *Rhodobacter sphaeroides* strain 2.4.1. *Journal of Bacteriology*, *191*(12), 3842–3851.

Gray, M. J., Tavares, N. K., & Escalante-Semerena, J. C. (2008). The genome of *Rhodobacter sphaeroides* strain 2.4.1 encodes functional cobinamide salvaging systems of archaeal and bacterial origins. *Molecular Microbiology*, *70*(4), 824–836.

Guimarães, D. H., Weber, A., Klaiber, I., Vogler, B., & Renz, P. (1994). Guanylcobamide and hypoxanthylcobamide-corrinoids formed by *Desulfovibrio vulgaris*. *Archives of Microbiology*, *162*(4), 272–276.

Hazra, A. B., Han, A. W., Mehta, A. P., Mok, K. C., Osadchiy, V., Begley, T. P., et al. (2015). Anaerobic biosynthesis of the lower ligand of vitamin B_{12}. *Proceedings of the National Academy of Sciences*, *112*(34), 10792–10797.

Hazra, A. B., Tran, J. L. A., Crofts, T. S., & Taga, M. E. (2013). Analysis of substrate specificity in CobT homologs reveals widespread preference for DMB, the lower axial ligand of vitamin B_{12}. *Chemistry & Biology*, *20*(10), 1275–1285.

Heal, K. R., Qin, W., Ribalet, F., Bertagnolli, A. D., Coyote-Maestas, W., Hmelo, L. R., et al. (2017). Two distinct pools of B_{12} analogs reveal community interdependencies in the ocean. *Proceedings of the National Academy of Sciences*, *114*(2), 364–369.

Helliwell, K. E., Lawrence, A. D., Holzer, A., Kudahl, U. J., Sasso, S., Kräutler, B., et al. (2016). Cyanobacteria and eukaryotic algae use different chemical variants of vitamin B_{12}. *Current Biology*, *26*(8), 999–1008.

Hendlin, D., & Ruger, M. L. (1950). The effect of cobalt on the microbial synthesis of LLD-active substances. *Science*, *111*, 541–542.

Hodgkin, D. C., Kamper, J., Mackay, M., Pickworth, J., Trueblood, K. N., & White, J. G. (1956). Structure of vitamin B_{12}. *Nature*, *178*(4524), 64–66.

Hoffmann, B., Oberhuber, M., Stupperich, E., Bothe, H., Buckel, W., Konrat, R., et al. (2000). Native corrinoids from *Clostridium cochlearium* are adeninylcobamides: Spectroscopic analysis and identification of pseudovitamin B_{12} and factor A. *Journal of Bacteriology*, *182*(17), 4773–4782.

Höllriegl, V., Lamm, L., Rowold, J., Hörig, J., & Renz, P. (1982). Biosynthesis of vitamin B_{12}. Different pathways in some aerobic and anaerobic microorganisms. *Archives of Microbiology*, *132*(2), 155–158.

Irion, E., & Ljungdahl, L. (1965). Isolation of factor III_m coenzyme and cobyric acid coenzyme plus other B_{12} factors from *Clostridium thermoaceticum*. *Biochemistry*, *4*(12), 2780–2790.

Jeter, V. L., & Escalante-Semerena, J. C. (2021). Insights into the relationship between Cobamide synthase and the cell membrane. *mBio*, *12*(2), e00215-21.

Jeter, V. L., Mattes, T. A., Beattie, N. R., & Escalante-Semerena, J. C. (2019). A new class of phosphoribosyltransferases involved in cobamide biosynthesis is found in methanogenic archaea and cyanobacteria. *Biochemistry*, *58*(7), 951–964.

Keck, B., & Renz, P. (2000). *Salmonella typhimurium* forms adenylcobamide and 2-methyladenylcobamide, but no detectable cobalamin during strictly anaerobic growth. *Archives of Microbiology*, *173*(1), 76–77.

Kelly, C. J., Alexeev, E. E., Farb, L., Vickery, T. W., Zheng, L., Eric, L. C., et al. (2019). Oral vitamin B_{12} supplement is delivered to the distal gut, altering the corrinoid profile and selectively depleting Bacteroides in C57BL/6 mice. *Gut Microbes*, *10*(6), 654–662.

Kirmiz, N., Galindo, K., Cross, K. L., Luna, E., Rhoades, N., Podar, M., et al. (2020). Comparative genomics guides elucidation of vitamin B_{12} biosynthesis in novel human-associated *Akkermansia* strains. *Applied and Environmental Microbiology*, *86*(3), e2117–19.

Kohler, H.-P. E. (1988). Isolation of cobamides from *Methanothrix soehngenii*: 5-methylbenzimidazole as the α-ligand of the predominant cobamide. *Archives of Microbiology*, *150*(3), 219–223.

Kräutler, B., Fieber, W., Ostermann, S., Fasching, M., Ongania, K.-H., Gruber, K., et al. (2003). The cofactor of tetrachloroethene reductive dehalogenase of *Dehalospirillum multivorans* is norpseudo-B_{12}, a new type of a natural corrinoid. *Helvetica Chimica Acta*, *86*(11), 3698–3716.

Kräutler, B., Kohler, H.-P. E., & Stupperich, E. (1988). 5′-Methylbenzimidazolyl-cobamides are the corrinoids from some sulfate-reducing and sulfur-metabolizing bacteria. *European Journal of Biochemistry*, *176*(2), 461–469.

Leviton, A., & Hargrove, R. E. (1952). Microbiological synthesis of vitamin B_{12} by propionic acid bacteria. *Industrial & Engineering Chemistry*, *44*(11), 2651–2655.

Lezius, A. G., & Barker, H. A. (1965). Corrinoid compounds of *Methanobacillus omelianskii*. I. Fractionation of the corrinoid compounds and identification of factor III and factor III coenzyme. *Biochemistry*, *4*(3), 510–518.

Lu, X., Heal, K. R., Ingalls, A. E., Doxey, A. C., & Neufeld, J. D. (2020). Metagenomic and chemical characterization of soil cobalamin production. *The ISME Journal*, *14*(1), 53–66.

Ma, A. T., & Beld, J. (2021). Direct cobamide remodeling via additional function of cobamide biosynthesis protein CobS from *Vibrio cholerae*. *Journal of Bacteriology*, *203*(15), e00172-21.

Ma, A. T., Beld, J., & Brahamsha, B. (2017). An amoebal grazer of cyanobacteria requires cobalamin produced by heterotrophic bacteria. *Applied and Environmental Microbiology*, *83*(10), e00035-17.

Ma, A. T., Tyrell, B., & Beld, J. (2020). Specificity of cobamide remodeling, uptake and utilization in *Vibrio cholerae*. *Molecular Microbiology*, *113*(1), 89–102.

Maggio-Hall, L. A., & Escalante-Semerena, J. C. (1999). In vitro synthesis of the nucleotide loop of cobalamin by *Salmonella typhimurium* enzymes. *Proceedings of the National Academy of Sciences*, *96*(21), 11798–11803.

Martens, J.-H., Barg, H., Warren, M. J., & Jahn, D. (2002). Microbial production of vitamin B_{12}. *Applied Microbiology and Biotechnology*, *58*(3), 275–285.
Mattes, T. A., & Escalante-Semerena, J. C. (2017). *Salmonella enterica* synthesizes 5,6-dimethylbenzimidazolyl-(DMB)-α-riboside. Why some Firmicutes do not require the canonical DMB activation system to synthesize adenosylcobalamin. *Molecular Microbiology*, *103*(2), 269–281.
McGoldrick, H., Deery, E., Warren, M., & Heathcote, P. (2002). Cobalamin (vitamin B_{12}) biosynthesis in *Rhodobacter capsulatus*. *Biochemical Society Transactions*, *30*(4), 646–648.
Men, Y., Seth, E. C., Yi, S., Allen, R. H., Taga, M. E., & Alvarez-Cohen, L. (2014). Sustainable growth of *Dehalococcoides mccartyi* 195 by corrinoid salvaging and remodeling in defined lactate-fermenting consortia. *Applied and Environmental Microbiology*, *80*(7), 2133–2141.
Men, Y., Seth, E. C., Yi, S., Crofts, T. S., Allen, R. H., Taga, M. E., et al. (2015). Identification of specific corrinoids reveals corrinoid modification in dechlorinating microbial communities. *Environmental Microbiology*, *17*(12), 4873–4884.
Miyamoto, E., Tanioka, Y., Nakao, T., Barla, F., Inui, H., Fujita, T., et al. (2006). Purification and characterization of a corrinoid-compound in an edible cyanobacterium *Aphanizomenon flos-aquae* as a nutritional supplementary food. *Journal of Agricultural and Food Chemistry*, *54*(25), 9604–9607.
Mok, K. C., Sokolovskaya, O. M., Nicolas, A. M., Hallberg, Z. F., Deutschbauer, A., Carlson, H. K., et al. (2020). Identification of a novel cobamide remodeling enzyme in the beneficial human gut bacterium *Akkermansia muciniphila*. *mBio*, *11*(6), e02507–20.
Mok, K. C., & Taga, M. E. (2013). Growth inhibition of *Sporomusa ovata* by incorporation of benzimidazole bases into cobamides. *Journal of Bacteriology*, *195*(9), 1902–1911.
Renz, P. (1999). Biosynthesis of the 5, 6-dimethylbenzimidazole moiety of cobalamin and of the other bases found in natural corrinoids. In R. Banerjee (Ed.), *Chemistry and biochemistry of B_{12}* (pp. 557–575). Wiley.
Renz, P., Blickle, S., & Friedrich, W. (1987). Two new vitamin B-12 factors from sewage sludge containing 2-methylsulfinyladenine or 2-methylsulfonyladenine as base component. *European Journal of Biochemistry*, *163*(1), 175–179.
Rickes, E. L., Brink, N. G., Koniuszy, F. R., Wood, T. R., & Folkers, K. (1948). Comparative data on vitamin B_{12} from liver and from a new source, *Streptomyces griseus*. *Science*, *108*, 634–635.
Saito, Y., Sato, T., Nomoto, K., & Tsuji, H. (2018). Identification of phenol-and p-cresol-producing intestinal bacteria by using media supplemented with tyrosine and its metabolites. *FEMS Microbiology Ecology*, *94*(9), fiy125.
Santos, F., Vera, J. L., Lamosa, P., de Valdez, G. F., de Vos, W. M., Santos, H., et al. (2007). Pseudovitamin B_{12} is the corrinoid produced by *Lactobacillus reuteri* CRL1098 under anaerobic conditions. *FEBS Letters*, *581*(25), 4865–4870.
Saunders, A. P., Otto, R. H., & Sylvester, J. C. (1952). The production of vitamin B_{12} by various strains of actinomycetes. *Journal of Bacteriology*, *64*(5), 725–728.
Schulze, B., Vogler, B., & Renz, P. (1998). Biosynthesis of vitamin B_{12} in anaerobic bacteria: Experiments with *Eubacterium limosum* on the transformation of 5-hydroxy-6-methylbenzimidazole, its nucleoside, its cobamide, and of 5-hydroxybenzimidazolylcobamide in vitamin B_{12}. *European Journal of Biochemistry*, *254*(3), 620–625.
Shelton, A. N., Seth, E. C., Mok, K. C., Han, A. W., Jackson, S. N., Haft, D. R., et al. (2019). Uneven distribution of cobamide biosynthesis and dependence in bacteria predicted by comparative genomics. *The ISME Journal*, *13*(3), 789–804.
Sokolovskaya, O. M., Shelton, A. N., & Taga, M. E. (2020). Sharing vitamins: Cobamides unveil microbial interactions. *Science*, *369*(6499), eaba0165.
Stadtman, T. C. (1960). Synthesis of adenine-B_{12} coenzyme by *Clostridium sticklandii*: Relationship to one-carbon metabolism. *Journal of Bacteriology*, *79*(6), 904–905.

Stupperich, E., & Eisinger, H. J. (1989). Biosynthesis of Para-cresolyl cobamide in *Sporomusa ovata*. *Archives of Microbiology*, *151*(4), 372–377.

Stupperich, E., Eisinger, H. J., & Kräutler, B. (1988). Diversity of corrinoids in acetogenic bacteria: P-Cresolylcobamide from *Sporomusa ovata*, 5-methoxy-6-methylbenzimidazolylcobamide from *Clostridium formicoaceticum* and vitamin B_{12} from *Acetobacterium woodii*. *European Journal of Biochemistry*, *172*(2), 459–464.

Stupperich, E., Eisinger, H. J., & Kräutler, B. (1989). Identification of phenolyl cobamide from the homoacetogenic bacterium *Sporomusa ovata*. *European Journal of Biochemistry*, *186*(3), 657–661.

Stupperich, E., & Kräutler, B. (1988). Pseudo vitamin B_{12} or 5-hydroxybenzimidazolylcobamide are the corrinoids found in methanogenic bacteria. *Archives of Microbiology*, *149*(3), 268–271.

Taga, M. E., Larsen, N. A., Howard-Jones, A. R., Walsh, C. T., & Walker, G. C. (2007). BluB cannibalizes flavin to form the lower ligand of vitamin B_{12}. *Nature*, *446*(7134), 449–453.

Tanioka, Y., Yabuta, Y., Yamaji, R., Shigeoka, S., Nakano, Y., Watanabe, F., et al. (2009). Occurrence of pseudovitamin B_{12} and its possible function as the cofactor of cobalamin-dependent methionine synthase in a cyanobacterium *Synechocystis* sp. PCC6803. *Journal of Nutritional Science and Vitaminology*, *55*(6), 518–521.

Torres, A. C., Vannini, V., Bonacina, J., Font, G., Saavedra, L., & Taranto, M. P. (2016). Cobalamin production by *Lactobacillus coryniformis*: Biochemical identification of the synthetized corrinoid and genomic analysis of the biosynthetic cluster. *BMC Microbiology*, *16*, 240.

Trzebiatowski, J. R., & Escalante-Semerena, J. C. (1997). Purification and characterization of CobT, the nicotinate-mononucleotide: 5, 6-dimethylbenzimidazole phosphoribosyltransferase enzyme from *Salmonella typhimurium* LT2. *Journal of Biological Chemistry*, *272*(28), 17662–17667.

Woodson, J. D., & Escalante-Semerena, J. C. (2004). CbiZ, an amidohydrolase enzyme required for salvaging the coenzyme B_{12} precursor cobinamide in archaea. *Proceedings of the National Academy of Sciences*, *101*(10), 3591–3596.

Woodson, J. D., & Escalante-Semerena, J. C. (2006). The *cbiS* gene of the archaeon *Methanopyrus kandleri* AV19 encodes a bifunctional enzyme with adenosylcobinamide amidohydrolase and α-ribazole-phosphate phosphatase activities. *Journal of Bacteriology*, *188*(12), 4227–4235.

Yan, J., Bi, M., Bourdon, A. K., Farmer, A. T., Wang, P.-H., Molenda, O., et al. (2018). Purinyl-cobamide is a native prosthetic group of reductive dehalogenases. *Nature Chemical Biology*, *14*(1), 8–14.

Yi, S., Seth, E. C., Men, Y.-J., Stabler, S. P., Allen, R. H., Alvarez-Cohen, L., et al. (2012). Versatility in corrinoid salvaging and remodeling pathways supports corrinoid-dependent metabolism in *Dehalococcoides mccartyi*. *Applied and Environmental Microbiology*, *78*(21), 7745–7752.

CHAPTER FOUR

Cubilin, the intrinsic factor-vitamin B12 receptor

Renata Kozyraki*, Pierre Verroust, and Olivier Cases
Centre de Recherche des Cordeliers, Sorbonne Université, INSERM, Université de Paris, Paris, France
*Corresponding author: e-mail address: renata.kozyraki@inserm.fr

Contents

Abstract

Cubilin (CUBN), the intrinsic factor-vitamin B12 receptor is a large endocytic protein involved in various physiological functions: vitamin B12 uptake in the gut; reabsorption of albumin and maturation of vitamin D in the kidney; nutrient delivery during

Vitamins and Hormones, Volume 119
ISSN 0083-6729
https://doi.org/10.1016/bs.vh.2022.01.005

embryonic development. Cubilin is an atypical receptor, peripherally associated to the plasma membrane. The transmembrane proteins amnionless (AMN) and Lrp2/Megalin are the currently known molecular partners contributing to plasma membrane transport and internalization of Cubilin. The role of Cubilin/Amn complex in the handling of vitamin B12 in health and disease has extensively been studied and so is the role of the Cubilin-Lrp2 tandem in renal pathophysiology. Accumulating evidence strongly supports a role of Cubilin in some developmental defects including impaired closure of the neural tube. Are these defects primarily caused by the dysfunction of a specific Cubilin ligand or are they secondary to impaired vitamin B12 or protein uptake? We will present the established Cubilin functions, discuss the developmental data and provide an overview of the emerging implications of Cubilin in the field of cardiovascular disease and cancer pathogenesis.

Abbreviations

ABCD4 ATP Binding Cassette Subfamily D member IV
AMN Amnionless
ARH autosomal recessive hypercholesterolemia
CNC cephalic neural crest
CUBN Cubilin
DAB2 disabled homolog 2
DBP vitamin D-binding protein
DKD diabetic kidney disease
DOCK8 dedicator of cytokinesis 8
EGF epidermal growth factor
ER endoplasmic reticulum
FGF8 fibroblast growth factor 8
GFR glomerular filtration rate
IF-B12 intrinsic factor-vitamin B12
IGS Immerslund-Gräsbeck syndrome
LDL low-density lipoprotein
LRP2 Megalin/Lrp2
NTD neural tube defect
OCRL inositol polyphosphate-5-phosphatase
PCT proximal convoluted tubule
SNX8 sorting nexin 8
TC transcobalamin
TLR4 toll-like receptor 4
VE visceral endoderm

1. Introduction

Eukaryotes have evolved highly regulated endocytic transport pathways to allow nutrient uptake and communication with the extracellular compartments. Clathrin-mediated endocytosis of cargo molecules from the plasma membrane to endosomes is the most studied process of vesicular transport to date. The cargo is mainly transmembrane proteins/receptors and their ligands. The cellular uptake of receptor-bound proteins, vitamins, iron, lipids and other factors is necessary for cellular homeostasis, growth control and cell proliferation.

In this pathway, ligand-receptor complexes are entrapped by deformation of the plasma membrane. The inward budding and detachment of the vesicles is followed by a complex sorting process involving organelle fusions and delivery of the ligands to the endolysosomal compartments. The ligands may either be degraded in the lysosomes or exported for further use during a closely monitored multi-step process (Mettlen, Chen, Srinivasan, Danuser, & Schmid, 2018). Disruption of the process at any step ultimately results in organ dysfunction and disease. The multiligand receptors Cubilin (CUBN) and Megalin/Lrp2 (LRP2) and their associated protein Amnionless (AMN) take this pathway. Dysfunction of these proteins due to inherited or acquired disorders reveal their important role in nutrient uptake, renal homeostasis and embryonic growth.

2. Expression sites of cubilin, Lrp2 and Amn

Cubilin and Lrp2 have close or overlapping expression patterns during development and in the adult (Assémat, Châtelet, et al., 2005; Assémat, Vinot, et al., 2005). During early development both receptors are found in the rodent blastocyst as soon as the eight-cell stage. Their initial vesicular distribution changes to an apical localization in the trophectoderm of the 32-cell stage embryo. In the early post-implantation embryo, both in man and mouse the receptors are co-expressed in numerous developing epithelia including the neuroepithelium, lung, thyroid, sensory, intestinal and renal epithelia as well as the cephalic neural crest cells (Cases et al., 2013; CNC).

In the adult both proteins are strongly expressed in the absorptive epithelia of the renal proximal convoluted tubule (PCT) as well as the glomerular podocytes, the reproductive organs, the choroid plexus, the lung alveoli, the gallbladder epithelium, the ciliary and retinal pigmentary epithelia of the eye and the auditory epithelia of the inner ear (Assémat, Châtelet, et al., 2005; Erranz et al., 2004; Tauris et al., 2009; Tsaroucha et al., 2008; Verroust, 2002; Willnow & Christ, 2017). Recently Cubilin and Lrp2 have also been described in the tear fluid producing lacrimal and accessory glands as well as the adipocytes (Bonnet et al., 2018; Lu et al., 2015). Cubilin has not been detected in some Lrp2-expressing sites such as astrocytes, ischemic neurons or brain endothelial cells (Bartolome et al., 2017; Gomes et al., 2016; Wicher & Aldskogius, 2008).

Early reports indicate that Amnionless is expressed in the heart, thymus, spleen, lung, colon and blood leucocytes (Tanner et al., 2003; Wang et al., 1996). Additionally, Amnionless is co-expressed with Cubilin in sites such as the developing and adult intestinal and renal epithelia, the mouse testis and the adipose tissue (Bonnet et al., 2018; Kozyraki & Cases, 2013; Oh, Seo, Ahn, & Gye, 2016; Ramanujam, Seetharam, & Seetharam, 1993; Strope, Rivi, Metzger, Manova, & Lacy, 2004). Finally, drosophila nephrocytes, highly endocytic cells sharing similarities with both mammalian podocytes and renal PCT cells express both Cubilin and Amn but not Lrp2 (Zhang, Zhao, Chao, Muir, & Han, 2013).

Cubilin, Amn and Lrp2 are co-localized at the maternal-fetal interfaces of the trophectoderm, the placenta, the visceral endoderm and its derivative the visceral yolk sac (Argraves & Morales, 2004; Assémat, Châtelet, et al., 2005; Cases et al., 2013; Müller, Nykjaer, & Willnow, 2003; Sahali et al., 1988; Willnow & Christ, 2017). In Teleosts *Cubilin*, *Amn* and *Lrp2* have also been found in the trophotaeniae, i.e., hindgut-derived pseudoplacental structures of Teleosts (Iida et al., 2021).

3. Cubilin, the intrinsic factor-vitamin B12 receptor

Cubilin, the intrinsic factor-vitamin B12 receptor (also known as gp280) was initially identified as the target of teratogenic antibodies (Sahali et al., 1988). Cubilin is a 460kDa species conserved, peripheral membrane protein encoded by the *CUBN* gene (10p12.33-p13) (Kozyraki et al., 1998; Moestrup et al., 1998). Mutations in *CUBN* cause the autosomal recessive Immerslund-Gräsbeck syndrome (IGS).

Cubilin encompasses a short amino-terminal region (100 residues) involved in protein trimerization and membrane anchoring; an epidermal growth factor (EGF) region composed of eight type B EFG repeats and a cluster of 27 contiguous CUB (Complement C1r/C1s, Uegf, BMP1) domains.

Cubilin contains a furin cleavage site (Mahmood, Shao, & Alpers, 2003) and 42 potential glycosylation Asn-X-Ser/Thr sites. Glycosylation is an important post-translational modification and plays many roles in intracellular protein sorting, cell communication or cell adhesion. At least

21 N-glycopeptides containing complex type glycans have been identified in Cubilin; some of them are also core and/or antennae fucosylated (Saraswat et al., 2015). Tunicamycin, a blocker of N-glycans impairs apical sorting of Cubilin confirming the importance of this modification in receptor processing (Coudroy et al., 2005). The biological significance of fucosylation is unclear but fucosylation of glycan structures has been associated with inflammation and cancer progression (Lau et al., 2007).

The CUB domains are involved in a wide range of biological functions (Bork, 1991). Their accumulation is unique to Cubilin, 85% of the entire protein, and accounts for the capacity of Cubilin to bind numerous, structurally unrelated proteins (Table 1) (Fig. 1). Some Cubilin ligands such as albumin, vitamin D-binding protein (DBP), hemoglobin or cystatin C (Jensen et al., 2017; Kozyraki & Cases, 2013) (Table 1) are common to LRP2. Others, including the intrinsic factor-vitamin B12 complex (IF-B12), transferrin, apolipoprotein A-I/HDL or Fgf8 are specific to Cubilin.

The binding sites for IF-B12 on Cubilin are the best described and comprise two distinct regions. The first encompasses CUB domains 5–8: crystallographic analysis showed that these domains function collectively to allow high-affinity binding of the complex. CUB6 and CUB8 harbor the binding sites whereas CUB5 and CUB7 structurally organize the two others to specifically recognize the bound form of IF, in a pH- and calcium-dependent manner (Andersen, Madsen, Storm, Moestrup, & Andersen, 2010; Kristiansen et al., 1999) (Fig. 1). Electrostatic interactions are the main binding force and N-glycosylation of IF may contribute to the binding. A second binding site for IF-B12 has been identified within the N-terminal stretch of Cubilin (Yammani, Seetharam, & Seetharam, 2001). This site would also bind albumin (Yammani et al., 2001) but no crystal structure is currently available.

Apical plasma membrane targeting and endocytosis of Cubilin also depends on the transmembrane proteins AMN, LRP2 and possibly additional yet to be identified partners.

4. Amnionless

Amnionless (38–50 kDa) is a type I trans-membrane protein of 458 amino acids encoded by *AMN* (14q32). Mutations in AMN cause IGS; CUBN and AMN are currently the only known causative genes for this syndrome. The *AMN* gene encodes at least five protein products, probably produced by excision of the N-terminus. It is possible that the truncated

Table 1 List of CUBN and LRP2 ligands.

Cubilin	**LRP2**
Vitamin carrier proteins	*Vitamin carrier proteins*
Intrinsic factor vitamin B_{12}	Folate-binding protein
Vitamin D–binding protein	Retinol-binding protein
	Transcobalamin–vitamin B_{12}
	Vitamin D–binding protein
Other carrier proteins	*Other carrier proteins*
Albumin	Albumin
Hemoglobin	Hemoglobin
Myoglobin	Lactoferrin
Transferrin	Liver-type fatty acid–binding protein
	Metallothionein
	Myoglobin
	Neutrophil gelatinase–associated lipocalin
	Odorant-binding protein
	Selenoprotein P
	Sex hormone–binding globulin
	Transthyretin
Lipoproteins	*Lipoproteins*
Apolipoprotein A-I	Apolipoprotein B
High-density lipoprotein	Apolipoprotein E
	Apolipoprotein J/clusterin
	Apolipoprotein H
	Apolipoprotein M
Hormones and signaling proteins	*Hormones and signaling proteins*
Fibroblast growth factor 8	Angiotensin II
Survivin	Bone morphogenic protein 4
	Connective tissue growth factor
	Epidermal growth factor

Table 1 List of CUBN and LRP2 ligands.—cont'd

Cubilin	LRP2
	Insulin
	Insulin-like growth factor
	Leptin
	Parathyroid hormone
	Prolactin
	Sonic hedgehog protein
	Survivin
	Thyroglobulin
Enzymes and enzyme inhibitors	*Enzymes and enzyme inhibitors*
Cystatin C	Cathepsin B
Recombinant activated factor VIIa	Cystatin C
	Lipoprotein lipase
	Lysozyme
	Plasminogen
	Plasminogen activator inhibitor type I
	Plasminogen activator inhibitor type I–tissue plasminogen activator
	Plasminogen activator inhibitor type I–urokinase
	Pro-urokinase
	α-Amylase
	α-Galactosidase A
	Recombinant activated factor VIIa
Immune- and stress-related proteins	***Immune- and stress-related proteins***
Ig light chains	Ig light chains
Clara cell secretory protein	Pancreatitis-associated protein 1
α_1-Microglobulin	α_1-Microglobulin
	β_2-Microglobulin

Continued

Table 1 List of CUBN and LRP2 ligands.—cont'd

Immune- and stress-related proteins	***Immune- and stress-related proteins***
Drugs and toxins	*Drugs and toxins*
Aminoglycosides	Aminoglycosides
	Aprotinin
	Colistin
	Polymyxin B
	Trichosanthin
Receptors	*Receptors*
LRP2	Cubilin
	Transcobalamin II–B_{12} receptor
Others	*Others*
Coagulation factor VII	Ca^{2+}
Receptor-associated protein	Coagulation factor VII
	Coagulation factor VIII
	Cytochrome C
	Receptor-associated protein
	Seminal vesicle secretory protein II

protein products are involved in embryonic development. Amnionless was indeed initially identified as a protein essential for gastrulation and amnion formation in mice (Tomihara-Newberger et al., 1998). Via its N-terminus Amnionless interacts with Cubilin in the ileum, PCT, the visceral endoderm and most likely the testis (Coudroy et al., 2005; Oh et al., 2016; Strope et al., 2004). Amnionless has no known related proteins; nonetheless, its extracellular domain comprises a stretch of 70 amino acids containing a cysteine-rich module present in the BMP inhibitors chordin and procollagen IIA. Amnionless has only one potential site for N-glycosylation (NXT/S). It is an endocytic receptor harboring two functionally redundant FXNPXF signals within its cytosolic domain that may associate with the Clathrin-associated proteins Disabled-2 (DAB2) and autosomal recessive hypercholesterolemia (ARH), known for its interactions with low density lipoprotein receptor (LDLR) family (Fig. 1).

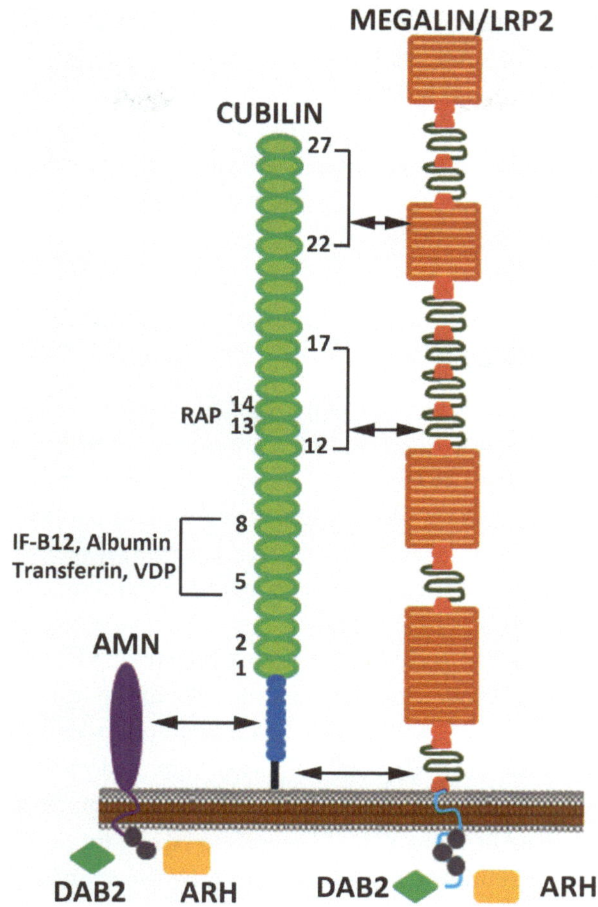

Fig. 1 Schematic representation of Cubilin and its coreceptor Amnionless and Lrp2/Megalin: Cubilin (Cubn) is a peripheral membrane protein with 27 cub domains and a sequence with amphipathic helical characteristics located in the amino-terminal region that has been implicated in cubn plasma membrane anchoring/association. Cubn requires Amnionless (Amn) or Lrp2/megalin for its internalization. Amn is a 50 kDa, type 1 transmembrane protein with a stretch of 70 amino acids in the extracellular region that has similarity to cysteine-rich modules present in a small group of proteins known to function as BMP inhibitors: chordin, *Short in gastrulation* and procollagen IIA. Amn contains endocytic motifs that interact with Dab2 and Arh. Megalin shares common structural motifs with other members of the LDLR family, including ligand-binding repeats, epidermal growth factor repeats, YWTD spacer domains, a single transmembrane domain and a short cytoplasmic domain containing conserved endocytic motifs that interacts with clathrin, Dab2 and Arh. The position of the IF-Cobolamin (IF-B12) binding site (CUB domains 5–8) is indicated. CUB domains 5–8 also correspond to binding of albumin, transferrin and vitamin D binding protein. CUB domains 13–14 are important in binding to receptor-associated protein (RAP). Cubn is shown associating with megalin via its N-terminal domain and CUB domains 12–17 and 22–27. Cubn also interacts with Amnionless via its EGF-like domains.

In addition to Cubilin, Amnionless interacts with the calcium release-activated channel (ORAI1) and the nuclear valosin-containing protein-like 2 (NVL2) (Urae et al., 2020; Zeng et al., 2017). ORAI1 is a four-spanning membrane protein expressed in the PCT and activated upon depletion of internal calcium stores. Activated ORAI1 would interact with AMN and possibly CUBN and LRP2 to allow concomitant ORAI1 and calcium endocytosis (Zeng et al., 2017).

5. Megalin/Lrp2

The Low-Density Lipoprotein (LDL) Receptor-related protein 2 (LRP2) or Megalin is a single-pass, transmembrane receptor of the LDL receptor gene family (Kerjaschki & Farquhar, 1982). It is encoded by *LRP2* (2q24-q31) and is the largest known mammalian single-chain receptor. Mutations in *LRP2* cause the autosomal recessive Donnai-Barrow syndrome (also known as facio-oculo-acoustico-renal syndrome) (Kantarci et al., 2007).

Initially described as gp330 or the autoantigen in Heymann nephritis, LRP2 consists of an extracellular domain of 4398 amino acids, a short transmembrane domain of 23 amino acids and a 209 amino acid long cytoplasmic tail. The extracellular domain comprises four regions of cysteine-rich complement-type repeats involved in ligand and calcium binding; four growth factor repeats spaced by eight YWTD repeats necessary for dissociation of the ligand in acidic conditions; a single epidermal growth factor-like repeat in the extracellular juxtamembrane region (Fig. 1). LRP2 is O- and N-glycosylated and glycosylation is necessary for ligand binding (Hirano, Totani, Fukuda, Gu, & Suzuki, 2017). The intracellular domain contains two (F/Y)XNPXY internalization motifs which interact with Clathrin, the adaptors AP-2, DAB2, ARH and other proteins involved in Clathrin-mediated endocytosis (Nagai, Meerloo, Takeda, & Farquhar, 2003; Oleinikov, Zhao, & Makker, 2000). The third, VENQNY, motif could mediate apical sorting of the receptor.

Like Cubilin, LRP2 has a furin cleavage site involved in the shedding of the extracellular part of LRP2 in the urine. Lrp2 has numerous ligands (Table 1). Among them is the chaperone LDL related protein associated protein 1 (LRPAP1), involved in the intracellular folding of newly synthesized LRP2 and Cubilin (Kristiansen et al., 1999).

6. The receptor complex CUBN/AMN, LRP2 and their partners (Fig. 2)

Cubilin interacts with other membrane proteins for plasma membrane localization and endocytosis.

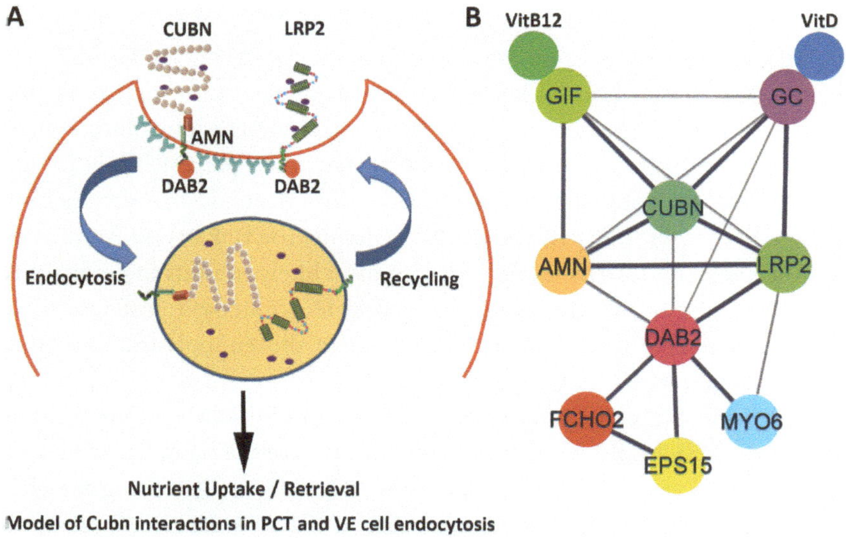

Fig. 2 (A) After endocytosis in the apical endocytic compartments, the receptor-ligand complexes dissociate and the receptors are recycled back to the plasma membrane via the dense apical tubules. The released ligands are generally delivered to lysosomes for degradation. A small fraction may escape this route and be directly released into the blood but this transcytotic pathway is of rather minor importance (Christensen & Nielsen, 2007). FcRn (not shown) may represent this route by directing apically internalized albumin to the basal membrane of tubular cells and releasing it in the interstitium due to the increased pH (Tenten et al., 2013). Receptor-mediated endocytosis shifts along the PCT. The uptake normally occurs in the first two segments (S1, S2) which strongly express Cubilin and LRP2 along the microvilli and the intermicrovillar spaces. The third segment (S3) would internalize proteins when the other two are overloaded (Christensen, Verroust, & Nielsen, 2009; Christensen et al., 2021; Devuyst & Luciani, 2015; Polesel & Hall, 2019; Verroust, 2002). Accordingly, the expression of Cubilin, LRP2 and DAB2 is fainter in S3. It is possible that in this segment Clathrin-dependent fluid-phase uptake takes over in case of abnormally increased protein levels in the ultrafiltrate. (B) STRING interactome of Cubilin/Amn and Lrp2 in vitamin B12 and vitamin D in DAB2 mediated endocytosis. EPS15, Epidermal growth factor receptor substrate 15; FCHO2, F-BAR domain only 2; GC, vitamin D-binding protein; GIF, Gastric intrinsic factor; MYO6, Unconventional myosin-VI.

The mutual Cubilin and Amnionless interdependence for apical membrane localization has been shown in the embryonic visceral endoderm (VE) and the adult kidney (Strope et al., 2004). In the VE of *Amn* null embryos or in the PCT of dogs expressing a dysfunctional *Amn* the apical membrane distribution of Cubilin is lost (Strope et al., 2004). Likewise, in *Cubn* null embryos or kidney-specific *Cubn* mice Amnionless remains blocked within the VE or PCT cells respectively (Amsellem et al., 2010; Coudroy et al., 2005; Storm et al., 2011).

Additional evidence on the interdependence between Cubilin and Amnionless in the human kidney comes from the study of the *CUBN* G653R mutation reported in a subject with macrocytic anemia, proteinuria, interstitial fibrosis and tubular atrophy. In this patient Cubilin and Amnionless, but not LRP2, were trapped within the endoplasmic reticulum (ER) of PCT cells (Udagawa et al., 2018) despite the lack of any Amnionless dysfunction. The *CUBN* G653R variant most likely inhibits the interaction between Cubilin and the glucose glycoprotein glucosyltransferase 1, an enzyme catalyzing the addition of glucosyl in the endoplasmic reticulum and allowing normal Cubilin folding.

At least in the kidney the three proteins form a multireceptor complex with most likely LRP2 driving the internalization of the complex and bound ligands (Ahuja et al., 2008; Amsellem et al., 2010) (Fig. 2B). Using the HEK 293 cell model Urae et al. suggested that LRP2 may be not required for CUBN-ligand endocytosis in the kidney. Nuclear valosin-containing protein like-2 (NVL2), would, through its interaction with Amnionless, facilitate ARH-dependent Cubilin endocytosis in the PCT and concomitant albumin uptake (Urae et al., 2020). However, ARH has not been detected in Cubilin and LRP2 expressing PCT segments (Fang, Garuti, Kim, Wade, & Welling, 2009; Wang, Yue, Sun, & Lin, 2010). ARH is indeed essentially expressed in the distal nephron. ARH deficiency causes hypercholesterolemia and perturbs the function of the renal potassium secretory channel but not Cubilin endocytosis (Fang et al., 2009; Wilund et al., 2002).

The functional relationship between Cubilin and LRP2, first suggested by the overlapping expression of the receptors in the PCT, has been confirmed in several contexts. Cubilin and LRP2 are important for the activation of vitamin D, a process that requires two hydroxylation steps. 25-hydroxyvitamin D3, 25-(OH)D3 produced in the liver binds to its serum transporter DBP. DBP-bound 25(OH)D3 enters the tubular lumen and is taken up by Cubilin and LRP2 in the PCT. This internalization step requires DAB2 and is the prerequisite for the formation of the active 1,25-(OH)D3 in the mitochondria, site of 1-alpha-hydroxylase activity (Chesney, 2016; Nykjaer et al., 2001). Active vitamin D binds to its receptor (VDR) and is taken up by the cell nucleus, a step that requires vitamin A sufficiency. Retinol binding protein (RBP)-bound retinoic acid is freely filtered and taken up by LRP2 in the PCT (Christensen & Willnow, 1999). The role of LRP2 and Cubilin in this process is highlighted by the high urinary excretion of vitamin A and vitamin D respectively and the vitamin D deficiency observed in patients with the Donnai-Barrow or Imerslund-Grasbeck syndromes. Urinary excretion of both vitamins and vitamin D deficiency are also observed in dogs and mice with *cubilin* and/or *lrp2* dysfunction.

Under normal conditions the Cubilin-LRP2 interaction is also critical for the endocytosis of albumin, and the other common Cubilin and LRP2 ligands, in the PCT and podocytes and loss of either one results in proteinuria (Amsellem et al., 2010; Nykjaer et al., 2001; Ren et al., 2020; Weyer et al., 2018). Even more important in Donnai-Barrow patients the dysfunction of LRP2 results in increased urinary shedding of Cubilin in the absence of any tubular or glomerular abnormality (Dachy et al., 2015). It is interesting to note that the residual expression of LRP2 is compatible with the apical distribution of Cubilin in the PCT and the most likely unmodified expression of Amnionless (Ahuja et al., 2008; Amsellem et al., 2010; Assémat, Châtelet, et al., 2005; Christensen et al., 2009; Hammad, Barth, Knaak, & Argraves, 2000a; Kozyraki et al., 1999; Verroust, 2002). Nevertheless, this expression is not sufficient to allow Cubilin endocytosis providing a clear demonstration of the primordial role of Cubilin and Lrp2 interaction for normal renal function.

With respect to extra-renal tissues, the observations from human and animal studies are contradictory. In the human terminal ileum LRP2 has not been detected (Jensen, Andersen, Hager, & Madsen, 2014). Lrp2 is detected in the rodent gut; in this site the expression of Amn is differentially regulated with age, does not depend on either Cubilin or Lrp2 (Pannérec et al., 2018) and is not modified by the lack of Cubilin (Aseem, Barth, Klatt, Smith, & Argraves, 2013; Cases et al., 2013). In the cephalic neural crest Lrp2 is the only known Cubilin partner expressed (Cases et al., 2013).It is clear today that the endocytic function of Cubilin is site/ligand dependent. Future studies will show whether additional site-specific endocytic Cubilin partners may exist.

7. Cubilin in the kidney

In the kidney, coordinate function of the nephrons controls body fluid composition, activates vitamin D, regulates hormone synthesis, allows excretion of toxins. The glomerular filter, composed of endothelial cells, a basement membrane and podocytes retains the majority of plasma proteins allowing the passage of water and small molecules. Due to higher sieving coefficients or charges a small fraction of proteins escapes the glomerular barrier and reaches the tubular lumen. Among these proteins are albumin and carriers for vitamins A, B, D et E. Their reabsorption in the PCT allows to recover these molecules from the ultrafiltrate and prevents urinary loss.

Cubilin and LRP2, abundantly expressed at the apical surface of epithelial PCT cells, retrieve the filtered proteins. More than 50 plasma proteins, including albumin, are reabsorbed by CUBN/AMN and LRP2, both interacting with DAB2.

The same molecular machinery is expressed in the podocytes and most likely, mesangial cells (Bryniarski et al., 2020), suggesting that receptor-mediated endocytosis may be the pathway for albumin uptake also in these cells. In podocytes, albumin overload increases the expression of Cubilin/Amn at the mRNA but not the protein level whereas neither Lrp2 nor Dab2 are modified (Gianesello et al., 2017). It is possible that the Cubilin/Amn pathway is an important entry point for proteins in human podocytes where it most likely cooperates with the other albumin binder, the neonatal Fc receptor (FcRn).

8. Cubilin and receptor-mediated endocytosis

The concept of receptor-mediated endocytosis was established by Brown and Goldstein (1986) during their studies on LDLR. Since then, many other endocytic receptors have been identified and the molecular mechanism of endocytosis has been characterized.

The PCT cells possess a well-developed apical endocytic apparatus, supporting a high degradative activity (Moestrup & Verroust, 2001; Nielsen et al., 2007; Zhuo & Li, 2013). Clathrin-dependent apical endocytosis seems to be the exclusive route for biomolecule uptake from the human PCT lumen (Zhuang, Marshansky, Breton, & Brown, 2011). Indeed, treatment of renal cells with chloropromazine, an inhibitor of Clathrin-dependent endocytosis, blocks albumin uptake whereas inhibition of Caveolin, the central constituent of caveolae, via Filipin or nystatin has no effect (Decorti, Malusà, Furlan, Candussio, & Klugmann, 1999; Raghavan, Rbaibi, Pastor-Soler, Carattino, & Weisz, 2014).

Clustering of Cubilin/Amn and Lrp2 in Clathrin-coated regions and interaction with Dab2, an endocytic adaptor, facilitates incorporation of ligand-receptor complexes in clathrin-coated pits and vesicles (Mishra et al., 2002). These first steps are followed by the uncoating of Clathrin-coated vesicles, fusion with subapical endosomal tubules and formation of large apical vacuoles. The subsequent dissociation of Cubilin-Lrp2-ligand complexes, receptor recycling and ligand retrieval requires a sustained vesicular acidification from early endosomes to lysosomes. Endosomal acidification in the PCT is driven by the vacuolar H+-ATPase and facilitated by the Chloride-proton antiporter ClC5, the chloride-channel CFTR and the sodium hydrogen exchanger NHE3 which collectively decrease the pH from around 7.4 in the tubular lumen to around 5.5 in the PCT endosomal compartments (Devuyst & Luciani, 2015; Gekle et al., 2004; Hurtado-Lorenzo et al., 2006; Reed et al., 2010; Scheel, Zdebik, Lourdel, & Jentsch, 2005; Stauber & Jentsch, 2013). The lower pH in the latter favors binding of albumin, and possibly other Cubilin/Lrp2 ligands, to FcRN (Anderson et al., 2006). It is thus possible that some ligands may switch receptors within an endosomal compartment, gain the endocytosis route and avoid default lysosomal degradation. Lysosomal degradation is attributed to cathepsins B, H and L and can be increased in case of proteinuria (Olbricht, 1992).

LRP2 is a constitutive recycling receptor and a marker of the complex apical recycling pathway in epithelial cells. In these cells the endosomal apparatus consists of separate apical and basolateral sorting endosomes (ASE, BSE), perinuclear Rab11-negative common recycling endosomes (CRE) and subapical Rab11-positive endosomes also known as the apical recycling endosomes/DAT (Bomsel, Prydz, Parton, Gruenberg, & Simons, 1989). After endocytosis Lrp2 traffics through the ASE, CRE and DAT, sorting that depends on the small GTPase Rab11 (Perez Bay et al., 2016). DAB2 is essential for LRP2 endocytosis whereas ARH retains LRP2 in the recycling endosomes, retarding or blocking its reappearance on the plasma membrane (Shah, Baterina, Taupin, & Farquhar, 2013).

The ability of LRP2 to carry out its functions in the kidney and other epithelia requires that ligand-free LRP2 is delivered to the plasma membrane with the same rate as ligand-bound LRP2 is internalized. The recycling pathway carries out this task. Fast recycling requires a rapid dissociation of Lrp2-lignad complexes in the acidic endosomes and depends on the voltage-gated chloride channel 5 (CLC5) and vacuolar proton ATPase (Hryciw et al., 2012).

Cubilin/Amn and Lrp2 endocytosis also depends on Myosin, microtubule and Actin dynamics. Lrp2 interacts with the heavy chain of nonmuscle Myosin II via Dab2 which can also interact with Myosin VI. Indeed, blocking of myosin II or microtubule polymerization via blebbistatin or nocodazole respectively, drastically decreases uptake of Albumin and other Cubilin and Lrp2 ligands (Gekle et al., 2004; Hosaka et al., 2009; Maurer & Cooper, 2005; Morris, Tallquist, Rock, & Cooper, 2002). Furthermore, knockout of *myosin VI* decreases the association of Dab2 to the plasma membrane, reduces the number of apical vacuoles and impairs PCT endocytosis (Gotoh et al., 2010). Finally, disruption of F-actin using cytochalasin D almost completely blocks Albumin endocytosis *in vitro* (Dance et al., 2004; Gekle, Mildenberger, Freudinger, Schwerdt, & Silbernagl, 1997).

Conversely, reduced or absent expression of the receptor complexes impairs the formation of apical endocytic compartments and affects both receptor-mediated and fluid-phase endocytosis in mice and men (Dachy et al., 2015; Leheste et al., 1999; Perea-Gomez et al., 2019).

9. Cubilin/Amn, Lrp2, DAB2

In renal, and extra-renal, tissues *in vivo* Cubilin/Amn and Lrp2 internalization requires Dab2 (Morris et al., 2002). Dab2 is a phosphoprotein particularly abundant in the kidney, notably the PCT and its first two segments (Moore, Cai, Tao, Smith, & Xu, 2013; Tao, Moore, Smith, & Xu, 2016). *Dab2* ablation in the mice leads to embryo-lethality but conditional inactivation in the embryonic tissues reduces endocytic vesicle formation in the PCT, the expression of Lrp2 and Cubilin and consequently increases proteinuria (Morris et al., 2002). In the PCT, Dab2 interacts with the protein kinase B/Akt. Akt-mediated Dab2 phosphorylation at Ser-448 and Ser-449 seems to be physiological relevant as mutation of these two Ser inhibits albumin endocytosis (Koral & Erkan, 2012; Koral et al., 2014). It has been proposed that Ser phosphorylation tethers Dab2 to the plasma membrane and brings it to close proximity to Clathrin, facilitating Lrp2 and other receptor-mediated endocytosis (Koral et al., 2014). A similar Lrp2-Cubilin-Dab2-dependent, but not Amn, mechanism may operate in mesangial cells to scavenge low amounts of albumin and preserve the integrity of the glomerular barrier (Bryniarski et al., 2020).

The VE is a polarized epithelial tissue harboring functional and morphological similarities with the PCT cells. The VE surrounds the developing mammalian embryo throughout gestation and is critical for embryonic survival until E7.5–8.5. VE supports nutrient supply and conveys signals necessary for embryonic patterning and growth. Similarly to yolk sac and placenta, the VE expresses Cubilin/Amn and Lrp2 (Burke, Jauniaux, Burton, & Cindrova-Davies, 2013). In *Dab2* null embryos Cubilin and Lrp2 accumulate at the apical plasma membrane of the VE; endocytosis is severely compromised and causes a developmental arrest around gastrulation (E7.5), a phenotype reminiscent of *Amn* and *Cubn* null mice (Maurer & Cooper, 2005; Smith et al., 2006; Tomihara-Newberger et al., 1998).

Cubilin-ligand endocytosis is thus linked to Dab2 both in the kidney and the VE. Observations based on mice lacking *Cubn-*, *Amn-* or *Lrp2* suggest that in the kidney Dab2 is the endocytic motor for receptor endocytosis. In the VE the function of Lrp2 may be dispensable but that of Dab2 is only partially compensated by ARH protein (Moore et al., 2013) and the latter is not sufficient to rescue Cubilin and Amn distribution in the VE. It is thus possible that another Dab2 interacting transmembrane protein contributes to Cubilin/Amn function in the VE.

Another site where Cubilin/Amn, Lrp2 and Dab2 may cooperate is the intestine of the suckling rat, a transitory epithelium with high endocytic capacity (Vázquez-Carretero et al., 2014). This epithelium has a complex apical endocytic apparatus which at weaning begins to undergo structural and functional changes to eventually be replaced by the adult intestinal cells.

These are absorptive cells with a greatly reduced endocytic apparatus (Fujita, Baba, Shimamoto, Sakuma, & Fujimoto, 2007). Dab2 co-localizes with Lrp2 and Cubilin/Amn this interaction would allow a co-operation during the suckling period. However, differential regulation of the four molecules after weaning suggests that this multiligand complex may not function in a similar way in the adult (Vázquez-Carretero et al., 2014). Supporting this possibility, no LRP2 has yet been detected in the human adult ileum (Jensen et al., 2014).

10. Regulation of cubilin expression

The expression of Cubilin is regulated by both environmental and genetic factors. Apical endocytosis in the PCT is particularly flexible and adjustable to the endocytic load. Changes in the endocytic load can alter the plasma membrane expression of Lrp2 and accordingly that of Cubilin.

The fluid shear stress can also activate endocytosis; this effect is not specific to Cubilin and Lrp2 but impacts both fluid-phase and receptor-mediated endocytosis (Ferrell, Ricci, Groszek, Marmerstein, & Fissell, 2012). Retinoic acid induces differentiation of various cell types, including VE, PCT and tumor cells, and is a positive regulator of Cubilin, Lrp2 and Dab2 (Alvarado-Morales et al., 2021; Chlon, Taffany, Welsh, & Rowling, 2008; Hammad, Barth, Knaak, & Argraves, 2000b). In mammary cells exposure to RA does not modify *Cubilin* mRNA levels but does increase *Lrp2* and *Dab2* transcripts indicating that modulating the traffic of Cubilin, but not its synthesis, is a rapid means to adjust the rate of endocytosis.

The apical plasma membrane expression of Cubilin and Lrp2 together with the formation and function of the primary cilium and brush border are markers of terminal epithelial PCT differentiation (Lima et al., 2010). The transcription factor ZONAB (zonula occludens 1-associated nucleic acid binding protein) binds to the promoter regions of *Cubn* and *Lrp2* and is inversely correlated with *Cubilin* and *Lrp2* expression and brush border formation (Lima et al., 2010). Indeed, overexpression of ZONAB in proximal tubule cells *in vitro* represses the expression of *Cubn* and *Lrp2* and promotes cell proliferation.

Hepatocyte nuclear factor 1 alpha (HNF1a) a transcription factor expressed in the liver pancreas and PCT, is a causal gene for autosomal dominant diabetes mellitus and PCT dysfunction. HNF1a functionally interacts with the promoters of *LRP2* and *CUBN* and is essential for the constitutive expression of Cubilin and Lrp2. Loss of this factor results in defective endocytosis whereas its overexpression promotes the expression of the receptors in proximal tubule cells (Terryn et al., 2016).

Transforming growth factor-b1 (TGF-b1) is a cytokine involved in various physiological and pathological effects including apoptosis, fibrogenesis and epithelial-mesenchymal transition. Gekle et al. showed that in proximal tubule cells increased TGF-b1 levels reduced albumin endocytosis; the overexpressed TGFb1 appears to repress *Lrp2* transcription via activation of the canonical Smad-dependent TGF-b1 pathway. The decrease in Cubilin expression and albumin reabsorption observed by Gekle et al. may thus be secondary to the downregulation of Lrp2 (Cabezas, Farfán, & Marzolo, 2019; Gekle et al., 2003).

The expression of Lrp2 is regulated by Peroxisome Proliferator-Activated Receptors (PPARs), transcription factors of the nuclear receptor superfamily, histone acetylation and methylation and DNA methylation (Kondo, Shen, Yan, Huang, & Issa, 2004). The functional interactions between Cubilin and Lrp2 suggested similar regulations for Cubilin.

In the mouse, *Cubilin* undergoes an allelic inactivation, i.e., downregulation or silencing of the maternal or paternal allele (Aseem et al., 2013). This phenomenon is called monoallelic expression and may be due to random gene inactivation, autosomal gene imprinting or chromosomal inactivation (Gimelbrant, Hutchinson, Thompson, & Chess, 2007). DNA methylation and histone deacytelation may explain, at least partly, monoallelic expression of *Cubn* gene but a role of noncoding RNAs is not excluded (Aseem et al., 2013). Additionally, *Cubn*, as *Lrp2*, is a PPAR alpha and gamma responsive gene. Histone deacetylation contributes to the pathogenesis of diseases like gastric and colorectal cancer, also characterized by impaired *Cubilin* expression. This observation may indicate similar mechanisms of *Cubilin* regulation in man and mice. It may also suggest that the improvement of albuminuria by PPAR agonists is mediated by increased Cubilin and Lrp2 levels.

Maternal protein restriction upregulates genes involved in lipid uptake and endocytosis including Cubn (Daniel, Swali, Emes, & Langley-Evans, 2016). The strong increase in Cubn is also observed when suboptimal fat supply is provided (Draycott et al., 2020). The underlying mechanism is not clear but may correspond to an adaptative response preserving fetal growth.

1,25(OH)-D3 is an essential micronutrient; it activates the vitamin D receptor (VDR) and alters the transcription rate of genes bearing the vitamin D response element in their promoter regions. It is believed that almost all cells in the body can convert 25(OH)-D3 to its active form 1,25(OH)-D3 including the ocular tissues or tumor cells insofar as they express Cubilin, Lrp2 and Dab2 (Alsalem et al., 2014; Holick, 2013). Currently, variants specifically associated with vitamin D deficiency, including during pregnancy, have been identified in LRP2 but not CUBN (Alharazy et al., 2021; Wang et al., 2021). Supplementation of pregnant women with vitamin D/cholecalcipherol increases the amounts of placental 25(OH)-D3 and 1,25(OH)-D3 and the blood levels of Cubilin and Lrp2 suggesting a role of these receptors in the modulation of vitamin D metabolites by the placenta (Park et al., 2017). On the contrary, in the kidney and white adipose tissue cholecalciferol supplementation downregulates the expression of Cubilin at the mRNA and/or protein levels. In the liver as well in adipocytes the same treatment results in the specific downregulation of Cubn mRNA without affecting the levels of Lrp2, Amn and Dab2. The meaning of this finding is currently unclear but demonstrates the importance of Cubilin in vitamin D handling and the tight tissue-specific regulation of the vitamin.

Cubilin and Lrp2 are expressed in the gallbladder in humans and mice; the expression of Lrp2 but not Cubilin is regulated by bile acids, most likely via the farnesoid X receptor. Genetic data suggest a role of Lrp2 in murine gallstone disease and a lithogenic diet increases Lrp2 levels (Erranz et al., 2004). The expression of Cubilin is not apparently modified but it is possible that increased Cubilin recycling, due to the elevation of Lrp2, contributes to the disease.

11. Cubilin in vitamin B12 homeostasis vitamin B12 (Fig. 3)

Vitamin B12 (cobalamin, B12) is an essential water-soluble molecule that functions as a coenzyme for the enzymes methionine synthase and methylmalonyl-CoA mutase. The first is a cytoplasmic enzyme involved in nucleotide synthesis by catalyzing the methylation of homocysteine to methionine. The latter allows the mitochondrial conversion of methylmalonyl-CoA to succinyl-CoA and is necessary for the digestion

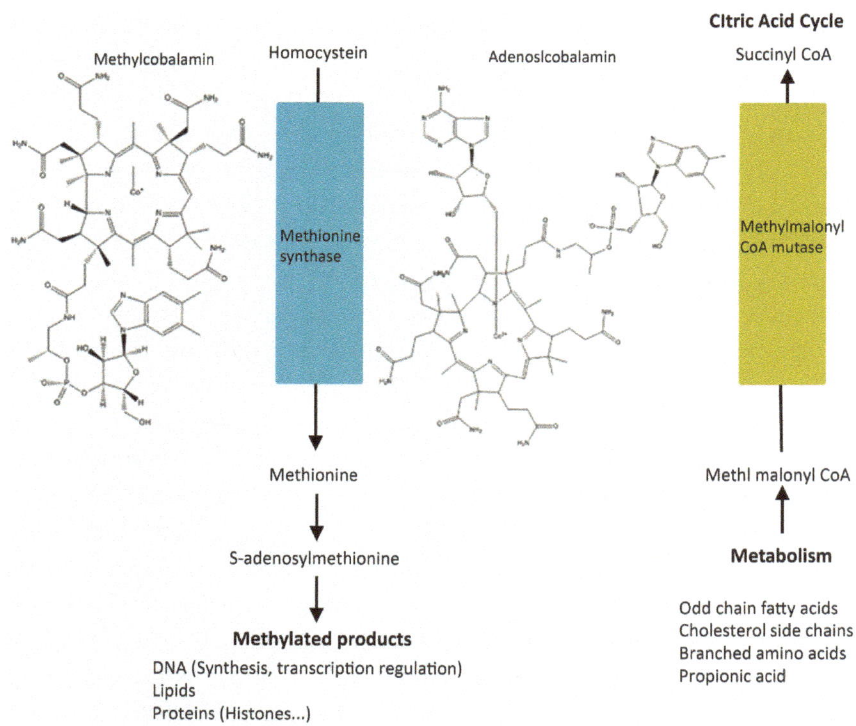

Fig. 3 Vitamin B12 an essential coenzyme for methionine and succinyl-CoA synthesis.

of various organic compounds including odd-chain fatty acids. Humans do not synthesize B12 and entirely depend on the consumption of food of animal origin. The daily requirements vary between 0.4 and 2.4 mg depending on age and status and the total body storage, mainly due to the hepatic and renal uptake of B12, ranges from 2 to 5 mg. Moderate B12 deficiency is common worldwide, especially in the elderly. Overt B12 deficiency is a severe disease characterized by hematological and neurological disorders (Allen, 2009; Carmel, 2008; Stabler, 2013). These include anemia, macrocytosis, leukopenia and thrombocytopenia, demyelinating disorder of the central nervous system, paresthesia and gait problems, sensory deficits (Allen, Stabler, & Lindenbaum, 1998; Green et al., 2017; Smith & Refsum, 2009). The hematological defects can be reversed by adequate B12 supplementation.

12. Cobalamin absorption and transport (Fig. 4)

In the extracellular fluids transport of B12 depends on three carriers: haptocorrin (or R-protein or transcobalamin I), gastric intrinsic factor and transcobalamin (or transcobalamin II). These proteins are homologous;

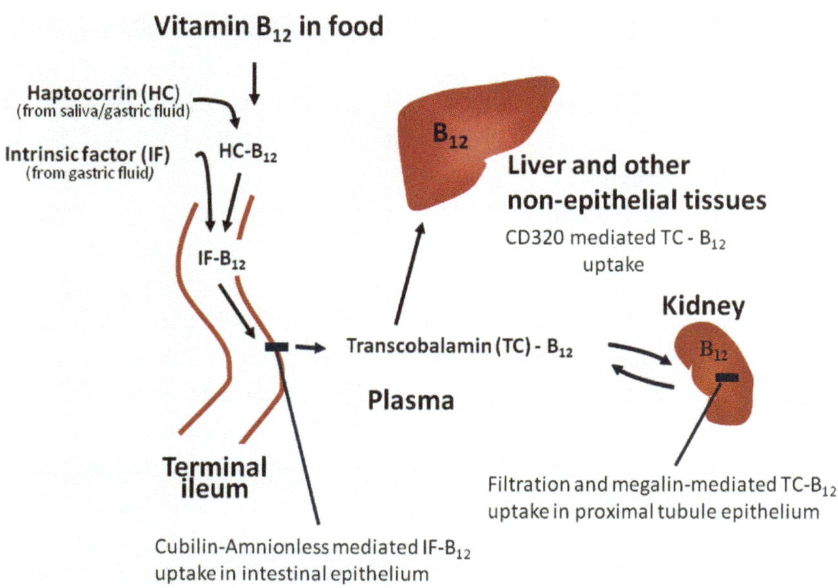

Fig. 4 B12 homeostasis.

each one is linked to a single vitamin B12 molecule (Fedosov, 2012; Morkbak, Poulsen, & Nexo, 2007; Quadros, 2010; Wuerges, Geremia, Fedosov, & Randaccio, 2007).

In the acidic environment of the stomach dietary B12 is released and associates with haptocorrin (HC, $K_d \sim 0.01$ pM) which protects B12 from degradation. In the duodenum HC is digested by pancreatic proteases and B12 binds the gastric intrinsic factor (IF, $K_d \sim 1$ pM) essentially produced by the gastric parietal cells (Gräsbeck, 2013; Kozyraki & Cases, 2013; Nakamura, Sagawa, & Mori, 1993; Russell-Jones & Alpers, 1999).

IF is a highly glycosylated protein, resistant to intestinal enzymes. IF displays a two-domain structure and B12 binds at the interface of these domains. B12-associated IF is crucial for the interaction with Cubilin expressed together with Amn in the ileal enterocytes. Several B12-analogues can bind IF but are unable to assemble the two domains and bind to Cubilin (Fedosov, Fedosova, Kräutler, Nexø, & Petersen, 2007). IF can oligomerize *in vitro*. Such a mechanism would allow a more efficient interaction with Cubilin trimers and increase B12 intestinal absorption *in vivo* (Fedosov et al., 2004).

Cubilin/Amn-mediated endocytosis guarantees endocytosis of the minute doses of ingested vitamin whereas receptor-independent uptake accounts for only 1% of the overall absorption. Within the enterocyte IF is degraded in the lysosomes and the released vitamin B12 enters the cytosol most likely via LMBD1, the lysosomal cobalamin transport escort protein and ABCD4, an ATP-binding cassette transporter (Coelho et al., 2012; Rutsch et al., 2009). The exact cellular handling of vitamin B12 after its lysosomal exit and its usage as a coenzyme or export from the cell is beyond the scope of this review.

Upon its exit B12 binds transcobalamin (TC, K_d 0.005 pM), the plasma carrier that transports 20–30% of the total vitamin. Further delivery of B12 to the body tissues depends on the interaction of the TC-B12 complex with CD320 and Lrp2, members of the LDL receptor family (Quadros, Nakayama, & Sequeira, 2009). After lysosomal degradation of TC, B12 is either stored (mainly in the liver and kidney) or partially secreted back to the circulation associated with HC or TC. Despite structural similarities between TC and HC, differences in the aminoacid sequence (Lys105 and Lys114 in TC vs Asp92 and Asn102 in HC) result in electrostatic potentials that favor TC binding to CD320 (Alam et al., 2016).

HC associated with all B12 analogues is taken up by the liver via the asialoglycoprotein receptor. HC is generally degraded and B12 compounds are secreted into the bile (Velkova et al., 2017). B12 secreted into the bile undergoes enterohepatic circulation (5–10 mg/day) and is reabsorbed via the IF-vitamin B12-Cubilin/Amn pathway.

13. Renal handling of vitamin B12

CD320 and Lrp2 are involved in renal B12 uptake. CD320 is expressed in virtually all tissues and is tightly regulated according to the proliferative/differentiation status of the cell and the cellular levels of B12. Increased expression of CD320 is observed in high B12 demanding dividing cells (Amagasaki, Green, & Jacobsen, 1990). High CD320 expression may also explain the accumulation of B12 in the kidney during times of high circulating levels of B12. Conversely, downregulation of CD320 would increase the levels of circulating TC-vitamin B12 complexes (Moestrup et al., 1996).

Lrp2 is the main receptor for TC-B12 reuptake in the kidney (Moestrup et al., 1996). Lrp2 binds TC with high affinity and prevents urinary loss of the vitamin. After internalization in the PCT cell, B12 can be locally used, stored or released back to the bloodstream. How vitamin B12 exits the tubular cell is unknown. During periods of high plasma B12 levels, downregulation of CD320 and efficient Lrp2-mediated uptake may explain the accumulation of vitamin B12 in the renal tissues. No HC/B12 analogues uptake has been described; due to its heavy glycosylation HC is less likely to be filtered.

14. IGS or juvenile megaloblastic anemia (Fig. 5)

IGS or selective B12 malabsorption or megaloblastic anemia 1 (MGA1: OMIM ID #261100) is a rare autosomal recessive disorder characterized by B12 deficiency due to intestinal B12 malabsorption, usually resulting in megaloblastic anemia. IGS appears in childhood and is generally diagnosed before the age of 14. Failure to thrive, frequent mild low molecular weight proteinuria and neurological symptoms of various severity are described (Grasbeck, Gordin, Kantero, & Kuhlback, 1960; Gräsbeck & Tanner, 2011). Rarely, pancytopenia and pseudo-thrombotic microangiopathy may occur (Andrès, Affenberger, Federici, & Korganow, 2006). The mechanism of hemolysis is unclear but may involve intra- and extramedullary destruction of erythrocytes (Acharya et al., 2008; Gurlek Gokcebay, Akpinar Tekgunduz, & Cavdarli, 2020). Proteinuria is

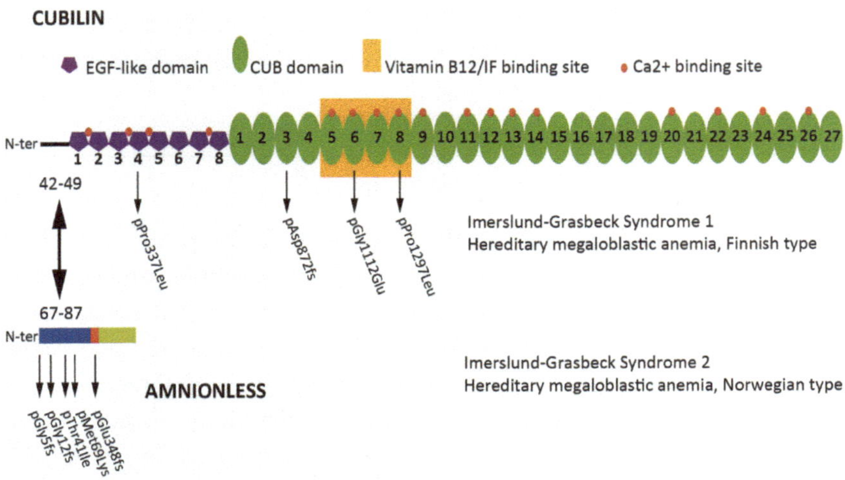

Fig. 5 Mutations in CUBN and AMN causing Imerslund-Grasbeck syndrome.

not progressive and less than 1 g/day. In IGS minor glomerular changes have been observed and the renal function is globally preserved (Broch, Imerslund, Monn, Hovig, & Seip, 1984; Storm et al., 2013; Udagawa et al., 2018). Congenital anomalies are rare. Nonetheless, urinary tract anomalies have been described. IGS occurs worldwide and has a higher prevalence in middle eastern countries, Norway and Finland.

To date 300 cases are described with 37 different mutations in *CUBN* and 30 in *AMN* (Aminoff et al., 1999; Tanner et al., 2003; Tanner, Sturm, Baack, Liyanarachchi, & de la Chapelle, 2012). Finnish- and Norvegian-type IGS are mapped to the *CUBN* and *AMN* locus respectively and are examples of founder effects. Both *CUBN* and *AMN* mutations are identified in consanguineous mediterranean families; compound heterozygous mutations in *AMN* associated with a *CUBN* missense variant have also been reported (Montgomery et al., 2015). In some cases both *CUBN* and *AMN* are excluded (Tanner et al., 2004); this may be due to misdiagnosis of patients suffering from the clinically close genetic deficiency in gastric IF. The mutations in *CUBN* may affect the recycling or glycosylation of the receptor, i.e., G653R (Udagawa et al., 2018) or the binding site of IF-B12, i.e., the FM1 mutation in CUB8 modifying Pro1297 to leucine. This proline is not directly involved in IF binding but interacts with Pro1299 to stabilize the loop of CUB8 and IF binding (Andersen et al., 2010; Kristiansen et al., 2000). All *AMN* mutations perturb its interaction with CUBN (Storm et al., 2013).

15. Cubilin in pathology

15.1 Renal pathology

Impairment in the PCT function prevents normal reabsorption of filtered proteins and results in their excretion into the urine. This sign is referred to as low molecular weight proteinuria or tubular proteinuria and is characterized by the preferential excretion of proteins with a MW $<$ 50 kDa and a rather high sieving coefficient. A dysfunction of the glomerular filter results in a different urinary profile with higher MW proteins caused by the overwhelming of the PCT reabsorption capacity. Albumin is the predominant protein in the urine in that case. CUBN is a high affinity, low-capacity receptor for Albumin whereas LRP2 carries low affinity, high-capacity sites suggesting that in higher concentrations of Albumin LRP2 is preferentially required. However, under normal conditions loss of either Cubilin or Lrp2 similarly results in tubular proteinuria indicating a requirement for both receptors for protein reabsorption (Amsellem et al., 2010; Nykjaer et al., 2001).

The glomerular filtration rate (GFR) varies in humans. GFR reflects the response of the kidney to normal conditions and variations. A protein-rich meal for instance will induce a relative hyperfiltration during which the abundance of filtered protein per unit time will increase without modifications in protein concentration in the ultrafiltrate. The PCT cells are able to adapt to these modifications by increasing protein uptake (Raghavan et al., 2014). Changes in tubular flow could alter the endocytic capacity to maintain efficient uptake by the PCT cells and an impaired response would be a cause of tubular proteinuria. *In vitro* studies showed that Opossum Kidney cells (OK) cells growing under continuous shear stress were able to increase by at least fivefold their endocytic capacity. In addition to the increased apical expression of Cubilin and Lrp2, stressed OK cells expanded the endolysosomal and mitochondrial compartments and had more numerous microvilli (Park et al., 2020; Ren et al., 2019). Whether the PCT cells can respond acutely to flow changes *in vivo* requires further investigation. What is clearly established however is that dysfunction of *Cubn* or *Lrp2* in humans and in mice prevents efficient uptake of Albumin and other proteins resulting in tubular proteinuria (Amsellem et al., 2010; Bedin et al., 2020; Weyer et al., 2011). If the glomerular filter is breached, albumin levels increase rapidly in the ultrafiltrate and overwhelm the capacity of Cubilin and possibly that of Lrp2. In this case, fluid-phase endocytosis could limit albumin excretion. If, as it is currently thought, Lrp2 is necessary for fluid-phase endocytosis the loss of Lrp2 would have a greater effect than that of Cubilin on Albumin retrieval.

Morphological modifications of the PCT cells, including reduced brush border microvilli and endosomal vesicle formation, could be important in human pathology during tubular damage and involve the mammalian target of rapamycin complex 1 (mTORC1) (Long et al., 2017; Raghavan et al., 2014).

mTORC1 and 2 are intracellular multiprotein complexes co-expressed together with Cubilin and Lrp2 in the two first segments of the RPT. They integrate various signals, i.e., growth factor signaling, amino acid and cellular energy content or cellular stress to regulate cell survival, growth and division. Specific inactivation of TORC1 and/or TORC2 in the mouse leads to a renal Fanconi syndrome characterized by a low molecular weight proteinuria, aminoaciduria, glucosuria and phosphaturia (Grahammer et al., 2017). In these mice both fluid-phase and receptor-mediated endocytosis are impaired, the length of apical microvilli and the number of endocytic vesicles are reduced, the length of the S1 and S2 segments is shortened. Reduction of DOCK8, an activator of Cdc42, a member of the Rho family of GTPases and key regulator of cell polarity and apical differentiation (Harada et al., 2012) may contribute to the morphological defects. The concomitant reduction of SNX8, a member of the sorting nexin protein family and Rab10, involved in vesicular trafficking in the mutants would impair endocytosis and vesicle transport (Shi & Grant, 2013). Surprisingly in these mice the expression of Cubilin and Lrp2 is preserved. This observation is inconsistent with reduced receptor-mediated endocytosis and previous data showing that: (a) apical dedifferentiation results in the re-distribution of Cubilin and Lrp2 in endocytic compartments and (b) lack of Lrp2 is sufficient to induce dramatic changes in the apical endocytic pathway of the PCT cell (Leheste et al., 1999). In this context it is interesting to note that the transcriptional down-regulation of Lrp2 was previously reported in mice treated with the mTOR inhibitor rapamycin (Gleixner et al., 2014).

Renal damage of various origins may cause urinary loss of albumin between 30–300 mg/day or microalbuminuria; 300–3500 mg/day or macroalbuminuria; more than 3500 mg/day known as nephrotic-range albuminuria. Exposure of PCT cells to serum proteins may trigger toxic responses and damages the tubular cells and/or impair lysosomal degradation as it was shown *in vitro* (Long, Rbaibi, Gliozzi, Ren, & Weisz, 2020).

The underlying mechanisms involved ER stress, NLRP3 inflammasome (cryopyrin) activation and tubule interstitial injury most likely resulting from the increased uptake of albumin-bound fatty acids and oxidized lipids (Fang et al., 2013; Khan et al., 2018; Liu et al., 2015; Ruggiero et al., 2014; Theilig et al., 2007). Knockdown of *Cubn* or *Lrp2* blocked these effects in vitro (Liu et al., 2015). Whether Cubilin and Lrp2-mediated endocytosis of albumin and other proteins protects from or favors the progression of renal damages and disease is an open question. It is also unclear how increased protein load impacts the plasma membrane expression of Cubilin and Lrp2 (Kim, Moradi, Yuan, Norris, & Vaziri, 2009).

Diabetic kidney disease (DKD) is the main cause of end-stage kidney disease. The pathogenesis of DKD involves PCT cell dysfunction in a high glucose environment (Gilbert, 2017; Tang & Lai, 2012). It was initially thought that in diabetic kidneys excessive amount of plasma proteins, and albumin, were filtered through the damaged glomerulus. Protein overload conditions could thus cause PCT cell death, inflammation and tubulo-interstitial fibrosis. Nonetheless, several studies showed that the glomerular filtration rate remained unchanged; the increased amount of albumin in the ultrafiltrate would result from reduced albumin reabsorption in the PCT via CUBN and LRP2 (Mori et al., 2017; Russo et al., 2009). Indeed, urinary excretion of CUBN and LRP2 was increased in diabetic patients (Thrailkill et al., 2009) and plasma membrane expression of Cubilin is lower in diabetic mice (Jheng et al., 2015) suggesting that this mechanism may protect from protein overload. A set of experimental data support this hypothesis.

High glucose is known to activate TLR4, a Toll-like receptor critical in diabetic nephropathy, and its potential ligand HSP70. TLR4 promotes inflammation and tubular cell injury and can be further stimulated by albumin (Jheng et al., 2015). In a *TLR4* null mouse the tubulointerstitial injury and albuminuria were inversely associated with the tubular expression of Cubilin (Jheng et al., 2015). The authors proposed that reduced plasma membrane expression of Cubilin and increased albuminuria could be due to the upregulation of TLR4. Supporting this mechanism, patients with diabetic nephropathy had higher levels of TLR4 and HSP70 than controls (Jheng et al., 2015). The protective role of Cubilin and Lrp2 from protein overload was further suggested in mice treated with ipragiflozin, an inhibitor of sodium-glucose cotransporter 2, used to increase glucosuria and lower blood glucose levels. Increased renal glucose uptake in diabetic patients stimulates O-glycosylation of various PCT proteins including Lrp2 and

Cubilin. Blocking O-glycosylation in mice decreased plasma membrane expression of Lrp2 suggesting that this effect contributed to the renal protective effect of these drugs (Otomo et al., 2020).

In patients with proteinuric diseases (IgA nephropathy, minimal change nephrotic syndrome) large amounts of proteins enter the tubular lumen. The renal investigation of these patients as well as of non-albuminuric, micro-and macro-albuminuric individuals showed a differential distribution of Cubilin and Lrp2. The two receptors were strongly expressed at the PCT plasma membrane of non-albuminuric individuals whereas in all albuminuric patients Cubilin, but not Lrp2 accumulated within the cell (Sun et al., 2017). Despite the limited number of individuals examined, these observations suggest that the tubular protein load regulates the apical membrane expression of Cubilin and Lrp2. Their re-distribution in intra-cellular compartments in proteinuric conditions may reduce RPT overload and prevent and/or limit renal damage. Although more studies are required to demonstrate this hypothesis, this adaptive mechanism is compatible with the unique plasticity of RPT cells to rapidly increase or decrease its endocytic capacity.

Nephrotic syndrome (NS) is a chronic kidney disease characterized by massive proteinuria, hypoalbuminemia, edema and hyperlipidemia. Steroid resistant nephrotic syndrome (SRNS) is the second most frequent cause of chronic kidney disease and is due to mutations of 39 monogenic genes including *CUBN* (Vivante et al., 2017). The combined approach of whole genome homozygosity mapping with human exome capture revealed an homozygous 1-bp deletion (c.8355delA; p.S2785fsX19) in exon 53 of *CUBN* in two patients with intermittent nephrotic-range proteinuria and epilepsia without anemia (Ovunc et al., 2011). The pathogenesis of the SRNS is still poorly understood and no curative treatment is currently available. The identification of *CUBN* and other single-gene mutations may be important to orient the treatment in the patients that do not respond to steroids.

Another genetic *CUBN* variant was identified as a risk variant for renal function in end-stage renal disease and graft loss in kidney transplantation (Böger et al., 2011; Reznichenko et al., 2012). Nonetheless, some *CUBN* variants may be indicators of normal renal function despite early onset (Bedin et al., 2020). The latter are generally located after the IF-B12 binding site (residues 1487–3618) and may reflect a means to reduce PCT protein/albumin uptake and preserve renal function.

16. Cubilin dysfunction in Fanconi-Bickel, Dent and Lowe syndromes

The Fanconi-Bickel syndrome (Santer, Schneppenheim, Suter, Schaub, & Steinmann, 1998) is a rare autosomal recessive disorder caused by mutations in the gene *SLC2A2* encoding the glucose transporter protein 2 (GLUT2) also expressed in the liver, pancreas and gut. GLUT2 is located in the first segment of the PCT and is necessary for glucose reabsorption. The renal manifestations include generalized proximal dysfunction and impaired utilization of glucose and galactose. The expression of Cubilin, Lrp2 and the beta2 subunit of the H+-ATPase, involved in the reabsorption of bicarbonate, were strongly reduced (Mihout et al., 2014). It is unclear however whether the downregulation of these proteins is secondary to chronic glycogen accumulation or additional mechanisms.

Impaired Cubilin and Lrp2 function is reported in Dent disease and Lowe syndrome, rare X-linked recessive disorders. In Dent disease, characterized by low molecular weight proteinuria, hypercalciuria and nephrocalcinosis, the loss of the endosome-associated chloride channel CLC-5 and the ensuing modification in Cl content, impairs endocytosis and recycling of Cubilin and Lrp2 (Devuyst, 2010). Knock out of CLC-5 in mice largely summarizes the human disease showing decreased expression of Cubilin and Lrp2 (Christensen et al., 2003). It is possible that reduced Cubilin/Lrp2-mediated ligand uptake may, in addition to proteinuria, contribute to the development of hypercalciuria and nephrocalcinosis. In these mice the reduced expressions of Cubilin and Lrp2 do not result from decreased mRNA levels but from defective recycling. However, the discordance between Cubilin and Lrp2 levels along the PCT segments is surprising and cannot be explained currently.

Some cases of Dent disease called Dent disease 2 are caused by mutations in the phosphatidylinositol phosphatase encoding *OCRL* gene which also causes the Lowe syndrome. The Lowe syndrome is characterized by vision defects, intellectual disability and renal Fanconi syndrome including aminoaciduria, proteinuria, tubular acidosis and renal rickets (Loi, 2006). Like in Dent disease, Lowe patients have reduced levels of Cubilin and Lrp2 at the renal brush border and excrete lower amounts of the receptors (Santo et al., 2004). OCRL1 has numerous roles in cytoskeletal dynamics and membrane trafficking but its precise mode of action remains unclear. *Ocrl1* knockout mice do not have any phenotype, most likely due to redundancy with Inpp5b, another phosphatidylinositol phosphatase. Confirming this hypothesis mutants lacking the murine form but expressing human INPP5B develop a tubulopathy and have impaired Lrp2 recycling (Bothwell et al., 2011; Jänne et al., 1998).

17. Cubilin in diabetes, the vitamin D/micronutrient connection

Diabetes mellitus is a multifactorial disorder with genetic and environmental causes (Kahn, Cooper, & Del Prato, 2014). The vitamin D biological network, including the tandem Cubilin and Lrp2, has been involved in the pathogenesis of both type 2 and type 1 diabetes by modulating among others insulin resistance or pancreatic beta-cell function. Vitamin D deficiency may have a role in the development of vascular complications and vitamin D supplementation improves glycemic control in diabetes.

A relationship between Cubilin and diabetes has been suggested by the co-existence of IGS and type 1 diabetes in the same subjects (Madhavan, Vijayakumar, Rajajee, & Nammalwar, 2009; Schröder, Witt, & Eber, 2000). In a study investigating the potential association between *CUBN* and type 2 diabetes it was suggested that CUBN variants rs11254375, rs7071576, rs6602175, rs1801224 and rs4366393 were associated with the disease. It is unclear however whether these intronic polymorphisms modify Cubilin activity.

Similarly, the *CUBN* polymorphism rs3740165 was associated with an increased risk for type 1 diabetes (Ramos-Lopez et al., 2010). While a relationship between some *CUBN* variants and dysglycemia becomes evident, other studies associate *CUBN* locus with vascular complications and microalbuminuria in diabetic subjects. Similarly, the variant rs1801239 has been documented in patients with end-stage renal disease, peripheral nephropathy and type 2 diabetes (Ahluwalia et al., 2019; Ma et al., 2016; Teumer et al., 2016; Tsekmekidou et al., 2020). It is thus possible that some CUBN variants play a role in the genetic susceptibility for diabetes and its complications.

In addition to its relationship with diabetes vitamin D deficiency has been associated with cardiovascular diseases and increased mortality (Holick, 2007; Jorde, Wilsgaard, & Grimnes, 2019). In the Tromsoe study on 11987 subjects followed for 22 years, a possible relationship between *CUBN* rs12785878, but not *LRP2*, and mortality was highlighted. This variant was not related to lower vitamin D levels suggesting that the association of *CUBN* with mortality is independent of its function as a vitamin D receptor.

Additional CUBN polymorphisms associated with micronutrient status have been reported. Genome-wide association studies showed that B12 levels could be influenced by variants in Cubilin, Transcobalamin-I and Fucosyltransferase 2 genes. In various independent studies on healthy individuals the *CUBN* rs1801222, rs4748353, rs11254363, rs12243895 variants have been either associated with a higher vitamin B12 status or a higher risk of vitamin B12 deficiency (Hazra et al., 2009; Pangilinan et al., 2012). The intronic variant rs11254363, associated with increased B12 levels would in addition provide a protection from congenital heart disease, observation which is consistent with a role of B12 in the embryonic heart development (Wang et al., 2014).

Lower blood iron levels are attributed to lower iron absorption. Iron loss is estimated to less than 0.1% of body iron due to desquamation or blood loss. Urinary iron excretion is not taken into account and no validated tests for urinary iron are currently available. Iron deficiency, as well as iron overload have been considered in multiple sclerosis (MS) pathogenesis since iron supplementation has been beneficial in some cases and iron deficiency affects myelin parameters (Herbert et al., 2018; van Rensburg, Kotze, & van Toorn, 2012). Vitamin D deficiency, associated with MS risk and lower iron levels are comorbidities in inflammatory bowel disease, a chronic intestinal condition causing nutrition-related complications (Syed et al., 2017). Another risk factor for demyelination and disability in MS is deficiencies in the vitamin B12 metabolism pathway. Not surprisingly, whole exome sequencing identified several *CUBN* variants (rs74431427, rs557482398, rs41289305, rs2271462, rs116114483, rs7905349) in patients with MS which were not relevant for megaloblastic anemia. This observation suggests that Cubilin-mediated renal iron re-uptake as well as iron and vitamin D excretion may be important parameters to evaluate in MS patients (van Rensburg et al., 2019).

18. Cubilin in neural tube defects

Neural tube defects (NTD) are congenital birth defects detected annually in approximately 300,000 neonates worldwide. They are complex disorders caused by both genetic and environmental factors. Anencephaly and spina bifida, corresponding to the failure of neural tube closure in the anterior and posterior ends respectively, are rather common.

Maternal risk factors linked to NTD include maternal iron and folate status, diabetes mellitus or drugs. Preventing micronutrient deficiency is important; indeed food fortification programs reduce the incidence of NTD (Czeizel, 2009; Liu et al., 2016; Sabatino, Stokes, & Zohn, 2017). B12 has been associated with NTD and vitamin B12 deficiency seems to be more prevalent in women with affected infants or both mothers and infants (Ayaz & Asoglu, 2020). Additionally, low maternal B12-TCII levels are lower in mothers of NTD-affected infants (Godbole et al., 2011).

LRP2 variants have also been associated with NTD risk (Rebekah Prasoona, Sunitha, Srinadh, Muni Kumari, & Jyothy, 2018). It is possible that these variants affect *CUBN* function. Anyhow polymorphisms in *CUBN* (rs7070148, rs2273737, rs1801222) are independently associated with maternal NTD risk and lower circulating vitamin B12 (Hazra et al., 2009; Pangilinan et al., 2012). Additionally, the *CUBN* polymorphism (rs1907362) was shown to be associated with a decreased risk of spina bifida (Franke et al., 2009). It is unclear whether Cubilin dysfunction may directly affect NT closure but IF has been detected in the amniotic fluid of the human fetus suggesting a role for Cubilin in vitamin B12 uptake by the developing neuroepithelium (Guéant et al., 1989).

19. Cubilin as a risk variant and/or prognostic factor in cancer progression

Several reports involve Cubilin in the development and progression of colorectal, cervical, renal, gastric cancers or osteosarcoma. CUBN may be a predictor of both good and bad prognosis.

Cervical cancer is a common cancer in women worldwide and cervical adenocarcinoma is one of its subtypes. Upregulation of *CUBN* inhibits the development of the disease via a mechanism that involves a circular RNA (circSPIDR). CircSPIDR derives from the human scaffold protein involved in DNA repair and is a cell-specific regulatory RNA with tumor suppressor activity (Xu & Lu, 2021). Indeed, *CUBN* is negatively regulated by the upregulated in tumor tissues miR-431-5p and positively regulated by circSPIDR, which inhibits miR-431-5p activity. *CUBN* expression has thus a good diagnostic potential in this cancer (Xu & Lu, 2021).

Cubilin expression is specific to renal clear cell carcinoma and has been validated as a marker for the classification of patients into low- and high-risk groups (Gremel et al., 2017). It is interesting to note that ZONAB, a regulator of *CUBN* expression is another marker of this cancer

(Lima et al., 2010). *CUBN* expressing tumors have better response to anti-angiogenetic treatments and a statistically better survival. Conversely, *CUBN* downregulation is associated with higher metastatic activity and a less favorable outcome (Gremel et al., 2017). *CUBN* is considered as a diagnostic marker in renal clear cell carcinoma (Gremel et al., 2017).

Genome-wide association studies (GWAS) identified the *CUBN* intronic variant (rs10904849) as a risk variant for colorectal cancer (Al-Tassan et al., 2015). In this cancer a higher *CUBN* expression is observed in cancer tissues and is correlated with shorter overall and disease-free survival (Wu & Xu, 2020). Patients under radiation therapy tend to express lower amounts of *CUBN* suggesting that *CUBN* may be used to predict the efficacy of the treatment. Bioinformatic analysis suggested that alteration in *CUBN* may induce abnormal transcription of various factors including miRNA-517B or FOXO4, a transcription factor overexpressed in colorectal cancer of poor prognosis (Wu & Xu, 2020). Alternatively this variant may modify the interactions between Cubilin and Fgf8, a positive regulator of colorectal cancer (Cases et al., 2013; Liu et al., 2015). Downregulation of Amn is also thought be involved in colorectal cancer progression (Galamb et al., 2009) indicating differential regulation of CUBN and AMN in disease.

Gastric cancer has been associated with impaired one-carbon metabolism and aberrant vitamin B12 levels may contribute to gastric cancer progression (Zhao, Wei, Song, & Li, 2016). GWAS showed that variants within the vitamin B12 pathway genes including *CUBN* or *TCN1* (Transcobalamin) could influence B12 levels. Whereas some *CUBN* variants had an increased risk for developing a gastric cancer, others were protective, and this could partially be explained by higher plasma B12 levels in the latter. The statistical power of these observations is rather low due to the low frequency of the *CUBN* variants.

Mutations in *CUBN* have been found in osteosarcomas, primary bone malignancies mainly diagnosed in adolescence. The significance of this finding is unclear as not correlated with the clinical data of the patients (Choy et al., 2012).

20. Cubilin in animal studies

20.1 Cubilin in embryonic nutrition and development (Fig. 6)

The long-suspected role of Cubilin in maternal-to fetal transport of essential maternal nutrients was first demonstrated in 2005 (Le Panse et al., 1994;

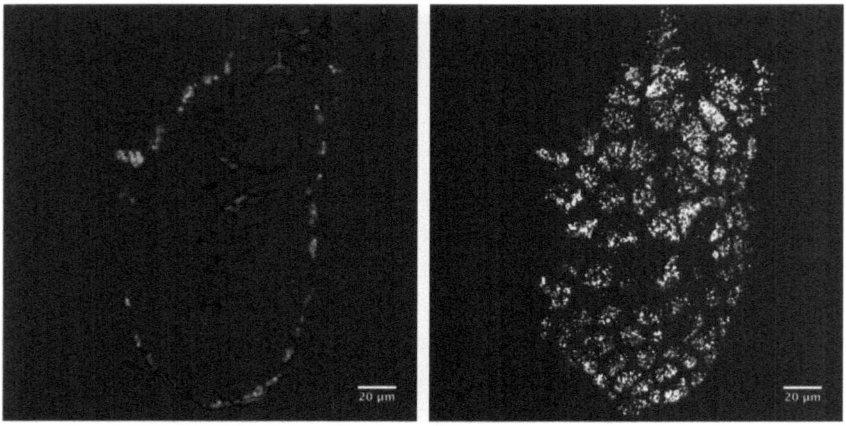

Fig. 6 Cubilin mediated uptake of transferrin (green) in the VE of a gastrulating embryo. Left panel is a sagittal cross-section of the embryo. Right panel is a surface view. Phalloidin stains in red.

Assémat, Vinot, et al., 2005). Using rodent blastocysts and extraembryonic cultures of rat conceptuses Assemat et al. showed a Cubilin-dependent uptake of ApoA-I/HDL. Upregulation of Cubilin in the rodent placenta during malnutrition may represent a compensatory mechanism to allow fetal growth (Daniel et al., 2016). A similar mechanism was proposed for lipid uptake in the developing neuroepithelium (Assémat, Vinot, et al., 2005). Subsequent work identified Cubilin as an essential and conserved factor for nutrient uptake in suckling mammals and stomach-less fish.

VE: The visceral endoderm (VE) is an extraembryonic tissue essential for early embryonic development in rodents. VE appears prior to gastrulation, i.e., mesoderm formation. VE supports embryonic growth by coordinating maternal nutrient uptake and transport and providing developmental cues to the early embryo (Bielinska, Narita, & Wilson, 1999). In the VE endocytosis is supported by the expression of specific receptors and large vacuolar structures, the apical vacuoles, specialized in nutrient processing and transcellular signaling. Defective VE endocytosis affects both embryonic patterning and growth. In the VE, Cubilin/Amn, Lrp2 and Dab2 mediated endocytosis is essential for lipid and other nutrient uptake and possibly Fibroblast growth factor signaling (Assémat, Vinot, et al., 2005; Maurer & Cooper, 2005; Perea-Gomez et al., 2019). Factors that jeopardize expression and/or function of Cubilin in the visceral yolk sac include: (a) the Lrp2 and Amn interacting adaptor Dab2 and (b) Gpr107, a Clathrin-interacting protein involved in receptor recycling from the endocytic compartments to the plasma membrane. In Gpr107 null mutants, as in Dab2 null mice,

Cubilin/Lrp2 recycling is disrupted (Zhou, Na, Niedra, & Seed, 2014). The downregulation of these receptors may contribute to the embryonic lethality.

Intestinal tissues: In mammals, macromolecule absorption in the intestine is limited to the suckling period. During this period, the high endocytic activity of the intestinal epithelium enables the absorption of growth factors, immunoglobulins and other molecules in maternal milk. Specialized intestinal cells, the lysosomal-rich enterocytes allow receptor-mediated and fluid phase endocytosis of nutrients. These cells express Cubilin/Amnionless, Lrp2 and Dab2 and have an elaborate but transient endocytic apparatus which gradually changes at weaning (Fujita et al., 2007; Park et al., 2019; Pedersen, Chakraborty, Steinhauser, Traub, & Madsen, 2010; Vázquez-Carretero et al., 2014). A similar endocytosis process has been reported in the chicken yolk-sac and the gastrointestinal tract of larval zebrafish (Bauer et al., 2013; Park et al., 2019).

In other species, i.e., stomachless fish, endocytosis-mediated absorption most likely continues for life in part of the intestine because of low digestive activity in the enteric cavity (Rombout, Lamers, Helfrich, Dekker, & Taverne-Thiele, 1985). Indeed, teleosts from the Goodeidae family possess trophotaeniae, i.e., hindgut-derived pseudoplacentas consisting of a single epithelial layer and connective tissues. These epithelial cells are like enterocytes and express Cubilin, Amnionless and Lrp2 suggesting a role for these proteins in maternal nutrient intake (Iida et al., 2021).

20.2 Cubilin inactivation in the mouse

Cubilin expressed in the VE is necessary for maternal nutrient supply of the developing embryo. Anti-cubilin antibodies interfering with VE endocytosis have a teratogenic effect (Marshall et al., 2015; Sahali et al., 1988) and impaired Cubilin distribution in the VE of mice lacking *Amn* or the reduced folate carrier (*RFC*1) or *Dab2* does not support embryonic growth (Assémat, Vinot, et al., 2005; Bauer et al., 2013; Gelineau-van Waes et al., 2008; Ishida, Hatae, Nishi, Araki, & Hamasaki, 2004; Maurer & Cooper, 2005; Strope et al., 2004).

In agreement with this central activity the gene-targeted inactivation of *Cubn* results in developmental arrest around gastrulation (Amsellem et al., 2010; Perea-Gomez et al., 2019; Smith et al., 2006) (Fig. 7). In these mutants the epithelial morphology of the VE cells is compromised and nutrient uptake is abolished (Smith et al., 2006) due to the deficient formation of the apical endocytic apparatus and the lack of apical vacuoles (Perea-Gomez et al., 2019). Although mesoderm formation and migration

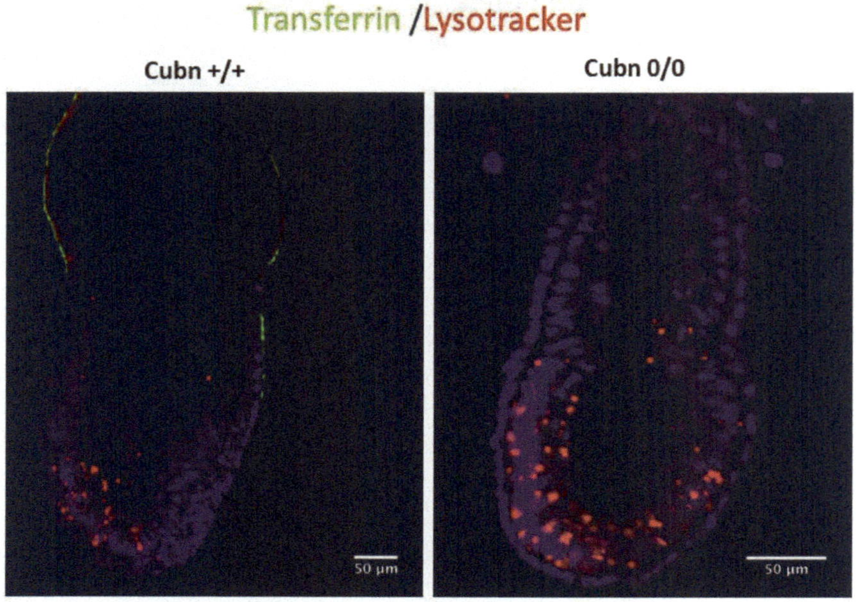

Fig. 7 Cubilin is required for VE transferrin uptake (green) in the gastrulating embryo.

are initiated in these mutants, further embryonic patterning and VE dispersal, essential for the formation of the definitive endoderm, are impaired. It is likely that deficient Fgf8 signaling, essential for normal VE movements, contributes to the mesodermal defects (Tam & Loebel, 2007).

Inactivation of *Cubn* exclusively in the epiblast does not interfere with mesoderm formation or neuroectodermal induction as evidenced by the use of the *Sox2-Cre* transgene (Cases et al., 2013). The morphogenesis of the head is however compromised due to the defective migration and possibly proliferation of the cephalic neural crest cells (CNC). CNC migration is indispensable for head, face and heart morphogenesis and requires a sustained Fgf8 signaling. The Fgf8 effector ERK1/2 is not phosphorylated in the *Sox2-Cubn* mutant head suggesting a role for Cubilin in this pathway (Cases et al., 2013). It is possible that deficiencies in the same signaling pathway contribute to heart defects and embryo-lethality of the *Sox2-Cubn* mutant after embryonic day 11.5 (E11.5).

20.3 Cubilin inactivation in the kidney

Various models of mosaic *Cubn* and/or *Lrp2* inactivation based on *Cre-Lox* recombination have been used to investigate the function of these receptors in vivo. *MORE-Cre* mediated inactivation of Cubilin and/or Lrp2 deleted both receptors in the PCT but embryo-lethality was observed in 20% of the mutants due to the ablation of extra-renal functions of Cubilin (Amsellem et al., 2010). This mouse model was the first to establish Cubilin as the physiological renal receptor for albumin (Amsellem et al., 2010) and confirm the role of Lrp2 as the molecular motor for Cubilin-albumin uptake. *Wnt4-Cre* was thereafter used to obtain an efficient, yet mosaic inactivation of *Cubn* and/or *Lrp2* in the PCT (Weyer et al., 2011). In *Cubn*- and *Lrp2*-deficient mice the amount of urinary albumin excretion was around 2 mg/d corresponding to the amount of the daily filtered albumin.

The apparent match between tubular levels of albumin and the capacity of Cubilin and Lrp2 to internalize albumin under physiological conditions raises two questions. First, whether Cubilin-Lrp2 mediated albumin uptake may contribute to the regulation of plasma albumin levels; second, why elevated levels of albumin are cytotoxic to PCT cells. In the *Wnt4-Cre* mediated inactivation of *Cubn* and *Lrp2* urinary albumin excretion increased 20-fold confirming the importance of the receptor complex in tubular albumin reabsorption (Weyer et al., 2018). Disruption of the glomerular filtration barrier via inactivation of *podocin*, a slit diaphragm protein, led to a nephrotic range albuminuria (almost 3000-fold increase compared to the base-line), which was further (1.5-fold) increased in the Cubn-Lrp2 renal mutants (Weyer et al., 2018). Despite the differences in albuminuria, the plasma levels of albumin were similarly decreased in the *podocin* mutants and the triple *Cubilin/Lrp2/podocin* mutants (Weyer et al., 2018). It is therefore likely that the function of Cubilin and Lrp2 contributes to the regulation of plasma albumin levels by a mechanism including FcRN. Albumin released from Cubilin could be retrieved by FcRN in the early acidic compartments where FcRn has the highest affinity for its ligand (Tenten et al., 2013).

Additional piece of evidence on the role of Cubilin in the regulation of circulating albumin comes from a rat model of adriamycin-induced model renal injury where the plasma level of albumin was reduced after silencing of *Cubn*. Finally, a decrease of almost 20% of plasma albumin was observed in heterozygous *Cubn* mice (Aseem et al., 2014).

Under nephrotic conditions the increased levels of albumin are likely to saturate the high affinity binding sites on Cubilin. Low affinity binding (via Lrp2) and fluid-phase uptake would be the remaining routes accounting for the accumulation of albumin as suggested by Park and Maack (1984). It is therefore possible that Lrp2 rather than Cubilin-mediated uptake of excessive levels of filtered proteins contributes to their cytotoxicity.

The mouse models of *Cubn* and/or *Lrp2* inactivation confirmed the in vivo interactions between Cubilin and Lrp2 with DBP, alpha-1-microglobulin, retinol binding protein (RBP) or the iron-transporting gelatinase-associated lipocalin (Hvidberg et al., 2005). More recently, the endogenous protease inhibitor and marker of tubular injury cystatin C (14.3kDa) was similarly identified as a common Cubilin and Lrp2 ligand (Jensen et al., 2017).

Drosophila was proposed as an alternative model to study renal function as it combines in the same cell type, the nephrocyte, podocyte and proximal tubule functions (Fu et al., 2017). Silencing of *Cubn* in the nephrocyte was shown to impair endocytosis (Hermle, Braun, Helmstädter, Huber, & Hildebrandt, 2017). Silencing of rabphilin, the orthologue of rabphilin 3A involved in proteinuria, reduced the expression of Cubilin, the number of endosomes and vesicular trafficking in the nephrocyte. Consequently, molecular reabsorption was decreased (Selma-Soriano et al., 2020). Microalbuminuria is observed in patients with mutations in RPH3A encoding rabphilin3A (Marrachelli et al., 2014); it could be explained by both cytoskeletal modifications and decreased expression of Cubilin in both the podocyte and the PCT.

21. Breed-specific cubilin mutations in dogs

In dogs, as in man, mutations in *Cubilin* or *Amn* result in vitamin B12 deficiency and proteinuria. These dogs develop a disorder analogue to IGS which typically manifests during the first year of life with intermittent diarrhea, inappetence, hematologic abnormalities including nonregenerative anemia, mild proteinuria and failure to grow (Gold, Scott, & Fyfe, 2015; Owczarek-Lipska et al., 2013). In contrast to humans neutropenia rather than macrocytic anemia is observed in canine IGS.

A genetic background for B12 deficiency is confirmed in Giant Schnauzers, Border Collies, Beagle dogs, and Australian Shepherd Dogs. The *Cubn*: c.8392delC and c786delC variants have for instance been identified as the causative defect in Border Collies and Beagles respectively (Fyfe, Hemker, Venta, Stebbing, & Giger, 2014; Owczarek-Lipska et al., 2013). In Border Collies the single base pair deletion is predicted to introduce a premature stop codon and a truncated Cubilin with a residual expression/function. It is interesting to note that in this breed the diagnosis of IGS is delayed. In Beagles the translational frameshift in exon 8 also causes a premature stop codon within the Cubilin/Amn interaction region suggesting that membrane expression of both Cubilin and Amn is impaired. Consistent with this hypothesis IGS is more severe in this breed (Fyfe et al., 2014). Recently, a homozygous pathogenic single-base change in c.8746 +1G>A, obliterating a splice donor consensus sequence in intron 55 of *Cubn* was reported in Komondor dogs. In total, four mutations in CUBN and 2 in AMN (Giant Schnauzers and Australian Shepherd Dogs) have been identified to date in dogs with selective B12 malabsorption.

22. Conclusion

The human and experimental data clearly establish the importance of Cubilin, Amn and Lrp2 in the intestinal absorption of vitamin B12 and the tubular reabsorption of filtered biomolecules, including albumin or vitamins.

In addition to these key functions growing evidence indicates implication of Cubilin in the progression and/or prognosis of cancer, cardiovascular and other diseases. Nonetheless future work will clarify the link(s) between Cubilin and these pathologies and identify the underlying protein network(s). Whether modulation of Cubilin expression or Cubilin-mediated endocytosis is a diagnostic, prognostic or even therapeutic tool remains to be established.

References

Acharya, U., Gau, J.-T., Horvath, W., Ventura, P., Hsueh, C.-T., & Carlsen, W. (2008). Hemolysis and hyperhomocysteinemia caused by cobalamin deficiency: Three case reports and review of the literature. *Journal of Hematology & Oncology*, *1*, 26. https://doi.org/10.1186/1756-8722-1-26.

Ahluwalia, T. S., Schulz, C.-A., Waage, J., Skaaby, T., Sandholm, N., van Zuydam, N., et al. (2019). A novel rare CUBN variant and three additional genes identified in Europeans with and without diabetes: Results from an exome-wide association study of albuminuria. *Diabetologia*, *62*, 292–305. https://doi.org/10.1007/s00125-018-4783-z.

Ahuja, R., Yammani, R., Bauer, J. A., Kalra, S., Seetharam, S., & Seetharam, B. (2008). Interactions of cubilin with megalin and the product of the amnionless gene (AMN): Effect on its stability. *The Biochemical Journal*, *410*, 301–308. https://doi.org/10.1042/BJ20070919.

Alam, A., Woo, J.-S., Schmitz, J., Prinz, B., Root, K., Chen, F., et al. (2016). Structural basis of transcobalamin recognition by human CD320 receptor. *Nature Communications*, *7*, 12100. https://doi.org/10.1038/ncomms12100.

Alharazy, S., Naseer, M. I., Alissa, E., Robertson, M. D., Lanham-New, S., & Chaudhary, A. G. (2021). Whole-exome sequencing for identification of genetic variants involved in vitamin D metabolic pathways in families with vitamin D deficiency in Saudi Arabia. *Frontiers in Genetics*, *12*, 677780. https://doi.org/10.3389/fgene.2021.677780.

Allen, L. H. (2009). How common is vitamin B-12 deficiency? *The American Journal of Clinical Nutrition*, *89*, 693S–696S. https://doi.org/10.3945/ajcn.2008.26947A.

Allen, R. H., Stabler, S. P., & Lindenbaum, J. (1998). Relevance of vitamins, homocysteine and other metabolites in neuropsychiatric disorders. *European Journal of Pediatrics*, *157*(Suppl. 2), S122–S126.

Alsalem, J. A., Patel, D., Susarla, R., Coca-Prados, M., Bland, R., Walker, E. A., et al. (2014). Characterization of vitamin D production by human ocular barrier cells. *Investigative Ophthalmology & Visual Science*, *55*, 2140–2147. https://doi.org/10.1167/iovs.13-13019.

Al-Tassan, N. A., Whiffin, N., Hosking, F. J., Palles, C., Farrington, S. M., Dobbins, S. E., et al. (2015). A new GWAS and meta-analysis with 1000Genomes imputation identifies novel risk variants for colorectal cancer. *Scientific Reports*, *5*, 10442. https://doi.org/10.1038/srep10442.

Alvarado-Morales, I., Olivares-Illana, V., Arenas-Huertero, C., Reynaga-Hernández, E., Layseca-Espinosa, E., Tokar, E. J., et al. (2021). Human prostate epithelial cells and prostate-derived stem cells malignantly transformed in vitro with sodium arsenite show impaired toll like receptor −3 (TLR3)-associated anti-tumor pathway. *Toxicology Letters*, *350*, 185–193. https://doi.org/10.1016/j.toxlet.2021.07.013.

Amagasaki, T., Green, R., & Jacobsen, D. W. (1990). Expression of transcobalamin II receptors by human leukemia K562 and HL-60 cells. *Blood*, *76*, 1380–1386.

Aminoff, M., Carter, J. E., Chadwick, R. B., Johnson, C., Gräsbeck, R., Abdelaal, M. A., et al. (1999). Mutations in CUBN, encoding the intrinsic factor-vitamin B12 receptor, cubilin, cause hereditary megaloblastic anaemia 1. *Nature Genetics*, *21*, 309–313. https://doi.org/10.1038/6831.

Amsellem, S., Gburek, J., Hamard, G., Nielsen, R., Willnow, T. E., Devuyst, O., et al. (2010). Cubilin is essential for albumin reabsorption in the renal proximal tubule. *Journal of the American Society of Nephrology*, *21*, 1859–1867. https://doi.org/10.1681/ASN.2010050492.

Andersen, C. B. F., Madsen, M., Storm, T., Moestrup, S. K., & Andersen, G. R. (2010). Structural basis for receptor recognition of vitamin-B(12)-intrinsic factor complexes. *Nature*, *464*, 445–448. https://doi.org/10.1038/nature08874.

Anderson, C. L., Chaudhury, C., Kim, J., Bronson, C. L., Wani, M. A., & Mohanty, S. (2006). Perspective—FcRn transports albumin: Relevance to immunology and medicine. *Trends in Immunology*, *27*, 343–348. https://doi.org/10.1016/j.it.2006.05.004.

Andrès, E., Affenberger, S., Federici, L., & Korganow, A. S. (2006). Pseudo-thrombotic microangiopathy related to' cobalamin deficiency. *The American Journal of Medicine*, *119*, e3. https://doi.org/10.1016/j.amjmed.2006.02.001.

Argraves, W. S., & Morales, C. R. (2004). Immunolocalization of cubilin, megalin, apolipoprotein J, and apolipoprotein A-I in the uterus and oviduct. *Molecular Reproduction and Development*, *69*, 419–427. https://doi.org/10.1002/mrd.20174.

Aseem, O., Barth, J. L., Klatt, S. C., Smith, B. T., & Argraves, W. S. (2013). Cubilin expression is monoallelic and epigenetically augmented via PPARs. *BMC Genomics*, *14*, 405. https://doi.org/10.1186/1471-2164-14-405.

Aseem, O., Smith, B. T., Cooley, M. A., Wilkerson, B. A., Argraves, K. M., Remaley, A. T., et al. (2014). Cubilin maintains blood levels of HDL and albumin. *Journal of the American Society of Nephrology*, *25*, 1028–1036. https://doi.org/10.1681/ASN.2013060671.

Assémat, E., Châtelet, F., Chandellier, J., Commo, F., Cases, O., Verroust, P., et al. (2005). Overlapping expression patterns of the multiligand endocytic receptors cubilin and megalin in the CNS, sensory organs and developing epithelia of the rodent embryo. *Gene Expression Patterns*, *6*, 69–78. https://doi.org/10.1016/j.modgep.2005.04.014.

Assémat, E., Vinot, S., Gofflot, F., Linsel-Nitschke, P., Illien, F., Châtelet, F., et al. (2005). Expression and role of cubilin in the internalization of nutrients during the peri-implantation development of the rodent embryo. *Biology of Reproduction*, *72*, 1079–1086. https://doi.org/10.1095/biolreprod.104.036913.

Ayaz, R., & Asoglu, M. R. (2020). Neural tube defects in eastern Turkey; is low folate status or vitamin B12 deficiency or both associated with a high rate of NTDs? *The Journal of Maternal-Fetal & Neonatal Medicine*, *33*, 3835–3840. https://doi.org/10.1080/14767058.2019.1623778.

Bartolome, F., Antequera, D., Tavares, E., Pascual, C., Maldonado, R., Camins, A., et al. (2017). Obesity and neuroinflammatory phenotype in mice lacking endothelial megalin. *Journal of Neuroinflammation*, *14*, 26. https://doi.org/10.1186/s12974-017-0800-2.

Bauer, R., Plieschnig, J. A., Finkes, T., Riegler, B., Hermann, M., & Schneider, W. J. (2013). The developing chicken yolk sac acquires nutrient transport competence by an orchestrated differentiation process of its endodermal epithelial cells. *The Journal of Biological Chemistry*, *288*, 1088–1098. https://doi.org/10.1074/jbc.M112.393090.

Bedin, M., Boyer, O., Servais, A., Li, Y., Villoing-Gaudé, L., Tête, M.-J., et al. (2020). Human C-terminal CUBN variants associate with chronic proteinuria and normal renal function. *The Journal of Clinical Investigation*, *130*, 335–344. https://doi.org/10.1172/JCI129937.

Bielinska, M., Narita, N., & Wilson, D. B. (1999). Distinct roles for visceral endoderm during embryonic mouse development. *The International Journal of Developmental Biology*, *43*, 183–205.

Böger, C. A., Chen, M.-H., Tin, A., Olden, M., Köttgen, A., de Boer, I. H., et al. (2011). CUBN is a gene locus for albuminuria. *Journal of the American Society of Nephrology*, *22*, 555–570. https://doi.org/10.1681/ASN.2010060598.

Bomsel, M., Prydz, K., Parton, R. G., Gruenberg, J., & Simons, K. (1989). Endocytosis in filter-grown Madin-Darby canine kidney cells. *The Journal of Cell Biology*, *109*, 3243–3258. https://doi.org/10.1083/jcb.109.6.3243.

Bonnet, L., Karkeni, E., Couturier, C., Astier, J., Dalifard, J., Defoort, C., et al. (2018). Gene expression pattern in response to cholecalciferol supplementation highlights Cubilin as a major protein of 25(OH)D uptake in adipocytes and male mice White adipose tissue. *Endocrinology*, *159*, 957–966. https://doi.org/10.1210/en.2017-00650.

Bork, P. (1991). Complement components C1r/C1s, bone morphogenic protein 1 and Xenopus laevis developmentally regulated protein UVS.2 share common repeats. *FEBS Letters*, *282*, 9–12. https://doi.org/10.1016/0014-5793(91)80433-4.

Bothwell, S. P., Chan, E., Bernardini, I. M., Kuo, Y.-M., Gahl, W. A., & Nussbaum, R. L. (2011). Mouse model for Lowe syndrome/dent disease 2 renal tubulopathy. *Journal of the American Society of Nephrology*, *22*, 443–448. https://doi.org/10.1681/ASN.2010050565.

Broch, H., Imerslund, O., Monn, E., Hovig, T., & Seip, M. (1984). Imerslund-Gräsbeck anemia. A long-term follow-up study. *Acta Paediatrica Scandinavica*, *73*, 248–253.

Brown, M. S., & Goldstein, J. L. (1986). A receptor-mediated pathway for cholesterol homeostasis. *Science*, *232*, 34–47. https://doi.org/10.1126/science.3513311.

Bryniarski, M. A., Yee, B. M., Chaves, L. D., Stahura, C. M., Yacoub, R., & Morris, M. E. (2020). Megalin-mediated albumin endocytosis in cultured murine mesangial cells. *Biochemical and Biophysical Research Communications*, *529*, 740–746. https://doi.org/10.1016/j.bbrc.2020.05.166.

Burke, K. A., Jauniaux, E., Burton, G. J., & Cindrova-Davies, T. (2013). Expression and immunolocalisation of the endocytic receptors megalin and cubilin in the human yolk sac and placenta across gestation. *Placenta*, *34*, 1105–1109. https://doi.org/10.1016/j.placenta.2013.08.003.

Cabezas, F., Farfán, P., & Marzolo, M.-P. (2019). Participation of the SMAD2/3 signalling pathway in the down regulation of megalin/LRP2 by transforming growth factor beta (TGF-ß1). *PLoS One*, *14*, e0213127. https://doi.org/10.1371/journal.pone.0213127.

Carmel, R. (2008). Efficacy and safety of fortification and supplementation with vitamin B12: Biochemical and physiological effects. *Food and Nutrition Bulletin*, *29*, S177–S187. https://doi.org/10.1177/15648265080292S121.

Cases, O., Perea-Gomez, A., Aguiar, D. P., Nykjaer, A., Amsellem, S., Chandellier, J., et al. (2013). Cubilin, a high affinity receptor for fibroblast growth factor 8, is required for cell survival in the developing vertebrate head. *The Journal of Biological Chemistry, 288*, 16655–16670. https://doi.org/10.1074/jbc.M113.451070.

Chesney, R. W. (2016). Interactions of vitamin D and the proximal tubule. *Pediatric Nephrology, 31*, 7–14. https://doi.org/10.1007/s00467-015-3050-5.

Chlon, T. M., Taffany, D. A., Welsh, J., & Rowling, M. J. (2008). Retinoids modulate expression of the endocytic partners megalin, cubilin, and disabled-2 and uptake of vitamin D-binding protein in human mammary cells. *The Journal of Nutrition, 138*, 1323–1328. https://doi.org/10.1093/jn/138.7.1323.

Choy, E., Hornicek, F., MacConaill, L., Harmon, D., Tariq, Z., Garraway, L., et al. (2012). High-throughput genotyping in osteosarcoma identifies multiple mutations in phosphoinositide-3-kinase and other oncogenes. *Cancer, 118*, 2905–2914. https://doi.org/10.1002/cncr.26617.

Christensen, E. I., Devuyst, O., Dom, G., Nielsen, R., Van der Smissen, P., Verroust, P., et al. (2003). Loss of chloride channel ClC-5 impairs endocytosis by defective trafficking of megalin and cubilin in kidney proximal tubules. *Proceedings of the National Academy of Sciences of the United States of America, 100*, 8472–8477. https://doi.org/10.1073/pnas.1432873100.

Christensen, E. I., Kristoffersen, I. B., Grann, B., Thomsen, J. S., Andreasen, A., & Nielsen, R. (2021). A well-developed endolysosomal system reflects protein reabsorption in segment 1 and 2 of rat proximal tubules. *Kidney International, 99*, 841–853. https://doi.org/10.1016/j.kint.2020.11.015.

Christensen, E. I., & Nielsen, R. (2007). Role of megalin and cubilin in renal physiology and pathophysiology. *Review of Physiology, Biochemistry and Pharmacology, 158*, 1–22. https://doi.org/10.1007/112_0604.

Christensen, E. I., Verroust, P. J., & Nielsen, R. (2009). Receptor-mediated endocytosis in renal proximal tubule. *Pflügers Archiv, 458*, 1039–1048. https://doi.org/10.1007/s00424-009-0685-8.

Christensen, E. I., & Willnow, T. E. (1999). Essential role of megalin in renal proximal tubule for vitamin homeostasis. *Journal of the American Society of Nephrology, 10*, 2224–2236. https://doi.org/10.1681/ASN.V10102224.

Coelho, D., Kim, J. C., Miousse, I. R., Fung, S., du Moulin, M., Buers, I., et al. (2012). Mutations in ABCD4 cause a new inborn error of vitamin B12 metabolism. *Nature Genetics, 44*, 1152–1155. https://doi.org/10.1038/ng.2386.

Coudroy, G., Gburek, J., Kozyraki, R., Madsen, M., Trugnan, G., Moestrup, S. K., et al. (2005). Contribution of cubilin and amnionless to processing and membrane targeting of cubilin-amnionless complex. *Journal of the American Society of Nephrology, 16*, 2330–2337. https://doi.org/10.1681/ASN.2004110925.

Czeizel, A. E. (2009). Periconceptional folic acid and multivitamin supplementation for the prevention of neural tube defects and other congenital abnormalities. *Birth Defects Research, Part A: Clinical and Molecular Teratology, 85*, 260–268. https://doi.org/10.1002/bdra.20563.

Dachy, A., Paquot, F., Debray, G., Bovy, C., Christensen, E. I., Collard, L., et al. (2015). In-depth phenotyping of a Donnai-Barrow patient helps clarify proximal tubule dysfunction. *Pediatric Nephrology, 30*, 1027–1031. https://doi.org/10.1007/s00467-014-3037-7.

Dance, A. L., Miller, M., Seragaki, S., Aryal, P., White, B., Aschenbrenner, L., et al. (2004). Regulation of myosin-VI targeting to endocytic compartments. *Traffic, 5*, 798–813. https://doi.org/10.1111/j.1600-0854.2004.00224.x.

Daniel, Z., Swali, A., Emes, R., & Langley-Evans, S. C. (2016). The effect of maternal undernutrition on the rat placental transcriptome: Protein restriction up-regulates cholesterol transport. *Genes & Nutrition, 11*, 27. https://doi.org/10.1186/s12263-016-0541-3.

Decorti, G., Malusà, N., Furlan, G., Candussio, L., & Klugmann, F. B. (1999). Endocytosis of gentamicin in a proximal tubular renal cell line. *Life Sciences*, *65*, 1115–1124.

Devuyst, O. (2010). Dent's disease: Chloride-proton exchange controls proximal tubule endocytosis. *Nephrology, Dialysis, Transplantation*, *25*, 3832–3835. https://doi.org/10.1093/ndt/gfq556.

Devuyst, O., & Luciani, A. (2015). Chloride transporters and receptor-mediated endocytosis in the renal proximal tubule. *Journal of Physiology (London)*, *593*, 4151–4164. https://doi.org/10.1113/JP270087.

Draycott, S. A. V., Daniel, Z., Khan, R., Muhlhausler, B. S., Elmes, M. J., & Langley-Evans, S. C. (2020). Expression of cholesterol packaging and transport genes in human and rat placenta: Impact of obesity and a high-fat diet. *Journal of Developmental Origins of Health and Disease*, *11*, 222–227. https://doi.org/10.1017/S2040174419000606.

Erranz, B., Miquel, J. F., Argraves, W. S., Barth, J. L., Pimentel, F., & Marzolo, M.-P. (2004). Megalin and cubilin expression in gallbladder epithelium and regulation by bile acids. *Journal of Lipid Research*, *45*, 2185–2198. https://doi.org/10.1194/jlr.M400235-JLR200.

Fang, L., Garuti, R., Kim, B.-Y., Wade, J. B., & Welling, P. A. (2009). The ARH adaptor protein regulates endocytosis of the ROMK potassium secretory channel in mouse kidney. *The Journal of Clinical Investigation*, *119*, 3278–3289. https://doi.org/10.1172/JCI37950.

Fang, L., Xie, D., Wu, X., Cao, H., Su, W., & Yang, J. (2013). Involvement of endoplasmic reticulum stress in albuminuria induced inflammasome activation in renal proximal tubular cells. *PLoS One*, *8*, e72344. https://doi.org/10.1371/journal.pone.0072344.

Fedosov, S. N. (2012). Physiological and molecular aspects of cobalamin transport. *Sub-Cellular Biochemistry*, *56*, 347–367. https://doi.org/10.1007/978-94-007-2199-9_18.

Fedosov, S. N., Fedosova, N. U., Berglund, L., Moestrup, S. K., Nexø, E., & Petersen, T. E. (2004). Assembly of the intrinsic factor domains and oligomerization of the protein in the presence of cobalamin. *Biochemistry*, *43*, 15095–15102. https://doi.org/10.1021/bi048924c.

Fedosov, S. N., Fedosova, N. U., Kräutler, B., Nexø, E., & Petersen, T. E. (2007). Mechanisms of discrimination between cobalamins and their natural analogues during their binding to the specific B12-transporting proteins. *Biochemistry*, *46*, 6446–6458. https://doi.org/10.1021/bi062063l.

Ferrell, N., Ricci, K. B., Groszek, J., Marmerstein, J. T., & Fissell, W. H. (2012). Albumin handling by renal tubular epithelial cells in a microfluidic bioreactor. *Biotechnology and Bioengineering*, *109*, 797–803. https://doi.org/10.1002/bit.24339.

Franke, B., Vermeulen, S. H. H. M., Steegers-Theunissen, R. P. M., Coenen, M. J., Schijvenaars, M. M. V. A. P., Scheffer, H., et al. (2009). An association study of 45 folate-related genes in spina bifida: Involvement of cubilin (CUBN) and tRNA aspartic acid methyltransferase 1 (TRDMT1). *Birth Defects Research, Part A: Clinical and Molecular Teratology*, *85*, 216–226. https://doi.org/10.1002/bdra.20556.

Fu, Y., Zhu, J.-Y., Richman, A., Zhao, Z., Zhang, F., Ray, P. E., et al. (2017). A Drosophila model system to assess the function of human monogenic podocyte mutations that cause nephrotic syndrome. *Human Molecular Genetics*, *26*, 768–780. https://doi.org/10.1093/hmg/ddw428.

Fujita, M., Baba, R., Shimamoto, M., Sakuma, Y., & Fujimoto, S. (2007). Molecular morphology of the digestive tract; macromolecules and food allergens are transferred intact across the intestinal absorptive cells during the neonatal-suckling period. *Medical Molecular Morphology*, *40*, 1–7. https://doi.org/10.1007/s00795-006-0346-3.

Fyfe, J. C., Hemker, S. L., Venta, P. J., Stebbing, B., & Giger, U. (2014). Selective intestinal cobalamin malabsorption with proteinuria (Imerslund-Gräsbeck syndrome) in juvenile beagles. *Journal of Veterinary Internal Medicine*, *28*, 356–362. https://doi.org/10.1111/jvim.12284.

Galamb, O., Sipos, F., Spisák, S., Galamb, B., Krenács, T., Valcz, G., et al. (2009). Potential biomarkers of colorectal adenoma-dysplasia-carcinoma progression: mRNA expression profiling and in situ protein detection on TMAs reveal 15 sequentially upregulated and 2 downregulated genes. *Cellular Oncology*, *31*, 19–29. https://doi.org/10.3233/clo-2009-0458.

Gekle, M., Knaus, P., Nielsen, R., Mildenberger, S., Freudinger, R., Wohlfarth, V., et al. (2003). Transforming growth factor-beta1 reduces megalin- and cubilin-mediated endocytosis of albumin in proximal-tubule-derived opossum kidney cells. *Journal of Physiology (London)*, *552*, 471–481. https://doi.org/10.1113/jphysiol.2003.048074.

Gekle, M., Mildenberger, S., Freudinger, R., Schwerdt, G., & Silbernagl, S. (1997). Albumin endocytosis in OK cells: Dependence on actin and microtubules and regulation by protein kinases. *The American Journal of Physiology*, *272*, F668–F677. https://doi.org/10.1152/ajprenal.1997.272.5.F668.

Gekle, M., Völker, K., Mildenberger, S., Freudinger, R., Shull, G. E., & Wiemann, M. (2004). NHE3 Na+/H+ exchanger supports proximal tubular protein reabsorption in vivo. *American Journal of Physiology. Renal Physiology*, *287*, F469–F473. https://doi.org/10.1152/ajprenal.00059.2004.

Gelineau-van Waes, J., Maddox, J. R., Smith, L. M., van Waes, M., Wilberding, J., Eudy, J. D., et al. (2008). Microarray analysis of E9.5 reduced folate carrier (RFC1; Slc19a1) knockout embryos reveals altered expression of genes in the cubilin-megalin multiligand endocytic receptor complex. *BMC Genomics*, *9*, 156. https://doi.org/10.1186/1471-2164-9-156.

Gianesello, L., Priante, G., Ceol, M., Radu, C. M., Saleem, M. A., Simioni, P., et al. (2017). Albumin uptake in human podocytes: A possible role for the cubilin-amnionless (CUBAM) complex. *Scientific Reports*, 7, 13705.

Gilbert, R. E. (2017). Proximal tubulopathy: Prime mover and key therapeutic target in diabetic kidney disease. *Diabetes*, *66*, 791–800. https://doi.org/10.2337/db16-0796.

Gimelbrant, A., Hutchinson, J. N., Thompson, B. R., & Chess, A. (2007). Widespread monoallelic expression on human autosomes. *Science*, *318*, 1136–1140. https://doi.org/10.1126/science.1148910.

Gleixner, E. M., Canaud, G., Hermle, T., Guida, M. C., Kretz, O., Helmstädter, M., et al. (2014). V-ATPase/mTOR signaling regulates megalin-mediated apical endocytosis. *Cell Reports*, *8*, 10–19. https://doi.org/10.1016/j.celrep.2014.05.035.

Godbole, K., Gayathri, P., Ghule, S., Sasirekha, B. V., Kanitkar-Damle, A., Memane, N., et al. (2011). Maternal one-carbon metabolism, MTHFR and TCN2 genotypes and neural tube defects in India. *Birth Defects Research. Part A: Clinical and Molecular Teratology*, *91*, 848–856. https://doi.org/10.1002/bdra.20841.

Gold, A. J., Scott, M. A., & Fyfe, J. C. (2015). Failure to thrive and life-threatening complications due to inherited selective cobalamin malabsorption effectively managed in a juvenile Australian shepherd dog. *The Canadian Veterinary Journal*, *56*, 1029–1034.

Gomes, J. R., Nogueira, R. S., Vieira, M., Santos, S. D., Ferraz-Nogueira, J. P., Relvas, J. B., et al. (2016). Transthyretin provides trophic support via megalin by promoting neurite outgrowth and neuroprotection in cerebral ischemia. *Cell Death and Differentiation*, *23*, 1749–1764. https://doi.org/10.1038/cdd.2016.64.

Gotoh, N., Yan, Q., Du, Z., Biemesderfer, D., Kashgarian, M., Mooseker, M. S., et al. (2010). Altered renal proximal tubular endocytosis and histology in mice lacking myosin-VI. *Cytoskeleton (Hoboken)*, *67*, 178–192. https://doi.org/10.1002/cm.20435.

Grahammer, F., Ramakrishnan, S. K., Rinschen, M. M., Larionov, A. A., Syed, M., Khatib, H., et al. (2017). mTOR regulates endocytosis and nutrient transport in proximal tubular cells. *Journal of the American Society of Nephrology*, *28*, 230–241. https://doi.org/10.1681/ASN.2015111224.

Gräsbeck, R. (2013). Hooked to vitamin B12 since 1955: A historical perspective. *Biochimie*, *95*, 970–975. https://doi.org/10.1016/j.biochi.2012.12.007.

Grasbeck, R., Gordin, R., Kantero, I., & Kuhlback, B. (1960). Selective vitamin B12 malabsorption and proteinuria in young people. A syndrome. *Acta Medica Scandinavica, 167*, 289–296.

Gräsbeck, R., & Tanner, S. M. (2011). Juvenile selective vitamin B_{12} malabsorption: 50 years after its description-10 years of genetic testing. *Pediatric Research, 70*, 222–228. https://doi.org/10.1203/PDR.0b013e3182242124.

Green, R., Allen, L. H., Bjørke-Monsen, A.-L., Brito, A., Guéant, J.-L., Miller, J. W., et al. (2017). Vitamin B12 deficiency. *Nature Reviews. Disease Primers, 3*, 17040. https://doi.org/10.1038/nrdp.2017.40.

Gremel, G., Djureinovic, D., Niinivirta, M., Laird, A., Ljungqvist, O., Johannesson, H., et al. (2017). A systematic search strategy identifies cubilin as independent prognostic marker for renal cell carcinoma. *BMC Cancer, 17*, 9. https://doi.org/10.1186/s12885-016-3030-6.

Guéant, J. L., Hambaba, L., Vidailhet, M., Schaefer, C., Wahlstedt, V., & Nicolas, J. P. (1989). Concentration and physicochemical characterisation of unsaturated cobalamin binding proteins in amniotic fluid. *Clinica Chimica Acta, 181*, 151–161.

Gurlek Gokcebay, D., Akpinar Tekgunduz, S., & Cavdarli, B. (2020). Imerslund-Gräsbeck syndrome presenting with microangiopathic hemolytic anemia in a child. *European Journal of Medical Genetics, 63*, 103880. https://doi.org/10.1016/j.ejmg.2020.103880.

Hammad, S. M., Barth, J. L., Knaak, C., & Argraves, W. S. (2000a). Megalin acts in concert with cubilin to mediate endocytosis of high density lipoproteins. *The Journal of Biological Chemistry, 275*, 12003–12008.

Hammad, S. M., Barth, J. L., Knaak, C., & Argraves, W. S. (2000b). Megalin acts in concert with Cubilin to mediate endocytosis of high density lipoproteins. *The Journal of Biological Chemistry, 16*, 275.

Harada, Y., Tanaka, Y., Terasawa, M., Pieczyk, M., Habiro, K., Katakai, T., et al. (2012). DOCK8 is a Cdc42 activator critical for interstitial dendritic cell migration during immune responses. *Blood, 119*, 4451–4461. https://doi.org/10.1182/blood-2012-01-407098.

Hazra, A., Kraft, P., Lazarus, R., Chen, C., Chanock, S. J., Jacques, P., et al. (2009). Genome-wide significant predictors of metabolites in the one-carbon metabolism pathway. *Human Molecular Genetics, 18*, 4677–4687. https://doi.org/10.1093/hmg/ddp428.

Herbert, E., Engel-Hills, P., Hattingh, C., Fouche, J.-P., Kidd, M., Lochner, C., et al. (2018). Fractional anisotropy of white matter, disability and blood iron parameters in multiple sclerosis. *Metabolic Brain Disease, 33*, 545–557. https://doi.org/10.1007/s11011-017-0171-5.

Hermle, T., Braun, D. A., Helmstädter, M., Huber, T. B., & Hildebrandt, F. (2017). Modeling monogenic human nephrotic syndrome in the Drosophila Garland cell nephrocyte. *Journal of the American Society of Nephrology, 28*, 1521–1533. https://doi.org/10.1681/ASN.2016050517.

Hirano, M., Totani, K., Fukuda, T., Gu, J., & Suzuki, A. (2017). N-Glycoform-dependent interactions of megalin with its ligands. *Biochimica et Biophysica Acta - General Subjects, 1861*, 3106–3118. https://doi.org/10.1016/j.bbagen.2016.10.015.

Holick, M. F. (2007). Vitamin D deficiency. *The New England Journal of Medicine, 357*, 266–281. https://doi.org/10.1056/NEJMra070553.

Holick, M. F. (2013). Bioavailability of vitamin D and its metabolites in black and white adults. *The New England Journal of Medicine, 369*, 2047–2048. https://doi.org/10.1056/NEJMe1312291.

Hosaka, K., Takeda, T., Iino, N., Hosojima, M., Sato, H., Kaseda, R., et al. (2009). Megalin and nonmuscle myosin heavy chain IIA interact with the adaptor protein Disabled-2 in proximal tubule cells. *Kidney International, 75*, 1308–1315. https://doi.org/10.1038/ki.2009.85.

Hryciw, D. H., Jenkin, K. A., Simcocks, A. C., Grinfeld, E., McAinch, A. J., & Poronnik, P. (2012). The interaction between megalin and ClC-5 is scaffolded by the Na^+-H^+ exchanger regulatory factor 2 (NHERF2) in proximal tubule cells. *The International Journal of Biochemistry & Cell Biology, 44*, 815–823. https://doi.org/10.1016/j.biocel.2012.02.007.

Hurtado-Lorenzo, A., Skinner, M., El Annan, J., Futai, M., Sun-Wada, G.-H., Bourgoin, S., et al. (2006). V-ATPase interacts with ARNO and Arf6 in early endosomes and regulates the protein degradative pathway. *Nature Cell Biology, 8*, 124–136. https://doi.org/10.1038/ncb1348.

Hvidberg, V., Jacobsen, C., Strong, R. K., Cowland, J. B., Moestrup, S. K., & Borregaard, N. (2005). The endocytic receptor megalin binds the iron transporting neutrophil- gelatinase-associated lipocalin with high affinity and mediates its cellular uptake. *FEBS Letters, 579*, 773–777. https://doi.org/10.1016/j.febslet.2004.12.031.

Iida, A., Sano, K., Inokuchi, M., Nomura, J., Suzuki, T., Kuriki, M., et al. (2021). Cubam receptor-mediated endocytosis in hindgut-derived pseudoplacenta of a viviparous teleost (Xenotoca eiseni). *The Journal of Experimental Biology, 224*, jeb242613. https://doi.org/10.1242/jeb.242613.

Ishida, T., Hatae, T., Nishi, N., Araki, N., & Hamasaki, M. (2004). Immunocytochemical analysis of cubilin-mediated endocytosis of high density lipoproteins (HDL) in epithelial cells of the rat visceral yolk sac. *Cell and Tissue Research, 318*, 533–543. https://doi.org/10.1007/s00441-004-0962-y.

Jänne, P. A., Suchy, S. F., Bernard, D., MacDonald, M., Crawley, J., Grinberg, A., et al. (1998). Functional overlap between murine Inpp5b and Ocrl1 may explain why deficiency of the murine ortholog for OCRL1 does not cause Lowe syndrome in mice. *The Journal of Clinical Investigation, 101*, 2042–2053. https://doi.org/10.1172/JCI2414.

Jensen, L. L., Andersen, R. K., Hager, H., & Madsen, M. (2014). Lack of megalin expression in adult human terminal ileum suggests megalin-independent cubilin/amnionless activity during vitamin B12 absorption. *Physiological Reports, 2*, e12086. https://doi.org/10.14814/phy2.12086.

Jensen, D., Kierulf-Lassen, C., Kristensen, M. L. V., Nørregaard, R., Weyer, K., Nielsen, R., et al. (2017). Megalin dependent urinary cystatin C excretion in ischemic kidney injury in rats. *PLoS One, 12*, e0178796. https://doi.org/10.1371/journal.pone.0178796.

Jheng, H.-F., Tsai, P.-J., Chuang, Y.-L., Shen, Y.-T., Tai, T.-A., Chen, W.-C., et al. (2015). Albumin stimulates renal tubular inflammation through an HSP70-TLR4 axis in mice with early diabetic nephropathy. *Disease Models & Mechanisms, 8*, 1311–1321. https://doi.org/10.1242/dmm.019398.

Jorde, R., Wilsgaard, T., & Grimnes, G. (2019). Polymorphisms in the vitamin D system and mortality—the Tromsø study. *The Journal of Steroid Biochemistry and Molecular Biology, 195*, 105481. https://doi.org/10.1016/j.jsbmb.2019.105481.

Kahn, S. E., Cooper, M. E., & Del Prato, S. (2014). Pathophysiology and treatment of type 2 diabetes: Perspectives on the past, present, and future. *Lancet, 383*, 1068–1083. https://doi.org/10.1016/S0140-6736(13)62154-6.

Kantarci, S., Al-Gazali, L., Hill, R. S., Donnai, D., Black, G. C. M., Bieth, E., et al. (2007). Mutations in LRP2, which encodes the multiligand receptor megalin, cause Donnai-Barrow and facio-oculo-acoustico-renal syndromes. *Nature Genetics, 39*, 957–959. https://doi.org/10.1038/ng2063.

Kerjaschki, D., & Farquhar, M. G. (1982). The pathogenic antigen of Heymann nephritis is a membrane glycoprotein of the renal proximal tubule brush border. *Proceedings of the National Academy of Sciences of the United States of America, 79*, 5557–5561.

Khan, S., Cabral, P. D., Schilling, W. P., Schmidt, Z. W., Uddin, A. N., Gingras, A., et al. (2018). Kidney proximal tubule Lipoapoptosis is regulated by fatty acid Transporter-2 (FATP2). *Journal of the American Society of Nephrology, 29*, 81–91. https://doi.org/10.1681/ASN.2017030314.

Kim, H. J., Moradi, H., Yuan, J., Norris, K., & Vaziri, N. D. (2009). Renal mass reduction results in accumulation of lipids and dysregulation of lipid regulatory proteins in the remnant kidney. *American Journal of Physiology. Renal Physiology*, *296*, F1297–F1306. https://doi.org/10.1152/ajprenal.90761.2008.

Kondo, Y., Shen, L., Yan, P. S., Huang, T. H.-M., & Issa, J.-P. J. (2004). Chromatin immunoprecipitation microarrays for identification of genes silenced by histone H3 lysine 9 methylation. *Proceedings of the National Academy of Sciences of the United States of America*, *101*, 7398–7403. https://doi.org/10.1073/pnas.0306641101.

Koral, K., & Erkan, E. (2012). PKB/Akt partners with Dab2 in albumin endocytosis. *American Journal of Physiology. Renal Physiology*, *302*, F1013–F1024. https://doi.org/10.1152/ajprenal.00289.2011.

Koral, K., Li, H., Ganesh, N., Birnbaum, M. J., Hallows, K. R., & Erkan, E. (2014). Akt recruits Dab2 to albumin endocytosis in the proximal tubule. *American Journal of Physiology. Renal Physiology*, *307*, F1380–F1389. https://doi.org/10.1152/ajprenal.00454.2014.

Kozyraki, R., & Cases, O. (2013). Vitamin B12 absorption: Mammalian physiology and acquired and inherited disorders. *Biochimie*, *95*, 1002–1007. https://doi.org/10.1016/j.biochi.2012.11.004.

Kozyraki, R., Fyfe, J., Kristiansen, M., Gerdes, C., Jacobsen, C., Cui, S., et al. (1999). The intrinsic factor-vitamin B12 receptor, cubilin, is a high-affinity apolipoprotein A-I receptor facilitating endocytosis of high-density lipoprotein. *Nature Medicine*, *5*, 656–661. https://doi.org/10.1038/9504.

Kozyraki, R., Kristiansen, M., Silahtaroglu, A., Hansen, C., Jacobsen, C., Tommerup, N., et al. (1998). The human intrinsic factor-vitamin B12 receptor, cubilin: Molecular characterization and chromosomal mapping of the gene to 10p within the autosomal recessive megaloblastic anemia (MGA1) region. *Blood*, *91*, 3593–3600.

Kristiansen, M., Aminoff, M., Jacobsen, C., de La Chapelle, A., Krahe, R., Verroust, P. J., et al. (2000). Cubilin P1297L mutation associated with hereditary megaloblastic anemia 1 causes impaired recognition of intrinsic factor-vitamin B(12) by cubilin. *Blood*, *96*, 405–409.

Kristiansen, M., Kozyraki, R., Jacobsen, C., Nexø, E., Verroust, P. J., & Moestrup, S. K. (1999). Molecular dissection of the intrinsic factor-vitamin B12 receptor, cubilin, discloses regions important for membrane association and ligand binding. *The Journal of Biological Chemistry*, *274*, 20540–20544.

Lau, K. S., Partridge, E. A., Grigorian, A., Silvescu, C. I., Reinhold, V. N., Demetriou, M., et al. (2007). Complex N-glycan number and degree of branching cooperate to regulate cell proliferation and differentiation. *Cell*, *129*, 123–134. https://doi.org/10.1016/j.cell.2007.01.049.

Le Panse, S., Ayani, E., Mulliez, N., Chatelet, F., Cywiner-Golenzer, C., Galceran, M., et al. (1994). Antibodies to the 280-kd coated pit protein, target of teratogenic antibodies, produce alterations in the traffic of internalized proteins. *The American Journal of Pathology*, *145*, 1526–1536.

Leheste, J. R., Rolinski, B., Vorum, H., Hilpert, J., Nykjaer, A., Jacobsen, C., et al. (1999). Megalin knockout mice as an animal model of low molecular weight proteinuria. *The American Journal of Pathology*, *155*, 1361–1370.

Lima, W. R., Parreira, K. S., Devuyst, O., Caplanusi, A., N'kuli, F., Marien, B., et al. (2010). ZONAB promotes proliferation and represses differentiation of proximal tubule epithelial cells. *Journal of the American Society of Nephrology*, *21*, 478–488. https://doi.org/10.1681/ASN.2009070698.

Liu, D., Wen, Y., Tang, T.-T., Lv, L.-L., Tang, R.-N., Liu, H., et al. (2015). Megalin/Cubulin-lysosome-mediated albumin reabsorption is involved in the tubular cell activation of NLRP3 inflammasome and tubulointerstitial inflammation. *The Journal of Biological Chemistry*, *290*, 18018–18028. https://doi.org/10.1074/jbc.M115.662064.

Liu, J., Zhang, L., Li, Z., Jin, L., Zhang, Y., Ye, R., et al. (2016). Prevalence and trend of neural tube defects in five counties in Shanxi province of northern China, 2000 to 2014. *Birth Defects Research, Part A: Clinical and Molecular Teratology*, *106*, 267–274. https://doi.org/10.1002/bdra.23486.

Loi, M. (2006). Lowe syndrome. *Orphanet Journal of Rare Diseases*, *1*, 16. https://doi.org/10.1186/1750-1172-1-16.

Long, K. R., Rbaibi, Y., Gliozzi, M. L., Ren, Q., & Weisz, O. A. (2020). Differential kidney proximal tubule cell responses to protein overload by albumin and its ligands. *American Journal of Physiology. Renal Physiology*, *318*, F851–F859. https://doi.org/10.1152/ajprenal.00490.2019.

Long, K. R., Shipman, K. E., Rbaibi, Y., Menshikova, E. V., Ritov, V. B., Eshbach, M. L., et al. (2017). Proximal tubule apical endocytosis is modulated by fluid shear stress via an mTOR-dependent pathway. *Molecular Biology of the Cell*, *28*, 2508–2517. https://doi.org/10.1091/mbc.E17-04-0211.

Lu, X., Elizondo, R. A., Nielsen, R., Christensen, E. I., Yang, J., Hammock, B. D., et al. (2015). Vitamin D in tear fluid. *Investigative Ophthalmology & Visual Science*, *56*, 5880–5887. https://doi.org/10.1167/iovs.15-17177.

Ma, J., Guan, M., Bowden, D. W., Ng, M. C. Y., Hicks, P. J., Lea, J. P., et al. (2016). Association analysis of the cubilin (CUBN) and megalin (LRP2) genes with ESRD in African Americans. *Clinical Journal of the American Society of Nephrology*, *11*, 1034–1043. https://doi.org/10.2215/CJN.12971215.

Madhavan, S., Vijayakumar, M., Rajajee, S., & Nammalwar, B. R. (2009). Imerslund-Grasbeck syndrome: Association with diabetes mellitus. *Indian Pediatrics*, *46*, 251–253.

Mahmood, A., Shao, J., & Alpers, D. H. (2003). Rat enterocytes secrete SLPs containing alkaline phosphatase and cubilin in response to corn oil feeding. *American Journal of Physiology. Gastrointestinal and Liver Physiology*, *285*, G433–G441. https://doi.org/10.1152/ajpgi.00466.2002.

Marrachelli, V. G., Monleon, D., Rentero, P., Mansego, M. L., Morales, J. M., Galan, I., et al. (2014). Genomic and metabolomic profile associated to microalbuminuria. *PLoS One*, *9*, e98227. https://doi.org/10.1371/journal.pone.0098227.

Marshall, V. A., Johnson, K. J., Moore, N. P., Rasoulpour, R. J., Tornesi, B., & Carney, E. W. (2015). Comparative response of rat and rabbit conceptuses in vitro to inhibitors of histiotrophic nutrition. *Birth Defects Research, Part A: Clinical and Molecular Teratology*, *104*, 1–10. https://doi.org/10.1002/bdrb.21134.

Maurer, M. E., & Cooper, J. A. (2005). Endocytosis of megalin by visceral endoderm cells requires the Dab2 adaptor protein. *Journal of Cell Science*, *118*, 5345–5355. https://doi.org/10.1242/jcs.02650.

Mettlen, M., Chen, P.-H., Srinivasan, S., Danuser, G., & Schmid, S. L. (2018). Regulation of clathrin-mediated endocytosis. *Annual Review of Biochemistry*, *87*, 871–896. https://doi.org/10.1146/annurev-biochem-062917-012644.

Mihout, F., Devuyst, O., Bensman, A., Brocheriou, I., Ridel, C., Wagner, C. A., et al. (2014). Acute metabolic acidosis in a GLUT2-deficient patient with Fanconi-Bickel syndrome: New pathophysiology insights. *Nephrology, Dialysis, Transplantation*, *29 Suppl 4*, iv113-116. https://doi.org/10.1093/ndt/gfu018.

Mishra, S. K., Keyel, P. A., Hawryluk, M. J., Agostinelli, N. R., Watkins, S. C., & Traub, L. M. (2002). Disabled-2 exhibits the properties of a cargo-selective endocytic clathrin adaptor. *The EMBO Journal*, *21*, 4915–4926. https://doi.org/10.1093/emboj/cdf487.

Moestrup, S. K., Birn, H., Fischer, P. B., Petersen, C. M., Verroust, P. J., Sim, R. B., et al. (1996). Megalin-mediated endocytosis of transcobalamin-vitamin-B12 complexes suggests a role of the receptor in vitamin-B12 homeostasis. *Proceedings of the National Academy of Sciences of the United States of America*, *93*, 8612–8617.

Moestrup, S. K., Kozyraki, R., Kristiansen, M., Kaysen, J. H., Rasmussen, H. H., Brault, D., et al. (1998). The intrinsic factor-vitamin B12 receptor and target of teratogenic antibodies is a megalin-binding peripheral membrane protein with homology to developmental proteins. *The Journal of Biological Chemistry, 273*, 5235–5242.

Moestrup, S. K., & Verroust, P. J. (2001). Megalin- and cubilin-mediated endocytosis of protein-bound vitamins, lipids, and hormones in polarized epithelia. *Annual Review of Nutrition, 21*, 407–428. https://doi.org/10.1146/annurev.nutr.21.1.407.

Montgomery, E., Sayer, J. A., Baines, L. A., Hynes, A. M., Vega-Warner, V., Johnson, S., et al. (2015). Novel compound heterozygous mutations in AMN cause Imerslund-Gräsbeck syndrome in two half-sisters: A case report. *BMC Medical Genetics, 16*, 35. https://doi.org/10.1186/s12881-015-0181-2.

Moore, R., Cai, K. Q., Tao, W., Smith, E. R., & Xu, X.-X. (2013). Differential requirement for Dab2 in the development of embryonic and extra-embryonic tissues. *BMC Developmental Biology, 13*, 39. https://doi.org/10.1186/1471-213X-13-39.

Mori, K. P., Yokoi, H., Kasahara, M., Imamaki, H., Ishii, A., Kuwabara, T., et al. (2017). Increase of total nephron albumin filtration and reabsorption in diabetic nephropathy. *Journal of the American Society of Nephrology, 28*, 278–289. https://doi.org/10.1681/ASN.2015101168.

Morkbak, A. L., Poulsen, S. S., & Nexo, E. (2007). Haptocorrin in humans. *Clinical Chemistry and Laboratory Medicine, 45*, 1751–1759. https://doi.org/10.1515/CCLM.2007.343.

Morris, S. M., Tallquist, M. D., Rock, C. O., & Cooper, J. A. (2002). Dual roles for the Dab2 adaptor protein in embryonic development and kidney transport. *The EMBO Journal, 21*, 1555–1564. https://doi.org/10.1093/emboj/21.7.1555.

Müller, D., Nykjaer, A., & Willnow, T. E. (2003). From holoprosencephaly to osteopathology: Role of multifunctional endocytic receptors in absorptive epithelia. *Annals of Medicine, 35*, 290–299.

Nagai, M., Meerloo, T., Takeda, T., & Farquhar, M. G. (2003). The adaptor protein ARH escorts megalin to and through endosomes. *Molecular Biology of the Cell, 14*, 4984–4996. https://doi.org/10.1091/mbc.e03-06-0385.

Nakamura, K., Sagawa, N., & Mori, T. (1993). The sources and biochemical characteristics of cobalamin-binders in human amniotic fluid. *Asia-Oceania Journal of Obstetrics and Gynaecology, 19*, 343–353.

Nielsen, R., Courtoy, P. J., Jacobsen, C., Dom, G., Lima, W. R., Jadot, M., et al. (2007). Endocytosis provides a major alternative pathway for lysosomal biogenesis in kidney proximal tubular cells. *Proceedings of the National Academy of Sciences of the United States of America, 104*, 5407–5412. https://doi.org/10.1073/pnas.0700330104.

Nykjaer, A., Fyfe, J. C., Kozyraki, R., Leheste, J. R., Jacobsen, C., Nielsen, M. S., et al. (2001). Cubilin dysfunction causes abnormal metabolism of the steroid hormone 25(OH) vitamin D(3). *Proceedings of the National Academy of Sciences of the United States of America, 98*, 13895–13900. https://doi.org/10.1073/pnas.241516998.

Oh, Y. S., Seo, J. T., Ahn, H. S., & Gye, M. C. (2016). Expression of cubilin in mouse testes and Leydig cells. *Andrologia, 48*, 325–332. https://doi.org/10.1111/and.12450.

Olbricht, C. J. (1992). Distribution of cathepsins B and L in the kidney and their role in tubular protein absorption. *European Journal of Clinical Chemistry and Clinical Biochemistry, 30*, 675–681.

Oleinikov, A. V., Zhao, J., & Makker, S. P. (2000). Cytosolic adaptor protein Dab2 is an intracellular ligand of endocytic receptor gp600/megalin. *The Biochemical Journal, 347*(Pt. 3), 613–621.

Otomo, H., Nara, M., Kato, S., Shimizu, T., Suganuma, Y., Sato, T., et al. (2020). Sodium-glucose cotransporter 2 inhibition attenuates protein overload in renal proximal tubule via suppression of megalin O-GlcNacylation in progressive diabetic nephropathy. *Metabolism, 113*, 154405. https://doi.org/10.1016/j.metabol.2020.154405.

Ovunc, B., Otto, E. A., Vega-Warner, V., Saisawat, P., Ashraf, S., Ramaswami, G., et al. (2011). Exome sequencing reveals cubilin mutation as a single-gene cause of proteinuria. *Journal of the American Society of Nephrology*, *22*, 1815–1820. https://doi.org/10.1681/ASN.2011040337.

Owczarek-Lipska, M., Jagannathan, V., Drögemüller, C., Lutz, S., Glanemann, B., Leeb, T., et al. (2013). A frameshift mutation in the cubilin gene (CUBN) in border collies with Imerslund-Gräsbeck syndrome (selective cobalamin malabsorption). *PLoS One*, *8*, e61144. https://doi.org/10.1371/journal.pone.0061144.

Pangilinan, F., Molloy, A. M., Mills, J. L., Troendle, J. F., Parle-McDermott, A., Signore, C., et al. (2012). Evaluation of common genetic variants in 82 candidate genes as risk factors for neural tube defects. *BMC Medical Genetics*, *13*, 62. https://doi.org/10.1186/1471-2350-13-62.

Pannérec, A., Migliavacca, E., De Castro, A., Michaud, J., Karaz, S., Goulet, L., et al. (2018). Vitamin B12 deficiency and impaired expression of amnionless during aging. *Journal of Cachexia, Sarcopenia and Muscle*, *9*, 41–52. https://doi.org/10.1002/jcsm.12260.

Park, H. J., Fan, Z., Bai, Y., Ren, Q., Rbaibi, Y., Long, K. R., et al. (2020). Transcriptional programs driving shear stress-induced differentiation of kidney proximal tubule cells in culture. *Frontiers in Physiology*, *11*, 587358. https://doi.org/10.3389/fphys.2020.587358.

Park, J., Levic, D. S., Sumigray, K. D., Bagwell, J., Eroglu, O., Block, C. L., et al. (2019). Lysosome-rich enterocytes mediate protein absorption in the vertebrate gut. *Developmental Cell*, *51*, 7–20.e6. https://doi.org/10.1016/j.devcel.2019.08.001.

Park, C. H., & Maack, T. (1984). Albumin absorption and catabolism by isolated perfused proximal convoluted tubules of the rabbit. *The Journal of Clinical Investigation*, *73*, 767–777. https://doi.org/10.1172/JCI111270.

Park, H., Wood, M. R., Malysheva, O. V., Jones, S., Mehta, S., Brannon, P. M., et al. (2017). Placental vitamin D metabolism and its associations with circulating vitamin D metabolites in pregnant women. *The American Journal of Clinical Nutrition*, *106*, 1439–1448. https://doi.org/10.3945/ajcn.117.153429.

Pedersen, G. A., Chakraborty, S., Steinhauser, A. L., Traub, L. M., & Madsen, M. (2010). AMN directs endocytosis of the intrinsic factor-vitamin B(12) receptor Cubam by engaging ARH or Dab2. *Traffic*, *11*, 706–720.

Perea-Gomez, A., Cases, O., Lelièvre, V., Pulina, M. V., Collignon, J., Hadjantonakis, A.-K., et al. (2019). Loss of Cubilin, the intrinsic factor-vitamin B12 receptor, impairs visceral endoderm endocytosis and endodermal patterning in the mouse. *Scientific Reports*, *9*, 10168. https://doi.org/10.1038/s41598-019-46559-0.

Perez Bay, A. E., Schreiner, R., Benedicto, I., Paz Marzolo, M., Banfelder, J., Weinstein, A. M., et al. (2016). The fast-recycling receptor Megalin defines the apical recycling pathway of epithelial cells. *Nature Communications*, 7, 11550. https://doi.org/10.1038/ncomms11550.

Polesel, M., & Hall, A. M. (2019). Axial differences in endocytosis along the kidney proximal tubule. *American Journal of Physiology. Renal Physiology*, *317*(6), 1526–1530. https://doi.org/10.1152/ajprenal.00459.2019.

Quadros, E. V. (2010). Advances in the understanding of cobalamin assimilation and metabolism. *British Journal of Haematology*, *148*, 195–204. https://doi.org/10.1111/j.1365-2141.2009.07937.x.

Quadros, E. V., Nakayama, Y., & Sequeira, J. M. (2009). The protein and the gene encoding the receptor for the cellular uptake of transcobalamin-bound cobalamin. *Blood*, *113*, 186–192. https://doi.org/10.1182/blood-2008-05-158949.

Raghavan, V., Rbaibi, Y., Pastor-Soler, N. M., Carattino, M. D., & Weisz, O. A. (2014). Shear stress-dependent regulation of apical endocytosis in renal proximal tubule cells mediated by primary cilia. *Proceedings of the National Academy of Sciences of the United States of America*, *111*, 8506–8511. https://doi.org/10.1073/pnas.1402195111.

Ramanujam, K. S., Seetharam, S., & Seetharam, B. (1993). Regulated expression of intrinsic factor-cobalamin receptor by rat visceral yolk sac and placental membranes. *Biochimica et Biophysica Acta, 1146*, 243–246.

Ramos-Lopez, E., Lange, B., Penna-Martinez, M., Brück, P., Swiech, K., Mauf, S., et al. (2010). The role of cubilin gene polymorphisms and their influence on 25(OH)D3 and 1,25(OH)2D3 plasma levels in type 1 diabetes patients. *The Journal of Steroid Biochemistry and Molecular Biology, 121*, 442–444. https://doi.org/10.1016/j.jsbmb.2010.03.087.

Rebekah Prasoona, K., Sunitha, T., Srinadh, B., Muni Kumari, T., & Jyothy, A. (2018). LRP2 gene variants and their haplotypes strongly influence the risk of developing neural tube defects in the fetus: A family-triad study from South India. *Metabolic Brain Disease, 33*, 1343–1352. https://doi.org/10.1007/s11011-018-0242-2.

Reed, A. A. C., Loh, N. Y., Terryn, S., Lippiat, J. D., Partridge, C., Galvanovskis, J., et al. (2010). CLC-5 and KIF3B interact to facilitate CLC-5 plasma membrane expression, endocytosis, and microtubular transport: Relevance to pathophysiology of Dent's disease. *American Journal of Physiology. Renal Physiology, 298*, F365–F380. https://doi.org/10.1152/ajprenal.00038.2009.

Ren, Q., Gliozzi, M. L., Rittenhouse, N. L., Edmunds, L. R., Rbaibi, Y., Locker, J. D., et al. (2019). Shear stress and oxygen availability drive differential changes in opossum kidney proximal tubule cell metabolism and endocytosis. *Traffic, 20*, 448–459. https://doi.org/10.1111/tra.12648.

Ren, Q., Weyer, K., Rbaibi, Y., Long, K. R., Tan, R. J., Nielsen, R., et al. (2020). Distinct functions of megalin and cubilin receptors in recovery of normal and nephrotic levels of filtered albumin. *American Journal of Physiology. Renal Physiology, 318*, F1284–F1294. https://doi.org/10.1152/ajprenal.00030.2020.

Reznichenko, A., Snieder, H., van den Born, J., de Borst, M. H., Damman, J., van Dijk, M. C. R. F., et al. (2012). CUBN as a novel locus for end-stage renal disease: Insights from renal transplantation. *PLoS One*, 7, e36512. https://doi.org/10.1371/journal.pone.0036512.

Rombout, J. H., Lamers, C. H., Helfrich, M. H., Dekker, A., & Taverne-Thiele, J. J. (1985). Uptake and transport of intact macromolecules in the intestinal epithelium of carp (Cyprinus carpio L.) and the possible immunological implications. *Cell and Tissue Research, 239*, 519–530. https://doi.org/10.1007/BF00219230.

Ruggiero, C., Elks, C. M., Kruger, C., Cleland, E., Addison, K., Noland, R. C., et al. (2014). Albumin-bound fatty acids but not albumin itself alter redox balance in tubular epithelial cells and induce a peroxide-mediated redox-sensitive apoptosis. *American Journal of Physiology. Renal Physiology, 306*, F896–F906. https://doi.org/10.1152/ajprenal.00484.2013.

Russell-Jones, G. J., & Alpers, D. H. (1999). Vitamin B12 transporters. *Pharmaceutical Biotechnology, 12*, 493–520.

Russo, L. M., Sandoval, R. M., Campos, S. B., Molitoris, B. A., Comper, W. D., & Brown, D. (2009). Impaired tubular uptake explains albuminuria in early diabetic nephropathy. *Journal of the American Society of Nephrology, 20*, 489–494. https://doi.org/10.1681/ASN.2008050503.

Rutsch, F., Gailus, S., Miousse, I. R., Suormala, T., Sagné, C., Toliat, M. R., et al. (2009). Identification of a putative lysosomal cobalamin exporter altered in the cblF defect of vitamin B12 metabolism. *Nature Genetics, 41*, 234–239. https://doi.org/10.1038/ng.294.

Sabatino, J. A., Stokes, B. A., & Zohn, I. E. (2017). Prevention of neural tube defects in Lrp2 mutant mouse embryos by folic acid supplementation. *Birth Defects Research, 109*, 16–26. https://doi.org/10.1002/bdra.23589.

Sahali, D., Mulliez, N., Chatelet, F., Dupuis, R., Ronco, P., & Verroust, P. (1988). Characterization of a 280-kD protein restricted to the coated pits of the renal brush border and the epithelial cells of the yolk sac. Teratogenic effect of the specific monoclonal antibodies. *The Journal of Experimental Medicine, 167*, 213–218.

Santer, R., Schneppenheim, R., Suter, D., Schaub, J., & Steinmann, B. (1998). Fanconi-Bickel syndrome- -the original patient and his natural history, historical steps leading to the primary defect, and a review of the literature. *European Journal of Pediatrics*, *157*, 783–797.

Santo, Y., Hirai, H., Shima, M., Yamagata, M., Michigami, T., Nakajima, S., et al. (2004). Examination of megalin in renal tubular epithelium from patients with dent disease. *Pediatric Nephrology*, *19*, 612–615. https://doi.org/10.1007/s00467-004-1445-9.

Saraswat, M., Joenväära, S., Musante, L., Peltoniemi, H., Holthofer, H., & Renkonen, R. (2015). N-linked (N-) Glycoproteomics of urinary exosomes. *Molecular & Cellular Proteomics*, *14*, 2298. https://doi.org/10.1074/mcp.A114.040345.

Scheel, O., Zdebik, A. A., Lourdel, S., & Jentsch, T. J. (2005). Voltage-dependent electrogenic chloride/proton exchange by endosomal CLC proteins. *Nature*, *436*, 424–427. https://doi.org/10.1038/nature03860.

Schröder, A., Witt, O., & Eber, S. (2000). Diabetes mellitus type I, celiac disease and Imerslund-Gräsbeck syndrome: Only an incidental combination of rare diseases? *Klinische Pädiatrie*, *212*, 22–25. https://doi.org/10.1055/s-2000-9646.

Selma-Soriano, E., Llamusi, B., Fernández-Costa, J. M., Ozimski, L. L., Artero, R., & Redón, J. (2020). Rabphilin involvement in filtration and molecular uptake in Drosophila nephrocytes suggests a similar role in human podocytes. *Disease Models & Mechanisms*, *13*, dmm041509. https://doi.org/10.1242/dmm.041509.

Shah, M., Baterina, O. Y., Taupin, V., & Farquhar, M. G. (2013). ARH directs megalin to the endocytic recycling compartment to regulate its proteolysis and gene expression. *The Journal of Cell Biology*, *202*, 113–127. https://doi.org/10.1083/jcb.201211110.

Shi, A., & Grant, B. D. (2013). Interactions between Rab and Arf GTPases regulate endosomal phosphatidylinositol-4,5-bisphosphate during endocytic recycling. *Small GTPases*, *4*, 106–109. https://doi.org/10.4161/sgtp.23477.

Smith, B. T., Mussell, J. C., Fleming, P. A., Barth, J. L., Spyropoulos, D. D., Cooley, M. A., et al. (2006). Targeted disruption of cubilin reveals essential developmental roles in the structure and function of endoderm and in somite formation. *BMC Developmental Biology*, *6*, 30. https://doi.org/10.1186/1471-213X-6-30.

Smith, A. D., & Refsum, H. (2009). Vitamin B-12 and cognition in the elderly. *The American Journal of Clinical Nutrition*, *89*, 707S–711S. https://doi.org/10.3945/ajcn.2008.26947D.

Stabler, S. P. (2013). Clinical practice. Vitamin B12 deficiency. *The New England Journal of Medicine*, *368*, 149–160. https://doi.org/10.1056/NEJMcp1113996.

Stauber, T., & Jentsch, T. J. (2013). Chloride in vesicular trafficking and function. *Annual Review of Physiology*, *75*, 453–477. https://doi.org/10.1146/annurev-physiol-030212-183702.

Storm, T., Emma, F., Verroust, P. J., Hertz, J. M., Nielsen, R., & Christensen, E. I. (2011). A patient with cubilin deficiency. *The New England Journal of Medicine*, *364*, 89–91. https://doi.org/10.1056/NEJMc1009804.

Storm, T., Zeitz, C., Cases, O., Amsellem, S., Verroust, P. J., Madsen, M., et al. (2013). Detailed investigations of proximal tubular function in Imerslund-Gräsbeck syndrome. *BMC Medical Genetics*, *14*, 111. https://doi.org/10.1186/1471-2350-14-111.

Strope, S., Rivi, R., Metzger, T., Manova, K., & Lacy, E. (2004). Mouse amnionless, which is required for primitive streak assembly, mediates cell-surface localization and endocytic function of cubilin on visceral endoderm and kidney proximal tubules. *Development*, *131*, 4787–4795. https://doi.org/10.1242/dev.01341.

Sun, J., Hultenby, K., Axelsson, J., Nordström, J., He, B., Wernerson, A., et al. (2017). Proximal tubular expression patterns of Megalin and Cubilin in Proteinuric nephropathies. *Kidney International Reports*, *2*, 721–732. https://doi.org/10.1016/j.ekir.2017.02.012.

Syed, S., Michalski, E. S., Tangpricha, V., Chesdachai, S., Kumar, A., Prince, J., et al. (2017). Vitamin D status is associated with hepcidin and hemoglobin concentrations in children with inflammatory bowel disease. *Inflammatory Bowel Diseases, 23*, 1650–1658. https://doi.org/10.1097/MIB.0000000000001178.

Tam, P. P. L., & Loebel, D. A. F. (2007). Gene function in mouse embryogenesis: Get set for gastrulation. *Nature Reviews. Genetics, 8*, 368–381. https://doi.org/10.1038/nrg2084.

Tang, S. C. W., & Lai, K. N. (2012). The pathogenic role of the renal proximal tubular cell in diabetic nephropathy. *Nephrology, Dialysis, Transplantation, 27*, 3049–3056. https://doi.org/10.1093/ndt/gfs260.

Tanner, S. M., Aminoff, M., Wright, F. A., Liyanarachchi, S., Kuronen, M., Saarinen, A., et al. (2003). Amnionless, essential for mouse gastrulation, is mutated in recessive hereditary megaloblastic anemia. *Nature Genetics, 33*, 426–429. https://doi.org/10.1038/ng1098.

Tanner, S. M., Li, Z., Bisson, R., Acar, C., Oner, C., Oner, R., et al. (2004). Genetically heterogeneous selective intestinal malabsorption of vitamin B12: Founder effects, consanguinity, and high clinical awareness explain aggregations in Scandinavia and the Middle East. *Human Mutation, 23*, 327–333. https://doi.org/10.1002/humu.20014.

Tanner, S. M., Sturm, A. C., Baack, E. C., Liyanarachchi, S., & de la Chapelle, A. (2012). Inherited cobalamin malabsorption. Mutations in three genes reveal functional and ethnic patterns. *Orphanet Journal of Rare Diseases, 7*, 56. https://doi.org/10.1186/1750-1172-7-56.

Tao, W., Moore, R., Smith, E. R., & Xu, X.-X. (2016). Endocytosis and physiology: Insights from Disabled-2 deficient mice. *Frontiers in Cell and Development Biology, 4*, 129. https://doi.org/10.3389/fcell.2016.00129.

Tauris, J., Christensen, E. I., Nykjaer, A., Jacobsen, C., Petersen, C. M., & Ovesen, T. (2009). Cubilin and megalin co-localize in the neonatal inner ear. *Audiology & Neuro-Otology, 14*, 267–278. https://doi.org/10.1159/000199446.

Tenten, V., Menzel, S., Kunter, U., Sicking, E.-M., van Roeyen, C. R. C., Sanden, S. K., et al. (2013). Albumin is recycled from the primary urine by tubular transcytosis. *Journal of the American Society of Nephrology, 24*, 1966–1980. https://doi.org/10.1681/ASN.2013010018.

Terryn, S., Tanaka, K., Lengelé, J.-P., Olinger, E., Dubois-Laforgue, D., Garbay, S., et al. (2016). Tubular proteinuria in patients with HNF1α mutations: HNF1α drives endocytosis in the proximal tubule. *Kidney International, 89*, 1075–1089.

Teumer, A., Tin, A., Sorice, R., Gorski, M., Yeo, N. C., Chu, A. Y., et al. (2016). Genome-wide association studies identify genetic loci associated with albuminuria in diabetes. *Diabetes, 65*, 803–817. https://doi.org/10.2337/db15-1313.

Theilig, F., Kriz, W., Jerichow, T., Schrade, P., Hähnel, B., Willnow, T., et al. (2007). Abrogation of protein uptake through megalin-deficient proximal tubules does not safeguard against tubulointerstitial injury. *Journal of the American Society of Nephrology, 18*, 1824–1834. https://doi.org/10.1681/ASN.2006111266.

Thrailkill, K. M., Nimmo, T., Bunn, R. C., Cockrell, G. E., Moreau, C. S., Mackintosh, S., et al. (2009). Microalbuminuria in type 1 diabetes is associated with enhanced excretion of the endocytic multiligand receptors megalin and cubilin. *Diabetes Care, 32*, 1266–1268. https://doi.org/10.2337/dc09-0112.

Tomihara-Newberger, C., Haub, O., Lee, H. G., Soares, V., Manova, K., & Lacy, E. (1998). The amn gene product is required in extraembryonic tissues for the generation of middle primitive streak derivatives. *Developmental Biology, 204*, 34–54. https://doi.org/10.1006/dbio.1998.9034.

Tsaroucha, A. K., Chatzaki, E., Lambropoulou, M., Despoudi, K., Laftsidis, P., Charsou, C., et al. (2008). Megalin and cubilin in the human gallbladder epithelium. *Clinical and Experimental Medicine, 8*, 165–170. https://doi.org/10.1007/s10238-008-0174-y.

Tsekmekidou, X., Tsetsos, F., Koufakis, T., Karras, S. N., Georgitsi, M., Papanas, N., et al. (2020). Association between CUBN gene variants, type 2 diabetes and vitamin D concentrations in an elderly Greek population. *The Journal of Steroid Biochemistry and Molecular Biology*, *198*, 105549. https://doi.org/10.1016/j.jsbmb.2019.105549.

Udagawa, T., Harita, Y., Miura, K., Mitsui, J., Ode, K. L., Morishita, S., et al. (2018). Amnionless-mediated glycosylation is crucial for cell surface targeting of cubilin in renal and intestinal cells. *Scientific Reports*, *8*, 2351. https://doi.org/10.1038/s41598-018-20731-4.

Urae, S., Harita, Y., Udagawa, T., Ode, K. L., Nagahama, M., Kajiho, Y., et al. (2020). A cellular model of albumin endocytosis uncovers a link between membrane and nuclear proteins. *Journal of Cell Science*, *133*, jcs242859. https://doi.org/10.1242/jcs.242859.

van Rensburg, S. J., Kotze, M. J., & van Toorn, R. (2012). The conundrum of iron in multiple sclerosis—time for an individualised approach. *Metabolic Brain Disease*, *27*, 239–253. https://doi.org/10.1007/s11011-012-9290-1.

van Rensburg, S. J., Peeters, A. V., van Toorn, R., Schoeman, J., Moremi, K. E., van Heerden, C. J., et al. (2019). Identification of an iron-responsive subtype in two children diagnosed with relapsing-remitting multiple sclerosis using whole exome sequencing. *Molecular Genetics and Metabolism Reports*, *19*, 100465. https://doi.org/10.1016/j.ymgmr.2019.100465.

Vázquez-Carretero, M. D., Palomo, M., García-Miranda, P., Sánchez-Aguayo, I., Peral, M. J., Calonge, M. L., et al. (2014). Dab2, megalin, cubilin and amnionless receptor complex might mediate intestinal endocytosis in the suckling rat. *Journal of Cellular Biochemistry*, *115*, 510–522. https://doi.org/10.1002/jcb.24685.

Velkova, A., Diaz, J. E. L., Pangilinan, F., Molloy, A. M., Mills, J. L., Shane, B., et al. (2017). The FUT2 secretor variant p.Trp154Ter influences serum vitamin B12 concentration via holo-haptocorrin, but not holo-transcobalamin, and is associated with haptocorrin glycosylation. *Human Molecular Genetics*, *26*, 4975–4988. https://doi.org/10.1093/hmg/ddx369.

Verroust, P. J. (2002). Pathophysiology of cubilin: Of rats, dogs and men. *Nephrology, Dialysis, Transplantation*, *17*(Suppl. 9), 55–56.

Vivante, A., Hwang, D.-Y., Kohl, S., Chen, J., Shril, S., Schulz, J., et al. (2017). Exome sequencing discerns syndromes in patients from consanguineous families with congenital anomalies of the kidneys and urinary tract. *Journal of the American Society of Nephrology*, *28*, 69–75. https://doi.org/10.1681/ASN.2015080962.

Wang, X., Bornslaeger, E. A., Haub, O., Tomihara-Newberger, C., Lonberg, N., Dinulos, M. B., et al. (1996). A candidate gene for the amnionless gastrulation stage mouse mutation encodes a TRAF-related protein. *Developmental Biology*, *177*, 274–290. https://doi.org/10.1006/dbio.1996.0162.

Wang, S., Xin, X., Luo, W., Mo, M., Si, S., Shao, B., et al. (2021). Association of vitamin D and gene variants in the vitamin D metabolic pathway with preterm birth. *Nutrition*, *89*, 111349. https://doi.org/10.1016/j.nut.2021.111349.

Wang, W.-H., Yue, P., Sun, P., & Lin, D.-H. (2010). Regulation and function of potassium channels in aldosterone-sensitive distal nephron. *Current Opinion in Nephrology and Hypertension*, *19*, 463–470. https://doi.org/10.1097/MNH.0b013e32833c34ec.

Wang, J., Zhao, J.-Y., Wang, F., Peng, Q.-Q., Hou, J., Sun, S.-N., et al. (2014). A genetic variant in vitamin B12 metabolic genes that reduces the risk of congenital heart disease in Han Chinese populations. *PLoS One*, *9*, e88332. https://doi.org/10.1371/journal.pone.0088332.

Weyer, K., Andersen, P. K., Schmidt, K., Mollet, G., Antignac, C., Birn, H., et al. (2018). Abolishment of proximal tubule albumin endocytosis does not affect plasma albumin during nephrotic syndrome in mice. *Kidney International*, *93*, 335–342. https://doi.org/10.1016/j.kint.2017.07.024.

Weyer, K., Storm, T., Shan, J., Vainio, S., Kozyraki, R., Verroust, P. J., et al. (2011). Mouse model of proximal tubule endocytic dysfunction. *Nephrology, Dialysis, Transplantation, 26*, 3446–3451. https://doi.org/10.1093/ndt/gfr525.

Wicher, G., & Aldskogius, H. (2008). Megalin deficiency induces critical changes in mouse spinal cord development. *Neuroreport, 19*, 559–563. https://doi.org/10.1097/WNR.0b013e3282f94267.

Willnow, T. E., & Christ, A. (2017). Endocytic receptor LRP2/megalin-of holoprosencephaly and renal Fanconi syndrome. *Pflügers Archiv, 469*, 907–916. https://doi.org/10.1007/s00424-017-1992-0.

Wilund, K. R., Yi, M., Campagna, F., Arca, M., Zuliani, G., Fellin, R., et al. (2002). Molecular mechanisms of autosomal recessive hypercholesterolemia. *Human Molecular Genetics, 11*, 3019–3030. https://doi.org/10.1093/hmg/11.24.3019.

Wu, Y., & Xu, Y. (2020). Bioinformatics for the prognostic value and function of cubilin (CUBN) in colorectal cancer. *Medical Science Monitor, 26*, e922447. https://doi.org/10.12659/MSM.922447.

Wuerges, J., Geremia, S., Fedosov, S. N., & Randaccio, L. (2007). Vitamin B12 transport proteins: Crystallographic analysis of beta-axial ligand substitutions in cobalamin bound to transcobalamin. *IUBMB Life, 59*, 722–729. https://doi.org/10.1080/15216540701673413.

Xu, J., & Lu, W. (2021). CircSPIDR acts as a tumour suppressor in cervical adenocarcinoma by sponging miR-431-5p and regulating SORCS1 and CUBN expression. *Aging (Albany NY), 13*, 18340–18359. https://doi.org/10.18632/aging.203283.

Yammani, R. R., Seetharam, S., & Seetharam, B. (2001). Identification and characterization of two distinct ligand binding regions of cubilin. *The Journal of Biological Chemistry, 276*, 44777–44784. https://doi.org/10.1074/jbc.M106419200.

Zeng, B., Chen, G.-L., Garcia-Vaz, E., Bhandari, S., Daskoulidou, N., Berglund, L. M., et al. (2017). ORAI channels are critical for receptor-mediated endocytosis of albumin. *Nature Communications, 8*, 1–12. https://doi.org/10.1038/s41467-017-02094-y.

Zhang, F., Zhao, Y., Chao, Y., Muir, K., & Han, Z. (2013). Cubilin and amnionless mediate protein reabsorption in Drosophila nephrocytes. *Journal of the American Society of Nephrology, 24*, 209–216. https://doi.org/10.1681/ASN.2012080795.

Zhao, L., Wei, Y., Song, A., & Li, Y. (2016). Association study between genome-wide significant variants of vitamin B12 metabolism and gastric cancer in a Han Chinese population. *IUBMB Life, 68*, 303–310.

Zhou, G. L., Na, S.-Y., Niedra, R., & Seed, B. (2014). Deficits in receptor-mediated endocytosis and recycling in cells from mice with Gpr107 locus disruption. *Journal of Cell Science, 127*, 3916–3927. https://doi.org/10.1242/jcs.135269.

Zhuang, Z., Marshansky, V., Breton, S., & Brown, D. (2011). Is caveolin involved in normal proximal tubule function? Presence in model PT systems but absence in situ. *American Journal of Physiology. Renal Physiology, 300*, F199–F206. https://doi.org/10.1152/ajprenal.00513.2010.

Zhuo, J. L., & Li, X. C. (2013). Proximal nephron. *Comprehensive Physiology, 3*, 1079–1123. https://doi.org/10.1002/cphy.c110061.

CHAPTER FIVE

Membrane transport of cobalamin

Mark Nijland[†], Jose M. Martínez Felices[†], Dirk J. Slotboom*, and Chancievan Thangaratnarajah
University of Groningen, Faculty of Science and Engineering, Groningen Biomolecular Sciences and Biotechnology, Membrane Enzymology Group, Groningen, Netherlands
*Corresponding author e-mail address: d.j.slotboom@rug.nl

Contents

Abstract

A wide variety of organisms encode cobalamin-dependent enzymes catalyzing essential metabolic reactions, but the cofactor cobalamin (vitamin B12) is only synthesized by a subset of bacteria and archaea. The biosynthesis of cobalamin is complex and energetically costly, making cobalamin variants and precursors metabolically valuable. Auxotrophs for these molecules have evolved uptake mechanisms to compensate for the lack of a synthesis pathway. Bacterial transport of cobalamin involves the passage over one or two lipidic membranes in Gram-positive and -negative bacteria, respectively. In higher eukaryotes, a complex system of carriers, receptors and transporters facilitates the delivery of the essential molecule to the tissues. Biochemical and genetic approaches have identified different transporter families involved in cobalamin transport. The majority of the characterized cobalamin transporters are active transport systems that belong to the ATP-binding cassette (ABC) superfamily of transporters. In this chapter, we describe the different cobalamin transport systems characterized to date that are present in bacteria and humans, as well as yet-to-be-identified transporters.

[†] Contributed equally.

Vitamins and Hormones, Volume 119
ISSN 0083-6729
https://doi.org/10.1016/bs.vh.2022.01.008

Abbreviations

ABC	ATP-binding cassette
ATP	adenosine triphosphate
Cryo-EM	cryogenic electron microscopy
DMB	5′,6′-dimethylbenzimidazole
DNA	deoxyribonucleic acid
ECF	energy coupling factor
HC	haptocorrin
IF	intrinsic factor
NBD	nucleotide-binding domain
RNA	ribonucleic acid
TBDT	TonB-dependent outer membrane transporter
TC	transcobalamin
TMD	transmembrane domain
VDAC	voltage-dependent anion channel

1. Introduction

Cobalamin (vitamin B12) is a complex organometallic cofactor belonging to the group of cobamides, and is essential for a large number of organisms from almost all domains of life, with the notable exception of plants (Roth, Lawrence, & Bobik, 1996). Also fungi were thought not to make use of cobalamin, but a recent publication showed that a few early branching fungi may make use of the cofactor (Orłowska, Steczkiewicz, & Muszewska, 2021). The structural core of cobamides consists of a corrin ring with a chelated cobalt ion in the center, which is additionally coordinated by a "lower" ligand at the α-axial position and an "upper" ligand at the β-axial position (Fig. 1) (Gude & Taga, 2020). The lower ligand is usually a covalently attached side chain of the corrin ring, and folds over to coordinate the cobalt ion. Different lower ligands give rise to distinctive cobamides, and include organic moieties that fall into the group of benzimidazoles, purines or phenols (Fig. 1) (Seth & Taga, 2014). Characteristic for cobalamin is the presence of a 5′6′-dimethylbenzimidazole (DMB) base as lower ligand (Gruber, Puffer, & Kräutler, 2011). While this moiety is generally coordinated to the cobalt ion (base-on conformation) in free cobalamin under physiological conditions (Fig. 1), it can dissociate (base-off conformation) depending on temperature, pH, oxidation state of the cobalt ion and protein environment (Fig. 1) (Obeid, Fedosov, & Nexo, 2015). In the base-off

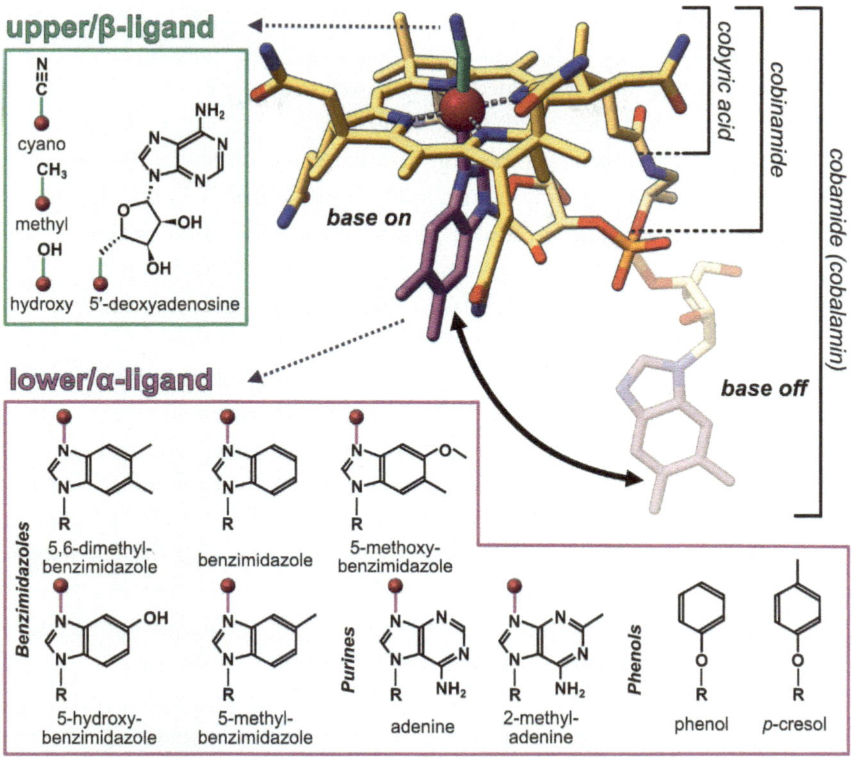

Fig. 1 Structure of cyanocobalamin in the base-on and -off conformation, the variants of the upper ligand to form different chemical forms and lower ligands to form distinct cobamides. Cobalamin is shown in yellow stick representation and illustrates a composite of the base-on (PDB: 1n2z) and -off conformation (PDB: 3bul). Oxygen, nitrogen and phosphate atoms within are colored in red, blue and orange, respectively. The coordinated cobalt ion within the corrin ring is shown as a wine-colored sphere. The upper and lower ligands forming a covalent interaction with the cobalt ion are colored in green and purple, respectively. Chemical structures of variants of the upper (green box) and lower (purple box) ligands are shown with the position forming the covalent interaction with the cobalt ion indicated by a wine-colored sphere and a solid line. Parts of the molecule indicating cobyric acid, cobinamide and cobamide are highlighted.

conformation, the cobalt coordination can be taken over by an external ligand, such as a protein side chain (Gruber et al., 2011; Rempel, Colucci, de Gier, Guskov, & Slotboom, 2018). The upper ligand of the cobalt ion in naturally occurring cobalamin is not covalently linked to the corrin ring, and may be a hydroxy-, methyl- or 5′-deoxyadenosyl-group (Fig. 1) (Giedyk, Goliszewska, & Gryko, 2015). Methylcobalamin and adenosylcobalamin represent the biological active cofactor forms that are used by distinctive cobalamin-dependent enzymes (Giedyk et al., 2015). These cofactors are involved in various metabolic processes, including DNA and RNA synthesis, amino acid metabolism, carbon and nitrogen metabolism and secondary metabolite synthesis (Sokolovskaya, Shelton, & Taga, 2020). In the process of industrial production of cobalamin, which involves bacterial fermentation, a cyano-group is added as the upper ligand. Cyanocobalamin is commonly used in pharmaceuticals, supplements and food additives, and requires removal of the cyano-group for conversion into the active cofactors methylcobalamin or adenosylcobalamin (Rizzo et al., 2016).

Despite its widespread use as cofactor, *de novo* biosynthesis of cobalamin is confined to only a limited number of bacterial and archaeal species. This complex synthesis route involves approximately 30 enzymatic steps and can occur by two alternative pathways, differing in oxygen requirements and timing of cobalt insertion (for details, see review Fang, Kang, & Zhang, 2017; Warren, Raux, Schubert, & Escalante-Semerena, 2002). Organisms that make use of cobalamin-dependent enzymes, but lack the capacity to synthesize the cofactors themselves, have to acquire cobalamin from external sources (auxotrophs). Humans can only meet their cobalamin requirement by absorbing the complete cofactor from their diet, while various bacteria can also convert precursor molecules ("salvaging") or alternative cobamides ("remodeling") taken up from the environment into cobalamin (Seth & Taga, 2014; Shelton et al., 2019). Often cobalamin producers can also import the cofactor, which is energetically more favorable than the biosynthesis (Fang et al., 2017; Shelton et al., 2019). While knowledge of the biosynthetic pathway of cobalamin and its chemical properties is extensive, insight in how cobalamin or its derivatives or precursors are taken up into the cell is limited. Here, we provide an overview of the characterized membrane transport systems that enable the uptake of cobalamin and its derivatives into the cell. We will be focusing on the systems that are present in bacteria and humans.

2. Membrane transport systems in bacteria

Cobalamin is used by bacteria for various metabolic processes and genomic analysis has shown that more than three quarters of the bacteria harbor cobalamin-dependent enzymes (Shelton et al., 2019; Zhang, Rodionov, Gelfand, & Gladyshev, 2009). However, more than half of this subset of bacteria lack the genetic capacity to synthesize the cofactor (Shelton et al., 2019; Zhang et al., 2009). Moreover, it is not clear whether species that have the genetic capacity, actually produce cobalamin in their natural habitat (Shelton et al., 2019). For the model organism *Escherichia coli*, it is estimated that only approximately 20 cobalamin molecules per cell are already sufficient to support growth in cobalamin-dependent conditions (Di Girolamo, Kadner, & Bradbeer, 1971), which means that high-capacity transport is not needed. Bacteria have evolved various membrane transport systems to mediate the uptake of cobalamin from the environment. The currently identified transporters are described in more detail below.

2.1 Transport across the outer membrane in Gram-negative bacteria

A key feature of Gram-negative bacteria is the presence of a double membrane system. In addition to the plasma membrane, Gram-negative bacteria also have an outer membrane, complicating the uptake of molecules into the cell (Faraldo-Gómez & Sansom, 2003). Outer membrane proteins play a crucial role in the transport of nutrients across this membrane. Porins and substrate-specific channels are passive transporters that facilitate the diffusion of small molecules (up to ~600 Da) into the periplasm (Vergalli et al., 2020). However, larger molecules or those present in low concentrations in the extracellular environment, such as cobalamin, require active transport systems to cross the outer membrane (Noinaj, Guillier, Barnard, & Buchanan, 2010). Gram-negative species have evolved TonB-dependent outer membrane transporters (TBDTs) to transport a variety of scarce nutrients, including cobalamin (Noinaj et al., 2010).

2.1.1 BtuB

BtuB is a member of the TonB-dependent outer membrane transport family, widespread among Gram-negative bacteria, and mediates the active transport of cobalamin and variants into the periplasm (Fig. 2) (Heller, Mann, & Kadner, 1985). This transporter exhibits a conserved architecture among the TonB-dependent outer membrane transport family, consisting of a 22-stranded β-barrel and an inserted plug domain (Noinaj et al., 2010).

Transport activity for BtuB has not been demonstrated *in vitro* using purified protein reconstituted in liposomes, which hampers mechanistic understanding. Although the exact transport mechanism is not well understood, X-ray crystal structures of members of the TonB-dependent outer membrane transport family in different conformational states provide a glimpse of the structural mechanism involved in the translocation of the bound substrate to the periplasm (Noinaj et al., 2010). The binding site

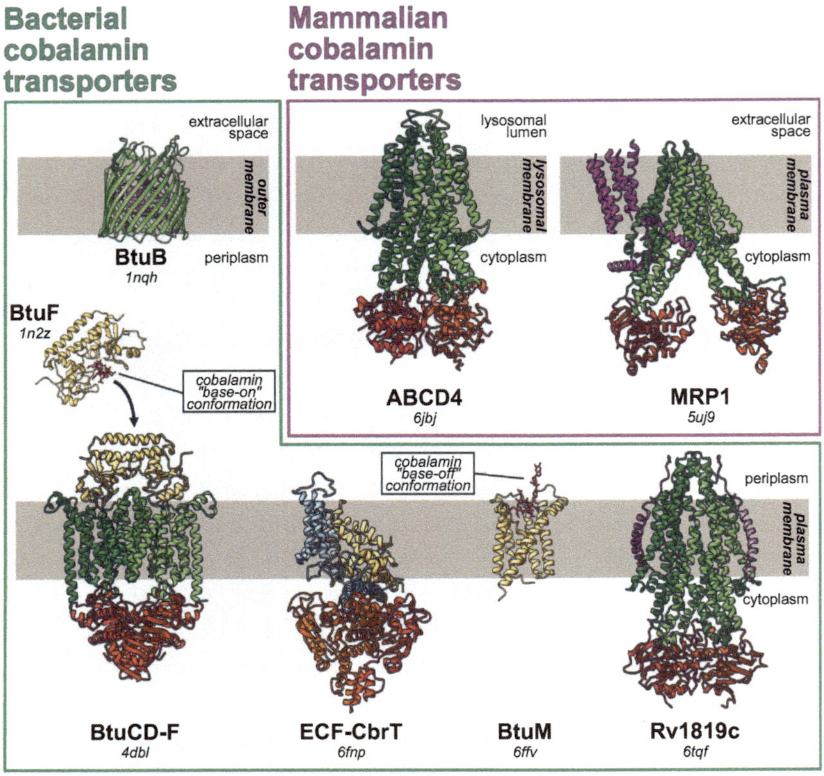

Fig. 2 Structures of bacterial and mammalian cobalamin transporters. Models of bacterial (green box) and mammalian (purple box) cobalamin transporters are shown in cartoon representation, and are colored according to common structural features: green for substrate translocation pathway, red or orange for nucleotide-binding domains/proteins, yellow for substrate binding proteins, blue for the scaffold protein in ECF transporters, purple for structural features of unknown function. Lighter shades indicate the different protomers within the homo-dimeric assembly. Bound cobalamin molecules are shown in stick representation and the base-on and -off conformations are indicated with labels. The lipid membrane is indicated as a gray bar and the type and orientation of the lipid membrane is indicated. PDB codes of the individual models are indicated under the protein labels.

for cobalamin is formed by residues located on the extracellular side of the so-called plug domain in the center of the β-barrel, the extracellular loops and the interior wall of the β-barrel (Chimento, Mohanty, Kadner, & Wiener, 2003). Crucially, co-operative binding of calcium ions is essential to support binding of cobalamin to BtuB within nanomolar affinity (Table 1) (Bradbeer et al., 1986; Cadieux, Barekzi, & Bradbeer, 2007). The proton motive force across the inner membrane is used as an energy source to

Table 1 Available functional data of cobalamin transporters.

Transporter	Organism	Cobalamin substrates	Kinetics	Futile ATP hydrolysis	References
Bacterial outer membrane transporters					
BtuB	*Escherichia coli* (Gram-negative)	**CN-Cbl**: Kd ~ 0.3 nM (on bacterial membranes (1), depletion of Ca^{2+} reduces affinity by 50- to 100-fold (1). **CN-Cbi**: Ki ~ 0.8 nM (competitive transport assay against labeled CN-Cbl (1))	**CN-Cbl**: $k_{off} = 8.6 \times 10^{-3} \pm 0.7\,min^{-1}$ (2) $k_{off} = 1.5 \times 10^{-3} \pm 0.5\,min^{-1}$ (in presence of $TonB_{\Delta\ 1-31}$) (2)	No NBDs (TonB-dependent)	(1) Bradbeer, Reynolds, Bauler, and Fernandez (1986) (2) Mills, Le, and Duong (2016)
Bacterial cytoplasmic membrane transporters					
BtuCDF	*E. coli* (Gram-negative)	**CN-Cbl**: $K_d \sim 9.1–15\,nM$ (BtuF in solution) (1, 2) **CN-H_2O-Cbi**: $K_d = 40 \pm 10\,nM$ (BtuF in solution) (2)	**CN-Cbl**: $V_{max} \sim 0.3\,nmol\,mg^{-1}\,min^{-1}$ (2)	ATPase rate = $180\,nmol\,min^{-1}\,mg^{-1}$ (BtuCD) (2) ATPase rate = $440\,nmol\,min^{-1}\,mg^{-1}$ (in presence of 20 μM BtuF) (1)	(1) Borths, Poolman, Hvorup, Locher, and Rees (2005) (2) Mireku et al. (2017)

Continued

Table 1 Available functional data of cobalamin transporters.—cont'd

Transporter	Organism	Cobalamin substrates	Kinetics	Futile ATP hydrolysis	References
CbrT	*Lactobacillus delbrueckii* (Gram-positive)	**CN-Cbl**: $K_d = 9.2 \pm 4.5$ nM (ITC on bacterial membranes) (1) **OH-Cbl**: $K_d = 9.6 \pm 6.9$ nM (ITC on bacterial membranes) (1) **Met-Cbl**: $K_d = 4.5 \pm 0.3$ nM (ITC on bacterial membranes) (1) **diCN-Cbi**: $K_d = 36 \pm 15$ nM (ITC on bacterial membranes) (1)	**CN-Cbl**: $K_m = 2.1 \pm 0.4$ nM (1) $V_{max} = 0.06 \pm 0.01$ nmol mg^{-1} s^{-1} (1)	Not directly tested (futile ATPase activity on the related folate ECF transporter FolT2) (2)	(1) Santos et al. (2018) (2) Swier, Guskov, and Slotboom (2016)
BtuM	*Thiobacillus denitrificans* (Gram-negative)	**CN-Cbl** (Growth assay) (1) **OH-Cbl** (Growth assay) (1) **diCN-Cbi**: $K_d \sim 0.58$–0.65 µM (ITC on solubilized BtuM) (1) $K_d = 5.6 \pm 2.8$ µM (ITC on solubilized BtuMC80S mutant) (1)	**diCN-Cbi**: $t_{1/2} = 12 \pm 0.7$ min (decyanation) (1)	No NBDs	(1) Rempel et al. (2018)

rv1819c	*Mycobacterium tuberculosis* (Gram-negative)	**Cobalamin** (growth assay, CN-Cbl) (1)	Not determined	Yes (1)	(1) Rempel et al. (2020)
Human membrane transporters					
ABCD4	*Homo sapiens*	**Cobalamin** (variant not specified) (1,2)	$K_m = \sim 426\,\mu M$ (1) $V_{max} = \sim 667\,pmol\,min^{-1}\,mg^{-1}$ (1)	ATPase rate $\sim 4.5–11.8\,nmol\,mg^{-1}\,min^{-1}$ (1, 2) ATPase rate $\sim 8\,nmol\,mg^{-1}\,min^{-1}$ (+1 mM cobalamin) (1)	(1) Kitai et al. (2021) (2) Xu et al. (2019)
MRP1	*H. sapiens*	**OH-Cbl** (1) **CN-Cbl** (competitive inhibition against E217βG) (1) **GS-Cbl**: $K_i = 6.2 \pm 1.6\,\mu M$ (competitive inhibition against E217βG) (1)	**OH-Cbl**: $K_m \sim 23\,\mu M$ (1) $V_{max} \sim 76\,pmol\,mg^{-1}\,min^{-1}$ (1)	K_m (ATP) $= 104 \pm 22\,\mu M$ (2) $V_{max} \sim 5–10\,nmol\,min^{-1}\,mg^{-1}$ (30–50% increase in presence of known substrates other than cobalamin) (2) K_m (ATP) $= 400 \pm 60\,\mu M$ (3) $V_{max} = 27 \pm 1\,nmol\,mg^{-1}\,min^{-1}$ (~2- to 3-fold different in presence of known substrates other than cobalamin) (digitonin) (3)	(1) Beedholm-Ebsen et al. (2010) (2) Mao, Leslie, Deeley, and Cole (1999) (3) Johnson and Chen (2017)

Abbreviations: CN-Cbl, cyanocobalamin; CN-H_2O-Cbl, cyano-aquocobinamide; diCN-Cbi, dicyanocobinamide; Met-Cbl, methylcobalamin; OH-Cbl, hydroxocobalamin. Experiments in artificial membranes were preferentially selected, unless otherwise stated.

facilitate the active transport process, which is transmitted to BtuB *via* the energy-transducing TonB complex (Noinaj et al., 2010). This is achieved by the physical interaction of the Ton box, a conserved motif at the periplasmic amino-terminal region of the plug domain, with the TonB complex (Shultis, Purdy, Banchs, & Wiener, 2006). Although it is known that this interaction is essential to release cobalamin into the periplasm, the mechanism of passage through BtuB remains to be elucidated.

Gram-negative bacteria usually utilize a single BtuB transport system to facilitate cobalamin transport (Chimento et al., 2003). However, nearly all members of the Bacteroides species inhabiting the human gut harbor multiple (up to four) BtuB homologs that are encoded on separate loci (Degnan, Barry, Mok, Taga, & Goodman, 2014). Interestingly, they display distinct preferences toward corrinoids that differ in their lower ligand (Degnan et al., 2014). Moreover, each homolog contributes differently to the relative bacterial fitness in competitive environments *in vitro* as well as in a germ-free mice model (Degnan et al., 2014). Whether the presence of multiple corrinoid transport systems simply provides the bacteria with competitive advantages or may play a bigger role in communication between bacterial species remains to be elucidated (Degnan et al., 2014).

Bacteroides species also encode the accessory protein BtuG within the genetic loci where BtuB is also encoded (Wexler et al., 2018). BtuG is a cell-surface exposed lipoprotein that binds cobalamin and cobinamide with femtomolar affinity (187 and 193 fM, respectively), and associates with BtuB to support efficient uptake of the substrate by the accompanying BtuB homolog in *Bacteroides thetaiotaomicron* (Wexler et al., 2018). Although BtuB is capable of importing cobalamin on its own, BtuG-mediated cobalamin scavenging is critical for *in vivo* growth of *B. thetaiotaomicron* in germ-free mice (Wexler et al., 2018). Remarkably, it was shown that BtuG is able to extract cobalamin from the human carrier protein intrinsic factor (IF), which binds cobalamin with picomolar affinity (1 pM) (Fedosov, Berglund, Fedosova, Nexo, & Petersen, 2002), providing Bacteroides a competitive advantage for the cobalamin sources present in the human gut (*see* section on Transport of cobalamin in human) (Wexler et al., 2018).

2.2 Transport systems of the plasma membrane in Gram-negative and -positive bacteria

While BtuB is the only identified cobalamin transporter in the outer membrane, the repertoire of membrane transport systems for cobalamin in the plasma membrane of Gram-negative and -positive bacteria is more extensive. Comparative genomic analysis for cobalamin metabolism in bacteria revealed that diverse solutions for cobalamin transport across the plasma membrane exist (Gopinath et al., 2013; Rodionov et al., 2009; Rodionov, Vitreschak, Mironov, & Gelfand, 2003). All characterized cobalamin transporters in bacteria to date belong to the superfamily of adenosine triphosphate (ATP)-binding cassette (ABC) transporters (Fig. 2), with the exception of BtuM, which is nonetheless structurally related to a part of an ABC transporter. Core to ABC transporters is the presence of two well conserved nucleotide-binding domains or subunits (NBDs) exhibiting ATP binding and hydrolysis activity. The NBDs are connected to transmembrane domains or subunits (TMDs) that contain a substrate translocation pathway (Thomas & Tampé, 2020). The energy released upon ATP hydrolysis is used to power conformational changes in the membrane domains, ultimately leading to substrate translocation across the membrane (Thomas & Tampé, 2020). Based on high resolution structures obtained by X-ray crystallography and cryogenic electron microscopy (cryo-EM), ABC transporters have been classified into seven different types. While all types make use of related NBDs, the TMDs are structurally diverse and do not show obvious evolutionary relatedness (Thomas et al., 2020). The characterized bacterial cobalamin transporters belong to type II, III or IV of ABC transporters, which we will lay out in detail below.

2.2.1 BtuCD-F

The cobalamin transporter from *E. coli*, BtuCD-F (Fig. 2), is the best characterized among the cobalamin transporters, and belongs to the type II class of ABC importers (Thomas et al., 2020). Homologs are widely distributed among Gram-negative and -positive bacteria. They are present in 90% of thebacteria that do not encode enzymes for cobalamin synthesis, yet require the vitamin as evidenced by the presence of genes coding for enzymes that use the cofactor cobalamin (Zhang et al., 2009). Some organisms encode more than one homolog of BtuCD-F within their genome (Degnan et al., 2014). Moreover, a BtuCD-F homolog has been found in the archaeon *Halobacterium* sp. NRC-1, constituting the first characterized cobalamin transporter in archaea (Woodson, Reynolds, & Escalante-Semerena, 2005).

BtuCD-F from the Gram-negative bacterium *E. coli* mediates the translocation of cobalamin from the periplasm across the plasma membrane into the cytoplasm (Borths et al., 2005; Cadieux et al., 2002; de Veaux, Clevenson, Bradbeer, & Kadner, 1986). The core is formed by an oligomeric assembly consisting of two identical NBDs (BtuD) and two identical TMDs (BtuC) (Fig. 2) (Locher, Lee, & Rees, 2002). The substrate translocation pathway is found at the interface between the two TMDs BtuC, allowing the passage of cobalamin during the transport process across the membrane (Locher et al., 2002). Each copy of BtuD interacts with BtuC on the cytoplasmic side (Fig. 2), and hence the NBDs are accessible for ATP from the cytoplasm (Korkhov, Mireku, Veprintsev, & Locher, 2014). A single substrate-binding protein specific for cobalamin, BtuF, interacts with the core complex on the periplasmic side forming the full transporter complex (Fig. 2) (Hvorup et al., 2007; Korkhov, Mireku, & Locher, 2012). Cobalamin binds BtuF in the base-on conformation, and with nanomolar affinity (Fig. 2, Table 1) (Cadieux et al., 2002; Karpowich, Huang, Smith, & Hunt, 2003). Such a high binding affinity is an essential property given the scarcity of cobalamin in the extracellular environment. Binding of the cobalamin precursor cobinamide to BtuF and transport by the BtuCD-F complex has also been demonstrated, albeit at lower affinity and transport rates (Table 1) (Mireku et al., 2017). The wealth of information from crystallographic and functional studies allows a detailed description of the transport cycle of cobalamin.

In absence of cobalamin, the BtuCD-F complex rests in a conformation with the membrane pathway sealed on the cytoplasmic side ("outward-facing" state) (Locher et al., 2002), from which BtuF partly or completely dissociates revealing its substrate-binding site for cobalamin (Goudsmits, Slotboom, & van Oijen, 2017; Korkhov et al., 2014), thus allowing the capture of cobalamin with high affinity. In presence of ATP, BtuCD retrieves cobalamin from the loaded BtuF, and traps it in a cavity approximately in the center of the membrane by closing off the translocation pathway to both sides of the membrane (Korkhov et al., 2012). ATP hydrolysis triggers conformational changes within the TMDs whereupon the cytoplasmic gate opens and the translocation pathway collapses (Hvorup et al., 2007), resulting in cobalamin being pushed into the cytoplasm. Once released, the cytoplasmic gate closes and the periplasmic gate opens back ready to accept a new cobalamin molecule (Korkhov et al., 2014). A common property among type II ABC transporters is their specificity for low-abundant substrates, as well as a high level of basal or futile ATP hydrolysis. This means that hydrolysis of the majority of ATP molecules is not mechanistically coupled to cobalamin transport (Borths et al., 2005). Such futile hydrolysis may be a necessary trade-off for efficient high affinity scavenging of cobalamin from the environment.

2.2.2 ECF-CbrT

On the basis of comparative genomic analysis, cobalamin transporters belonging to type III ABC transporters (also named the energy coupling factor (ECF)-type) were identified in a subgroup of bacteria that lack a homolog of BtuCD-F (Rodionov et al., 2009). Notably, this study also identified ECF-CbrT in the archaeon *Methanosphaera stadtmanae*, representing, along with BtuCD-F, the only cobalamin transporters identified in archaea to date. ECF transporters constitute the phylogenetically, structurally and mechanistically distinct type III clade of ABC transporters (Rempel, Stanek, & Slotboom, 2019; Thomas et al., 2020). Like all ABC transporters, ECF transporters have two NBDs EcfA exhibiting ATPase activity (Fig. 2), which may be either encoded by the same (EcfA) or two homologous genes (EcfA and EcfA'), or as a fusion protein from a single gene (EcfA-A) (Rempel et al., 2019). The NBDs are associated with a membrane-embedded scaffold protein (EcfT) that is structurally distinct from the TMDs found in BtuCD-F (Fig. 2). Both binding and translocation of the substrate are conferred by a membrane embedded substrate-binding protein termed S-component (Fig. 2), which is structurally unrelated to the TMD EcfT. CbrT binds cobalamin in a site near the extracellular face of the membrane, and then moves the substrate to the cytosol in a unique toppling motion (Santos et al., 2018). EcfT and the two NBDs form together the ECF or energizing module, which ultimately powers substrate transport by the S-component (Slotboom, 2014).

ECF transporters can be divided into two subgroups based on their modular nature with respect to their assembly. In group I ECF transporters, the S-component associates exclusively with a dedicated ECF module (Rempel et al., 2019), while in group II ECF transporters the ECF module is shared among multiple S-components that have different substrate specificities (Rempel et al., 2019; Setyawati et al., 2020). It is of note that predicted cobalamin-specific ECF transporters can be found in both groups.

In vivo complementation assays with a cobalamin-dependent and BtuCD-F-defective *E. coli* strain, and *in vitro* transport assays with purified and into-liposomes-reconstituted protein, showed that the group II ECF transporter ECF-CbrT from the Gram-positive bacterium *Lactobacillus delbrueckii* is indeed a cobalamin transporter (Santos et al., 2018). The S-component CbrT binds cobalamin and cobinamide with nanomolar affinity *in vitro* (Table 1) in absence of the ECF module. However, the presence of the energizing module is essential to support transport of the substrates

(Santos et al., 2018), a property also observed for other S-components (Rempel et al., 2019; Setyawati et al., 2020).

A crystallographic structure of the full complex of ECF-CbrT was solved, in which the S-component CbrT assumes a toppled conformation (with the alpha-helices lying almost parallel to the membrane) in a substrate-free state (Santos et al., 2018). A similar conformation has also been found in structures of ECF transporters for other substrates (Swier et al., 2016; Wang et al., 2013; Xu et al., 2013; Zhang, Wang, & Shi, 2010). Based on this single conformation, it is difficult to derive a model for the transport mechanism. However, the mechanism is likely conserved among related ECF transporters for which more structural states have been resolved. Some of them, such as the folate-specific S-component FolT2, share the same ECF module as CbrT (Swier et al., 2016).

The structure of the folate-bound S-component FolT1 represents a conformation, in which the substrate binding site is occluded, and located close to the extracellular boundary of the lipid bilayer (Swier et al., 2016). Both the substrate-bound and -free S-component can dynamically associate with the ECF module, which powers the transport (Setyawati et al., 2020). Structural and functional data support the notion that the ECF module provides a favorable environment for the toppling of the S-component by bending and thinning of the membrane (Faustino et al., 2020). This toppling motion brings the bound substrate to the cytoplasmatic side of the cell membrane. Once toppled, the substrate is released form the substrate binding site, which is the conformation that has been observed in X-ray crystal structures of ECF transporters to date (Santos et al., 2018; Swier et al., 2016; Wang et al., 2013; Xu et al., 2013; Zhang et al., 2010). The binding and possibly also hydrolysis of ATP is likely to result in the reorientation of the substrate-free S-component from an inward-facing to the outward-facing state and release from the ECF module for a renewed transport cycle (Swier et al., 2016).

2.2.3 BtuM

Similar to ECF-CbrT, genomic analysis also helped to reveal the inner membrane protein BtuM as a predicted cobalamin transporter (Rodionov et al., 2003). Interestingly, BtuM is primarily distributed among bacterial species that do not encode the BtuCD-F homolog, nor any gene encoding the ECF module (Fig. 2) (Rempel et al., 2018). While all other previously characterized cobalamin transporters were ATP-powered ABC transporters, *in vivo* complementation assays in *E. coli* revealed that BtuM from *Thiobacillus denitrificans* alone is sufficient to mediate the uptake of cobalamin and cobinamide across the inner membrane (Rempel et al., 2018). Notably, the precursor cobinamide has been shown to bind BtuM within nanomolar affinity *in vitro* (Table 1) (Rempel et al., 2018). Although experimentally not verified, it is expected that cobalamin also binds with a comparable affinity to BtuM. Moreover, there are also no reports of successful *in vitro* transport assays using the purified protein in liposomes.

Surprisingly, a structure of BtuM determined by X-ray crystallography revealed that the overall fold resembles the S-component of ECF transporters (Fig. 2) (Rempel et al., 2018). BtuM has subsequently been classified based on the structural homology as a "solitary" S-component, a subgroup of ECF-type ABC transporters that appears to be able to transport substrates across the membrane without the need for an ECF module and ATP hydrolysis (Rempel et al., 2019). Notably, BtuM and CbrT share the same structural framework and overlap in their substrate specificity, but do not show significant similarity on the sequence level (Rempel et al., 2018).

Unprecedented among the cobalamin transporters is that the structure of BtuM revealed a bound cobalamin molecule in a base-off conformation (Fig. 2) (Rempel et al., 2018). This represents a binding mode that is generally observed for proteins with enzymatic activity (Drennan, Huang, Drummond, Matthews, & Lidwig, 1994; Gruber et al., 2011), which has not been observed for cobalamin transporters before. In BtuM, the DMB base that serves as α-axial ligand of the cobalt ion is replaced by a cysteine residue, which is highly unusual for cobalamin binding proteins (Rempel et al., 2018). The binding of cyanocobalamin to BtuM is accompanied by to the removal of the cyano-group from the β-axial position, which could be considered as enzymatic activity (Rempel et al., 2018). The cysteine ligation of the cobalt ion is essential to support the enzymatic removal of the β-ligand. The unexpected enzymatic activity of BtuM has also been confirmed biochemically for the substrate cobinamide (Rempel et al., 2018). The biological implications of this enzymatic modification are not understood yet.

Apart from its crystallographic structure, very little is known about the transport mechanism of BtuM. A speculative transport model described a facilitated transport that involves metabolic trapping of the substrate in the cytoplasm (Rempel et al., 2018). In this model BtuM would bind its substrate from the periplasm and topple over in the membrane. It is unclear how the substrate would be released into the cytoplasm, but since no NBDs are involved to power this process using ATP, it is possible that the release is very slow. To prevent rebinding to the transporter, the released cobalamin may be trapped by binding to enzymes, shifting the equilibrium toward the uptake. With this transport system it is expected that chemical modification of the substrate leads to lower transport rates compared to full ABC transporters, but survival of the cell would still be possible due to the minimal requirements of the vitamin for biological processes (Rempel et al., 2018).

2.2.4 Rv1819c

Random *in vivo* mutagenesis screening in the pathogenic bacterium *Mycobacterium tuberculosis* revealed that the disruption of the gene for Rv1819c, encoding an ABC transporter, leads to abolished cobalamin uptake (Gopinath et al., 2013). Rv1819c belongs to the Type IV ABC transporters that includes many members with exporter functions (*e.g.*, multidrug exporters), but recently was also found to contain members with importer functions (Thomas et al., 2020). Rv1819c is a homo-dimeric membrane complex, where each protomer consists of a TMD that is covalently linked to a NBD (Fig. 2) (Rempel et al., 2020). The main difference between Rv1819c and the other bacterial cobalamin-specific uptake systems BtuCD-F, ECF-CbrT and BtuM is that Rv1819c operates as a multi-solute transporter that facilitates the uptake of multiple unrelated hydrophilic substrates including cobalamin (Rempel et al., 2020; Slotboom, Ettema, Nijland, & Thangaratnarajah, 2020).

The poly-specific nature of the transport system arises from two important structural characteristics. First, in contrast to type I–III ABC importers (Thomas et al., 2020), Rv1819c does not make use of a substrate binding protein to confer substrate specificity on the transport system and to bind the substrate with high affinity (Rempel et al., 2020). Second, in the structure that was solved, an enormous occluded cavity with a volume over 7700 $Å^3$ was found that spans the entire thickness of the membrane (Rempel et al., 2020). The surface of this cavity consists of mainly polar and negatively charged residues that favors the interaction of hydrophilic molecules, which may explain its poly-specificity (Rempel et al., 2020). Noteworthy is that the cavity is the largest found among the characterized ABC transporter systems and is nearly seven times larger than the internal cavity found in the cobalamin transporter system BtuCD-F, which can only accommodate one cobalamin molecule (Korkhov et al., 2012). It was speculated that the cavity opens in an alternating fashion to the outside and the cytosol, such that cobalamin may diffuse across the membrane (Rempel et al., 2020). However, more work is needed to substantiate this model.

So far, it is evident that Rv1819c not only mediates cobalamin uptake, but also supports the uptake of the antimicrobial peptides bleomycin and Bac7, which are structurally unrelated substrates (Domenech, Kobayashi, LeVier, Walker, & Barry, 2009; Rempel et al., 2020). Consistent with the poly-specificity of the transporter, Rv1819c does not seem to have a high affinity binding site for cobalamin, which is an important aspect of the above-mentioned dedicated cobalamin transporters to support efficient uptake of the scarce vitamin (Rempel et al., 2020). Although Rv1819c is expected to transport cobalamin less efficiently than the dedicated uptake systems, physiologically it may allow uptake of cobalamin at a sufficient level for *M. tuberculosis*, which is an extremely slow-growing bacterium that only requires minute amounts of cobalamin.

2.2.5 BtuN

BtuN was identified by conserved positional linkage of the encoding gene to the *btuB* gene and has been implicated to be involved as a potential candidate in BtuCD-independent transport of cobalamin (Rodionov et al., 2003). While BtuN was predicted to contain four transmembrane helices (Rodionov et al., 2003), information on its structure and function are not known to date, and transport activity has not been confirmed.

2.3 Cobamides in microbial communities: Are exporters needed?

Bacteria are usually part of microbial communities, and their fitness is shaped by the flow of nutrients among each other within dynamic physical and chemical networks to fuel metabolism (Seth & Taga, 2014). Cobamides are considered as precious commodities placed at the core of microbial interactions, and comparative genomics analyses estimate that about 85% of bacteria utilize cobamides as coenzyme (Shelton et al., 2019). Notably, cobamides that differ in the lower ligand structure are not functionally interchangeable in bacteria and therefore the distribution of specific bacterial species within microbial communities and environments is modulated by the presence of diverse sets of cobamides that these bacteria utilize, and constrained by the presence of a subset of bacteria capable of cobamide *de novo* synthesis. These interactions are either shaped by competition of a micronutrient limited in availability, by utilizing metabolic partnerships between microbes to gain access to a micronutrient that none of the individual microbe is capable to metabolize on its own (syntropy), or by making intermediary or end products available to other microbes within a community (cross-feeding) (Seth & Taga, 2014). However, all types of interactions within a microbial community requires transport systems that enable the exchange of cobamides between its producers and consumers. While different systems for the import of exogenous cobamides have been identified and characterized in Gram-negative and -positive bacteria (see above), it is not known how synthesized cobamides are released into the surrounding environment. Membrane transport systems for the active export of cobalamin in bacteria have not been identified to date. The occasional lysis of bacteria mediated by the action of bacteriophages or bacteriocins and also by sporulation may therefore serve as mechanisms during which cobamides are made available to others within a microbial community (Danchin & Braham, 2017).

3. Transport of cobalamin in humans

The capacity to produce cobalamin has not made the transition to eukaryotes. It is an essential cofactor for only two enzymes in humans: the cobalamin-dependent methionine synthase that catalyzes the methylation of homocysteine in the cytoplasm, and the mitochondrial methylmalonyl-CoA mutase that catalyzes the conversion of methylmalonyl-CoA to succinyl-CoA (Banerjee & Matthews, 1990; Janata, Kogekar, & Fenton, 1997). Humans obtain cobalamin exclusively from their diet, where animal products such as meat, fish and dairy products are the primary natural source of the essential vitamin (Obeid et al., 2019). The absorption and subsequent distribution of cobalamin to the sites of utilization in the human body is more complex than the uptake systems described for bacteria. The uptake of cobalamin in humans involves multiple carrier proteins, receptors and membrane transport proteins (Fig. 3) (Nielsen, Rasmussen, Andersen, Nexø, & Moestrup, 2012; Rizzo & Laganà, 2020). The dysfunction of any of these proteins may result in cobalamin deficiency (Green et al., 2017). Here, we will focus on what is currently known about the membrane transport systems that mediate the translocation of cobalamin across membranes inside the cell. To provide some context of these transport systems, we briefly describe the absorption and distribution process for cobalamin in humans, but for details refer to extensive reviews (Green et al., 2017; Nielsen et al., 2012; Rizzo & Laganà, 2020).

In humans, three homologous carrier proteins, haptocorrin (HC), gastric intrinsic factor (IF) and transcobalamin (TC), play essential roles in escorting cobalamin through the extracellular fluids. HC and IF protect the vitamin from enzymatic degradation and non-catalyzed breakdown due to extreme pH values (Del Corral & Carmel, 1990). In the stomach, HC binds cobalamin after its release from food and aids in the passage toward the duodenum. HC is subsequently degraded by pancreatic proteases in the duodenum, where the difference in pH favors binding of cobalamin to

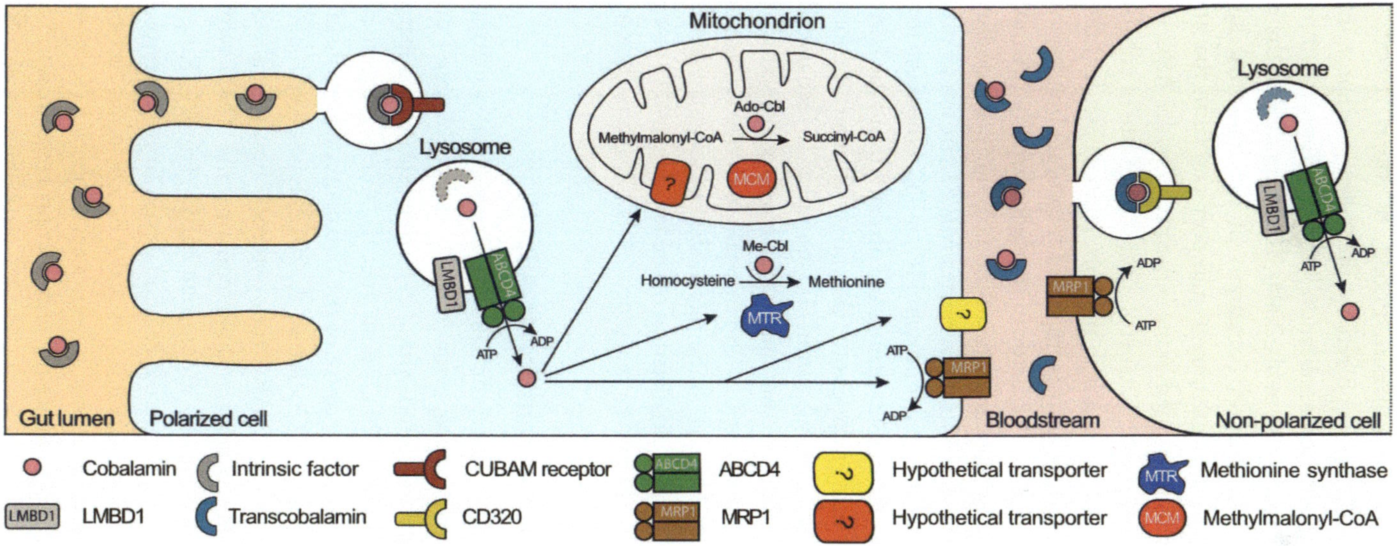

Fig. 3 Uptake of cobalamin by enterocytes from the distal ileum and posterior utilization and systemic transport. Cobalamin-bound IF is recognized by the CUBAM receptor of the enterocyte in the distal ileum. The endocytosed complexed IF is subsequently proteolyzed and free cobalamin imported to the cytoplasm by the transporter ABCD4. Cytoplasmic cobalamin can then be utilized in the enterocyte metabolism as a cofactor for MTR (in the methyl-cobalamin variant), transported to the mitochondrion *via* a yet-to-be-discovered mitochondrial transporter (where it is utilized by MCM as adenosyl-cobalamin) or exported toward the plasma by MRP1 or another unknown transporter. In the bloodstream, free cobalamin is predominantly captured by TC, which delivers the vitamin throughout the system. Cobalamin-bound TC is recognized by the CD320 receptor. Once imported, cytoplasmic cobalamin undergoes the same fate as in the enterocyte. Abbreviations: MTR, methionine synthase; MCM, methylmalonyl-CoA.

the carrier protein IF (Allen, Seetharam, Podell, & Alpers, 1978). The cobalamin-IF complex is then recognized by the cubam receptor in the distal ileum, which facilitates the internalization into the enterocytes through receptor-mediated endocytosis (Fyfe et al., 2004). IF is degraded in the lysosome of enterocytes, which liberates cobalamin in the lysosomal lumen (Nielsen et al., 2012). The actual membrane transport step from the lumen to the cytosol across the lysosomal membrane is mediated by the ABC transporter ABCD4 (Coelho et al., 2012; Kitai et al., 2021). The fate of cobalamin is threefold: either it serves as a cofactor in the cobalamin-dependent methionine synthase in the cytoplasm, or it is imported into mitochondria *via* an unknown transporter to serve as a cofactor in methylmalonyl-CoA mutase, or it is transported from the basolateral side of the cell into the bloodstream (Rizzo & Laganà, 2020) by the ABC transporter MRP1. In blood plasma, cobalamin is bound to either the carrier protein HC or TC, where the former is responsible for excretion of inactive versions of cobalamin, and the latter for cellular delivery to the cells throughout the body (Burger, Schneider, Mehlman, & Allen, 1975). The cobalamin-TC complex is recognized by the ubiquitous receptor CD320 and subsequently internalized by endocytosis, after which TC is degraded to eventually end up again with free cobalamin in the lysosome (Quadros, Nakayama, & Sequeira, 2009; Seligman & Allen, 1978). Only in the kidney, the internalization of the cobalamin-TC complex is mediated by megalin, an alternative receptor to CD320 (Birn, 2006).

3.1 Lysosomal cobalamin transporter ABCD4

Genetic analysis of cobalamin deficient patients revealed that mutations in the genes encoding the ABC transporter ABCD4 and the integral membrane protein LMBD1 are associated with a defective release of cobalamin from the lysosome (Coelho et al., 2012; Rutsch et al., 2009). The mutations in the genes are also known as the cause of the cobalamin metabolism disorders cblJ and cblF, respectively (Coelho et al., 2012; Rutsch et al., 2009).

ABCD4 belongs to the type IV ABC transporters (Thomas et al., 2020) and is a homo-dimer, where each protomer contains one TMD with six transmembrane helices followed by a cytosolic NBD (Fig. 2) (Xu et al., 2019). The protein imports the internalized cobalamin from the lysosomal lumen into the cytoplasm (Fig. 3). While purified ABCD4 alone was shown to mediate cobalamin transport across in liposomes (Table 1) (Kitai et al., 2021), LMBD1 is essential to facilitate the transport process across the lysosomal membrane *in vivo* (Coelho et al., 2012; Fettelschoss et al., 2017), where it plays an important role in lysosomal targeting of ABCD4. Defects in LMBD1 or ABCD4 that preclude interaction between the two proteins or mislocalization of LMBD1 affect transport of cobalamin across the lysosomal membrane (Fettelschoss et al., 2017; Kawaguchi, Okamoto, Morita, & Imanaka, 2016). Both proteins have been shown to interact *in vivo* and *in vitro* with sub-nanomolar dissociation constants (Deme et al., 2014; Fettelschoss et al., 2017; Kawaguchi et al., 2016; Kawaguchi & Morita, 2016). The structure of ABCD4 was resolved by cryo-EM (Fig. 2) (Xu et al., 2019), revealing a fold similar to the distant homolog Rv1819c of *M. tuberculosis* (Fig. 2) (Rempel et al., 2020; Thomas et al., 2020). The transporter was captured in an open conformation with the binding site exposed to the lysosomal lumen (Xu et al., 2019). Although more structures would be needed to elucidate the transport mechanism, the characterization of similar type IV ABC transporters suggest an alternating access transport mechanism transport (Johnson & Chen, 2017; Kodan et al., 2019). The structure of ABCD4 has not revealed a specific binding site for cobalamin (Xu et al., 2019), which opens the possibility that there are similarities in the transport mechanism with Rv1819c, but more work is needed to substantiate possible mechanistic similarity.

3.2 MRP1

Upon the transport of cobalamin from the lysosomal lumen into the cytoplasm, cobalamin is utilized within the cell or released from the basolateral side of the enterocyte into the bloodstream (Fig. 3) (Nielsen et al., 2012). No cobalamin-specific human transporter has been reported to perform the latter function. Instead, the multidrug resistant protein MRP1, which functions as a non-specific ABC exporter, has been shown to mediate the export of cobalamin into the medium of human and murine cultured cells (Beedholm-Ebsen et al., 2010). MRP or MRP-like proteins are present in all eukaryotic organisms, but are absent from bacteria and archaea (Cole, 2014).

MRP1 is a versatile transporter, which belongs to the type IV ABC transporters (Thomas et al., 2020). However, unlike the structurally related Rv1819c and ABCD4, which are importers, MRP1 functions as an exporter (Beedholm-Ebsen et al., 2010). MRP1 is encoded by a single peptide consisting of two NBDs, two TMDs forming the transport path, and an additional, membrane embedded N-terminal domain of unknown function consisting of five transmembrane helices (Fig. 2) (Johnson & Chen, 2017).

The substrates recognized by MRP1 are diverse. However, they share a common amphipathic organo-acidic nature, typically containing large hydrophobic groups (Johnson & Chen, 2017). In humans, MRP1 has a dual "*good-evil*" role regarding its function as a multidrug exporter. While its widespread presence in the tissue-blood interface provides protection against toxic molecules (Li et al., 2020; Zhao et al., 2019), its expression also impedes chemotherapy (Torres et al., 2018) and the treatment of a variety of brain disorders (Wijnholds et al., 2000; Zhao et al., 2019).

Structural and functional analysis of MRP1 revealed a substrate-driven transport mechanism, where the transporter rests in an inward-open state, with the NBDs separated from each other, which impedes ATP hydrolysis and provides access to the substrate-binding cavity between the two TMDs. Substrate interaction brings the TMDs together, which also leads to closer proximity of the NBDs, enabling ATP hydrolysis to export the bound substrate toward the extracellular space (Johnson & Chen, 2017).

In vivo, knockdown mice for the MRP1 homolog showed decreased cobalamin levels in blood and tissue (Beedholm-Ebsen et al., 2010). However, these mice did not present the characteristic metabolic markers for cobalamin deficiency (Beedholm-Ebsen et al., 2010). Consistent with these findings, sequence analysis of cobalamin deficient patients that could not be explained by malabsorption of cobalamin did not present severe mutations in the gene encoding for MRP1 either (Shah, Beech, Sturm, & Tanner, 2011). As suggested by Shah et al., these findings point toward at least another exporter implicated in cobalamin efflux from mammalian cells (Fig. 3) (Shah et al., 2011), preventing MRP1-defective individuals from developing cobalamin-related diseases and inducing cobalamin malabsorption-like phenotype even when MRP1 is fully operational (Beedholm-Ebsen et al., 2010). Further investigations are needed to clarify the exact role of this potential transporter.

3.3 Mitochondrial cobalamin transporter

It is apparent that cobalamin has to enter the mitochondrion as methylmalonyl-CoA mutase activity is dependent on cobalamin as cofactor (Takahashi-Iñiguez, García-Hernandez, Arreguín-Espinosa, & Flores, 2012). However, it is not clear how cobalamin is imported into mitochondria. While the passage of cobalamin across the mitochondrial outer membrane is likely to be mediated by the voltage-dependent anion channel (VDAC), which enables the exchange of small molecular-mass solutes up to 5 kDa between the cytoplasm and the mitochondrial intermembrane space (Benz, 1994), it is unclear how the molecule passes the inner membrane. As the mitochondrial inner membrane is impermeable to ions and molecules, the passage of cobalamin must be mediated by a transporter. Nonetheless, the identity of the transporter that facilitates the import of cobalamin is unknown. A candidate for cobalamin import into mitochondria has been described in knockout studies of multiple ABC transporters in the roundworm *Caenorhabditis elegans* (McDonald et al., 2017). The *wht-6* gene product encoding an ABC transporter has been suggested, but the disruption of the *wht-6* gene did not result in a complete loss of transport and direct evidence of mitochondrial localization and cobalamin transport is currently lacking (McDonald et al., 2017). As many of the transport steps of structurally diverse sets of molecules across the mitochondrial inner membrane are facilitated by members of the SLC25 family of mitochondrial carriers (Kunji, King, Ruprecht, & Thangaratnarajah, 2020), the import of cobalamin could potentially be mediated by one of the unidentified transporters within this family.

4. Concluding remarks

Cobalamin plays an essential role in metabolism as a cofactor in organisms from all domains of life. Especially in the last few years, great progress has been made in the structural and biochemical characterization of cobalamin transporters. These have remarkably advanced our understanding of cobalamin transport and highlight that the proteins share common features, yet have evolved different solutions that facilitate the uptake and transport of cobalamin and its derivatives across lipid membranes. More work is however required to fully understand and describe the transport mechanism of many cobalamin transporters. While almost all cobalamin transporters characterized to date are powered by ATP, other mechanisms of transport may be uncovered among the yet-to-be-identified cobalamin transporters in the future.

References

Allen, R. H., Seetharam, B., Podell, E., & Alpers, D. H. (1978). Effect of proteolytic enzymes on the binding of cobalamin to R protein and intrinsic factor. In vitro evidence that a failure to partially degrade R protein is responsible for cobalamin malabsorption in pancreatic insufficiency. *The Journal of Clinical Investigation*, *61*(1), 47–54.

Banerjee, R. V., & Matthews, R. G. (1990). Cobalamin-dependent methionine synthase. *The FASEB Journal*, *4*(5), 1450–1459.

Beedholm-Ebsen, R., van de Wetering, K., Hardlei, T., Nexø, E., Borst, P., & Moestrup, S. K. (2010). Identification of multidrug resistance protein 1 (MRP1/ABCC1) as a molecular gate for cellular export of cobalamin. *Blood*, *115*(8), 1632–1639.

Benz, R. (1994). Permeation of hydrophilic solutes through mitochondrial outer membranes: Review on mitochondrial porins. *Biochimica et Biophysica Acta*, *1197*(2), 167–196.

Birn, H. (2006). The kidney in vitamin B12 and folate homeostasis: Characterization of receptors for tubular uptake of vitamins and carrier proteins. *American Journal of Physiology. Renal Physiology*, *291*(1), F22–F36.

Borths, E. L., Poolman, B., Hvorup, R. N., Locher, K. P., & Rees, D. C. (2005). In vitro functional characterization of BtuCD-F, the *Escherichia coli* ABC transporter for vitamin B12 uptake. *Biochemistry*, *44*(49), 16301–16309.

Bradbeer, C., Reynolds, P. R., Bauler, G. M., & Fernandez, M. T. (1986). A requirement for calcium in the transport of cobalamin across the outer membrane of *Escherichia coli*. *The Journal of Biological Chemistry*, *261*(6), 2520–2523.

Burger, R. L., Schneider, R. J., Mehlman, C. S., & Allen, R. H. (1975). Human plasma R-type vitamin B12-binding proteins. II. The role of transcobalamin I, transcobalamin III, and the normal granulocyte vitamin B12-binding protein in the plasma transport of vitamin B12. *The Journal of Biological Chemistry*, *250*(19), 7707–7713.

Cadieux, N., Barekzi, N., & Bradbeer, C. (2007). Observations on the calcium dependence and reversibility of cobalamin transport across the outer membrane of *Escherichia coli*. *The Journal of Biological Chemistry*, *282*(48), 34921–34928.

Cadieux, N., Bradbeer, C., Reeger-Schneider, E., Köster, W., Mohanty, A. K., Wiener, M. C., et al. (2002). Identification of the periplasmic cobalamin-binding protein BtuF of *Escherichia coli*. *Journal of Bacteriology*, *184*(3), 706–717.

Chimento, D. P., Mohanty, A. K., Kadner, R. J., & Wiener, M. C. (2003). Substrate-induced transmembrane signaling in the cobalamin transporter BtuB. *Nature Structural Biology*, *10*(5), 394–401.

Coelho, D., Kim, J. C., Miousse, I. R., Fung, S., du Moulin, M., Buers, I., et al. (2012). Mutations in ABCD4 cause a new inborn error of vitamin B12 metabolism. *Nature Genetics*, *44*(10), 1152–1155.

Cole, S. P. C. (2014). Multidrug resistance protein 1 (MRP1, ABCC1), a "multitasking" ATP-binding cassette (ABC) transporter. *The Journal of Biological Chemistry*, *289*(45), 30880–30888.

Danchin, A., & Braham, S. (2017). Coenzyme B12 synthesis as a baseline to study metabolite contribution of animal microbiota. *Microbial Biotechnology*, *10*(4), 688–701.

de Veaux, L. C., Clevenson, D. S., Bradbeer, C., & Kadner, R. J. (1986). Identification of the btuCED polypeptides and evidence for their role in vitamin B12 transport in Escherichia coli. *Journal of Bacteriology*, *167*(3), 920–927.

Degnan, P. H., Barry, N. A., Mok, K. C., Taga, M. E., & Goodman, A. L. (2014). Human gut microbes use multiple transporters to distinguish vitamin B_{12} analogs and compete in the gut. *Cell Host & Microbe*, *15*(1), 47–57.

Del Corral, A., & Carmel, R. (1990). Transfer of cobalamin from the cobalamin-binding protein of egg yolk to R binder of human saliva and gastric juice. *Gastroenterology*, *98*(6), 1460–1466.

Deme, J. C., Hancock, M. A., Xia, X., Shintre, C. A., Plesa, M., Kim, J. C., et al. (2014). Purification and interaction analyses of two human lysosomal vitamin B12 transporters: LMBD1 and ABCD4. *Molecular Membrane Biology*, *31*(7–8), 250–261.

Di Girolamo, P. M., Kadner, R. J., & Bradbeer, C. (1971). Isolation of vitamin B 12 transport mutants of Escherichia coli. *Journal of Bacteriology*, *106*(3), 751–757.

Domenech, P., Kobayashi, H., LeVier, K., Walker, G. C., & Barry, C. E., 3rd. (2009). BacA, an ABC transporter involved in maintenance of chronic murine infections with *Mycobacterium tuberculosis*. *Journal of Bacteriology*, *191*(2), 477–485.

Drennan, C. L., Huang, S., Drummond, J. T., Matthews, R. G., & Lidwig, M. L. (1994). How a protein binds B12: A 3.0 A X-ray structure of B12-binding domains of methionine synthase. *Science (New York, N.Y.)*, *266*(5191), 1669–1674.

Fang, H., Kang, J., & Zhang, D. (2017). Microbial production of vitamin B12: A review and future perspectives. *Microbial Cell Factories*, *16*(1), 15.

Faraldo-Gómez, J. D., & Sansom, M. S. P. (2003). Acquisition of siderophores in gram-negative bacteria. *Nature Reviews. Molecular Cell Biology*, *4*(2), 105–116.

Faustino, I., Abdizadeh, H., Souza, P. C. T., Jeucken, A., Stanek, W. K., Guskov, A., et al. (2020). Membrane mediated toppling mechanism of the folate energy coupling factor transporter. *Nature Communications*, *11*(1), 1763.

Fedosov, S. N., Berglund, L., Fedosova, N. U., Nexo, E., & Petersen, T. E. (2002). Comparative analysis of cobalamin binding kinetics and ligand protection for intrinsic factor, transcobalamin, and haptocorrin. *The Journal of Biological Chemistry*, *277*(12), 9989–9996.

Fettelschoss, V., Burda, P., Sagné, C., Coelho, D., De Laet, C., Lutz, S., et al. (2017). Clinical or ATPase domain mutations in ABCD4 disrupt the interaction between the vitamin B12-trafficking proteins ABCD4 and LMBD1. *The Journal of Biological Chemistry*, *292*(28), 11980–11991.

Fyfe, J. C., Madsen, M., Højrup, P., Christensen, E. I., Tanner, S. M., de la Chapelle, A., et al. (2004). The functional cobalamin (vitamin B12)-intrinsic factor receptor is a novel complex of cubilin and amnionless. *Blood*, *103*(5), 1573–1579.

Giedyk, M., Goliszewska, K., & Gryko, D. (2015). Vitamin B12 catalysed reactions. *Chemical Society Reviews*, *44*(11), 3391–3404.

Gopinath, K., Venclovas, C., Ioerger, T. R., Sacchettini, J. C., McKinney, J. D., Mizrahi, V., et al. (2013). A vitamin B_{12} transporter in *Mycobacterium tuberculosis*. *Open Biology*, *3*(2), 120175.

Goudsmits, J. M. H., Slotboom, D. J., & van Oijen, A. M. (2017). Single-molecule visualization of conformational changes and substrate transport in the vitamin B(12) ABC importer BtuCD-F. *Nature Communications*, *8*(1), 1652.

Green, R., Allen, L. H., Bjørke-Monsen, A.-L., Brito, A., Guéant, J.-L., Miller, J. W., et al. (2017). Vitamin B12 deficiency. *Nature Reviews. Disease Primers*, *3*, 17040.

Gruber, K., Puffer, B., & Kräutler, B. (2011). Vitamin B12-derivatives-enzyme cofactors and ligands of proteins and nucleic acids. *Chemical Society Reviews*, *40*(8), 4346–4363.

Gude, S., & Taga, M. E. (2020). Multi-faceted approaches to discovering and predicting microbial nutritional interactions. *Current Opinion in Biotechnology*, *62*, 58–64.

Heller, K., Mann, B. J., & Kadner, R. J. (1985). Cloning and expression of the gene for the vitamin B12 receptor protein in the outer membrane of *Escherichia coli*. *Journal of Bacteriology*, *161*(3), 896–903.

Hvorup, R. N., Goetz, B. A., Niederer, M., Hollenstein, K., Perozo, E., & Locher, K. P. (2007). Asymmetry in the structure of the ABC transporter-binding protein complex BtuCD-BtuF. *Science (New York, N.Y.)*, *317*(5843), 1387–1390.

Janata, J., Kogekar, N., & Fenton, W. A. (1997). Expression and kinetic characterization of methylmalonyl-CoA mutase from patients with the mut-phenotype: Evidence for naturally occurring interallelic complementation. *Human Molecular Genetics*, *6*(9), 1457–1464.

Johnson, Z. L., & Chen, J. (2017). Structural basis of substrate recognition by the multidrug resistance protein MRP1. *Cell*, *168*(6), 1075–1085.e9.

Karpowich, N. K., Huang, H. H., Smith, P. C., & Hunt, J. F. (2003). Crystal structures of the BtuF periplasmic-binding protein for vitamin B12 suggest a functionally important reduction in protein mobility upon ligand binding. *The Journal of Biological Chemistry*, *278*(10), 8429–8434.

Kawaguchi, K., & Morita, M. (2016). ABC transporter subfamily D: Distinct differences in behavior between ABCD1-3 and ABCD4 in subcellular localization, function, and human disease. *BioMed Research International*, *2016*, 6786245.

Kawaguchi, K., Okamoto, T., Morita, M., & Imanaka, T. (2016). Translocation of the ABC transporter ABCD4 from the endoplasmic reticulum to lysosomes requires the escort protein LMBD1. *Scientific Reports*, *6*, 30183.

Kitai, K., Kawaguchi, K., Tomohiro, T., Morita, M., So, T., & Imanaka, T. (2021). The lysosomal protein ABCD4 can transport vitamin B(12) across liposomal membranes in vitro. *The Journal of Biological Chemistry*, *296*, 100654.

Kodan, A., Yamaguchi, T., Nakatsu, T., Matsuoka, K., Kimura, Y., Ueda, K., et al. (2019). Inward- and outward-facing X-ray crystal structures of homodimeric P-glycoprotein CmABCB1. *Nature Communications*, *10*(1), 88.

Korkhov, V. M., Mireku, S. A., & Locher, K. P. (2012). Structure of AMP-PNP-bound vitamin B12 transporter BtuCD-F. *Nature*, *490*(7420), 367–372.

Korkhov, V. M., Mireku, S. A., Veprintsev, D. B., & Locher, K. P. (2014). Structure of AMP-PNP-bound BtuCD and mechanism of ATP-powered vitamin B12 transport by BtuCD-F. *Nature Structural & Molecular Biology*, *21*(12), 1097–1099.

Kunji, E. R. S., King, M. S., Ruprecht, J. J., & Thangaratnarajah, C. (2020). The SLC25 carrier family: Important transport proteins in mitochondrial physiology and pathology. *Physiology (Bethesda, Md.)*, *35*(5), 302–327.

Li, N.-N., Guo, Y., Jiang, C.-J., Zhou, Y.-Y., Li, C.-H., Li, Z.-G., et al. (2020). Allyl isothiocyanate upregulates MRP1 expression through Notch1 signaling in human bronchial epithelial cells. *Canadian Journal of Physiology and Pharmacology*, *98*(5), 324–331.

Locher, K. P., Lee, A. T., & Rees, D. C. (2002). The *E. coli* BtuCD structure: A framework for ABC transporter architecture and mechanism. *Science (New York, N.Y.)*, *296*(5570), 1091–1098.

Mao, Q., Leslie, E. M., Deeley, R. G., & Cole, S. P. (1999). ATPase activity of purified and reconstituted multidrug resistance protein MRP1 from drug-selected H69AR cells. *Biochimica et Biophysica Acta*, *1461*(1), 69–82.

McDonald, M. K., Fritz, J.-A., Jia, D., Scheuchner, D., Snyder, F. F., Stanislaus, A., et al. (2017). Identification of ABC transporters acting in vitamin B12 metabolism in Caenorhabditis elegans. *Molecular Genetics and Metabolism*, *122*(4), 160–171.

Mills, A., Le, H.-T., & Duong, F. (2016). TonB-dependent ligand trapping in the BtuB transporter. *Biochimica et Biophysica Acta*, *1858*(12), 3105–3112.

Mireku, S. A., Ruetz, M., Zhou, T., Korkhov, V. M., Kräutler, B., & Locher, K. P. (2017). Conformational change of a tryptophan residue in BtuF facilitates binding and transport of cobinamide by the vitamin B12 transporter BtuCD-F. *Scientific Reports*, *7*, 41575.

Nielsen, M. J., Rasmussen, M. R., Andersen, C. B. F., Nexø, E., & Moestrup, S. K. (2012). Vitamin B12 transport from food to the body's cells- -a sophisticated, multistep pathway. *Nature Reviews. Gastroenterology & Hepatology*, *9*(6), 345–354.

Noinaj, N., Guillier, M., Barnard, T. J., & Buchanan, S. K. (2010). TonB-dependent transporters: Regulation, structure, and function. *Annual Review of Microbiology*, *64*, 43–60.

Obeid, R., Fedosov, S. N., & Nexo, E. (2015). Cobalamin coenzyme forms are not likely to be superior to cyano- and hydroxyl-cobalamin in prevention or treatment of cobalamin deficiency. *Molecular Nutrition & Food Research*, *59*(7), 1364–1372.

Obeid, R., Heil, S. G., Verhoeven, M. M. A., van den Heuvel, E. G. H. M., de Groot, L. C. P. G. M., & Eussen, S. J. P. M. (2019). Vitamin B12 intake from animal foods, biomarkers, and health aspects. *Frontiers in Nutrition*, *6*, 93.

Orłowska, M., Steczkiewicz, K., & Muszewska, A. (2021). Utilization of cobalamin is ubiquitous in early-branching fungal phyla. *Genome Biology and Evolution*, *13*(4).

Quadros, E. V., Nakayama, Y., & Sequeira, J. M. (2009). The protein and the gene encoding the receptor for the cellular uptake of transcobalamin-bound cobalamin. *Blood*, *113*(1), 186–192.

Rempel, S., Colucci, E., de Gier, J. W., Guskov, A., & Slotboom, D. J. (2018). Cysteine-mediated decyanation of vitamin B12 by the predicted membrane transporter BtuM. *Nature Communications*, *9*(1), 3038.

Rempel, S., Gati, C., Nijland, M., Thangaratnarajah, C., Karyolaimos, A., de Gier, J. W., et al. (2020). A mycobacterial ABC transporter mediates the uptake of hydrophilic compounds. *Nature*, *580*(7803), 409–412.

Rempel, S., Stanek, W. K., & Slotboom, D. J. (2019). ECF-type ATP-binding cassette transporters. *Annual Review of Biochemistry*, *88*, 551–576.

Rizzo, G., & Laganà, A. S. (2020). Chapter 6—A review of vitamin B12. In V. Patel (Ed.), *Molecular nutrition* (pp. 105–129). Academic Press.

Rizzo, G., Laganà, A. S., Rapisarda, A. M. C., La Ferrera, G. M. G., Buscema, M., Rossetti, P., et al. (2016). Vitamin B12 among vegetarians: Status, assessment and supplementation. *Nutrients*, *8*(12).

Rodionov, D. A., Hebbeln, P., Eudes, A., ter Beek, J., Rodionova, I. A., Erkens, G. B., et al. (2009). A novel class of modular transporters for vitamins in prokaryotes. *Journal of Bacteriology*, *191*(1), 42–51.

Rodionov, D. A., Vitreschak, A. G., Mironov, A. A., & Gelfand, M. S. (2003). Comparative genomics of the vitamin B12 metabolism and regulation in prokaryotes. *The Journal of Biological Chemistry*, *278*(42), 41148–41159.

Roth, J. R., Lawrence, J. G., & Bobik, T. A. (1996). Cobalamin (coenzyme B12): Synthesis and biological significance. *Annual Review of Microbiology*, *50*, 137–181.

Rutsch, F., Gailus, S., Miousse, I. R., Suormala, T., Sagné, C., Toliat, M. R., et al. (2009). Identification of a putative lysosomal cobalamin exporter altered in the cblF defect of vitamin B12 metabolism. *Nature Genetics*, *41*(2), 234–239.

Santos, J. A., Rempel, S., Mous, S. T., Pereira, C. T., Ter Beek, J., de Gier, J.-W., et al. (2018). Functional and structural characterization of an ECF-type ABC transporter for vitamin B12. *ELife*, 7.

Seligman, P. A., & Allen, R. H. (1978). Characterization of the receptor for transcobalamin II isolated from human placenta. *The Journal of Biological Chemistry*, *253*(6), 1766–1772.

Seth, E. C., & Taga, M. E. (2014). Nutrient cross-feeding in the microbial world. *Frontiers in Microbiology*, *5*, 350.

Setyawati, I., Stanek, W. K., Majsnerowska, M., Swier, L. J. Y. M., Pardon, E., Steyaert, J., et al. (2020). In vitro reconstitution of dynamically interacting integral membrane subunits of energy-coupling factor transporters. *ELife*, *9*.

Shah, N. P., Beech, C. M., Sturm, A. C., & Tanner, S. M. (2011). Investigation of the ABC transporter MRP1 in selected patients with presumed defects in vitamin B12 absorption. *Blood*, *117*(16), 4397–4398.

Shelton, A. N., Seth, E. C., Mok, K. C., Han, A. W., Jackson, S. N., Haft, D. R., et al. (2019). Uneven distribution of cobamide biosynthesis and dependence in bacteria predicted by comparative genomics. *The ISME Journal*, *13*(3), 789–804.

Shultis, D. D., Purdy, M. D., Banchs, C. N., & Wiener, M. C. (2006). Outer membrane active transport: Structure of the BtuB:TonB complex. *Science*, *312*(5778), 1396–1399.

Slotboom, D. J. (2014). Structural and mechanistic insights into prokaryotic energy-coupling factor transporters. *Nature Reviews. Microbiology*, *12*(2), 79–87.

Slotboom, D. J., Ettema, T. W., Nijland, M., & Thangaratnarajah, C. (2020). Bacterial multi-solute transporters. *FEBS Letters*, *594*(23), 3898–3907.

Sokolovskaya, O. M., Shelton, A. N., & Taga, M. E. (2020). Sharing vitamins: Cobamides unveil microbial interactions. *Science (New York, N.Y.)*, *369*(6499).

Swier, L. J. Y. M., Guskov, A., & Slotboom, D. J. (2016). Structural insight in the toppling mechanism of an energy-coupling factor transporter. *Nature Communications*, 7, 11072.

Takahashi-Iñiguez, T., García-Hernandez, E., Arreguín-Espinosa, R., & Flores, M. E. (2012). Role of vitamin B12 on methylmalonyl-CoA mutase activity. *Journal of Zhejiang University. Science. B*, *13*(6), 423–437.

Thomas, C., Aller, S. G., Beis, K., Carpenter, E. P., Chang, G., Chen, L., et al. (2020). Structural and functional diversity calls for a new classification of ABC transporters. *FEBS Letters*, *594*(23), 3767–3775.

Thomas, C., & Tampé, R. (2020). Structural and mechanistic principles of ABC transporters. *Annual Review of Biochemistry*, *89*, 605–636.

Torres, Á., Arriagada, V., Erices, J. I., de los Ángeles Toro, M., Rocha, J. D., Niechi, I., et al. (2018). FK506 attenuates the MRP1-mediated chemoresistant phenotype in glioblastoma stem-like cells. *International Journal of Molecular Sciences*, *19*(9).

Vergalli, J., Bodrenko, I. V., Masi, M., Moynié, L., Acosta-Gutiérrez, S., Naismith, J. H., et al. (2020). Porins and small-molecule translocation across the outer membrane of Gram-negative bacteria. *Nature Reviews. Microbiology*, *18*(3), 164–176.

Wang, T., Fu, G., Pan, X., Wu, J., Gong, X., Wang, J., et al. (2013). Structure of a bacterial energy-coupling factor transporter. *Nature*, *497*(7448), 272–276.

Warren, M. J., Raux, E., Schubert, H. L., & Escalante-Semerena, J. C. (2002). The biosynthesis of adenosylcobalamin (vitamin B12). *Natural Product Reports*, *19*(4), 390–412.

Wexler, A. G., Schofield, W. B., Degnan, P. H., Folta-Stogniew, E., Barry, N. A., & Goodman, A. L. (2018). Human gut Bacteroides capture vitamin B(12) via cell surface-exposed lipoproteins. *ELife*, *7*.

Wijnholds, J., DeLange, E. C., Scheffer, G. L., van den Berg, D. J., Mol, C. A., van der Valk, M., et al. (2000). Multidrug resistance protein 1 protects the choroid plexus epithelium and contributes to the blood-cerebrospinal fluid barrier. *The Journal of Clinical Investigation*, *105*(3), 279–285.

Woodson, J. D., Reynolds, A. A., & Escalante-Semerena, J. C. (2005). ABC transporter for corrinoids in Halobacterium sp. strain NRC-1. *Journal of Bacteriology*, *187*(17), 5901–5909.

Xu, D., Feng, Z., Hou, W.-T., Jiang, Y.-L., Wang, L., Sun, L., et al. (2019). Cryo-EM structure of human lysosomal cobalamin exporter ABCD4. *Cell Research*, *29*(12), 1039–1041.

Xu, K., Zhang, M., Zhao, Q., Yu, F., Guo, H., Wang, C., et al. (2013). Crystal structure of a folate energy-coupling factor transporter from *Lactobacillus brevis*. *Nature*, *497*(7448), 268–271.

Zhang, Y., Rodionov, D. A., Gelfand, M. S., & Gladyshev, V. N. (2009). Comparative genomic analyses of nickel, cobalt and vitamin B12 utilization. *BMC Genomics*, *10*, 78.

Zhang, P., Wang, J., & Shi, Y. (2010). Structure and mechanism of the S component of a bacterial ECF transporter. *Nature*, *468*(7324), 717–720.

Zhao, X., Li, Y., Du, K., Wu, Y., Liu, L., Cui, S., et al. (2019). Involvement of human and canine MRP1 and MRP4 in benzylpenicillin transport. *PLoS One*, *14*(11), e0225702.

CHAPTER SIX

Vitamin B_{12} photoreceptors

S. Padmanabhan[a,*], Ricardo Pérez-Castaño[b], Lucía Osete-Alcaraz[b], María Carmen Polanco[b], and Montserrat Elías-Arnanz[b,*]
[a]Instituto de Química Física "Rocasolano", Consejo Superior de Investigaciones Científicas, Madrid, Spain
[b]Departamento de Genética y Microbiología, Área de Genética (Unidad Asociada al IQFR-CSIC), Facultad de Biología, Universidad de Murcia, Murcia, Spain
*Corresponding authors: e-mail address: padhu@iqfr.csic.es; melias@um.es

Contents

Abstract

Photoreceptor proteins enable living organisms to sense light and transduce this signal into biochemical outputs to elicit appropriate cellular responses. Their light sensing is typically mediated by covalently or noncovalently bound molecules called chromophores, which absorb light of specific wavelengths and modulate protein structure and biological activity. Known photoreceptors have been classified into about ten families based on the chromophore and its associated photosensory domain in the protein. One widespread photoreceptor family uses coenzyme B_{12} or 5′-deoxyadenosylcobalamin, a biological form of vitamin B_{12}, to sense ultraviolet, blue, or green light, and its discovery revealed

Vitamins and Hormones, Volume 119
ISSN 0083-6729
https://doi.org/10.1016/bs.vh.2022.01.007

both a new type of photoreceptor and a novel functional facet of this vitamin, best known as an enzyme cofactor. Large strides have been made in our understanding of how these B_{12}-based photoreceptors function, high-resolution structural descriptions of their functional states are available, as are details of their unusual photochemistry. Additionally, they have inspired notable applications in optogenetics/optobiochemistry and synthetic biology. Here, we provide an overview of what is currently known about these B_{12}-based photoreceptors, their discovery, distribution, molecular mechanism of action, and the structural and photochemical basis of how they orchestrate signal transduction and gene regulation, and how they have been used to engineer optogenetic control of protein activities in living cells.

1. Introduction to photoreceptors and their chromophores

Living organisms must sense and respond to light, a crucial environmental factor and energy source that determines diverse biological processes. Thus, plants, algae and photosynthetic bacteria harness light for conversion to chemical energy (Croce & van Amerongen, 2014, 2020; Nelson & Junge, 2015); human vision relies on sensing visible light (wavelength range 380–760 nm) (Palczewski & Kiser, 2020); and circadian rhythms depend on light (Johnson, Zhao, Xu, & Mori, 2017; Koronowski & Sassone-Corsi, 2021). Cells have also developed protective responses to combat photooxidative damage of cellular components, like DNA, proteins and membranes, which can result from highly reactive species such as singlet oxygen ($^1O_2$) generated upon exposure to light (Di Mascio et al., 2019; Glaeser, Nuss, Berghoff, & Klug, 2011; Ziegelhoffer & Donohue, 2009). Central to these processes are specialized proteins called photoreceptors that sense and transduce the light signal to yield cellular cues and elicit the appropriate response. To sense visible light, photoreceptors use chromophores (Fig. 1), which may be part of the protein itself or, more often, a non-protein molecule supplied in *trans* and bound to the protein noncovalently or covalently usually via cysteine or lysine sidechains (Seong & Lin, 2021; Shcherbakova, Shemetov, Kaberniuk, & Verkhusha, 2015). Chromophores are often unsaturated molecules with conjugated π systems and delocalized electrons that absorb light at specific wavelengths, undergo electron transfer, *cis–trans* isomerization or bond formation-disruption events, and convert "dark" to "light" states with altered conformation and/or chemistry. The light-driven changes are relayed to a photosensory protein module in the photoreceptor and then to an effector module in the same or different protein(s) to modulate a specific function such as enzyme activity or interactions with DNA or proteins.

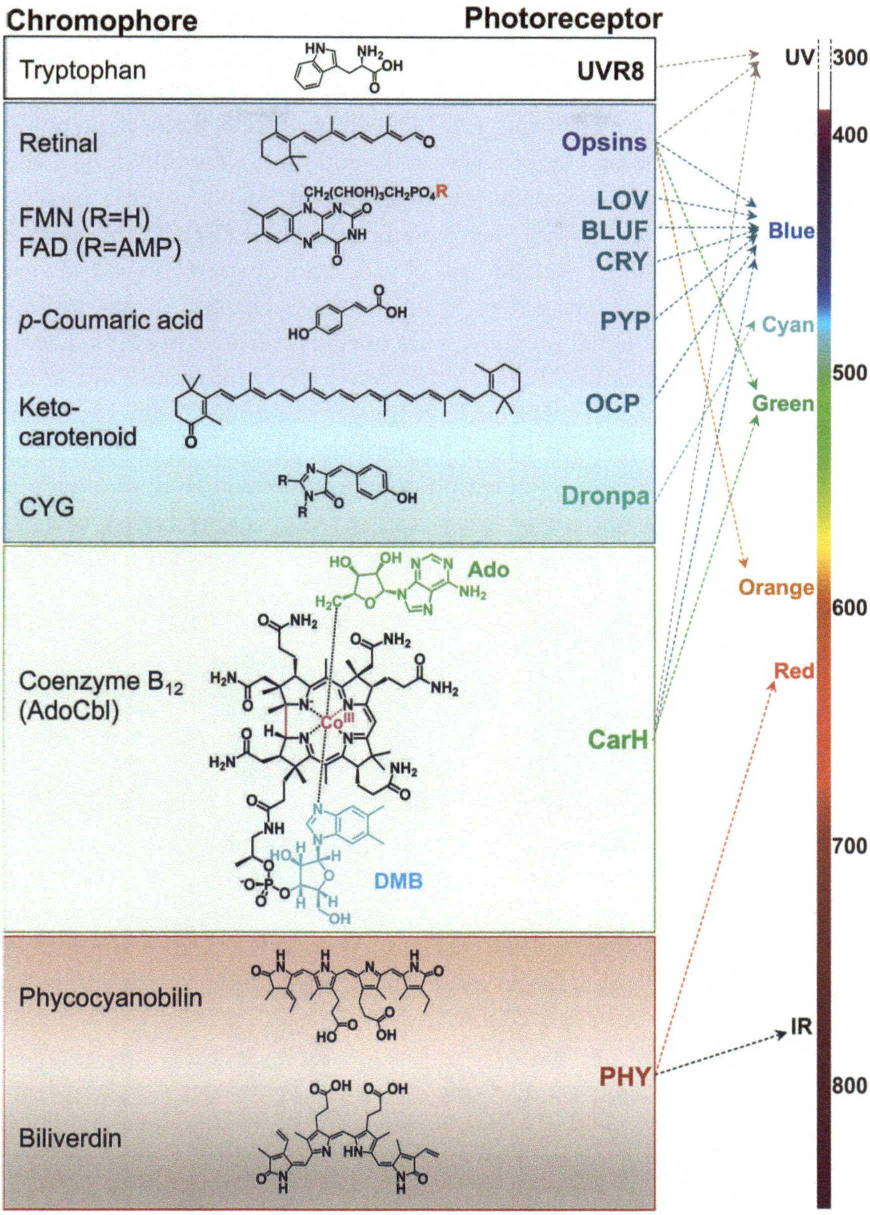

Fig. 1 Photoreceptors (PR) and their chromophores. Chromophore abbreviations (left): AdoCbl, 5′-deoxyadenosylcobalamin; AMP, adenosinemonophosphate; CYG, cysteine-tyrosine-glycine tripeptide; FAD, flavin adenine dinucleotide; FMN, flavin mononucleotide. Shown in AdoCbl are the upper axial 5′-deoxyadenosyl group (Ado) in green and the lower axial 5,6-dimethylbenzimidazole ribonucleoside (DMB in cyan). Photoreceptor abbreviations (right): BLUF, Blue-Light-Using FAD; CRY, cryptochrome; LOV: Light-Oxygen-Voltage-sensing; OCP, orange carotenoid protein; PHY, phytochrome; PYP, photoactive yellow protein. Arrows point to the light sensed by the photoreceptor.

Photoreceptors have been classified into about ten families based on the chromophore and its host photosensory protein domain (Fig. 1). Some exist exclusively in prokaryotes; others also occur in eukaryotes. Most known chromophores (for example, FMN, FAD, *p*-coumaric acid, keto-carotenoids, CYG) absorb in the blue light region (400-500 nm), a few like retinal and coenzyme B_{12} or 5′-deoxyadenosylcobalamin (AdoCbl) cover a broader range from ultraviolet (UV) to green light including blue (300–570 nm), while linear tetrapyrroles like biliverdin and phycocyanobilin perceive red/far-red light (620–750 nm). Light-induced chromophore changes in the photoreceptor "dark" to "light" photocycle are usually reversible, but light cleaves irreversibly a metal-carbon bond in the AdoCbl chromophore (Jost et al., 2015; Padmanabhan, Jost, Drennan, & Elías-Arnanz, 2017; Seong & Lin, 2021). Most photoreceptors have a modular design with the chromophore bound to an autonomously folded photosensory domain (Fig. 2), which is

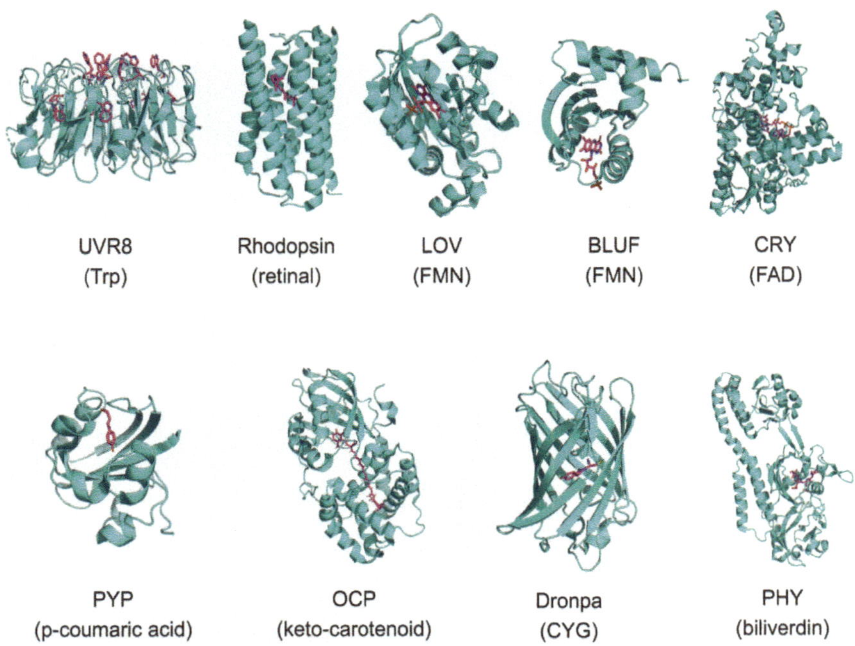

Fig. 2 Representative photoreceptor protomer structures. An example from each family is shown except for B_{12}-based CarH photoreceptors, which will be described later. Photoreceptor (in cyan) with its chromophore (in magenta) is indicated below the structure. PDB ID codes: 4D9S (UVR8), 1XIO (rhodopsin), 3P7N (LOV), 2BYC (BLUF), 1NP7 (CRY), 2PHY (PYP), 4XB5 (OCP), 2POX (Dronpa), 4R70 (PHY).

either the entire protein or part of a multidomain protein with additional effector modules for enzyme activity (kinase, cyclase, esterase) or interaction with specific biomolecules (proteins, DNA, RNA, lipids, etc.). Nature exploits this modularity to combine specific photosensory domains with diverse effector modules to generate a variety of light-sensing proteins and to evolve signaling and regulatory complexity (Mayer, 2015). Autonomous photosensory domains also serve as versatile optogenetics/optobiochemistry and synthetic biology tools to engineer artificial protein fusions amenable to very precise, optical spatiotemporal control of function (Seong & Lin, 2021; Shcherbakova et al., 2015).

This chapter surveys the photoreceptor family whose chromophore is AdoCbl, a biological form of vitamin B_{12} (Fig. 1), with some pertinent forays into aspects of the B_{12} molecule, its chemistry, photochemistry, and biology. Discovery of B_{12}-based photoreceptors and their mode of action in a seminal study a decade ago (Ortiz-Guerrero, Polanco, Murillo, Padmanabhan, & Elias-Arnanz, 2011) unveiled a new functional facet, that as a photoreceptor chromophore, of this fascinating vitamin described as "Nature's most beautiful cofactor" (Stubbe, 1994). It also overturned the notion that a basic property of AdoCbl, its light sensitivity, has no biological function.

2. Vitamin B_{12} structure, chemistry, photochemistry

Vitamin B_{12}, one of the most complex organometallic biomolecules, was discovered about a century ago as the antipernicious anemia factor (Minot & Murphy, 1926; Whipple & Robscheit-Robbins, 1925), and purified as bright red crystals shown to contain cobalt some 20 years later (Rickes, Brink, Koniuszy, Wood, & Folkers, 1948a, 1948b; Smith, 1948a, 1948b). The complex structure of this water-soluble vitamin determined by Dorothy Hodgkin and co-workers was a landmark in X-ray crystallography (Hodgkin et al., 1956), and its total chemical synthesis in the early 1970s by Woodward, Eschenmoser and coworkers a veritable tour de force (Eschenmoser & Wintner, 1977). Its biological role as a cofactor in enzymes susceptible to photoinactivation was also discovered early (Barker, Weissbach, & Smyth, 1958). Vitamin B_{12} and its derivatives, collectively described as cobalamins or just B_{12}, have since been intensely investigated revealing many intricate and exceptional details of their chemistry and biology.

2.1 B_{12}, a complex organometallic biomolecule

B_{12} is a tetrapyrrolic biological cofactor, like heme, chlorophyll or F_{430}, but with a central cobalt metal atom usually in the Co^{3+} or Co(III) oxidation state (Fig. 1) and a more complex molecular structure (Bryant, Hunter, & Warren, 2020). Unlike heme, the B_{12} tetrapyrrole core or corrin lacks a carbon between two of its pyrrole units and a fully conjugated structure, and hence is nonplanar, skewed and ring contracted with a smaller central metal coordination hole (Fig. 3A). Covalently attached methyl, acetamide and

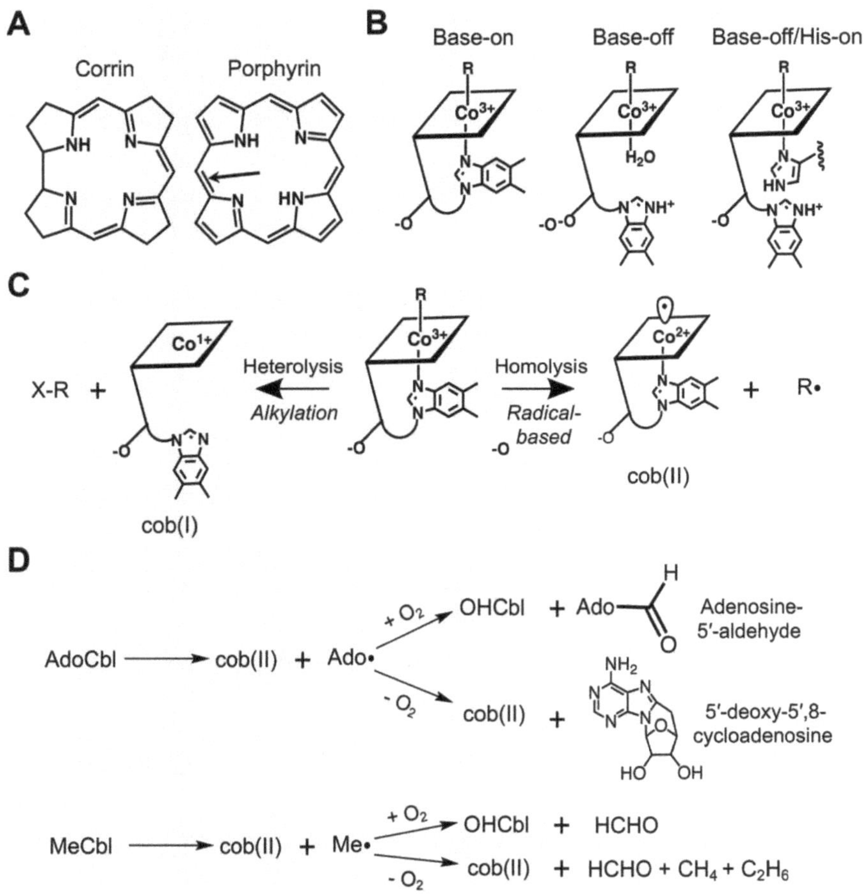

Fig. 3 Vitamin B_{12} chemistry and photochemistry. (A) B_{12} corrin and heme porphyrin tetrapyrrole rings (arrow in porphyrin points to the C missing in corrin). (B) Lower axial B_{12} ligand in "base-on", "base-off" and enzyme-bound "base-off/His-on" conformations. (C) Types of cobalamin Co-alkyl carbon bond cleavage and products. The cobalamin Co^{3+} upper axial group with alkyl carbon linkage, R, is Me in MeCbl and Ado in AdoCbl. X is a methyl acceptor. (D) Aerobic and anaerobic AdoCbl and MeCbl photolysis products.

propionamide groups decorate the corrin ring periphery, with one propionamide group having a pendant 5,6-dimethylbenzimidazole (DMB) ribonucleoside in cobalamins (Fig. 1). Analogs called cobamides have 5-hydroxy- or 5-methoxybenzimidazole or adenosine rather than DMB. The central Co^{3+} ion coordinates to four equatorial pyrrole nitrogens, to an imidazole nitrogen of DMB (or equivalent) as the lower (α) axial Co^{3+} ligand at neutral pH (base-on or DMB-on form; Figs. 1 and 3B), or to H_2O at low pH (base-off or DMB-off form). Many B_{12}-binding proteins have DMB displaced by a His imidazole side chain (base-off/His-on binding; Fig. 3B), which increases the molecular surface by ~300 Å^2, relieves geometric constraints of a base-on DMB loop, and likely alters cobalt activity (Gruber, Puffer, & Krautler, 2011). The upper (β) axial Co^{3+} ligand defines the B_{12} form. It is 5′-deoxyadenosyl (Ado) in AdoCbl (Fig. 1) or methyl (Me) in MeCbl, the two biological forms; cyano in vitamin B_{12} (CNCbl), hydroxyl in OHCbl, or H_2O in AqCbl. Rare examples of B_{12}-binding proteins with unusual axial Co ligands have also been reported. Thus, a light-exposed B_{12}-based photoreceptor has His as both upper and lower axial ligands (Jost, Fernández-Zapata, et al., 2015; Padmanabhan et al., 2017); the upper axial ligand in B_{12} bound base-on is His in a human B_{12} transporter (Randaccio, Geremia, Demitri, & Wuerges, 2010; Wuerges et al., 2006) and Cys in a human B_{12} processing chaperone (Banerjee, Gouda, & Pillay, 2021; Li et al., 2020); and Cys is the lower axial ligand of base-off B_{12} in a bacterial B_{12} transporter (Rempel, Colucci, de Gier, Guskov, & Slotboom, 2018). Diversity in the axial ligands is thus a striking feature in B_{12} chemistry and biology.

2.2 B_{12}, a rich chemistry and photochemistry

Why nature chose to recruit for biological use the B_{12} molecule, with a difficult-to-synthesize corrin coupled to relatively scarce Co and not another metal is a long-standing issue (Osman et al., 2021). A fundamental aspect of B_{12} chemistry in its two biological forms, AdoCbl and MeCbl, is an upper axial Co-C (specifically Co-alkyl) bond, with exceptional properties (Banerjee & Ragsdale, 2003; Gruber et al., 2011; Osman et al., 2021). The AdoCbl or MeCbl Co-C bond has relatively low dissociation energy (30–40 kcal mol^{-1}) and, though thermally stable, it is susceptible to homolytic or heterolytic cleavage (Fig. 3C) by light (near UV-visible range <530 nm, photon energies >40 kcal mol^{-1}) when free or bound to proteins, as exemplified in CarH photoreceptors (Jones, 2017; Padmanabhan et al., 2017;

Rury, Wiley, & Sension, 2015). Enzymes exploit Co-C bond cleavage to tackle diverse chemically exacting reactions (Banerjee & Ragsdale, 2003; Gruber et al., 2011; Osman et al., 2021). Its cleavage generates highly reactive species and radicals, with changes in cobalt oxidation, coordination geometry and ultimately reactivity from six-coordinate Co^{3+}/Co(III) to five-coordinate, paramagnetic Co^{2+}/Co(II), or four-coordinate Co^{1+}/ Co(I) (Fig. 3C), transitions empowered at least in part by the skewed ring-contracted corrin ring (Kieninger et al., 2019). This endows cobalamins with remarkable versatility and unique chemistry to carry out functions that, moreover, can be controlled and fine-tuned by proteins, probably explaining why nature has harnessed this cobalt-corrin pair.

An arresting feature of B_{12} derivatives is their remarkably bright and beautiful red colors. This owes to strong absorption in the UV-visible range mostly from π–π^* transitions, and the spectral and electronic features depend strongly on cobalt oxidation state and nature of the upper and lower axial ligands (Jones, 2017; Rury et al., 2015; Stich, Brooks, Buan, & Brunold, 2003). Both MeCbl and AdoCbl are photosensitive, and their photolysis mechanisms serve as models to understand enzymatic Co-C bond cleavage and reactivity (Schwartz & Frey, 2007). AdoCbl and MeCbl photolysis products and yields have been analyzed, and transient ultrafast spectroscopy has provided photophysical-photochemical insights into their dynamics, intermediates and electronic processes (Jones, 2017; Padmanabhan et al., 2017; Rury et al., 2015). Usually, homolytic Co–C bond photolysis occurs within 100 ps upon irradiation in 400–530 nm range with quantum yields that are wavelength-dependent for MeCbl (high at 400 nm, lower at 522 nm) but not for AdoCbl (near unit yields over this wavelength range), and transient caged alkyl radical:cob(II) pairs either dissociate or Co–C bond reversion reduces photolysis yields. Solvent cages around Cbl reduce photolysis, whereas reaction of radicals with molecules like oxygen curbs recombination. Various factors (size and geometry of the upper axial ligand, base-off or base-on lower axial coordination, conditions like pH, solvent viscosity, etc.) affect the stability and diffusion of the initial photolysis products, determining the reversion versus dissociation-escape outcome. AdoCbl and MeCbl differ in the photophysics and radical pair reaction dynamics, with some enigmatic MeCbl photolysis features attributed to the small Me• radical size and geometry (Cole et al., 2002; Jones, 2017; Sension, Harris, & Cole, 2005). Cob(II) from AdoCbl or MeCbl homolysis is stable in the absence of oxygen but oxidizes rapidly to OHCbl or AqCbl under aerobic conditions (Fig. 3D). Ado• radicals from AdoCbl homolysis yield 5′-deoxy-5′,8-cycloadenosine under anaerobic conditions (Hogenkamp, 1963; Law & Wood, 1973), or react with oxygen to produce 5′-peroxyadenosine, which decomposes to adenosine-5′-aldehyde and low levels of adenosine and adenine (Jost, Simpson, & Drennan, 2015; Schwartz & Frey, 2007). Me• radicals from MeCbl photolysis form formaldehyde, low levels of methanol, formic acid and carbon dioxide with oxygen or methane and ethane with no oxygen (Hogenkamp, 1966).

3. Vitamin B_{12}, a precious biological cofactor

3.1 B_{12} biosynthesis

B_{12} is made exclusively by bacteria and archaea (Roth, Lawrence, & Bobik, 1996; Zhang, Rodionov, Gelfand, & Gladyshev, 2009). Animals, including humans, and many prokaryotes require B_{12} but cannot make it, and rely on their diet or specific environmental niches and microbial communities for B_{12} (Degnan, Taga, & Goodman, 2014; Sokolovskaya, Shelton, & Taga, 2020). Plants, fungi, and many prokaryotes do not need B_{12} and resort to alternative, usually less efficient, B_{12}-independent enzymes and pathways. Biosynthesis of B_{12} *de novo* ranks among the most intricate in biology, with two genetically distinct pathways known involving over thirty enzymatic steps and a series of overall similar amidation, methylation and rearrangement reactions (Bryant et al., 2020). Metal insertion occurs early before ring contraction in the anaerobic "cobalt-early" pathway, but after ring contraction and various methylations in the aerobic "cobalt-late" route (Bryant et al., 2020; Osman et al., 2021). The earliest steps in both B_{12} pathways (and in heme and chlorophyll biosynthesis) are common and produce the uroporphyrinogen III tetrapyrrole precursor. They then branch into anaerobic and aerobic routes, where different reactions culminate in the same adenosylcobyrinic acid *a,c*-diamide intermediate, after which both pathways generate the Cbl product using the same final steps, including DMB addition. The sheer complexity of the two pathways, with many unique enzymes and required conditions, may explain why only prokaryotes can synthesize B_{12} *de novo*. Almost 90% of bacteria have an enzyme dependent on B_{12} but just a third can make it, usually via the anaerobic route; and even bacteria capable of B_{12} biosynthesis have enzymes that do not require B_{12} coexisting with ones that do require it for the same activity (Shelton et al., 2019). This divide between B_{12} producers and consumers is reflected in cooperative or conflictive inter-dependences within microbial communities or in niches like the human gut (Degnan et al., 2014; Rowley & Kendall, 2019; Sokolovskaya et al., 2020; Wexler et al., 2018).

3.2 B_{12}-binding proteins in B_{12} uptake and conversion to biologically active forms

B_{12} is very precious because it is scarce, and its complex synthesis incurs a steep genetic-metabolic cost. Consequently, prokaryotes as well as eukaryotes effectively trap, recover and reuse even minute amounts of available exogenous or intracellular B_{12} using active elaborate mechanisms, which have been reviewed elsewhere (Banerjee et al., 2021; Cracan & Banerjee, 2013; Gherasim, Lofgren, & Banerjee, 2013). To capture and import B_{12}, bacteria use highly regulated assembly lines of TonB-dependent ABC (or ATP-binding cassette) or ECF (energy coupling factor) membrane-bound transporters and associated proteins, usually with broad specificities and high affinities (nM–fM dissociation constants, K_D) for various B_{12} forms (Korkhov, Mireku, Veprintsev, & Locher, 2014; Wexler et al., 2018), although low affinity general multi-solute transporters are also known (Rempel et al., 2020). Similarly, eukaryotes absorb and mobilize B_{12} using factors that bind avidly to it ($K_D < 1\,pM$, with slight preferences for some forms or conformations) and defined transporters. These proteins have diverse molecular architectures and most bind to B_{12} forms base-on or variations like base-on/His-on (Gruber et al., 2011; Sukumar, 2013), although base-off B_{12}-binding, typical of a subset of B_{12}-dependent enzymes, occurs in some transporters with enzyme activity (Koutmos, Gherasim, Smith, & Banerjee, 2011; Rempel et al., 2018) (Fig. 4). Conditions within specific cellular compartments modulate transitions between the two B_{12}-binding modes, with a free DMB tail or a lower axial site possibly helping to shuttle B_{12} between sites and proteins (Banerjee et al., 2021). Dedicated enzymes and a corps of chaperone escorts process imported or recyclable B_{12} forms and convert them to AdoCbl or MeCbl for delivery to specific targets such as enzymes. Methyltransferases generate MeCbl for their own use. ATP-dependent cob(I)alamin adenosyltransferases produce AdoCbl for other factors, and three types occur in bacteria: CobA (BtuR or CobO) acts in *de novo* AdoCbl biosynthesis, while PduO (and its homolog in eukaryotes) and EutT are involved in salvaging and converting B_{12} to AdoCbl. In general, intricate interprotein communications through formation of stable or transient complexes, and novel chemistries fine-tuned by protein architecture and dynamics, direct B_{12} trafficking and assimilation (Banerjee et al., 2021).

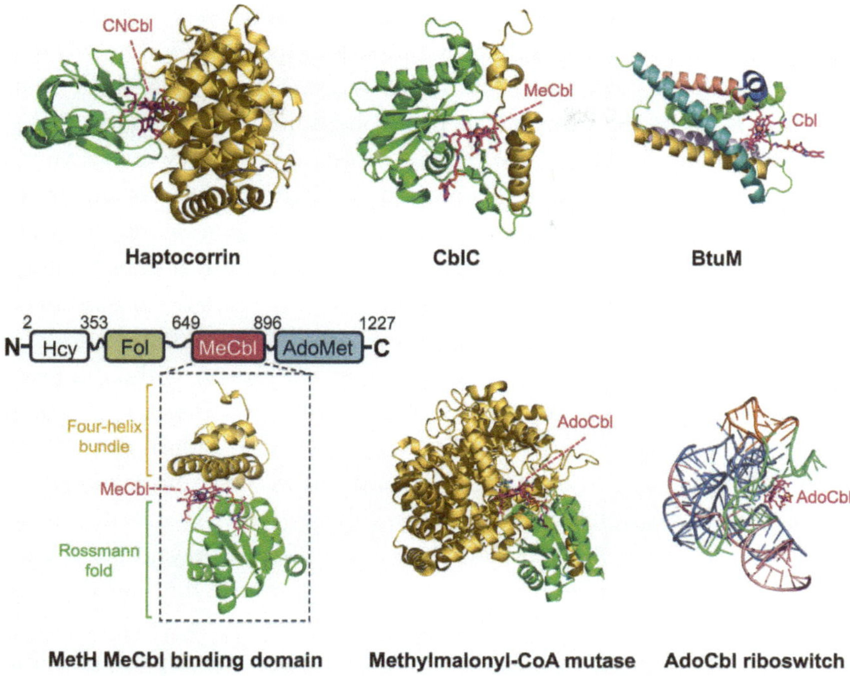

Fig. 4 Representative B_{12}-binding proteins and B_{12} riboswitch structures. PDB ID codes are: 4KKI (haptocorrin with CNCbl bound base-on), 3SC0 (human CblC B_{12} processing chaperone with MeCbl bound base-off); 6FFV (bacterial vitamin B_{12} transporter BtuM with Cbl bound base-off, with a lower axial Cys); 1BMT (MeCbl-binding domain of MetH methionine synthase with MeCbl bound base-off); 2XIJ (one chain of the homodimeric human methylmalonyl-CoA mutase with AdoCbl bound base-off); 4GXY (AdoCbl riboswitch with AdoCbl bound base-on). BtuM is a membrane protein, the rest are soluble, as is the riboswitch. The four domains of MetH are shown and their ligands indicated (Hcy: homocysteine; Fol: folate; AdoMet: S-adenosyl-methionine). Note the specific B_{12} ligand (in magenta, with the upper axial ligand in cyan) sandwiched between two modules in the structures.

3.3 B_{12}-dependent enzymes

The best known biological role of B_{12} as an enzyme cofactor, and the amazing range of activities, chemistries and mechanisms of these enzymes have been reviewed extensively (Banerjee & Ragsdale, 2003; Bridwell-Rabb & Drennan, 2017; Dowling, Croft, & Drennan, 2012; Giedyk, Goliszewska, & Gryko, 2015). Some salient general features pertinent to B_{12}-photoreceptors are noted here. Methyltransferases such as methionine synthase use MeCbl as cofactor; mutases, dehydratases, reductases, and deaminases require AdoCbl for radical-based rearrangements; and reductive dehalogenases, RNA-modifying reductases, and B_{12} decyanation/dealkylation enzymes rely on Cbl without an upper ligand. The enzymes bind to specific B_{12} forms base-on (DMB-on) or base-off (DMB-off) with affinities (K_D) ranging from nM to μM (sometimes mM; Sukumar, 2013). Base-off binding, mostly observed in a subset of proteins with enzymatic activity is often His-on, with His from a DxHxxG motif (x is any amino acid) replacing DMB as the lower axial ligand, and helps to sense cobalt oxidation/alkylation states and control protein conformations at different stages of the catalytic cycle (Gruber et al., 2011; Rempel et al., 2018). Interestingly, nonenzymatic B_{12}-based photoreceptors share this B_{12} binding mode, as described later. Some enzymes bind B_{12} base-off/His-off without a lower axial protein ligand at the cobalt, and a few use Cys as the lower axial protein ligand (Bridwell-Rabb & Drennan, 2017). The Ado• radical from Co-C bond homolysis in AdoCbl enzymes lies at the heart of various radical-based rearrangement reactions and recombines with cob(II) at the end of each catalytic round to reuse AdoCbl (Dowling et al., 2012; Toraya, 2003). Until recently, Ado• was too reactive to trap and characterize fully (Yang et al., 2019). Heterolytic Co–C bond cleavage in MeCbl enzymes like methyltransferases (Fig. 3C) yields the potent reductant and supernucleophile Co(I), and a methyl carbocation for transfer to a nucleophilic acceptor (Matthews, Koutmos, & Datta, 2008; Ragsdale, 2008). MeCbl methyltransferases usually have components to bind a methyl donor, a methyl acceptor, and Cbl (which forms MeCbl using methyl from the donor and transfers it to the acceptor) that can occur as separate domains in one polypeptide as in methionine synthase (Fig. 4), or as individual subunits of a multiprotein complex (Matthews et al., 2008; Ragsdale, 2008). Enzymes with Cbl lacking an upper ligand also rely on reactive species and unprecedented chemistry (Bridwell-Rabb & Drennan, 2017). In sum, B_{12}-dependent enzymes accomplish diverse difficult reactions by effectively leveraging the highly reactive species emanating from Co-C bond cleavage.

Clearly, B_{12}-based enzymes must properly manage the very reactive species they generate to both prevent unwanted reactions and protect their active sites from damage. They do this through ingeniously designed protein architectures that typically combine common folding units like the four-helix bundle (4HB), $(\beta/\alpha)_8$ triosephosphate isomerase (TIM) barrel, and α/β Rossmann fold (RF) domain (Dowling et al., 2012). Usually, the large B_{12} molecule is sandwiched between an upper protective "capping" domain and a lower B_{12}-binding domain (Fig. 4). This enables controlled Co-C bond homolysis in AdoCbl enzymes or heterolysis in

MeCbl enzymes to act on diverse substrates and shield against undesired chemistry. Base-off/His-on binding in many methyltransferases and AdoCbl enzymes usually involves an RF domain, comprising five parallel β-strands and helices, at the lower B_{12} face. It contains a $DxHxxGx_{41}S/TxLx_{26-28}GG$ motif, whose H supplies the lower axial ligand and forms a catalytic triad with D and S/T, while the two GG in a β-strand leave room for the displaced DMB tail, as was first visualized in the MeCbl-binding domain of MetH methionine synthase (Drennan, Huang, Drummond, Matthews, & Ludwig, 1994), and is interestingly mirrored in B_{12}-based photoreceptors (see below). Enzymes that bind B_{12} base-off/His-off lack a DxHxxG motif and use instead a hydrophobic helix or polar residues, like Arg or Glu, to block access of water or other small ligands to the lower axial position of B_{12} (Bridwell-Rabb & Drennan, 2017; Ragsdale, 2008). Many AdoCbl-based enzymes exhibit base-on binding using variable motifs, separate domains of β-strands and helices (not RF) or loops, and often occur as homodimers, heterodimers or tetramers (Dowling et al., 2012). Thus, the same small set of protein folds tailored differently encase specific B_{12} factors to modulate exquisitely controlled Co-C cleavage and act on specific substrates. Yet, despite built-in design features to protect against undesired reactions, agents that target the cofactor (like oxygen, substrates, and notably light) can irreversibly inactivate B_{12} enzymes, which often require reactivating chaperones to resuscitate the cofactor (Toraya, 2003, 2014). Indeed, ever since B_{12} was first unmasked as an enzyme cofactor, photoinactivation of B_{12} enzymes has been known (Barker et al., 1958), and B_{12} light-sensitivity was considered a hindrance of no biological relevance, until the discovery of B_{12}-based photoreceptors.

3.4 B_{12} riboswitches

A fascinating biological role of B_{12} in gene regulation emerged with the discovery of B_{12} riboswitches two decades ago. These ~200-nucleotide long RNA regulatory elements usually occur within long 5′-untranslated regions (UTRs) of mRNAs of bacterial genes associated with cobalt or B_{12} metabolism, such as for various B_{12}-dependent enzymes or proteins for cobalamin/cobalt uptake and processing (Mandal & Breaker, 2004; Nahvi et al., 2002; Nahvi, Barrick, & Breaker, 2004). The first such RNA-based mechanism discovered was in fact the AdoCbl riboswitch (Nahvi et al., 2002), after all attempts failed to link a protein factor with regulation of the *E. coli btuB* gene by its 5′-UTR and AdoCbl (Lundrigan, Köster, & Kadner, 1991). Crucial in this discovery was to not overlook

AdoCbl light-sensitivity, with special care taken to perform the experiments using adequate precautions against inadvertant exposure to light. The *btuB* 5′-UTR RNA binds preferentially to AdoCbl, but later work unearthed riboswitches specific for OHCbl or MeCbl, as well as promiscuous ones that bind all three B_{12} forms (Chan & Mondragón, 2020; Polaski, Webster, Johnson, & Batey, 2017). All have a conserved structural core but differ in the size and folding of peripheral structural elements, which determine tight binding affinities (K_D <10 to 250nM) and specificities (relative K_D values ranging from 2–10-fold to even ≥200-fold) for given B_{12} forms, and serve to classify B_{12} riboswitches into two classes and some subclasses (Polaski et al., 2017). The B_{12} riboswitch mechanism of gene regulation is based on conformational changes induced by base-on binding of a specific B_{12} form to ligand-free riboswitch RNA. They usually operate in *cis* at transcriptional or translational levels, but *trans*-acting B_{12} riboswitches that regulate transcription of non-coding RNAs or sequester specific proteins have been reported (DebRoy et al., 2014; Mellin et al., 2014). Whether and how light plays a role in B_{12} riboswitch function remains unexplored. Conceivably, inherent B_{12} light sensitivity can serve as a specificity switch between intact and photolyzed B_{12} forms.

4. B_{12}-based photoreceptors

4.1 B_{12}-binding proteins in gene regulation and the discovery of CarH photoreceptors

B_{12} can regulate gene expression not only through binding to a riboswitch RNA but also to protein regulatory factors. In early work, a bacterial AraC/XylS-type transcriptional factor named EutR was proposed to sense AdoCbl and ethanolamine by direct binding, in order to activate expression of the *Salmonella typhimurium eut* operon that encodes enzymes for ethanolamine utilization, some of which are B_{12}-dependent (Kaval & Garsin, 2018; Roof & Roth, 1992). Various studies linked EutR activity to AdoCbl, most recently using synthetic B_{12} mimic activity probes or EutR-based B_{12} biosensors (McNerney, Piorino, Michel, & Styczynski, 2020; Rosnow et al., 2018), but direct binding of AdoCbl to EutR remains to be shown experimentally. Arguably, the first experimental demonstration of direct and specific AdoCbl binding to a transcriptional repressor to regulate gene expression was the discovery of CarH photoreceptors ten years ago, which unmasked the new facet of vitamin B_{12} as a light sensor (Ortiz-Guerrero et al., 2011; Padmanabhan et al., 2017).

The first CarH photoreceptor was identified in the Gram-negative soil bacterium *Myxococcus xanthus* from studies on how light induces the biosynthesis of carotenoids, which are lipophilic isoprenoid pigments that protect cells against photooxidative damage (Padmanabhan et al., 2017, 2021). Extensive genetic, biochemical, and structural studies revealed two distinct modes of light sensing and signaling that trigger a transcriptional response leading to carotenoid biosynthesis in *M. xanthus*, and identified structural genes for the enzymes involved and several regulatory genes encoding novel factors (Padmanabhan et al., 2021; Pérez-Castaño et al., 2022). The first inkling of two B_{12}-binding transcription factors came from sequences of CarA and CarH, the products of the two most downstream genes in the light-inducible *carB* operon, which contains nine out of the ten structural genes for carotenogenesis (Fig. 5A). Both have similar domain architectures: N-terminal domains akin to the DNA-binding domain (DBD) of MerR family transcription factors (Botella, Murillo, & Ruiz-Vazquez, 1995), and C-terminal domains akin to the MetH MeCbl-binding domain (Fig. 4) with the base-off/His-on B_{12}-binding motif E/DxHxxGx_{41}S/ TxLx_{26-28}GG (Cervantes & Murillo, 2002). CarA and CarH were shown to block transcription of the *carB* operon by binding to the same DNA site at the *carB* P_{crt} promoter (Fig. 5A) (López-Rubio et al., 2004; López-Rubio, Elías-Arnanz, Padmanabhan, & Murillo, 2002) via their N-terminal DBDs, whose structure was indeed of the MerR-type (Navarro-Avilés et al., 2007; Pérez-Marín, López-Rubio, Murillo, Elías-Arnanz, & Padmanabhan, 2004). On the other hand, their C-terminal domains mediate oligomerization essential for function (López-Rubio et al., 2002; Ortiz-Guerrero et al., 2011). CarA could bind to B_{12} through its C-terminal domain but, puzzlingly, its activity neither required B_{12} nor an intact B_{12}-binding motif, whereas both were indispensable for CarH function (Ortiz-Guerrero et al., 2011; Pérez-Marín, Padmanabhan, Polanco, Murillo, & Elías-Arnanz, 2008). Moreover, light relieved CarA and CarH repression of P_{crt} *in vivo* using distinct mechanisms. Thus, CarA dimers bind cooperatively to its large operator site at P_{crt} to block access to RNA polymerase and initiate transcription in the dark (López-Rubio et al., 2004); and light inactivates CarA indirectly by inducing the expression of the antirepressor CarS, a DNA structural mimic that physically interacts with CarA to block its binding at P_{crt} and initiate transcription (León et al., 2010). Light sensing and signaling to induce CarS expression does not depend on B_{12} and is complex. It requires $^1O_2$ generated by blue light excitation of protoporphyrin IX (the heme precursor and a photosensitizer), glycerophospholipids called

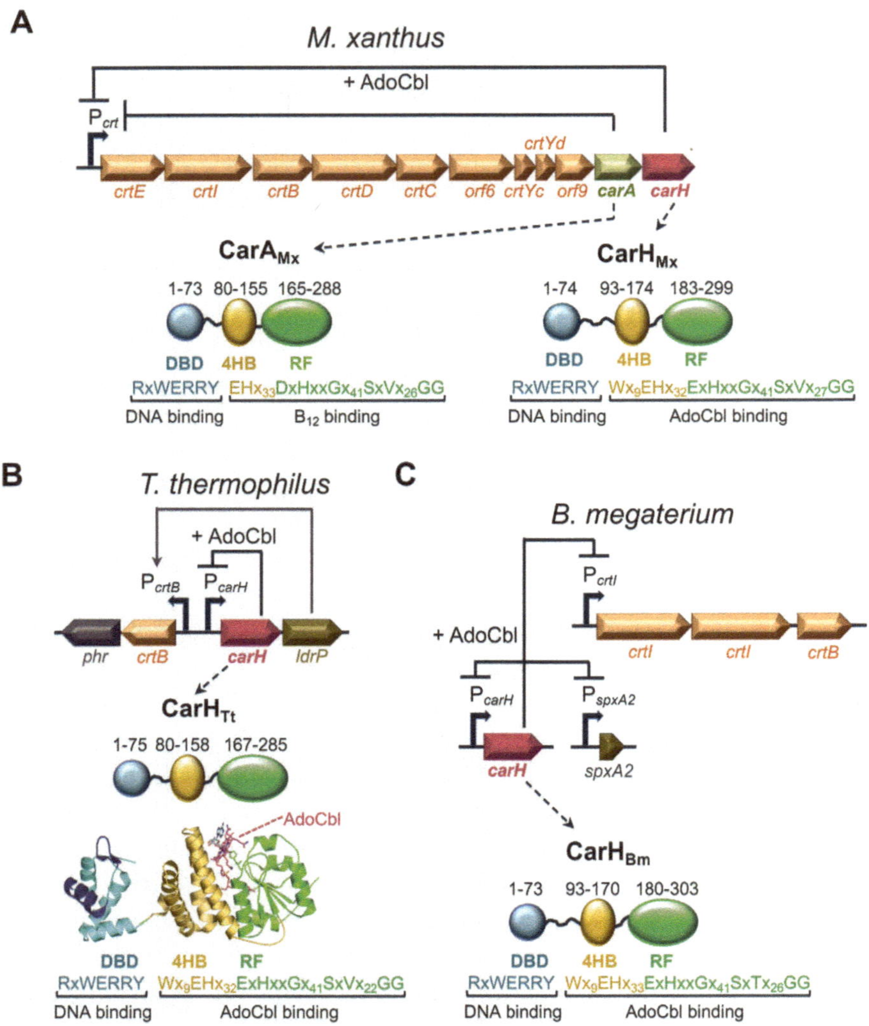

Fig. 5 CarH photoreceptor genes, genome context and domain architectures. (A) *M. xanthus* CarH ($CarH_{Mx}$) and CarA ($CarA_{Mx}$). Top: *carA*, *carH* and carotenoid biosynthesis genes (product enzymes in parentheses) *crtE* (farnesyltransferase), *crtI* (phytoene desaturase), *crtB* (phytoene synthase), *crtD* (hydroxyneurosporene dehydrogenase), *crtC* (neurosporene hydroxylase), *orf6* (putative carotenoid biosynthesis protein), *crtYc*, *crtYd* (hetero-dimeric lycopene cyclase), and *orf9* (putative acyltransferase). Bottom: CarA/CarH domain architectures indicating the predicted MerR-type winged-helix DBD and MetH-type B_{12}-binding domain, with its 4HB and RF subdomains, and their characteristic motifs. Domain delimiting residues are numbered. (B) Genome context of the gene for $CarH_{Tt}$ and protein domain architecture. Adjacent genes are *crtB*, *phr* (DNA photolyase), *ldrP* (CRP/FNR family transcription factor). Protomer crystal structure is shown below the protein domain architecture, with AdoCbl in magenta and its Ado group in cyan. (C) $CarH_{Bm}$ gene, the genes it regulates (*crtB*, *crtI*, as in A; *spxA2*, general oxidative stress regulator), and domain architecture (as in A and B).

plasmalogens, and many regulatory factors (Gallego-García et al., 2019; Padmanabhan et al., 2021). By contrast, light directly inactivates CarH bound specifically to AdoCbl (Ortiz-Guerrero et al., 2011).

Genetic analysis revealed that CarH oligomerizes in the dark with B_{12} present and represses expression from P_{crt} *in vivo*, but not without B_{12} or when exposed to light. CarH could not repress P_{crt} *in vivo* when AdoCbl formation in *M. xanthus* was blocked upon deleting its sole adenosyltransferase (PduO) gene, establishing AdoCbl as the specific B_{12} form that is required for CarH function *in vivo* (Ortiz-Guerrero et al., 2011). Although *M. xanthus* CarH has resisted purification as native soluble protein for *in vitro* analysis, two homologs (among several later identified in bacterial genomes over a wide taxonomical range) could be purified as soluble proteins and examined *in vitro*. These turned out to be AdoCbl-specific and light-dependent like *M. xanthus* CarH. They were $CarH_{Tt}$ from the Gram-negative *Thermus thermophilus* (Fig. 5B) and $CarH_{Bm}$ from Gram-positive *Bacillus megaterium* (Fig. 5C), two bacteria unrelated to *M. xanthus* that, unlike the latter, can synthesize B_{12} *de novo*. Both homologs control light-dependent expression of their own gene and of those for carotenoid biosynthesis (as in *M. xanthus*), a transcription factor and a photolyase ($CarH_{Tt}$), and a general oxidative stress factor ($CarH_{Bm}$) (Fig. 5B and C), although the target genes are organized differently in their respective genomes (Fernández-Zapata et al., 2018; Ortiz-Guerrero et al., 2011; Takano et al., 2011, 2015). Notably, unlike EutR or B_{12} riboswitches, which control genes for B_{12}-dependent enzymes or for proteins in B_{12} uptake or use, light-sensing CarH appear to regulate genes unrelated to B_{12} metabolism.

4.2 CarH molecular mechanism of action

Studies of $CarH_{Tt}$ and $CarH_{Bm}$ have provided invaluable molecular insights into CarH photoreceptors, on how they sense light and how this modulates conformational changes, protein oligomerization/quaternary structures and repressor activity. In particular, crystal structures of functionally relevant states of $CarH_{Tt}$ have been resolved (Jost, Fernández-Zapata, et al., 2015) and its photochemistry has been analyzed in detail (Jost, Simpson, & Drennan, 2015; Kutta et al., 2015; Miller et al., 2020). Crystal structures and photochemistry are not known for $CarH_{Bm}$, which shares various features with $CarH_{Tt}$ yet differs in others (Fernández-Zapata et al., 2018; Takano et al., 2015). Both proteins have significant secondary and tertiary structures in their apo forms and bind poorly to operator DNA; even so, apo$CarH_{Tt}$ is a monomer while apo$CarH_{Bm}$ is a loosely-packed tetramer with features typical of molten globules (Fernández-Zapata et al., 2018). Binding of $CarH_{Tt}$ or $CarH_{Bm}$ to AdoCbl in the dark (with $K_D \approx 200$–250 nM, comparable to those for various other proteins and riboswitches) yields tetramers that bind to operator DNA with high affinity ($K_D \approx 20$–100 nM) (Fernández-Zapata et al., 2018; Jost, Fernández-Zapata, et al., 2015; Kutta et al., 2015). CarH operators are large (~40–50 bp) and overlap with the −35 and/or −10 promoter elements recognized by RNA polymerase holoenzyme associated to the primary σ^A factor (Fernández-Zapata et al., 2018; Jost, Fernández-Zapata, et al., 2015; Ortiz-Guerrero et al., 2011). Hence, CarH binding denies promoter access to RNA polymerase to block transcription. Blue, green, or near-UV light cleaves the Co-C bond of CarH-bound AdoCbl and frees its upper axial Ado group. This disrupts active tetramers to monomers for $CarH_{Tt}$ but to dimers for $CarH_{Bm}$ (both remain bound to photolyzed AdoCbl), neither of which bind to operator, thereby allowing RNA polymerase to proceed with transcription. A new twist to this oligomeric plasticity was recently reported with CarH homologs from the genus *Cystobacter* (of the same bacterial suborder as *M. xanthus*). These are inactive dimers in the apo form but transition to AdoCbl-bound dimers that bind to a large ~ 50 bp DNA operator with an affinity and mode of recognition similar to the $CarH_{Tt}$ and $CarH_{Bm}$ tetramers, and become inactive monomers when exposed to light (Pérez-Castaño et al., 2022).

4.3 CarH structures

A far-reaching structural study of the free and DNA-bound tetramer and light-exposed monomer provided detailed snapshots of the molecular architectures and mechanism of action of these photoreceptors (Jost, Fernández-Zapata, et al., 2015). The $CarH_{Tt}$ protomer has an N-terminal MerR-type winged-helix module matching that of CarA, connected by a flexible linker to a C-terminal AdoCbl-binding domain (CBD) with a MetH B_{12}-binding domain-like structure but with AdoCbl, instead of MeCbl, sandwiched between the 4HB and the RF subdomain (Fig. 5B). The 4HB caps the upper axial Ado moiety of AdoCbl, whose lower face rests on the RF in the base-off/His-on mode, with DMB displaced by the imidazole side chain of a conserved His in an $ExHxxGx_{41}SxVx_{22}GG$ B_{12}-binding motif. Conserved in all CarH homologs, and absent in the MetH B_{12}-binding domain, is a Wx_9EH motif in the 4HB, whose Trp packs against the upper axial Ado group. This repurposes CarH to bind specifically to AdoCbl (rather than MeCbl, with a smaller upper ligand) in a more polar, spacious pocket. Mutating Trp, Glu or His of the Wx_9EH motif impairs AdoCbl binding, tetramerization and DNA binding (Fernández-Zapata et al., 2018; Jost, Fernández-Zapata, et al., 2015). Tellingly, B_{12}-independent CarA lacks this Trp and thus a Wx_9EH motif (Fig. 5A).

The AdoCbl-bound $CarH_{Tt}$ protomer adopts an extended upright conformation because the Ado moiety and Trp of the Wx_9EH motif "press" against each other. Two protomers pack head-to-tail, with the 4HB of one facing the RF of another, to form an extensive dimer interface with numerous interactions including salt bridges, H-bonds, with crucial contributions from the Ado group (Fig. 6A). Two such dimers rapidly assemble into a stable $CarH_{Tt}$ tetramer (Fig. 6B), with isolated dimers detected only upon disrupting the dimer-dimer interface by mutation (Díez et al., 2013; Jost, Fernández-Zapata, et al., 2015; Ortiz-Guerrero et al., 2011). Tetramer formation is closely tied to AdoCbl-binding and dimerization and can be impaired even by single mutations at the dimer or tetramer interfaces (Jost et al., 2015; Padmanabhan et al., 2017). Recent stopped-flow spectroscopy data support an intuitive stepwise mechanism for $CarH_{Tt}$ tetramerization (Camacho et al., 2021). Here, AdoCbl binds predominantly to $apoCarH_{Tt}$ monomers (apparent bimolecular rate constant $\sim 34\,s^{-1}\,mM^{-1}$) to drive AdoCbl-$CarH_{Tt}$ dimer formation, with rate-limiting structural changes at each step, followed by rapid assembly without further major structural changes to tetramers. Mass spectrometry data in this study showed this stepwise mechanism as predominant, but also hinted at minor pathways that may contribute to the formation of the final AdoCbl-bound tetramer.

The four DBDs of the $CarH_{Tt}$ tetramer, a dimer of two unusual head-to-tail dimers, radiate away from from a central core (Fig. 6B). The $CarH_{Tt}$-DNA complex structure revealed an unusual DNA-binding mode (Fig. 6C), validated by footprinting data and systematic analysis of mutant operators, where three out of the four DBDs contacted three tandem 11-bp direct repeats on the same face of operator DNA (Jost, Fernández-Zapata, et al., 2015). The DNA-bound and free tetramers have the same overall architecture but the three DBDs reorient to contact DNA. Each 11-bp direct repeat has a consensus nAnnTnnACAn sequence (n = any base), whose conserved T and AC make crucial contacts with the DNA recognition helix RxWERRY motif (Fig. 5) found in all CarH homologs (Padmanabhan et al., 2017). The less-conserved bases in the direct repeats also contribute to DNA binding (Fernández-Zapata et al., 2018).

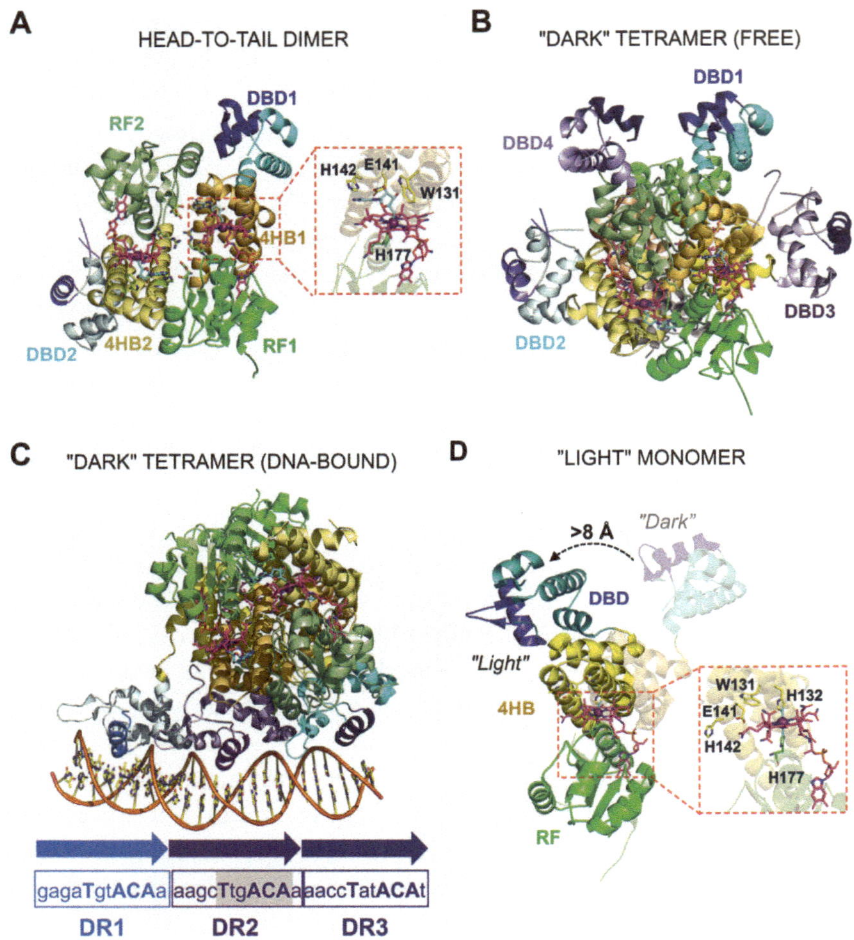

Fig. 6 Structures of free and DNA-bound dark and light-exposed forms of $CarH_{Tt}$. (A) Head-to-tail packed $CarH_{Tt}$ dimer with the DBD (blue), 4HB (yellow) and RF (green) modules indicated as "1" for one protomer and "2" for the other. AdoCbl is in magenta with its Ado group in cyan. Right: close-up of the AdoCbl binding site showing residues of the Wx_9EH motif (W131, E141, H142) that contact the upper axial Ado ligand (cyan), and the lower axial His ligand (H177). (B) Dark $CarH_{Tt}$ tetramer, a dimer of two head-to-tail dimers. Note the four splayed DBDs of the tetramer. (PDB ID code: 5C8D in A and B). (C) Dark CarH tetramer-DNA complex (PDB ID code 5C8E). Three of the four DBDs contact the same face of operator DNA with their recognition helices (darker colors), each of which contacts a conserved TnnACA sequence in the major groove of a 11-bp direct repeat (DR). Direct repeats are shown as thick arrows with sequences below and numbered. The -35 promoter element within the $CarH_{Tt}$ operator is shaded gray. (D) Light-exposed CarH monomer structure (PDB ID code 5C8F). The large shift (>8 Å) in the 4HB relative to the dark form (transparent), and a close-up of its photolyzed AdoCbl binding through bis-His Co coordination (right) are shown.

Curiously, structurally similar DBDs of MerR proteins use most of these same contacts to bind to (pseudo)palindromic DNA sites. The $CarH_{Bm}$ tetramer binds like $CarH_{Tt}$ but uses its four DBDs to contact four 11-bp direct repeats with the characteristic TnnACA motif (Fernández-Zapata et al., 2018), while *Cystobacter* CarH binds cooperatively as two dimers, hence effectively as a tetramer, to four such tandem 11-bp direct repeats (Pérez-Castaño et al., 2022). Interestingly, $CarH_{Bm}$ can recognize the smaller $CarH_{Tt}$ operator and *vice versa*, even though their direct repeats are not identical, indicating a notable plasticity in operator design and DNA binding. Possibly important in this is the linker connecting the CarH N- and C-terminal modules. This linker, which would transmit allosteric effects between domains (Papaleo et al., 2016), varies in size (4–120 residues) and composition among CarH homologs and can notably affect CarH properties (Fernández-Zapata et al., 2018).

How light inactivates the $CarH_{Tt}$ tetramer is clearly seen in the light-exposed AdoCbl-$CarH_{Tt}$ structure (Fig. 6D). It revealed bound Cbl lacking the Ado group and a large shift in the 4HB relative to the RF (compared to the dark form), where Trp of the Wx_9EH motif moves in to occupy the void left by the missing Ado group. This disrupts the dimer interface and, with it, the tetramer and its DNA binding. Interestingly, a His adjacent to the Trp (H132) also moves and becomes the upper axial Co^{3+} ligand in a very tight, unprecedented bis-His linkage (Jost, Fernández-Zapata, et al., 2015). This His is not conserved in many CarH homologs including *M. xanthus* CarH and $CarH_{Bm}$ (both with Glu), and mutating it does not affect AdoCbl binding, tetramer formation or DNA-binding. Although the dark and light state CarH structures are known, structural transitions between them are not; and to visualize such changes over the short timescales involved, CarH may be a promising candidate for future time-resolved serial femtosecond crystallography and related studies (Poddar et al., 2022). Moreover, light-exposed AdoCbl-$CarH_{Tt}$ is a monomer (as is Cystobacter CarH), while AdoCbl-$CarH_{Bm}$ is a dimer (of unknown structure), which indicates a plasticity even in the light states.

4.4 CarH photochemistry

A striking paradox posed by CarH function in a cellular response to curb photooxidative damage from highly reactive species is that AdoCbl photochemistry typically generates such species. Experimental data suggest that CarH resolves this dilemma by altering AdoCbl photochemistry. Analysis of

AdoCbl-CarH$_{Tt}$ photolysis products identified 4′,5′-anhydroadenosine (Fig. 7A), a harmless product not previously detected upon photolysis of free or enzyme-bound AdoCbl (Jost, Simpson, & Drennan, 2015). Also supportive of distinct photochemistry are transient ultrafast spectroscopic data using

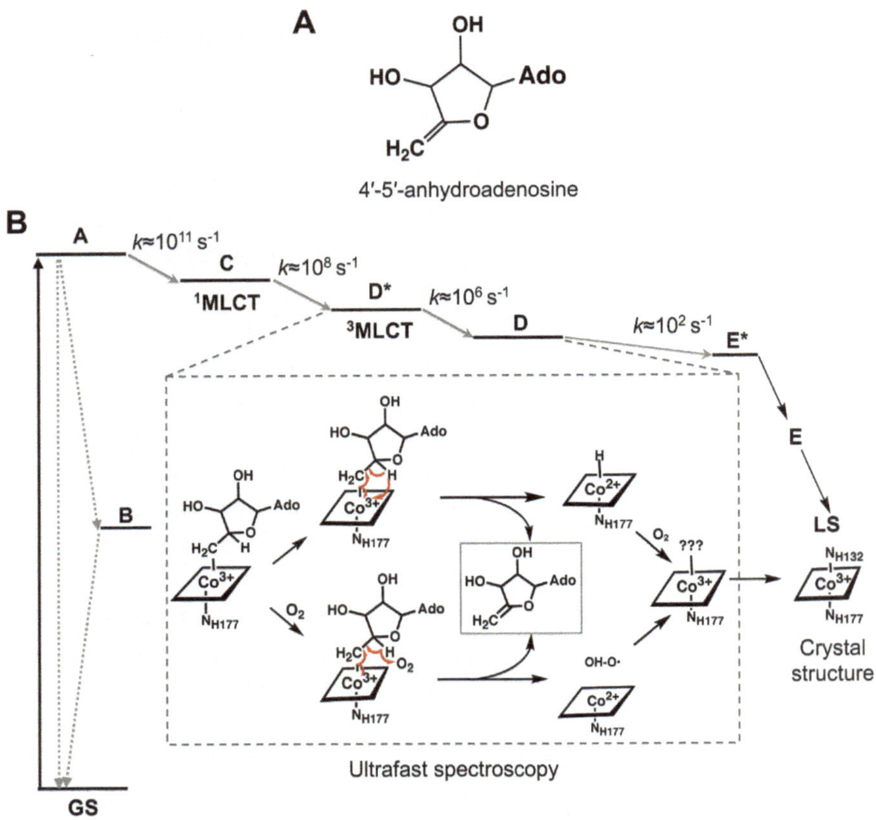

Fig. 7 CarH photochemistry. (A) Chemical structure of 4′,5′-anhydroadenosine, the AdoCbl-CarH$_{Tt}$ photolysis product. (B) Scheme for the proposed CarH photolysis mechanism indicating ground state (GS), final "light" state (LS), excited cob(III)alamin-like metal-to-ligand charge transfer singlet (^{1}MLCT, C) and triplet (^{3}MLCT, D*) states, and initial (A) and final intermediates (E*, E). The long-lived D* state formed with high quantum efficiency allows reactions (inset) yielding the 4′,5′-anhydroadenosine product directly via a cobalt hydride that reacts rapidly with O_2 (top), or by interaction with O_2 (bottom). An undefined cob(III)alamin species generates the final bis-His CarH product. In the earlier model (Kutta et al., 2015), the initial excited state A relaxes to GS (dotted grey arrows) directly (~90%) or through a non-productive intermediate B (~2%). Heterolytic Co-C cleavage at low quantum yield was proposed for the rest (~8%) in the D* state. *Part B: Adapted from Miller, N. A., Kaneshiro, A. K., Konar, A., Alonso-Mori, R., Britz, A., Deb, A., et al. (2020). The photoactive excited state of the B_{12}-based photoreceptor CarH.* The Journal of Physical Chemistry B, 124*(47), 10732–10738. 10.1021/acs.jpcb.0c09428.*

laser pulse photolysis of AdoCbl-$CarH_{Tt}$ (Kutta et al., 2015; Miller et al., 2020). Based on data obtained using near-UV/blue excitation (370 nm) pulses, it was proposed that ~90% of excited AdoCbl-$CarH_{Tt}$ reverted to the initial ground state, a small amount (~2%) from Co-C homolysis was not released, and the rest (~8%) underwent an unprecedented, heterolytic AdoCbl Co-C cleavage (Fig. 3D) to limit Ado• formation (Kutta et al., 2015). A theoretical QM/MM (quantum mechanics/molecular mechanics) study, based on an unsolvated monomer model generated starting from the $CarH_{Tt}$ tetramer-DNA crystal structure, suggested that rather than alter photochemistry, CarH controlled AdoCbl ribose moiety stereochemistry (Toda, Mamun, Lodowski, & Kozlowski, 2020). Another computational study using the same AdoCbl-bound CarH monomer model, but with solvent included, led to the conclusion that solvent exposure of the CarH Ado group and interactions of the Trp and Glu of the Wx_9EH motif contributed to stabilize the electronic states involving charge transfer and enable the proposed heterolytic cleavage (Cooper, Panitz, Edwards, & Goyal, 2021). The most recent transient ultrafast spectroscopy study, using green (520–565 nm) excitation pulses and transient X-ray absorption near edge structure (XANES) spectra to directly probe changes in the excited-state structure (Miller et al., 2020), proposed an alternative mechanism. Rather than Co–C bond heterolysis of low quantum yield, intersystem crossing generated a six-coordinate triplet excited state (D* in Fig. 7B), which was long-lived (as it persisted over μs) and provided time for excited-state reactions to form, with high quantum efficiency, the stable and safe 4′,5′-anhydroadenosine and bis-His Cbl-CarH photoproducts, bypassing bond homolysis and formation of highly reactive radicals. The issue of homolytic versus heterolytic Co–C bond cleavage is thus a matter of debate. Clearly, CarH somehow steers AdoCbl photochemistry for safe use as a chromophore, the probable key players being the protein environment around the Ado group, notably residues of the Wx_9EH motif, and factors like molecular oxygen. More experimental-theoretical work could help to better define the novel molecular mechanisms and reactive species behind CarH photochemistry.

4.5 Distribution of CarH photoreceptors and other B_{12}-binding light sensors

CarH homologs with all its functional domains and signature motifs (DBD with RxWERRY and CBD with Wx_9EH, or the less frequent $Wx_{10}EH$ found in *Streptomyces*, and $ExHxxGx_{41}SxVx_{22}GG$ motifs) are absent among eukaryotes, extremely rare in archaea, and widespread across diverse bacterial species, mostly belonging to Actinobacteria, Bacteroidetes, β-proteobacteria, Chloroflexi, δ-proteobacteria, Firmicutes, and Thermus-Deinococcus (Padmanabhan et al., 2017). Not all species within given bacterial phyla have CarH. Only a subset of those with CarH can synthesize B_{12} *de novo*, and others must rely on their environmental niche and/or their microbial consortia to acquire B_{12} for CarH activity. Most species generally have one *carH* gene, often amidst genes encoding proteins with light-related functions, such as those for carotenoid biosynthesis, DNA photolyase, and some regulatory proteins. Usually, CarH regulates its own gene and ones in its genomic vicinity; however, as shown in *B. megaterium*, CarH can also control genes at distant, unlinked genome loci. Genes for CarH and carotenogenesis frequently co-occur, suggesting a possible evolutionary relationship that needs to be explored. The *M. xanthus* CarH paralog CarA, which binds to B_{12} but does not need it for function, illustrates two aspects. A second CarH-like protein in the same species need not depend on B_{12}, and limited B_{12} availability might drive evolution of systems not requiring B_{12} to maintain a given protective cell response (Pérez-Castaño et al., 2022).

The photosensory CarH CBD is no exception to the finding that most photosensory domains are autonomous folded units that can be the entire protein or associate with diverse effector modules. Association with a MerR-type DBD as in CarH is most common, but genome data reveal many CBD standalone proteins as well as some with CBD linked to modules for heme/oxygen-sensing globins, histidine kinases, and GAF domains (cGMP-specific phosphodiesterases, adenylyl cyclases, transcription factor FhlA), among others (Cheng, Yamamoto, & Bauer, 2016; Padmanabhan et al., 2017; Padmanabhan, Pérez-Castaño, & Elías-Arnanz, 2019). These proteins have unknown functions and characterizing them are exciting prospects for future work. They may coexist with a *bona fide* CarH in some species, suggesting possible crosstalk between them and their functions. It must be cautioned that only a subset of the large number of putative B_{12}-binding proteins being identified may have light-sensing functions, because B_{12} binding can also occur to enzymes and proteins involved in transport whose functions are not based on light.

Standalone CBD proteins, which are the most numerous after CarH homologs, occur primarily among Firmicutes, Actinobacteria, and Deltaproteobacteria. Since they lack an obvious effector module, their B_{12}/light-dependent activities likely depend on heterotypic interactions with other effector proteins, in contrast to homotypic CarH oligomer interactions. An interesting, functionally characterized standalone

B_{12}-binding protein is AerR (or PpaA), found in purple photosynthetic α-proteobacteria. It has a E/DxHxxGx$_{41}$TxGx$_{22}$GG base-off/His-on B_{12}-binding motif but no Wx$_9$EH motif, and it binds to AqCbl generated from AdoCbl or MeCbl photolysis to control light-dependent gene expression (Cheng, Li, Hammad, Karty, & Bauer, 2014; Vermeulen & Bauer, 2015). AqCbl-AerR interacts directly with another protein, the redox-regulated CrtJ repressor, to modulate DNA binding by CrtJ and regulate gene expression under aerobic and anaerobic photosynthetic growth conditions (Fang & Bauer, 2017). CrtJ is a dimer (its homolog PpsR from some species is a tetramer (Winkler et al., 2013)) that forms with AerR a 1:2 AerR:CrtJ$_2$ complex, whose regulon includes photosystem genes for the synthesis of carotenoids (like CarH), bacteriochlorophyll, light-harvesting complex II structural proteins and many others (Fang & Bauer, 2017). Interestingly, AerR is expressed full-length (LAerR) or as a 40-residue shorter isoform (SAerR) from an alternative internal transcription start site, depending on growth conditions. LAerR binds tightly through a bis-His linkage to photolyzed AdoCbl/MeCbl, but this upper axial His is absent in SAerR, which binds promiscuously to various B_{12} derivatives in the dark or light (Yamamoto, Fang, Dragnea, & Bauer, 2018). Both forms interact with CrtJ but with opposite effects on gene expression. The distinct B_{12} and light-dependent modes of action of AerR and CarH thus presage novel and diverse mechanisms for discovery in future studies of putative B_{12}-based photoreceptors.

4.6 CarH photoreceptor use in optogenetics and synthetic biology

Ever since its discovery, the use of CarH as a versatile optogenetics/synthetic biology tool has grown steadily (Table 1), and its unusual photochemistry has even inspired development of B_{12}-based light-driven catalytic systems for mild olefin formation and other chemical transformations (Bam, Pollatos, Moser, & West, 2020; Yang et al., 2022). Most applications exploit the autonomous CarH CBD for light-controlled protein oligomerization and function. CarH CBD senses UV and blue light, but its major selling point is as one of the few suitable for green light optogenetics, and it is nontoxic and tolerated in the biological systems where it has been tested.

Even before molecular details were known, the CarH-vitamin B_{12} pair was used for conditional repression and photoactivated expression of essential genes in *M. xanthus*. Here, a single plasmid bears a constitutively expressing *carH* gene and the gene of interest under CarH operator-promoter control,

Table 1 Optogenetics-synthetic biology applications of B_{12}-based CarH photoreceptors.

Application	Tested in organism/cells	Reference
Light/B_{12}-dependent expression of essential genes in bacteria	*M. xanthus* (Gram-negative bacterium)	García-Moreno et al. (2009)
Green-light/B_{12}-dependent inactivation of receptor signaling	HEK293 (human embryonic kidney 293) cells; zebrafish (*Danio rerio*) embryos	Kainrath, Stadler, Reichhart, Distel, and Janovjak (2017)
SpyTag-SpyCatcher chemistry coupled with CarH CBD to generate B_{12}-dependent photoresponsive protein hydrogels for controlled stem cell/protein encapsulation, injectability and release	Mouse fibroblasts (NIH/3T3); human mesenchymal stem cells (hMSCs), mice	Wang, Yang, Luo, Hsing, and Sun (2017) Jiang et al. (2020) Narayan, Mu, Hasturk, and Kaplan (2021)
SpyTag-SpyCatcher chemistry coupled with CarH CBD to generate protein-based, photo- and thermoresponsive layer-by-layer nanofilm on a gold surface		Wei, Wu, Wang, Yang, and Zhang (2018)
B_{12}-/green-light directed control of transgene expression system in mammalian and plant cells	HEK293, human cervix carcinoma cells (HeLa); mouse fibroblasts (NIH/3T3); African green monkey kidney fibroblast-like cells (COS-7); *A. thaliana* protoplasts	Chatelle et al. (2018)
High precision green-light/B_{12}-dependent, spatiotemporal control of integrin-mediated cell adhesions to extracellular matrix	Human cancer cell line (MCF-7)	Xu, Ricken, and Wegner (2020)
Photoresponsive green-light/B_{12}-dependent cell-cell interactions for artificial cell adhesions	Human MDA-MB-231 epithelial cells	Nzigou Mombo, Bijonowski, Rasoulinejad, Mueller, and Wegner (2021)
B_{12}-based, green-light-operated, smart-watch-controlled ("Glow Control") mammalian gene switch for programmed percutaneous remote control of implanted human cells	Human cell lines, mouse	Mansouri et al. (2021)

whereby the latter is repressed in the dark with B_{12} present and activated by light (García-Moreno et al., 2009). Another application has CarH CBD fused to a membrane-tethered signaling receptor for AdoCbl-mediated interaction in the dark and its abrogation by light. It conferred robust temporal green-light control of signaling pathways in human cell lines and in zebrafish embryo development, a vertebrate model (Kainrath et al., 2017). An optogenetic system comprising a constitutively expressed CarH-transactivator domain fusion protein and a reporter plasmid with multiple CarH operators upstream of a minimal eukaryotic promoter has been engineered for controlled targeted gene activation (high ~350-fold induction, low leakiness, good reversibility) in the dark and green-light inactivation in mammalian cells and plant protoplasts (Chatelle et al., 2018). Recently, a wearable green-light controlled smart-watch device named Glow Control was designed using CarH CBD for robust, tunable, and reversible on-demand transgene expression in various mammalian and human cell types (Mansouri et al., 2021). It consists of membrane-tethered CarH CBD to sequester a synthetic cytosolic CarH CBD-fused transactivator protein, which is programmed for green-light triggered release, translocation into the nucleus, and transgene expression. The device works in various mammalian and human cell types and, as proof-of-concept, was tasked with programmed percutaneous control of implanted cells to produce and release human glucagon-like peptide-1 in a type-2 diabetes mouse model.

SpyTag/SpyCatcher technology (Keeble & Howarth, 2020) has been used with CarH CBD to produce AdoCbl/light-sensitive biomimetic and elastic protein hydrogels and nanofilms that have facile encapsulation, injectability and release properties for biomedical applications like drug and cell delivery, and tissue engineering (Jiang et al., 2020; Narayan et al., 2021; Wang et al., 2017; Wei et al., 2018). In the design, the two ends of CarH CBD-bridged elastin-like pentapeptide repeat (ELP) units are fused to the 13-residue SpyTag peptide, or separately the 113-residue SpyCatcher protein. These two complementary telechelic proteins are purified and mixed to produce protein polymers held covalently by the SpyTag-SpyCatcher isopeptide linkage. CBD-mediated interchain interactions in the dark with AdoCbl present form protein hydrogels, which disassemble rapidly when exposed to light. Spatiotemporal green-light control of cell-cell adhesions has also been achieved using CarH CBD (Nzigou Mombo et al., 2021; Xu et al., 2020). Basically, a layer of cell-adhesion peptides are masked in the dark by AdoCbl-CarH CBD tetramers, whose disruption by green light unmasks the peptides, which bind to cell integrins for adhesions (Xu et al., 2020). Alternatively, in the dark with AdoCbl present, CarH CBD-fused adhesion peptides immobilized with a metal affinity tag bind to integrins for cell adhesion and this is disrupted by green light. Cell-cell adhesions were also engineered by expressing CarH CBD as a fusion protein with an N-terminal secretion signal and a C-terminal membrane anchoring domain, which places CarH CBD on the outer cell surface (Nzigou Mombo et al., 2021). Distinct cells adhere in the dark with AdoCbl present through intercellular interactions between their surface CarH CBDs that can be disrupted by green light to separate the adherent cells. All these diverse optogenetic applications of the CarH CBD in the short time since its discovery augur well for new avenues to be pursued in this area.

5. Concluding remarks

This chapter has surveyed B_{12}-based photoreceptors from their discovery only a decade ago to the distribution, molecular basis of action, and increasing use as optogenetics tools over the last five years, with some asides on general aspects of photoreceptors and B_{12} chemistry and biology. Studies thus far have shown how nature has repurposed vitamin B_{12} and structural modules seen in enzymes and DNA-binding proteins to build a photosensory regulator, with an unusual photochemistry and a surprising plasticity in oligomerization and effector functions. Many details of these photoreceptors are understood, yet various questions remain to be answered. Future studies can be expected to deliver more surprises to add to the rich trove already known.

Acknowledgments

We thank our group and research collaborators who contributed to the work on B_{12} photoreceptors, currently supported by the Ministerio de Ciencia, Innovación y Universidades (MCIU)/Agencia Estatal de Investigación (AEI)-Spain/European Regional Development Fund (ERDF) grants PGC2018-094635-B-C21 to M.E.-A. and PGC2018-094635-B-C22 to S.P., both funded by MCIU/AEI 10.13039/501100011033 and by 'ERDF A way of making Europe'; Ph.D. contract to R. P.-C (BES-2016-076883, MINECO/AEI) and Fundación Séneca (Murcia)-Spain grant 20992/PI/18 to M.E.-A.

References

Bam, R., Pollatos, A. S., Moser, A. J., & West, J. G. (2020). Mild olefin formation via bio-inspired vitamin B_{12} photocatalysis. *Chemical Science*, *12*(5), 1736–1744. https://doi.org/10.1039/d0sc05925k.

Banerjee, R., Gouda, H., & Pillay, S. (2021). Redox-linked coordination chemistry directs vitamin B_{12} trafficking. *Accounts of Chemical Research*, *54*(8), 2003–2013. https://doi.org/10.1021/acs.accounts.1c00083.

Banerjee, R., & Ragsdale, S. W. (2003). The many faces of vitamin B_{12}: Catalysis by cobalamin-dependent enzymes. *Annual Review of Biochemistry*, *72*, 209–247. https://doi.org/10.1146/annurev.biochem.72.121801.161828.

Barker, H. A., Weissbach, H., & Smyth, R. D. (1958). A coenzyme containing pseudovitamin B_{12}. *Proceedings of the National Academy of Sciences of the United States of America*, *44*(11), 1093–1097. https://doi.org/10.1073/pnas.44.11.1093.

Botella, J. A., Murillo, F. J., & Ruiz-Vazquez, R. (1995). A cluster of structural and regulatory genes for light-induced carotenogenesis in *Myxococcus xanthus*. *European Journal of Biochemistry*, *233*(1), 238–248. http://www.ncbi.nlm.nih.gov/pubmed/7588751.

Bridwell-Rabb, J., & Drennan, C. L. (2017). Vitamin B_{12} in the spotlight again. *Current Opinion in Chemical Biology*, *37*, 63–70. https://doi.org/10.1016/j.cbpa.2017.01.013.

Bryant, D. A., Hunter, C. N., & Warren, M. J. (2020). Biosynthesis of the modified tetrapyrroles-the pigments of life. *Journal of Biological Chemistry*, *295*(20), 6888–6925. https://doi.org/10.1074/jbc.REV120.006194.

Camacho, I. S., Black, R., Heyes, D. J., Johannissen, L. O., Ramakers, L. A. I., Bellina, B., et al. (2021). Interplay between chromophore binding and domain assembly by the B_{12}-dependent photoreceptor protein, CarH. *Chemical Science*, *12*(24), 8333–8341. https://doi.org/10.1039/D1SC00522G.

Cervantes, M., & Murillo, F. J. (2002). Role for vitamin B_{12} in light induction of gene expression in the bacterium *Myxococcus xanthus*. *Journal of Bacteriology*, *184*(8), 2215–2224. https://doi.org/10.1128/JB.184.8.2215-2224.2002.

Chan, C. W., & Mondragón, A. (2020). Crystal structure of an atypical cobalamin riboswitch reveals RNA structural adaptability as basis for promiscuous ligand binding. *Nucleic Acids Research*, *48*(13), 7569–7583. https://doi.org/10.1093/nar/gkaa507.

Chatelle, C., Ochoa-Fernandez, R., Engesser, R., Schneider, N., Beyer, H. M., Jones, A. R., et al. (2018). A green-light-responsive system for the control of transgene expression in mammalian and plant cells. *ACS Synthetic Biology*, 7(5), 1349–1358. https://doi.org/10.1021/acssynbio.7b00450.

Cheng, Z., Li, K., Hammad, L. A., Karty, J. A., & Bauer, C. E. (2014). Vitamin B_{12} regulates photosystem gene expression via the CrtJ antirepressor AerR in *Rhodobacter capsulatus*. *Molecular Microbiology*, *91*(4), 649–664. https://doi.org/10.1111/mmi.12491.

Cheng, Z., Yamamoto, H., & Bauer, C. E. (2016). Cobalamin's (vitamin B_{12}) surprising function as a photoreceptor. *Trends Biochemical Sciences*, *41*(8), 647–650. https://doi.org/10.1016/j.tibs.2016.05.002.

Cole, A. G., Yoder, L. M., Shiang, J. J., Anderson, N. A., Walker, L. A., 2nd, Banaszak Holl, M. M., et al. (2002). Time-resolved spectroscopic studies of B_{12} coenzymes: a comparison of the primary photolysis mechanism in methyl-, ethyl-, n-propyl-, and 5'-deoxyadenosylcobalamin. *Journal of the American Chemical Society*, *124*(3), 434–441. https://doi.org/10.1021/ja011628s.

Cooper, C. L., Panitz, N., Edwards, T. A., & Goyal, P. (2021). Role of the CarH photoreceptor protein environment in the modulation of cobalamin photochemistry. *Biophys Journal*, *120*(17), 3688–3696. https://doi.org/10.1016/j.bpj.2021.07.020.

Cracan, V., & Banerjee, R. (2013). Cobalt and corrinoid transport and biochemistry. In L. Banci (Ed.), *Vol. 12. Metallomics and the Cell* (2013/04/19 ed., pp. 333–374). Springer. https://doi.org/10.1007/978-94-007-5561-1_10.

Croce, R., & van Amerongen, H. (2014). Natural strategies for photosynthetic light harvesting. *Nature Chemical Biology*, *10*(7), 492–501. https://doi.org/10.1038/nchembio.1555.

Croce, R., & van Amerongen, H. (2020). Light harvesting in oxygenic photosynthesis: Structural biology meets spectroscopy. *Science*, *369*(6506), eaay2058. https://doi.org/10.1126/science.aay2058.

DebRoy, S., Gebbie, M., Ramesh, A., Goodson, J. R., Cruz, M. R., van Hoof, A., et al. (2014). Riboswitches. A riboswitch-containing sRNA controls gene expression by sequestration of a response regulator. *Science*, *345*(6199), 937–940. https://doi.org/10.1126/science.1255091.

Degnan, P. H., Taga, M. E., & Goodman, A. L. (2014). Vitamin B_{12} as a modulator of gut microbial ecology. *Cell Metabolism*, *20*(5), 769–778. https://doi.org/10.1016/j.cmet.2014.10.002.

Di Mascio, P., Martinez, G. R., Miyamoto, S., Ronsein, G. E., Medeiros, M. H. G., & Cadet, J. (2019). Singlet molecular oxygen reactions with nucleic acids, lipids, and proteins. *Chemical Reviews*, *119*(3), 2043–2086. https://doi.org/10.1021/acs.chemrev.8b00554.

Díez, A. I., Ortiz-Guerrero, J. M., Ortega, A., Elías-Arnanz, M., Padmanabhan, S., & García de la Torre, J. (2013). Analytical ultracentrifugation studies of oligomerization and DNA-binding of TtCarH, a *Thermus thermophilus* coenzyme B_{12}-based photosensory regulator. *European Biophysics Journal*, *42*(6), 463–476. https://doi.org/10.1007/s00249-013-0897-x.

Dowling, D. P., Croft, A. K., & Drennan, C. L. (2012). Radical use of Rossmann and TIM barrel architectures for controlling coenzyme B_{12} chemistry. *Annual Review of Biophysics*, *41*, 403–427. https://doi.org/10.1146/annurev-biophys-050511-102225.

Drennan, C. L., Huang, S., Drummond, J. T., Matthews, R. G., & Ludwig, M. L. (1994). How a protein binds B_{12}: A 3.0 A X-ray structure of B_{12}-binding domains of methionine synthase. *Science*, *266*(5191), 1669–1674. https://doi.org/10.1126/science.7992050.

Eschenmoser, A., & Wintner, C. E. (1977). Natural product synthesis and vitamin B_{12}. *Science*, *196*(4297), 1410–1420. http://www.ncbi.nlm.nih.gov/pubmed/867037.

Fang, M., & Bauer, C. E. (2017). The vitamin B_{12}-dependent photoreceptor AerR relieves photosystem gene repression by extending the interaction of CrtJ with photosystem promoters. *mBio*, *8*(2), e00261-00217. https://doi.org/10.1128/mBio.00261-17.

Fernández-Zapata, J., Pérez-Castaño, R., Aranda, J., Colizzi, F., Polanco, M. C., Orozco, M., et al. (2018). Plasticity in oligomerization, operator architecture, and DNA binding in the mode of action of a bacterial B_{12}-based photoreceptor. *Journal of Biological Chemistry*, *293*(46), 17888–17905. https://doi.org/10.1074/jbc.RA118.004838.

Gallego-García, A., Monera-Girona, A. J., Pajares-Martínez, E., Bastida-Martínez, E., Pérez-Castaño, R., Iniesta, A. A., et al. (2019). A bacterial light response reveals an orphan desaturase for human plasmalogen synthesis. *Science*, *366*(6461), 128–132. https://doi.org/10.1126/science.aay1436.

García-Moreno, D., Polanco, M. C., Navarro-Avilés, G., Murillo, F. J., Padmanabhan, S., & Elías-Arnanz, M. (2009). A vitamin B_{12}-based system for conditional expression reveals *dksA* to be an essential gene in *Myxococcus xanthus*. *Journal of Bacteriology*, *191*(9), 3108–3119. https://doi.org/10.1128/JB.01737-08.

Gherasim, C., Lofgren, M., & Banerjee, R. (2013). Navigating the B_{12} road: Assimilation, delivery, and disorders of cobalamin. *Journal of Biological Chemistry*, *288*(19), 13186–13193. https://doi.org/10.1074/jbc.R113.458810.

Giedyk, M., Goliszewska, K., & Gryko, D. (2015). Vitamin B_{12} catalysed reactions. *Chemical Society Reviews*, *44*(11), 3391–3404. https://doi.org/10.1039/c5cs00165j.

Glaeser, J., Nuss, A. M., Berghoff, B. A., & Klug, G. (2011). Singlet oxygen stress in microorganisms. *Advances in Microbial Physiology*, *58*, 141–173. https://doi.org/10.1016/B978-0-12-381043-4.00004-0.

Gruber, K., Puffer, B., & Krautler, B. (2011). Vitamin B_{12}-derivatives-enzyme cofactors and ligands of proteins and nucleic acids. *Chemical Society Reviews*, *40*(8), 4346–4363. https://doi.org/10.1039/c1cs15118e.

Hodgkin, D. C., Kamper, J., Mackay, M., Pickworth, J., Trueblood, K. N., & White, J. G. (1956). Structure of vitamin B_{12}. *Nature*, *178*(4524), 64–66. http://www.ncbi.nlm.nih.gov/pubmed/13348621.

Hogenkamp, H. P. (1963). A cyclic nucleoside derived from coenzyme B_{12}. *Journal of Biological Chemistry, 238*(2), 477–480. http://www.ncbi.nlm.nih.gov/pubmed/5940929.

Hogenkamp, H. P. (1966). The photolysis of methylcobalamin. *Biochemistry, 5*(2), 417–422. http://www.ncbi.nlm.nih.gov/pubmed/5940929.

Jiang, B., Liu, X., Yang, C., Yang, Z., Luo, J., Kou, S., et al. (2020). Injectable, photoresponsive hydrogels for delivering neuroprotective proteins enabled by metal-directed protein assembly. *Science Advances, 6*(41), eabc4824. https://doi.org/10.1126/sciadv.abc4824.

Johnson, C. H., Zhao, C., Xu, Y., & Mori, T. (2017). Timing the day: What makes bacterial clocks tick? *Nature Reviews: Microbiology, 15*(4), 232–242. https://doi.org/10.1038/nrmicro.2016.196.

Jones, A. R. (2017). The photochemistry and photobiology of vitamin B_{12}. *Photochemical and Photobiological Sciences, 16*(6), 820–834. https://doi.org/10.1039/c7pp00054e.

Jost, M., Fernández-Zapata, J., Polanco, M. C., Ortiz-Guerrero, J. M., Chen, P. Y., Kang, G., et al. (2015). Structural basis for gene regulation by a B_{12}-dependent photoreceptor. *Nature, 526*(7574), 536–541. https://doi.org/10.1038/nature14950.

Jost, M., Simpson, J. H., & Drennan, C. L. (2015). The transcription factor CarH safeguards use of adenosylcobalamin as a light sensor by altering the photolysis products. *Biochemistry, 54*(21), 3231–3234. https://doi.org/10.1021/acs.biochem.5b00416.

Kainrath, S., Stadler, M., Reichhart, E., Distel, M., & Janovjak, H. (2017). Green-light-induced inactivation of receptor signaling using cobalamin-binding domains. *Angewandte Chemie (International Edition), 56*(16), 4608–4611. https://doi.org/10.1002/anie.201611998.

Kaval, K. G., & Garsin, D. A. (2018). Ethanolamine utilization in bacteria. *mBio, 9*(1), e00066-00018. https://doi.org/10.1128/mBio.00066-18.

Keeble, A. H., & Howarth, M. (2020). Power to the protein: Enhancing and combining activities using the Spy toolbox. *Chemical Science, 11*(28), 7281–7291. https://doi.org/10.1039/d0sc01878c.

Kieninger, C., Deery, E., Lawrence, A. D., Podewitz, M., Wurst, K., Nemoto-Smith, E., et al. (2019). The hydrogenobyric acid structure reveals the corrin ligand as an entatic state module empowering B_{12} cofactors for catalysis. *Angewandte Chemie (International Edition), 58*(31), 10756–10760. https://doi.org/10.1002/anie.201904713.

Korkhov, V. M., Mireku, S. A., Veprintsev, D. B., & Locher, K. P. (2014). Structure of AMP-PNP-bound BtuCD and mechanism of ATP-powered vitamin B_{12} transport by BtuCD-F. *Nature Structural & Molecular Biology, 21*(12), 1097–1099. https://doi.org/10.1038/nsmb.2918.

Koronowski, K. B., & Sassone-Corsi, P. (2021). Communicating clocks shape circadian homeostasis. *Science, 371*(6530), eabd0951. https://doi.org/10.1126/science.abd0951.

Koutmos, M., Gherasim, C., Smith, J. L., & Banerjee, R. (2011). Structural basis of multifunctionality in a vitamin B_{12}-processing enzyme. *Journal of Biological Chemistry, 286*(34), 29780–29787. https://doi.org/10.1074/jbc.M111.261370.

Kutta, R. J., Hardman, S. J., Johannissen, L. O., Bellina, B., Messiha, H. L., Ortiz-Guerrero, J. M., et al. (2015). The photochemical mechanism of a B_{12}-dependent photoreceptor protein. *Nature Communications, 6*, 7907. https://doi.org/10.1038/ncomms8907.

Law, P. Y., & Wood, J. M. (1973). The photolysis of 5'-deoxyadenosylcobalamin under anaerobic conditions. *Biochimica et Biophysica Acta, 331*(3), 451–454. http://www.ncbi.nlm.nih.gov/pubmed/4777666.

León, E., Navarro-Avilés, G., Santiveri, C. M., Flores-Flores, C., Rico, M., González, C., et al. (2010). A bacterial antirepressor with SH3 domain topology mimics operator DNA in sequestering the repressor DNA recognition helix. *Nucleic Acids Research, 38*(15), 5226–5241. https://doi.org/10.1093/nar/gkq277.

Li, Z., Mascarenhas, R., Twahir, U. T., Kallon, A., Deb, A., Yaw, M., et al. (2020). An inter-protein Co-S coordination complex in the B_{12}-trafficking pathway. *Journal of the American Chemical Society*, *142*(38), 16334–16345. https://doi.org/10.1021/jacs.0c06590.

López-Rubio, J. J., Elías-Arnanz, M., Padmanabhan, S., & Murillo, F. J. (2002). A repressor-antirepressor pair links two loci controlling light-induced carotenogenesis in *Myxococcus xanthus*. *Journal of Biological Chemistry*, *277*(9), 7262–7270. https://doi.org/10.1074/jbc.M110351200.

López-Rubio, J. J., Padmanabhan, S., Lázaro, J. M., Salas, M., Murillo, F. J., & Elías-Arnanz, M. (2004). Operator design and mechanism for CarA repressor-mediated down-regulation of the photoinducible *carB* operon in *Myxococcus xanthus*. *Journal of Biological Chemistry*, *279*(28), 28945–28953. https://doi.org/10.1074/jbc.M403459200.

Lundrigan, M. D., Köster, W., & Kadner, R. J. (1991). Transcribed sequences of the *Escherichia coli btuB* gene control its expression and regulation by vitamin B_{12}. *Proceedings of the National Academy of Sciences of the United States of America*, *88*(4), 1479–1483. https://doi.org/10.1073/pnas.88.4.1479.

Mandal, M., & Breaker, R. R. (2004). Gene regulation by riboswitches. *Nature Reviews: Molecular Cell Biology*, *5*(6), 451–463. https://doi.org/10.1038/nrm1403.

Mansouri, M., Hussherr, M.-D., Strittmatter, T., Buchmann, P., Xue, S., Camenisch, G., et al. (2021). Smart-watch-programmed green-light-operated percutaneous control of therapeutic transgenes. *Nature Communications*, *12*(1), 3388. https://doi.org/10.1038/s41467-021-23572-4.

Matthews, R. G., Koutmos, M., & Datta, S. (2008). Cobalamin-dependent and cobamide-dependent methyltransferases. *Current Opinion in Structural Biology*, *18*(6), 658–666. https://doi.org/10.1016/j.sbi.2008.11.005.

Mayer, B. J. (2015). The discovery of modular binding domains: Building blocks of cell signalling. *Nature Reviews: Molecular Cell Biology*, *16*(11), 691–698. https://doi.org/10.1038/nrm4068.

McNerney, M. P., Piorino, F., Michel, C. L., & Styczynski, M. P. (2020). Active analyte import improves the dynamic range and sensitivity of a vitamin B_{12} biosensor. *ACS Synthetic Biology*, *9*(2), 402–411. https://doi.org/10.1021/acssynbio.9b00429.

Mellin, J. R., Koutero, M., Dar, D., Nahori, M. A., Sorek, R., & Cossart, P. (2014). Riboswitches. Sequestration of a two-component response regulator by a riboswitch-regulated noncoding RNA. *Science*, *345*(6199), 940–943. https://doi.org/10.1126/science.1255083.

Miller, N. A., Kaneshiro, A. K., Konar, A., Alonso-Mori, R., Britz, A., Deb, A., et al. (2020). The photoactive excited state of the B_{12}-based photoreceptor CarH. *The Journal of Physical Chemistry B*, *124*(47), 10732–10738. https://doi.org/10.1021/acs.jpcb.0c09428.

Minot, G. R., & Murphy, W. P. (1926). Treatment of pernicious anemia by a special diet. *Journal of the American Medical Association*, *87*, 470–476.

Nahvi, A., Barrick, J. E., & Breaker, R. R. (2004). Coenzyme B_{12} riboswitches are widespread genetic control elements in prokaryotes. *Nucleic Acids Research*, *32*(1), 143–150. https://doi.org/10.1093/nar/gkh167.

Nahvi, A., Sudarsan, N., Ebert, M. S., Zou, X., Brown, K. L., & Breaker, R. R. (2002). Genetic control by a metabolite binding mRNA. *Chemical Biology*, *9*(9), 1043–1049. https://doi.org/10.1016/s1074-5521(02)00224-7.

Narayan, O. P., Mu, X., Hasturk, O., & Kaplan, D. L. (2021). Dynamically tunable light responsive silk-elastin-like proteins. *Acta Biomaterialia*, *121*, 214–223. https://doi.org/10.1016/j.actbio.2020.12.018.

Navarro-Avilés, G., Jiménez, M. A., Pérez-Marín, M. C., González, C., Rico, M., Murillo, F. J., et al. (2007). Structural basis for operator and antirepressor recognition by *Myxococcus xanthus* CarA repressor. *Molecular Microbiology*, *63*(4), 980–994. https://doi.org/10.1111/j.1365-2958.2006.05567.x.

Nelson, N., & Junge, W. (2015). Structure and energy transfer in photosystems of oxygenic photosynthesis. *Annual Review of Biochemistry*, *84*, 659–683. https://doi.org/10.1146/annurev-biochem-092914-041942.

Nzigou Mombo, B., Bijonowski, B. M., Rasoulinejad, S., Mueller, M., & Wegner, S. V. (2021). Spatiotemporal control over multicellular migration using green light reversible cell-cell interactions. *Advanced Biology*, *5*(5), e2000199. https://doi.org/10.1002/adbi.202000199.

Ortiz-Guerrero, J. M., Polanco, M. C., Murillo, F. J., Padmanabhan, S., & Elias-Arnanz, M. (2011). Light-dependent gene regulation by a coenzyme B_{12}-based photoreceptor. *Proceedings of the National Academy of Sciences of the United States of America*, *108*(18), 7565–7570. https://doi.org/10.1073/pnas.1018972108.

Osman, D., Cooke, A., Young, T. R., Deery, E., Robinson, N. J., & Warren, M. J. (2021). The requirement for cobalt in vitamin B_{12}: A paradigm for protein metalation. *Biochim Biophys Acta Mol Cell Res*, *1868*(1), 118896. https://doi.org/10.1016/j.bbamcr.2020.118896.

Padmanabhan, S., Jost, M., Drennan, C. L., & Elías-Arnanz, M. (2017). A new facet of vitamin B_{12}: Gene regulation by cobalamin-based photoreceptors. *Annual Review of Biochemistry*, *86*, 485–514. https://doi.org/10.1146/annurev-biochem-061516-044500.

Padmanabhan, S., Monera-Girona, A. J., Pérez-Castaño, R., Bastida-Martínez, E., Pajares-Martínez, E., Bernal-Bernal, D., et al. (2021). Light-triggered carotenogenesis in *Myxococcus xanthus*: New paradigms in photosensory signaling, transduction and gene regulation. *Microorganisms*, *9*(5), 1067. https://doi.org/10.3390/microorganisms9051067.

Padmanabhan, S., Pérez-Castaño, R., & Elías-Arnanz, M. (2019). B_{12}-based photoreceptors: From structure and function to applications in optogenetics and synthetic biology. *Current Opinion in Structural Biology*, *57*, 47–55. https://doi.org/10.1016/j.sbi.2019.01.020.

Palczewski, K., & Kiser, P. D. (2020). Shedding new light on the generation of the visual chromophore. *Proceedings of the National Academy of Sciences of the United States of America*, *117*(33), 19629–19638. https://doi.org/10.1073/pnas.2008211117.

Papaleo, E., Saladino, G., Lambrughi, M., Lindorff-Larsen, K., Gervasio, F. L., & Nussinov, R. (2016). The role of protein loops and linkers in conformational dynamics and allostery. *Chemical Reviews*, *116*(11), 6391–6423. https://doi.org/10.1021/acs.chemrev.5b00623.

Pérez-Castaño, R., Bastida-Martínez, E., Fernández-Zapata, J., Polanco, M. C., Galbis-Martínez, M. L., Iniesta, A. A., et al. (2022). Coenzyme B $_{12}$ -dependent and independent photoregulation of carotenogenesis across Myxococcales. *Environmental Microbiology*. https://doi.org/10.1111/1462-2920.15895.

Pérez-Marín, M. C., López-Rubio, J. J., Murillo, F. J., Elías-Arnanz, M., & Padmanabhan, S. (2004). The N-terminus of *Myxococcus xanthus* CarA repressor is an autonomously folding domain that mediates physical and functional interactions with both operator DNA and antirepressor protein. *Journal of Biological Chemistry*, *279*(32), 33093–33103. https://doi.org/10.1074/jbc.M405225200.

Pérez-Marín, M. C., Padmanabhan, S., Polanco, M. C., Murillo, F. J., & Elías-Arnanz, M. (2008). Vitamin B_{12} partners the CarH repressor to downregulate a photoinducible promoter in *Myxococcus xanthus*. *Molecular Microbiology*, *67*(4), 804–819. https://doi.org/10.1111/j.1365-2958.2007.06086.x.

Poddar, H., Heyes, D. J., Schirò, G., Weik, M., Leys, D., & Scrutton, N. S. (2022). A guide to time-resolved structural analysis of light-activated proteins. *FEBS Journal*, *289*(3), 576–595. https://doi.org/10.1111/febs.15880.

Polaski, J. T., Webster, S. M., Johnson, J. E., Jr., & Batey, R. T. (2017). Cobalamin riboswitches exhibit a broad range of ability to discriminate between methylcobalamin and adenosylcobalamin. *Journal of Biological Chemistry*, *292*(28), 11650–11658. https://doi.org/10.1074/jbc.M117.787176.

Ragsdale, S. W. (2008). Catalysis of methyl group transfers involving tetrahydrofolate and B_{12}. *Vitamins and Hormones*, *79*, 293–324. https://doi.org/10.1016/S0083-6729(08)00410-X.
Randaccio, L., Geremia, S., Demitri, N., & Wuerges, J. (2010). Vitamin B_{12}: Unique metalorganic compounds and the most complex vitamins. *Molecules*, *15*(5), 3228–3259. https://doi.org/10.3390/molecules15053228.
Rempel, S., Colucci, E., de Gier, J. W., Guskov, A., & Slotboom, D. J. (2018). Cysteine-mediated decyanation of vitamin B_{12} by the predicted membrane transporter BtuM. *Nature Communications*, *9*(1), 3038. https://doi.org/10.1038/s41467-018-05441-9.
Rempel, S., Gati, C., Nijland, M., Thangaratnarajah, C., Karyolaimos, A., de Gier, J. W., et al. (2020). A mycobacterial ABC transporter mediates the uptake of hydrophilic compounds. *Nature*, *580*(7803), 409–412. https://doi.org/10.1038/s41586-020-2072-8.
Rickes, E. L., Brink, N. G., Koniuszy, F. R., Wood, T. R., & Folkers, K. (1948a). Crystalline vitamin B_{12}. *Science*, *107*(2781), 396–397. https://doi.org/10.1126/science.107.2781.396.
Rickes, E. L., Brink, N. G., Koniuszy, F. R., Wood, T. R., & Folkers, K. (1948b). Vitamin B_{12}, a cobalt complex. *Science*, *108*(2797), 134. https://doi.org/10.1126/science.108.2797.134.
Roof, D. M., & Roth, J. R. (1992). Autogenous regulation of ethanolamine utilization by a transcriptional activator of the eut operon in *Salmonella typhimurium*. *Journal of Bacteriology*, *174*(20), 6634–6643. https://doi.org/10.1128/jb.174.20.6634-6643.1992.
Rosnow, J. J., Hwang, S., Killinger, B. J., Kim, Y. M., Moore, R. J., Lindemann, S. R., et al. (2018). A cobalamin activity-based probe enables microbial cell growth and finds new cobalamin-protein interactions across domains. *Applied and Environmental Microbiology*, *84*(18). https://doi.org/10.1128/aem.00955-18.
Roth, J. R., Lawrence, J. G., & Bobik, T. A. (1996). Cobalamin (coenzyme B_{12}): Synthesis and biological significance. *Annual Review of Microbiology*, *50*, 137–181. https://doi.org/10.1146/annurev.micro.50.1.137.
Rowley, C. A., & Kendall, M. M. (2019). To B_{12} or not to B_{12}: Five questions on the role of cobalamin in host-microbial interactions. *PLoS Pathogens*, *15*(1), e1007479. https://doi.org/10.1371/journal.ppat.1007479.
Rury, A. S., Wiley, T. E., & Sension, R. J. (2015). Energy cascades, excited state dynamics, and photochemistry in cob(III)alamins and ferric porphyrins. *Accounts of Chemical Research*, *48*(3), 860–867. https://doi.org/10.1021/ar5004016.
Schwartz, P. A., & Frey, P. A. (2007). 5'-Peroxyadenosine and 5'-peroxyadenosylcobalamin as intermediates in the aerobic photolysis of adenosylcobalamin. *Biochemistry*, *46*(24), 7284–7292. https://doi.org/10.1021/bi700077v.
Sension, R. J., Harris, D. A., & Cole, A. G. (2005). Time-resolved spectroscopic studies of B_{12} coenzymes: Comparison of the influence of solvent on the primary photolysis mechanism and geminate recombination of methyl-, ethyl-, n-propyl-, and 5'-deoxyadenosylcobalamin. *Journal of Physical Chemistry B*, *109*(46), 21954–21962. https://doi.org/10.1021/jp053202w.
Seong, J., & Lin, M. Z. (2021). Optobiochemistry: Genetically encoded control of protein activity by light. *Annual Review of Biochemistry*, *90*, 475–501. https://doi.org/10.1146/annurev-biochem-072420-112431.
Shcherbakova, D. M., Shemetov, A. A., Kaberniuk, A. A., & Verkhusha, V. V. (2015). Natural photoreceptors as a source of fluorescent proteins, biosensors, and optogenetic tools. *Annual Review of Biochemistry*, *84*, 519–550. https://doi.org/10.1146/annurev-biochem-060614-034411.
Shelton, A. N., Seth, E. C., Mok, K. C., Han, A. W., Jackson, S. N., Haft, D. R., et al. (2019). Uneven distribution of cobamide biosynthesis and dependence in bacteria predicted by comparative genomics. *ISME Journal*, *13*(3), 789–804. https://doi.org/10.1038/s41396-018-0304-9.

Smith, E. L. (1948a). Presence of cobalt in the anti-pernicious anaemia factor. *Nature, 162*(4108), 144. https://doi.org/10.1038/162144b0.
Smith, E. L. (1948b). Purification of anti-pernicious anaemia factors from liver. *Nature, 161*(4095), 638. http://www.ncbi.nlm.nih.gov/pubmed/18856623.
Sokolovskaya, O. M., Shelton, A. N., & Taga, M. E. (2020). Sharing vitamins: Cobamides unveil microbial interactions. *Science, 369*(6499), eaba0165. https://doi.org/10.1126/science.aba0165.
Stich, T. A., Brooks, A. J., Buan, N. R., & Brunold, T. C. (2003). Spectroscopic and computational studies of Co^{3+}-corrinoids: Spectral and electronic properties of the B_{12} cofactors and biologically relevant precursors. *Journal of the American Chemical Society, 125*(19), 5897–5914. https://doi.org/10.1021/ja029328d.
Stubbe, J. (1994). Binding site revealed of nature's most beautiful cofactor. *Science, 266*, 1663–1664. https://doi.org/10.1126/science.7992049.
Sukumar, N. (2013). Crystallographic studies on B_{12} binding proteins in eukaryotes and prokaryotes. *Biochimie, 95*(5), 976–988. https://doi.org/10.1016/j.biochi.2013.01.014.
Takano, H., Kondo, M., Usui, N., Usui, T., Ohzeki, H., Yamazaki, R., et al. (2011). Involvement of CarA/LitR and CRP/FNR family transcriptional regulators in light-induced carotenoid production in *Thermus thermophilus*. *Journal of Bacteriology, 193*(10), 2451–2459. https://doi.org/10.1128/JB.01125-10.
Takano, H., Mise, K., Hagiwara, K., Hirata, N., Watanabe, S., Toriyabe, M., et al. (2015). Role and function of LitR, an adenosyl B_{12}-bound light-sensitive regulator of *Bacillus megaterium* QM B1551, in regulation of carotenoid production. *Journal of Bacteriology, 197*(14), 2301–2315. https://doi.org/10.1128/JB.02528-14.
Toda, M. J., Mamun, A. A., Lodowski, P., & Kozlowski, P. M. (2020). Why is CarH photolytically active in comparison to other B_{12}-dependent enzymes? *Journal of Photochemistry and Photobiology. B: Biology, 209*, 111919. https://doi.org/10.1016/j.jphotobiol.2020.111919.
Toraya, T. (2003). Radical catalysis in coenzyme B_{12}-dependent isomerization (eliminating) reactions. *Chemical Reviews, 103*(6), 2095–2128. https://doi.org/10.1021/cr020428b.
Toraya, T. (2014). Cobalamin-dependent dehydratases and a deaminase: Radical catalysis and reactivating chaperones. *Archives of Biochemistry and Biophysics, 544*, 40–57. https://doi.org/10.1016/j.abb.2013.11.002.
Vermeulen, A. J., & Bauer, C. E. (2015). Members of the PpaA/AerR antirepressor family bind cobalamin. *Journal of Bacteriology, 197*(16), 2694–2703. https://doi.org/10.1128/JB.00374-15.
Wang, R., Yang, Z., Luo, J., Hsing, I. M., & Sun, F. (2017). B_{12}-dependent photoresponsive protein hydrogels for controlled stem cell/protein release. *Proceedings of the National Academy of Sciences of the United States of America, 114*(23), 5912–5917. https://doi.org/10.1073/pnas.1621350114.
Wei, J., Wu, W.-H., Wang, R., Yang, Z., & Zhang, W.-B. (2018). B_{12}-dependent protein oligomerization facilitates layer-by-layer growth of photo/thermal responsive nanofilms. *ACS Macro Letters*, 7(5), 514–518. https://doi.org/10.1021/acsmacrolett.8b00147.
Wexler, A. G., Schofield, W. B., Degnan, P. H., Folta-Stogniew, E., Barry, N. A., & Goodman, A. L. (2018). Human gut *Bacteroides* capture vitamin B_{12} via cell surface-exposed lipoproteins. *Elife*, 7, e37138. https://doi.org/10.7554/eLife.37138.
Whipple, G. H., & Robscheit-Robbins, F. S. (1925). Favourable influence of liver, heart and skeletal muscle in diet on blood regeneration in anemia. *American Journal of Physiology, 72*, 408–418.
Winkler, A., Heintz, U., Lindner, R., Reinstein, J., Shoeman, R. L., & Schlichting, I. (2013). A ternary AppA-PpsR-DNA complex mediates light regulation of photosynthesis-related gene expression. *Nature Structural & Molecular Biology, 20*(7), 859–867. https://doi.org/10.1038/nsmb.2597.

Wuerges, J., Garau, G., Geremia, S., Fedosov, S. N., Petersen, T. E., & Randaccio, L. (2006). Structural basis for mammalian vitamin B_{12} transport by transcobalamin. *Proceedings of the National Academy of Sciences of the United States of America*, *103*(12), 4386–4391. https://doi.org/10.1073/pnas.0509099103.

Xu, D., Ricken, J., & Wegner, S. (2020). Turning cell adhesions ON or OFF with high spatiotemporal precision using the green light responsive protein CarH. *Chemistry*. https://doi.org/10.1002/chem.202001238.

Yamamoto, H., Fang, M., Dragnea, V., & Bauer, C. E. (2018). Differing isoforms of the cobalamin binding photoreceptor AerR oppositely regulate photosystem expression. *Elife*, 7. https://doi.org/10.7554/eLife.39028.

Yang, H., McDaniel, E. C., Impano, S., Byer, A. S., Jodts, R. J., Yokoyama, K., et al. (2019). The elusive 5'-deoxyadenosyl radical: Captured and characterized by electron paramagnetic resonance and electron nuclear double resonance spectroscopies. *Journal of the American Chemical Society*, *141*(30), 12139–12146. https://doi.org/10.1021/jacs.9b05926.

Yang, X., Gerroll, B. H. R., Jiang, Y., Kumar, A., Zubi, Y. S., Baker, L. A., et al. (2022). Controlling non-native cobalamin reactivity and catalysis in the transcription factor CarH. *ACS Catalysis*, *12*(2), 935–942. https://doi.org/10.1021/acscatal.1c04748.

Zhang, Y., Rodionov, D. A., Gelfand, M. S., & Gladyshev, V. N. (2009). Comparative genomic analyses of nickel, cobalt and vitamin B_{12} utilization. *BMC Genomics*, *10*, 78. https://doi.org/10.1186/1471-2164-10-78.

Ziegelhoffer, E. C., & Donohue, T. J. (2009). Bacterial responses to photo-oxidative stress. *Nature Reviews Microbiology*, 7(12), 856–863. https://doi.org/10.1038/nrmicro2237.

CHAPTER SEVEN

Photolytic properties of B_{12}-dependent enzymes: A theoretical perspective

Arghya Pratim Ghosh, Megan J. Toda, and Pawel M. Kozlowski*
Department of Chemistry, University of Louisville, Louisville, KY, United States
*Corresponding author: e-mail address: pawel@louisville.edu

Contents

Abstract

The biologically active vitamin B_{12} derivates, methylcobalamin (MeCbl) and adenosylcobalamin (AdoCbl), are ubiquitous organometallic cofactors. In addition to their key roles in enzymatic catalysis, B_{12} cofactors have complex photolytic properties which have been the target of experimental and theoretical studies. With the recent discovery of B_{12}-dependent photoreceptors, there is an increased need to elucidate the underlying photochemical mechanisms of these systems. This book chapter summarizes the photolytic properties of MeCbl- and AdoCbl-dependent enzymes with particular emphasis on the effect of the environment of the cofactor on the excited state processes. These systems include isolated MeCbl and AdoCbl as well as the enzymes, ethanolamine ammonia-lyase (EAL), glutamate mutase (GLM), methionine synthase (MetH), and photoreceptor CarH. Central to determining the photodissociation mechanism of each system is the analysis of the lowest singlet excited state (S_1) potential energy surface (PES). Time-dependent density functional theory (TD-DFT), employing BP86/TZVPP, is widely used to construct such PESs. Regardless of the environment, the topology of the S_1 PES of AdoCbl or MeCbl is marked by characteristic features, namely the metal-to-ligand charge transfer (MLCT) and ligand field (LF) regions. Conversely, the relative energetics of these electronic states are affected by the environment. Applications and outlooks for Cbl photochemistry are also discussed.

Vitamins and Hormones, Volume 119
ISSN 0083-6729
https://doi.org/10.1016/bs.vh.2022.01.012

1. Introduction

Derivatives of vitamin B_{12}, commonly known as cobalamins (Cbls) (Fig. 1), are among nature's most complex organometallic compounds (Pratt & Thorp, 1966). Cbls belong to the porphyrinoid family, comprised

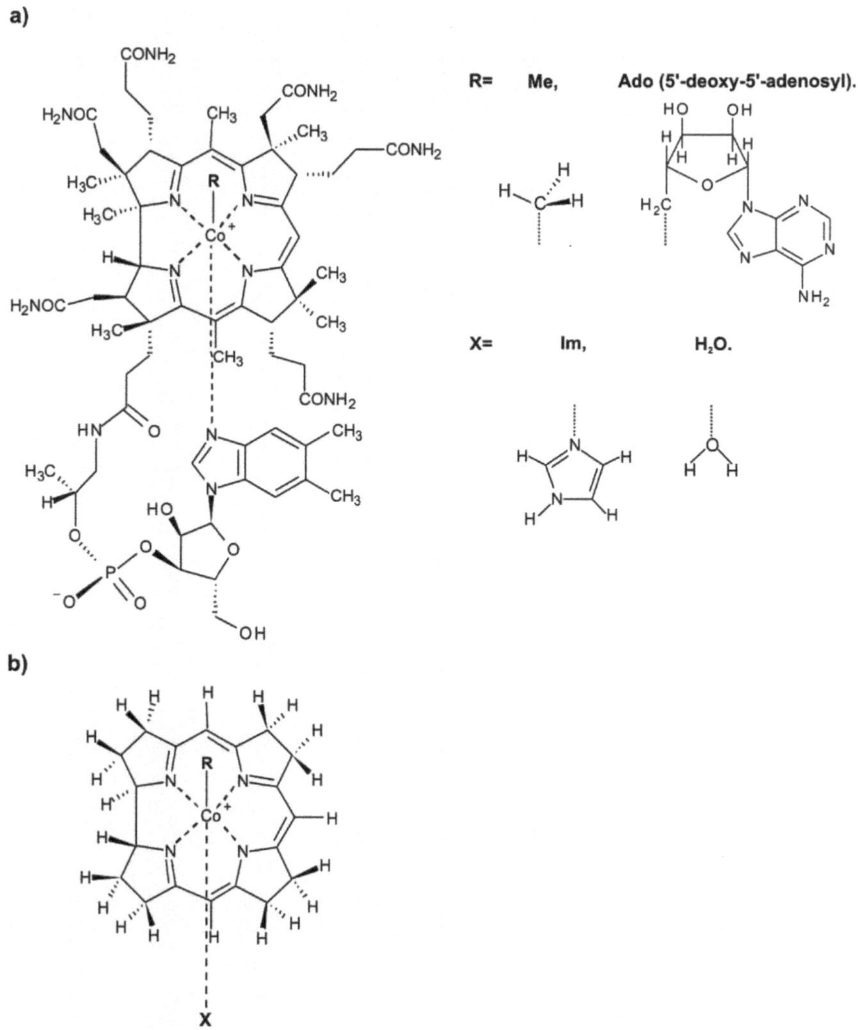

Fig. 1 (A) Left: General molecular structure of cobalamins. Right: Molecular structure of upper axial ligands for methylcobalamin (R=Me) and adenosylcobalamin (R=Ado). Molecular structure of lower axial ligands for base-on (X=Im) and base-off (X=H_2O) model structures. (B) Simplified model typically used in calculations.

of a corrin macrocycle with several side chains of acetamide and propionamide groups bound to the periphery of the corrin ring. Centralized in the corrin ring, is a cobalt (Co) ion, which is equatorially coordinated to four nitrogens of the macrocycle (Firth et al., 1968; Hogenkamp, Ladd, & Barker, 1962). In the Co(III) form, the Co is further coordinated to upper and lower axial ligands. In what is known as the base-on conformation, the metal center is ligated with 5,6-dimethylbenzimidazole (DBI) as the lower ligand. The upper axial position is then coordinated with various ligands, such as cyano (CN), hydroxyl (OH), methyl (Me), adenosyl (Ado), etc.

This book chapter will focus on Cbls with Me and Ado groups as the upper ligands, known as methylcobalamin (MeCbl) and adenosylcobalamin (AdoCbl), respectively. These are the biologically active forms of vitamin B_{12} (also known as cyanocobalamin, CNCbl) which is itself biologically inactive (Banerjee & Ragsdale, 2003). In addition to the base-on conformation, which is found in both solution and enzymes, Cbls can be found in different conformations depending on their environment. In the solution environment and in very acidic conditions (i.e., ~pH 2), the protonated DBI group can be replaced with a weakly coordinated water molecule. This is known as the base-off configuration. Upon binding with enzymes, the DBI base of the lower axial face can be detached and replaced with imidazole (Im) moiety of histidine (His) residue of the protein. This is known as base-off/His-on configuration, exemplified by methionine synthase (MetH), where the MeCbl cofactor possesses the base-off/His-on conformation (Drennan, Huang, Drummond, Matthews, & Lidwig, 1994).

MeCbl and AdoCbl act as cofactors for a series of enzymatic reactions that catalyze complex molecular transformations in many living organisms (Brown, 2005; Ludwig & Matthews, 1997; Matthews, 2001, 2009). MeCbl acts as a cofactor for the methyltransferase (MeTrs) class of enzymes, which catalyze the intermolecular methyl transfer reactions (Matthews, 2001; Matthews, Koutmos, & Datta, 2008). Among all MeTrs, MetH is one of the most well-studied MeCbl-dependent enzymes, due to its critical role in human metabolism (Drennan et al., 1994). MetH catalyzes the transfer of the Me group from methyltetrahydrofolate (CH_3-H_4 folate) to homocysteine (Hcy) to generate methionine (Met) and tetrahydrofolate (THF) (Goulding, Postigo, & Matthews, 1997; Matthews, 2001). On the other hand, AdoCbl-dependent enzymes, play a pivotal role in carrying out carbon skeleton rearrangements, heteroatom eliminations, and intermolecular amino group migration in various enzymes in the mutase and eliminase classes (Banerjee, 2003; Banerjee & Padmakumar, 1995; Frey, 2001; Frey & Reed, 2000; Gruber, Reitzer, & Kratky, 2001). In particular, homolytic cleavage of the Co-C bond of AdoCbl is the key catalytic step, which initiates the generation of the Ado radical which is necessary for these complex molecular transformations (Banerjee, 2001; Finke & Martin, 1990). Methymalonyl-CoA mutase (MCM) is one of the most well-studied AdoCbl-dependent enzymes primarily due to its role in human metabolism (Banerjee & Padmakumar, 1995; Chowdhury & Banerjee, 2000).

Apart from their role in enzymatic catalysis, Cbls also possess complex photolytic properties due to the light-sensitive Co-C bond (Barker et al., 1960; Barker, Weissbach, & Smyth, 1958). Initially, the photodegradation of Cbls was considered as an unwanted side reaction and an experimental nuisance with no physiological purpose. However, recent findings have proven this line of thinking to be outdated (Hogenkamp, 1963; Hogenkamp & Barker, 1961; Johnson, Mervyn, Shaw, & Smith, 1963; Schrauzer, Lee, & Sibert, 1970; Schrauzer, Sibert, & Windgassen, 1968). Cbls have been shown to be potential candidates for light-activated drug delivery and OH radical generation (Shell & Lawrence, 2011, 2015). In addition, a recently discovered photoreceptor, CarH, operates via photocleavage of the Co-C bond (Jost, Simpson, & Drennan, 2015; Ortiz-Guerrero, Polanco, Murillo, Padmanabhan, & Elías-Arnanz, 2011). This example of a physiological role for the light sensitivity of vitamin B_{12} derivates has enlivened the field of B_{12} chemistry. While applications in the area of optogenetics are explored, theoretical insights into the mechanism of photolytic pathways are of critical importance. This is discussed in more detail in the Outlook section.

The crucial step involved in the photolytic reactions of Cbls is the activation and, in some cases, the subsequent cleavage of the Co-C bond. Toward this, a molecular-level understanding of the electronically excited states is paramount and the photolytic properties of MeCbl and AdoCbl in various environments and conformations have been investigated. Both advanced spectroscopic techniques and state-of-the-art quantum mechanical calculations have been extensively employed to unravel the photochemical properties of Cbls (Cole et al., 2001; Andruniow, Kozlowski, & Zgierski, 2001; Andruniow, Zgierski and Kozlowski, 2000; Brooks, Vlasie, Banerjee, & Brunold, 2004; Ghosh, Lodowski, Bazarganpour, Leks, & Kozlowski, 2020; Harris, Stickrath, Carroll, & Sension, 2007; Sension et al., 2005; Walker et al., 1998). In particular, the photolytic properties of the biologically active forms of Cbls have been previously reviewed by Kozlowski and co-workers (Kozlowski, Garabato, Lodowski, & Jaworska, 2016; Toda, Lodowski, Mamun, Jaworska, & Kozlowski, 2019). The most important outcome of these computational investigations was the determination of potential energy surfaces (PESs) associated with low-lying electronically excited states of Cbls. The PES associated with the S_1 state, which consists of two energy minima, metal-to-ligand charge transfer (MLCT) and ligand-field (LF) states, is a reliable tool for understanding photolytic properties of Cbls. The pathways for photodissociation vary with the type of axial ligand and with the environment because the photochemistry is primarily determined by seams of intersection between MLCT/LF states (Toda, Lodowski, et al., 2019).

The purpose of this contribution is to summarize the influence of the environment on the photochemical and photophysical processes of MeCbl and AdoCbl cofactors. Recent contributions have shown that the photolytic properties of Cbls can be noticeably influenced by the cofactor's environment, whether it is in solution or inside an enzyme. While the enzymatic environment is impeding the photolysis of MeCbl inside MetH by de-stabilizing the LF state, the protein environment influences the photo-reaction of AdoCbl-dependent enzymes, such as glutamate mutase (GLM) and ethanolamine ammonia-lyase (EAL), by stabilizing the LF electronic state. These points will be delineated throughout the review.

2. Summary of experimental results

The light sensitivity of Cbls was first reported in 1958 when Barker et al. found that the pseudo vitamin B_{12} undergoes decomposition upon irradiation with visible light from a tungsten lamp (Barker et al., 1958). Although initially it was considered as an unwanted side reaction, in 1960, the same group of researchers conducted a detailed study to explore the photochemical cleavage of Cbl (Barker et al., 1960). In another study, Dolphin, Pratt, and co-workers showed that while the alkyl-Cbls were photolytically unstable, the non-alkyl Cbls are insensitive to light (Pratt, 1964). Since these important contributions, the photochemistry of Cbls gained interest among the scientific community.

In the 1990s, Matthews and co-workers used spectroscopic techniques such a UV-Vis spectroscopy and stopped-flow spectroscopy to investigate the photolytic cleavage of MeCbl inside methionine synthase (MetH) (Jarrett et al., 1996). It was shown that absorption of UV or visible light results in partitioning between relaxation through vibrational modes to the ground state and bond homolysis. The photolytic cleavage resulted in the formation of Co(II)/CH_3 radical pair (RP), which are surrounded by a cage provided by the protein residues. This prevents the diffusive loss of the methyl radical which results in the recombination of RPs. The quantum yield of ~~the~~ photolysis is dictated by the competition between radical escape and recombination of RPs. It was further suggested that a mutation of these residues, that surround the upper face of the MeCbl cofactor, can increase the exposure of methyl radical and increase the rate of photolysis of MeCbl (Jarrett et al., 1996).

With the advancement of laser spectroscopic techniques, the photochemistry of Cbls has continued to intrigue researchers more than ever before. Sension and co-workers have been using ultrafast transient absorption spectroscopy (TAS) to investigate the photolytic properties of Cbls since the late 1990's (Walker, Jarrett, et al., 1998; Walker, Shiang, Anderson, Pullen, & Sension, 1998). They demonstrated that the photolysis of MeCbl is sensitive to excitation wavelength (Shiang, Walker, Anderson, Cole, & Sension, 1999). Excitation at 400 nm results in the formation of two distinct photoproducts, bond homolysis, and a metastable state, which has an absorption spectrum consistent with the formation of cob(III)alamin. On the other hand, following excitation with 500 nm results in only the formation of metastable cob(III)alamin without prompt photolysis. They have also investigated the photolytic properties of AdoCbl in solution as well as to the protein environment in GLM to study how the protein-coenzyme interaction effect photoreactivity (Sension et al., 2005). It was shown that that unlike MeCbl, in the photolysis of AdoCbl, the quantum yields of RP formation are unaffected by the excitation wavelength. Based on the TAS, it appears that the enzymatic environment modulates the low-lying excited states and stabilizes the product of photolysis. This stabilization makes the recombination of RPs less favorable. It was reported that that the enzymatic environment reduces the rate constant for geminate recombination by ~30% in AdoCbl-dependent GLM when compared with the isolated cofactor.

The photolytic properties of EAL have been extensively studied experimentally. In a key study, Robertson and Warncke sought to look into RP formation that results from photolytic cleavage of the Co-$C_{5'}$ bond of AdoCbl in EAL using TAS (Robertson & Warncke, 2009; Warncke & Utada, 2001). The following conclusion is among the major findings of the study. It was shown that photolysis of AdoCbl in EAL leads to a quantum yield for cob(II)alamin that is threefold smaller than for AdoCbl in aqueous solution indicating that the protein binding site suppresses photoproduct, namely RP formation. It is also important to note that this study considered the role of the substrate in photolysis as it is well-known that substrate binding is a key step in the native catalytic cycle. Interestingly it was found that, unlike true substrates, the substrate analog for TAS experiments does not induce the RP stabilizing changes in EAL. In another study, Warncke and co-workers used time-resolved, full-spectrum electron paramagnetic resonance (EPR) spectroscopy to further probe the substrate-triggered Co-$C_{5'}$ cleavage and RP formation in AdoCbl-dependent EAL (Robertson, Wang, & Warncke, 2011). It was shown that substrate binding to holo-EAL did not convert the protein to a structural state that would stabilize the RP photoproduct. Accordingly, a major conclusion of the study was that a change in protein structure is not a basis of Co-$C_{5'}$ bond cleavage in native catalysis.

Scrutton and Jones also studied EAL, especially to determine how the substrate affects the native catalytic cycle, using pre-steady-state magnetic field effect (MFE) investigations (Jones, Hardman, Hay, & Scrutton, 2011; Jones, Hay, Woodward, & Scrutton, 2007). First, it should be noted that in all MFE experiments, RP formation has been shown to be singlet-born in Cbls. Stopped-flow MFE studies were insensitive to magnetically induced changes in the net forward rate of Co-$C_{5'}$ homolytic bond cleavage. A magnetic dependence in the continuous-wave Co-$C_{5'}$ photolysis of free AdoCbl in 75% glycerol was observed. However, magnetic dependence was not observed in the thermal Co-$C_{5'}$ homolysis in EAL with the substrate. It was concluded that the RP formation after homolysis of the Co-$C_{5'}$ suppressed geminate recombination to the extent that an MFE could not be observed. In another study, Scrutton and co-workers investigated the effect of protein dynamics on geminate RP by determining the influence of viscosity on the cw-photolysis rate, and its MFE (Jones, Levy, Hay, & Scrutton, 2013). This was achieved by using a specially configured MFE stopped-flow spectrophotometer to determine the magnitude of MFE of isolated AdoCbl and AdoCbl in EAL. It was found that the magnitude of the MFE increases with viscosity in isolated AdoCbl whereas the magnitude of the MFE remains constant, at around 18% when protein-bound. This seemed to indicate that viscosity, and hence protein dynamics, affect the transient RP dynamics in the EAL active site on a timescale likely to be not much more than a few ns. It was also shown that the majority of the geminate pairs have recombined within 3 ns using a 375 nm laser flash in a fs pump-probe spectrometer experiment. This is similar to GLM where there is also a marked delay in the appearance of the Cob(II)alamin photoproduct after photolysis (Sension et al., 2005).

In addition to the Cbl-dependent enzymes, recently, an entirely new field of Cbl chemistry, namely B_{12} photoreceptors, has emerged with the discovery and X-ray structure determination of CarH (Ortiz-Guerrero et al., 2011). CarH, a coenzyme B_{12}-dependent transcription regulator which employs light to perform its catalytic function. This was an exciting finding for the B_{12} community because, although the light sensitivity of the Co-$C_{5'}$ bond in coenzyme B_{12} has been known for five decades, only until CarH has photolytic properties been associated with controlled reactivity. CarH, present in bacteria, undergoes significant conformational changes in the presence of daylight to biosynthesize carotenoids (Jost et al., 2015; Jost, Simpson, & Drennan, 2015; Padmanabhan, Jost, Drennan, & Elías-Arnanz, 2017). In the dark, CarH, referred to as CarH-Dark State (CarH-DS), forms a tetramer upon AdoCbl binding. Each monomer contains a DNA binding domain, a helical bundle, and a Cbl binding domain. Monomers form head-to-tail dimers which are combined to form the tetramer. DNA binds to CarH-DS and transcription of biosynthetic genes is prevented in the absence of daylight. In the presence of light, the Co-$C_{5'}$ bond is broken and the Ado ligand is converted to the photoproduct, 4′,5′-anhydroadenosine, inducing a large-scale conformational change (Jost, Fernandez-Zapata, et al., 2015; Jost, Simpson, & Drennan, 2015). A His residue (His132) from the protein will take the place of the Ado ligand, forming a stable adduct with the cofactor known as *bis*-HisCbl (Jost, Fernandez-Zapata, et al., 2015; Jost, Simpson, & Drennan, 2015). Subsequently, the tetramer dissociates into monomers forming the CarH light state (CarH-LS). In CarH-LS, the cofactor retains His132 and His177 as upper and lower axial ligands, respectively (Padmanabhan et al., 2017). At this point, the transcription of carotenoid biosynthetic genes is initiated.

There are several proposals for how CarH forms 4′,5′-anhydroadenosine upon light-induced cleavage of the Co-$C_{5'}$ bond that has been put forth from at least three different research groups based on various experimental techniques (Jost, Fernandez-Zapata, et al., 2015; Jost, Simpson, & Drennan, 2015; Kutta et al., 2015; Miller et al., 2020). Thus, there is a clear need for theoretical studies to provide insights into the photoreactivity of CarH, especially considering the differences in experimental proposals. There are studies underway targeting this. However, another question has come to the forefront, namely, "why is CarH active photolytically active in comparison to many other B_{12}-dependent enzymes that rely on substrate binding to induce the catalytic cycle with the key step being cleavage of the Co-$C_{5'}$ bond?" This question was the motivation for a theoretical study of CarH which will be discussed in further detail in a proceeding section.

3. Computational approach

Due to the size and the complexity of Cbls, density functional theory (DFT) is the method of choice for investigating the ground-state properties of Cbls. One of the most important features of Cbl chemistry is the strength of the Co-C bond. The strength of the Co-C bond and accurate prediction of the bond dissociation energy (BDE) is of utmost importance in the exploration of the structure and reactivity of B_{12} derivatives. Accordingly, the accuracy of computed BDEs in comparison with the experiment is widely used to judge the efficacy of a particular functional for Cbl studies. For more details, the reader is referred to the appended citations (Kepp, 2014; Kozlowski et al., 2012; Toda, Kozlowski, & Andruniów, 2019). It was reported that the Co-C BDEs are very sensitive to the exchange-correlation functions. The widely used B3LYP hybrid functional is known to significantly underestimate the Co-C BDE. However, the use of pure GGA-type functionals can accurately predict BDEs while still providing computational efficiency. Among tested DFT functionals, it has been demonstrated, that BP86 with dispersion (D3) corrections can provide the most accurate BDE for the Co-C bond.

In terms of excited-state properties, time-dependent DFT (TD-DFT) is the most widely used method for the photolytic properties of Cbls. Again the reader may refer to Chapter 8 of the appended reference (Toda, Kozlowski, & Andruniów, 2019), for more details but in sum, TD-DFT has been benchmarked against the higher-level *ab initio* methods, such as complete active space self-consistent field theory (CASSCF) combined with second-order perturbation theory (CASSCF/MC-XQDPT2) and equation-of-motion coupled-cluster (EOM-CCSD) calculations to determine its reliability in describing the electronically excited states of Cbls (Kornobis et al., 2011, 2013). In the context of TD-DFT, the choice of functional is paramount for a proper description of low-lying excited states and the S_1 PES. At present, the BP86 functional is the most suitable functional for TD-DFT studies of Cbls especially because it accurately predicts the MLCT character of the S_1 state of MeCbl (Kornobis et al., 2013).

For enzymatic studies, the ground state and excited states properties of MeCbl- and AdoCbl-dependent enzymes, such as MetH, EAL, GLM, and CarH were computed based on the combined density functional theory/molecular mechanics (DFT/MM) and time-dependent DFT (TD-DFT/MM), respectively (Ghosh et al., 2018; Mamun, Toda, Lodowski, Jaworska, & Kozlowski, 2018; Toda, Mamun, Lodowski, & Kozlowski, 2020). For brevity of presentation, the reader is referred to the appended citations for the computational set-ups of these complex systems. The computed photochemical data of base-on MeCbl and AdoCbl were also compared with the corresponding enzymes, to investigate the influence of the environment on the photolysis of biologically active forms of vitamin B_{12}. For comparison, the base-on conformation of the cofactors was chosen over the base-off conformation because the lower axial bond (Co-N_{ax}) is more in line with the enzymatic environment. Both the base-on and base-off/His-on Cbl have a Co-N_{ax} bond to the lower axial ligand. Recall that the base-off conformation of the solution-based cofactor involves a weakly coordinated water molecule with a Co-O_{H2O} bond while the DBI is detached.

4. Importance of axial bonding in construction of S_1 PES

In terms of excitations, the changes associated with axial bond length are the most sensitive parameter to consider while studying the photochemical processes of Cbl derivatives (Miller et al., 2017; Rury, Wiley, & Sension, 2015). On the other hand, the corrin structure is comparatively less altered upon excitation. The comparison of BP86/DFT-based ground-state (S_0) and experimental axial bond lengths between MeCbl and AdoCbl base-on and base-off derivatives is depicted in Fig. 2. The DFT optimized axial bond lengths for MeCbl and AdoCbl base-on at S_0 state agree well with experimental x-ray bond lengths (Kozlowski et al., 2016). However, for the base-off Cbls, we must rely on the calculated data for predicting their structure as there is no structural data available. Based on XANES measurements of CNCbl, it was shown that upon excitation the axial bonds undergo the largest structural changes. In addition, based on TAS studies it is evident that the photolysis of Cbls occurs from the S_1 state (Walker, Jarrett, et al., 1998). Therefore, with these two points in mind, PESs as a function of two axial bonds, Co-R and Co-N_{ax}, are reliable tools to explore the photochemical and photophysical processes of Cbls.

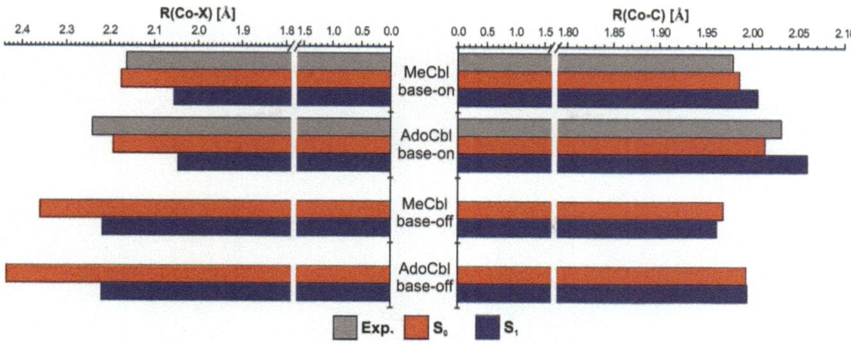

Fig. 2 Experimental and optimized axial bond lengths at the DFT/BP86 level, for S_0 and TD-DFT/BP86 for S_1 excited states of the cobalamins under consideration (X= Im and H_2O for *base-on* and *base-off* form, respectively). In all calculations the same basis sets, TZVPP for Co, C, N, and TZVP for H were used, employing the Conductor-like Screening Model (COSMO) solvent. *Adapted from Toda, M. J., Lodowski, P., Mamun, A. A., Jaworska, M., & Kozlowski, P. M. (2019). Photolytic properties of the biologically active forms of vitamin B_{12}.* Coordination Chemistry Review, 385, *20–43.*

5. Photolytic properties of cobalamins

5.1 Photolysis of AdoCbl

5.1.1 Isolated AdoCbl

To obtain a reliable description of the base-on AdoCbl photoreaction, a PES associated with the lowest vertical excited state has been depicted in Fig. 3. The S_0 and S_1 PESs of AdoCbl were constructed by the systematic elongation of the Co-C and Co-N_{Im} bond (Garabato, Lodowski, Jaworska, & Kozlowski, 2016; Kozlowski et al., 2016). The S_0 carries a single low-energy region, which is extended along the lower axial bond lengths. Unlike the S_0 state, the S_1 PES contains two low-energy regions, namely the MLCT and LF region, separated by a seam. The electronic excitation in the MLCT state is associated with the metal d to corrin π^* ($d+\pi \rightarrow \pi^*$) type transition, whereas the transition in the LF electronic state has a dominant $d+\pi \rightarrow \sigma^*$ (d_z^2) character. In the MLCT region, the energy minimum is localized at a Co-C bond length of 2.1 Å and Co-N_{Im} 2.0 Å (Fig. 3A). The second energy minima, which is associated with the LF region is located at a Co-C bond length of 2.5 Å and Co-N_{Im} bond length of 3.5 Å.

In this case, the mechanism of the base-on AdoCbl photoreaction is based on the two possible pathways, namely Path A and Path B, that have been identified for the photodissociation of the Co-C bond (Fig. 4). Path A

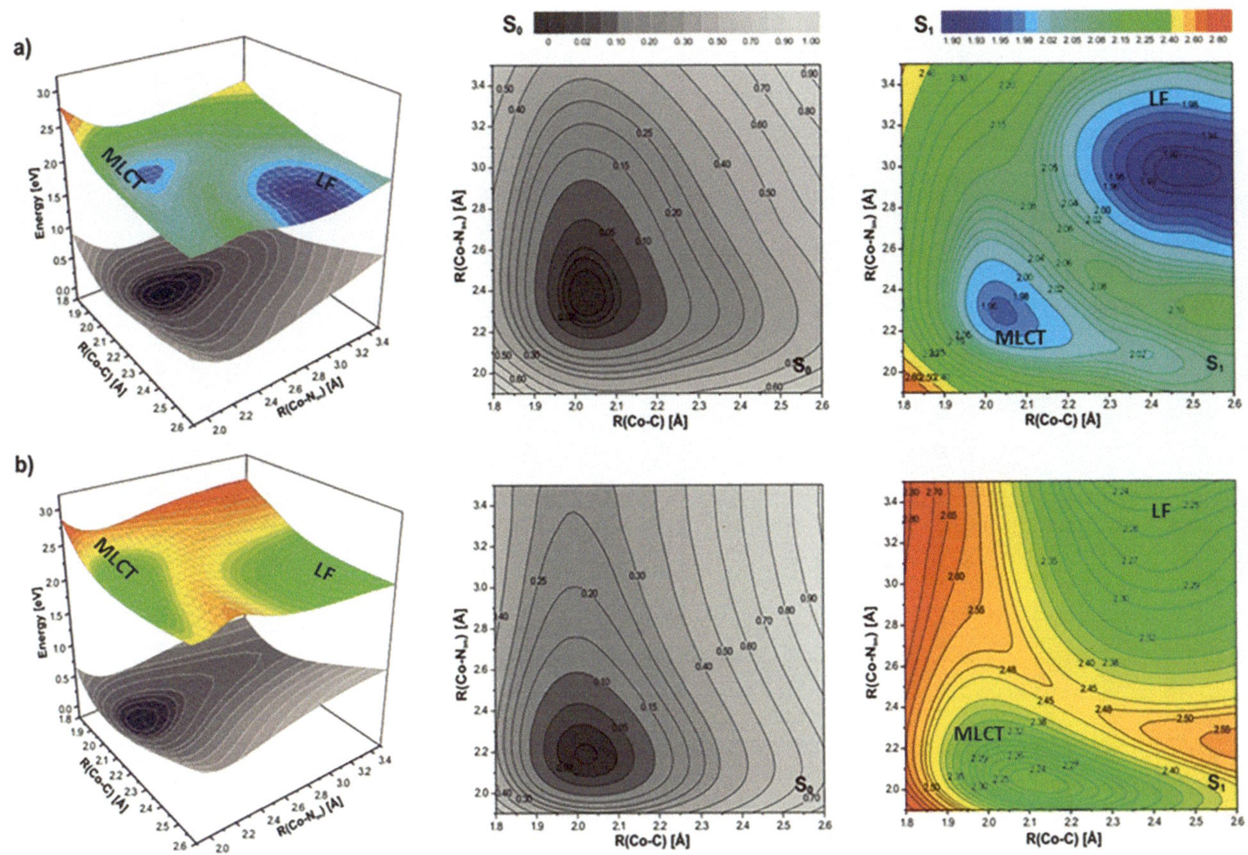

Fig. 3 Potential energy surfaces for the S_0 with vertical projections of the S_1 state plotted as functions of axial bond lengths along with the corresponding S_0 and S_1 contour PESs with color codes for (A) AdoCbl inside EAL (B) Im$-$[Co^{III}(corrin)]$-$Ado$^+$ base-on model complex in water. *Adapted from Mamun, A. A., Toda, M. J., Lodowski, P., Jaworska, M., & Kozlowski, P. M. (2018). Mechanism of light-induced radical pair formation in coenzyme B_{12}-dependent ethanolamine ammonia-lyase.* ACS Catalysis, 8, *7164–7178.*

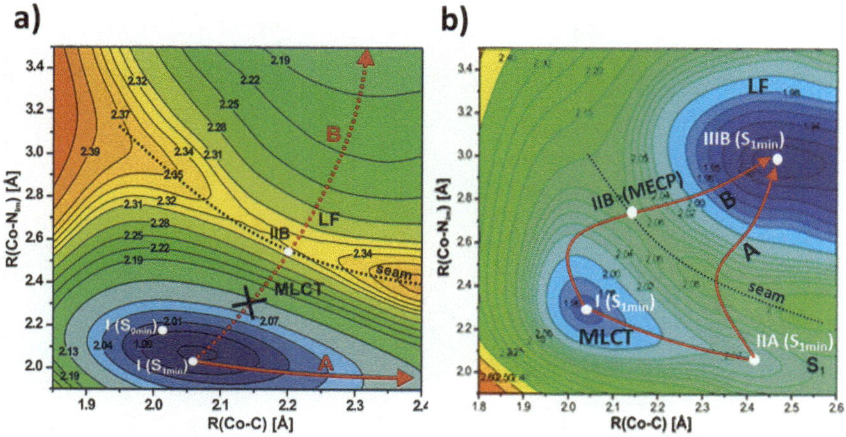

Fig. 4 Potential energy surfaces as a function of axial bond lengths with photoreaction paths for (A) Im-[Co^{III}(corrin)]-Ado^+, (B) AdoCbl inside EAL. The S_1 PES with minima regions, separated by a seam, where MLCT minimum is denoted as I (S_{1min}) and the LF minimum is denoted as IIIB (S_{1min}).

involves the initial elongation of the Co-C bond prior to the detachment of the Co-N_{Im} bond. On the other hand, Path B is associated with the initial elongation of the Co-N_{Im} bond from the MLCT, denoted as I (S_{1min}), followed by the displacement of the Co-C bond. However, the energetics profile of the two pathways is different which means that one would logically be favored over the other. Regardless of feasibility, from a theoretical point of view, either Path A or Path B will result in the same outcome, Co-C bond cleavage, and RP formation (Garabato et al., 2016).

The scheme of photoreaction is depicted in Fig. 5. Based on the energetics of computed S_1 PES of AdoCbl base-on, it is evident that Path A is energetically more favorable for the photodissociation of the Co-C bond. The energy difference between the I (S_{1min}) and IIB in S_1 PES of AdoCbl is 8.8 kcal/mol and is almost four times higher than the dissociation energy of the Co-C bond along with Path A. Therefore, the energy barrier associated with Path B, makes it energetically unfavorable and inactive in the photolysis of base-on AdoCbl (Garabato et al., 2016; Toda, Kozlowski, & Andruniów, 2019; Toda, Lodowski, et al., 2019). So, in base-on AdoCbl the photodissociation of Co-C bond initiate from the MLCT (I (S_{1min})) state and proceeds to the LF state by lengthening the Co-C bond, namely Path A. After reaching the LF excited state, the Co(II)/Ado RP is generated and associated with the partial or full dissociation of the axial base. In addition,

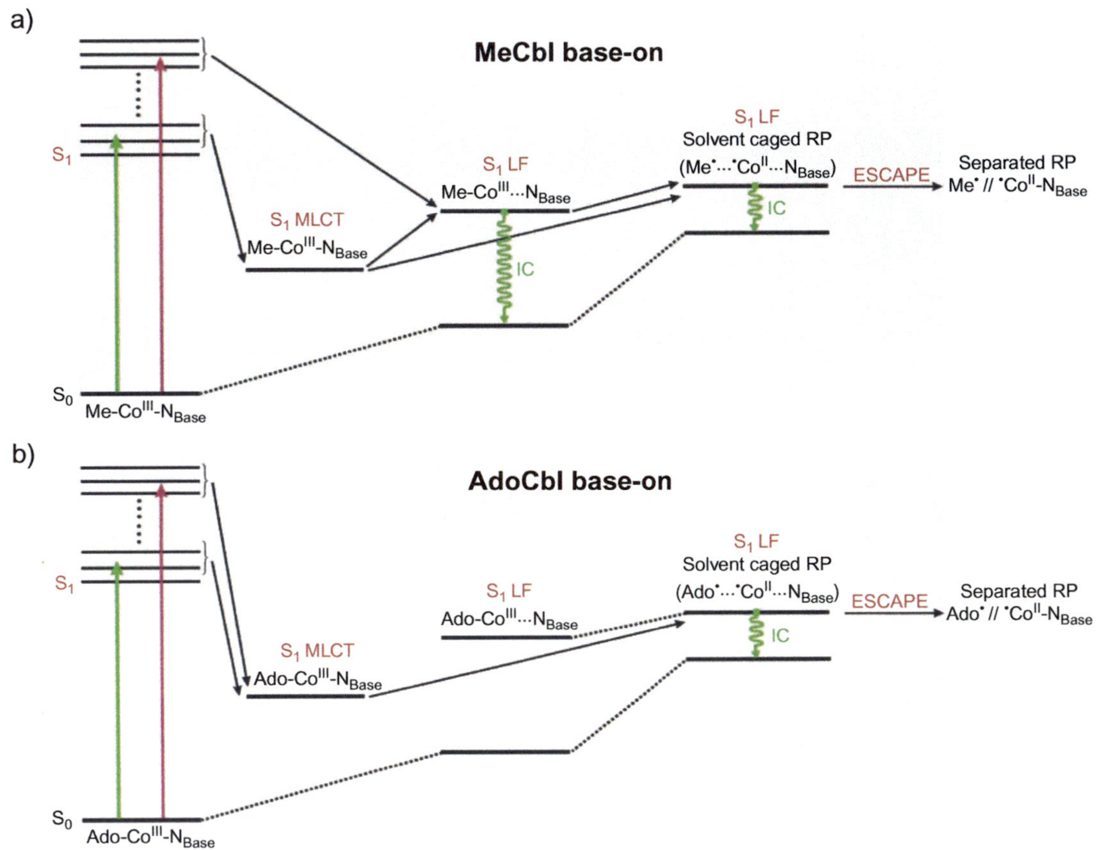

Fig. 5 Schemes of photoreaction mechanism for (A) base-on form of methylcobalamin (MeCbl) (B) base-on form of adenosylcobalamin (AdoCbl). *Adapted from Toda, M. J., Lodowski, P., Mamun, A. A., Jaworska, M., & Kozlowski, P. M. (2019). Photolytic properties of the biologically active forms of vitamin B_{12}.* Coordination Chemistry Review, 385, *20–43.*

once the RP is formed from the LF state, there is a competition between deactivation to the ground state through the geminate recombination and the radical escape from the solvent cage.

5.1.2 Photolysis of AdoCbl inside enzymes: EAL and GLM

In AdoCbl-dependent enzymes the activation of the Co-C bond is triggered by the binding of substrate, leading to the formation of the Co(II)/Ado RP (Banerjee & Padmakumar, 1995). This cleavage of the Co-C bond is kinetically coupled with the subsequent H-abstraction from the substrate (Banerjee & Padmakumar, 1995; Padmakumar, Padmakumar, & Banerjee, 1997). Unlike native catalysis, the Co-C bond in AdoCbl-dependent enzymes can also be photolytically cleaved to generate Co(II)/Ado RP (Fig. 5). The most significant aspect of the photolysis of AdoCbl-dependent enzymes is that it generates the same RP as in the case of native enzymatic reaction without the coupling of subsequent H-abstraction step. Thus, the investigation of the photolytic properties of enzyme-bound AdoCbl is of particular importance in AdoCbl-dependent enzymatic catalysis. It has been demonstrated that the photolysis of AdoCbl differs significantly in presence of an enzymatic environment when compared to that in an isolated cofactor in solution (Jones, 2017; Jones, Woodward, & Scrutton, 2009). Moreover, it has been suggested that the protein environment impedes the diffusive loss of Ado radical and increases the rate of geminate recombination of the photogenerated RP when compared with the solution (Sension et al., 2004, 2005; Walker, Shiang, et al., 1998). Therefore, it is needless to say that the protein environment plays a critical role in the photoactivation of the Co-C bond in the AdoCbl-dependent enzymes.

To explore the influence of the enzymatic environment on the photolysis of AdoCbl, the computed photochemical data of the isolated AdoCbl was compared with the AdoCbl-dependent GLM and AdoCbl-dependent EAL. In GLM the AdoCbl cofactor is in the base-off/His-on configuration and for EAL, the cofactor is in the base-on conformation. As discussed in Section 4, upon excitations, the most significant changes of Cbls occur in the axial ligands. Due to this, the photoreaction of EAL and GLM was explored by computing the PES of the low-lying excited state (S_1 excited state) as a function of both Co-C and Co-N_{axial} bond distances. The computational methodology of the construction of the PESs has been discussed in the previous section.

The major feature of the S_1 PES is the two energy minima, the MLCT and LF, which are consistent with base-on AdoCbl (Mamun et al., 2018). The MLCT minimum (I (S_{1min}), correspond to a shorter Co-C and Co-N_{Im} bond (Figs. 3 and 6). In the MLCT electronic state, the electronic

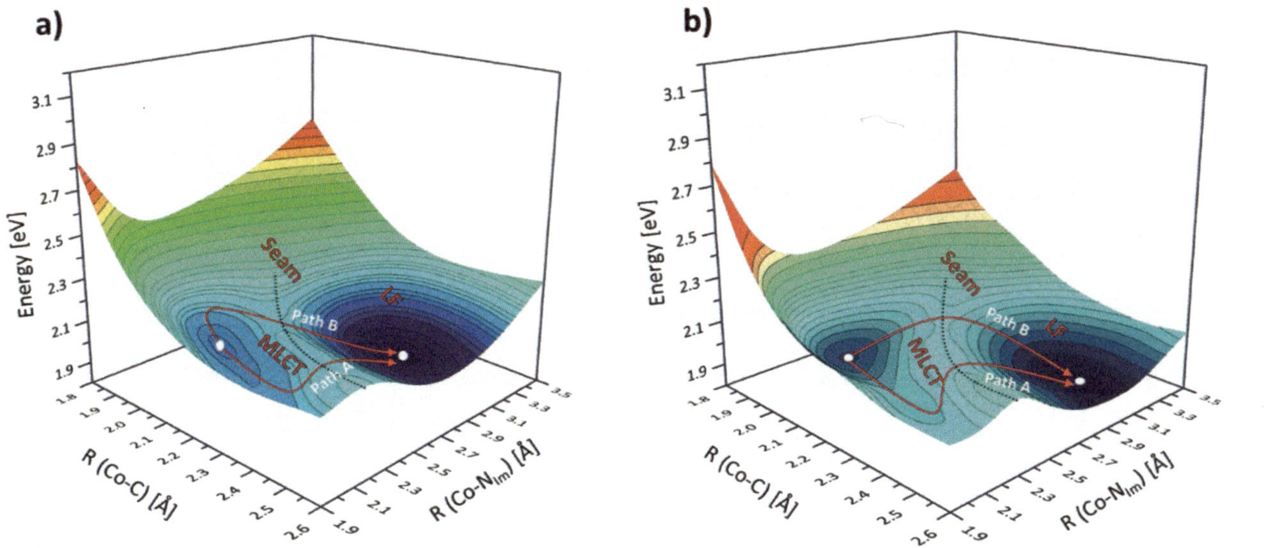

Fig. 6 S_1 PES as a function of axial bond lengths Co-C and Co-N_{Im} with the photodissociation pathways depicted by red arrows for (A) AdoCbl-dependent GLM and (B) AdoCbl-dependent EAL. *Adapted from Mamun, A. A.; Toda, M. J.; Lodowski, P.; Kozlowski, P. M. (2019). Photolytic cleavage of Co–C bond in coenzyme B_{12}-dependent glutamate mutase.* Journal of Physical Chemistry B, 123, *2585–2598.*

excitation is associated with the metal d to corrin π^* ($d+\pi \rightarrow \pi^*$) type transition. On the other hand, the LF minimum IIIB (S_{1min}) which is localized at an elongated Co-C and Co-N_{Im} bond distance has a dominant metal d orbital $(d_{yz}+\pi) \rightarrow \sigma^*(d_z^2)$ contribution. While the topology of the S_1 surface for both the enzymes and base-on AdoCbl is very similar, a noticeable difference can be seen in the relative energetics between the MLCT and LF excited states (Fig. 6). In the base-on AdoCbl inside the solution, where the MLCT state is energetically lower than LF excited state, whereas inside enzyme the LF minimum IIB (S_{1min}) is energetically more stable than the MLCT minimum I (S_{1min}) which is consistent with the experiment. In the case of GLM, the LF minimum IIB (S_{1min}) is 1.87 kcal/mol lower than the MLCT minima of S_1 PES (Mamun, Toda, Lodowski, & Kozlowski, 2019). On the other hand, in the EAL the LF state is stabilized by about 1.2 kcal/mol than the MLCT state (Fig. 7) (Mamun et al., 2018). These indicate that the enzymatic environment plays a role to stabilizes the LF electronic state.

Similar to the isolated AdoCbl, the photolysis of AdoCbl inside enzyme to generate Co(II)/Ado RP primarily occurs from the LF region of the S_1 state. The mechanism of the photoreaction for EAL and GLM is not significantly different from the isolated cofactor in the solution. Based on the

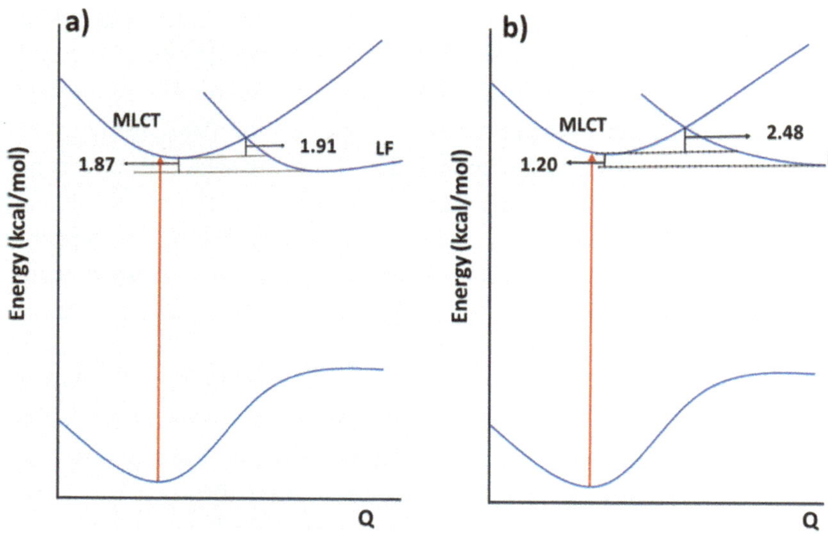

Fig. 7 Relative energetics of the MLCT and LF state in AdoCbl cofactor for (A) AdoCbl-dependent GLM and (B) AdoCbl-dependent EAL along the coordinate Q. *Adapted from Mamun, A. A., Toda, M. J., Lodowski, P., & Kozlowski, P. M. (2019). Photolytic cleavage of Co–C bond in coenzyme B_{12}-dependent glutamate mutase.* Journal of Physical Chemistry B, 123, *2585–2598.*

energetics of the S_1 state two pathways for the photodissociation have been identified, which are distinguished by whether the Co-C (Path A) bond or Co-N_{Im} (Path B) bond elongates first. Based on the energy profile of both pathways, it appears that in the case of EAL Path B is more energetically favorable for the photodissociation of the Co-C bond, where photoreaction initiates from MLCT state and proceeds through the elongation of Co-N_{Im} bond to LF state (Mamun et al., 2018). The energy barrier associated with Path B is 2.48 kcal/mol, which is about 1 kcal/mol lower than the energy barrier for Path A. On the other hand, in AdoCbl-bound GLM both Path A and Path B seem equally possible. The energetic barrier to cross the seam along Path A and Path B are 1.88 kcal/mol and 1.91 kcal/mol, respectively (Mamun et al., 2019). Whereas in base-on AdoCbl inside the solution, only Path A was active for the RP formation. Therefore, it appears that the enzymatic environment is affecting the photolysis of AdoCbl by stabilizing the LF electronic state.

5.1.3 Photolysis AdoCbl in B_{12}-dependent photoreceptors

As described in the previous section (Section 2), many details about the mode of action of CarH, especially those determined from crystal structures, seem to be generally understood. However, some questions remain. Among these is an important fundamental question, "what specifically makes CarH photolytically active in comparison to other AdoCbl-dependent enzymes which rely on substrate binding?" This question was the driving force for a recent study where the photolytic properties of AdoCbl-dependent enzymes GLM and EAL, were compared to those of CarH. Based on experimentally and computational analyses, it was found that the photo-cleavage of Co-$C_{5'}$ in GLM and EAL enzymes is suppressed and only a small portion undergo photodissociation. In stark contrast to CarH, excited state dynamics and photolytic studies of GLM and EAL have revealed that these enzymes were essentially photostable. Upon comparison of the topology of the S_1 PESs of the three enzymes, it was found that the topology of the S_1 state did not vary and for each, the S_1 state could be characterized as a crossing between the MLCT and LF regions (Fig. 8) (Toda et al., 2020). There was an interesting observation upon a comparison of the relative energetics. While, for GLM and EAL the ligand field was stabilized and its minimum energy point was lower in energy than the minimum of the MLCT region, the reverse was true for CarH (Fig. 8). The LF state in CarH was higher in energy than the MLCT state, although the barrier to the MECP from the MLCT minimum was small (3.85 kcal/mol) (Fig. 9).

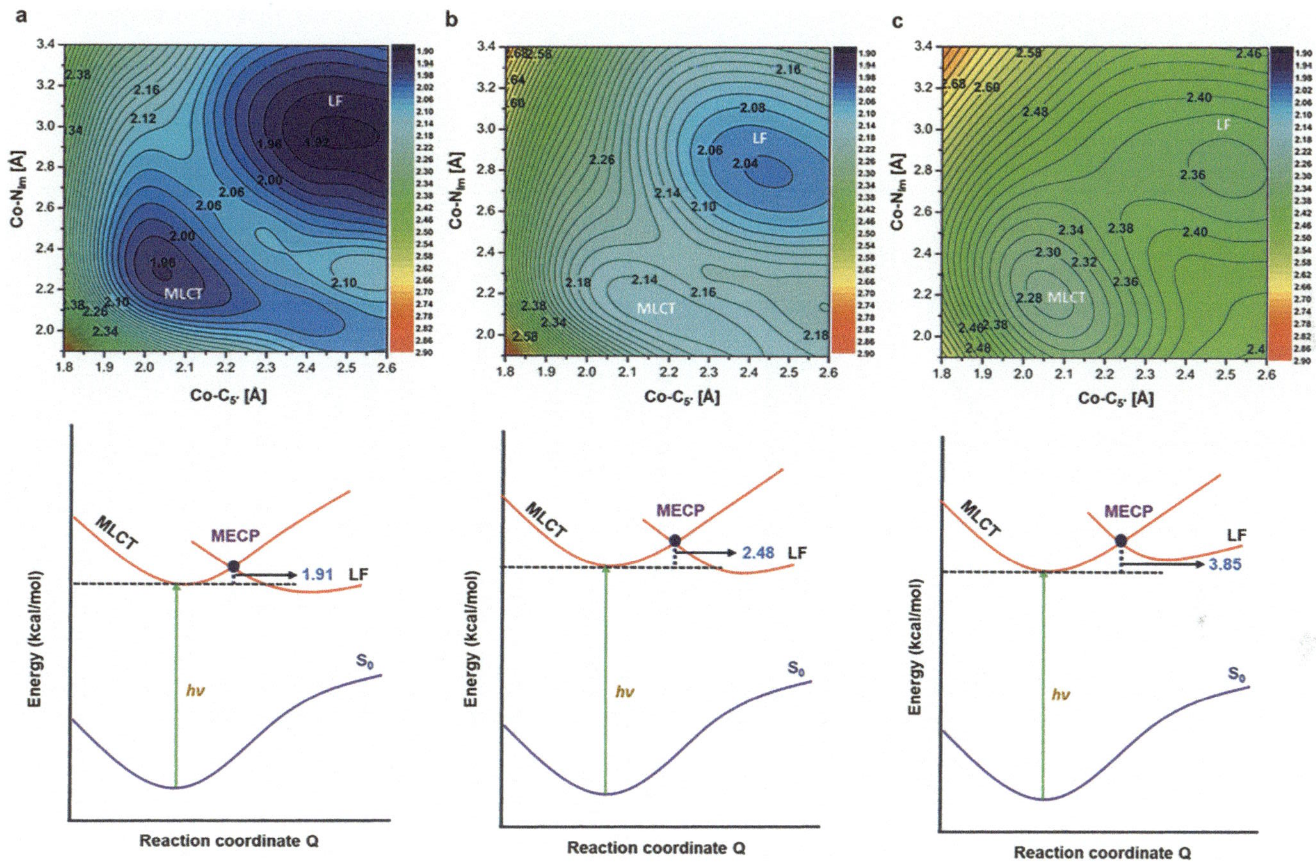

Fig. 8 Topologies of the S_1 excited states and relative energetics of the MLCT and LF state along the coordinate Q for (A) EAL, (B) GLM, and (C) CarH. *Adapted from Toda, M. J., Mamun, A. A., Lodowski, P., & Kozlowski, P. M. (2020). Why is CarH photolytically active in comparison to other B_{12}-dependent enzymes?* Journal of Photochemistry & Photobiology, B: Biology, 209, *111919.*

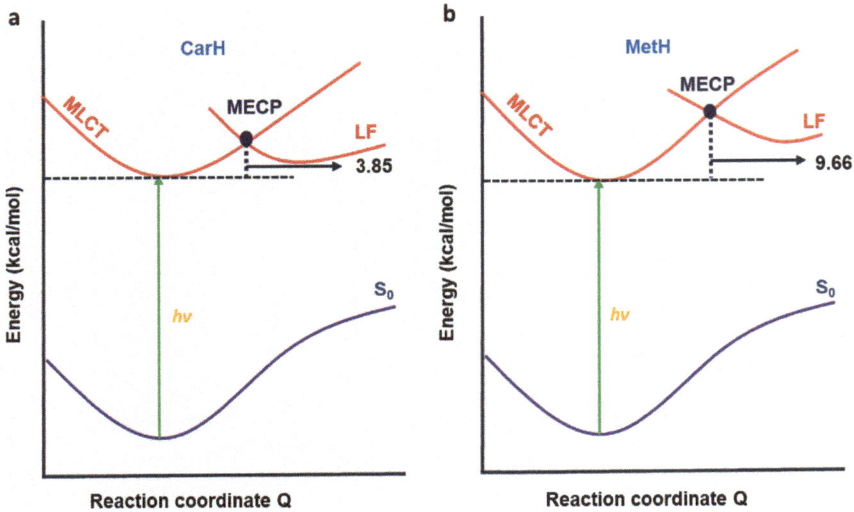

Fig. 9 Relative energetics of the MLCT and LF states for (A) CarH and (B) MetH. *Adapted from Toda, M. J., Mamun, A. A., Lodowski, P., & Kozlowski, P. M. (2020). Why is CarH photolytically active in comparison to other B_{12}-dependent enzymes?* Journal of Photochemistry & Photobiology, B: Biology, 209, *111919.*

Considering that the B_{12}-binding domain of MetH and CarH are remarkably similar, the relative energetics of the S_0 and S_1 states were compared. In both MetH and CarH the LF state is destabilized (Fig. 9). Thus, it is evident that the enzymatic environment alters the relative energetics between the minima regions on the S_1 PES.

The major conclusion from this study was based on a key observation that was supported by the crystal structures and QM/MM optimized structures of corresponding models. In terms of geometry, the orientation of the -$C_{5'}$ of the Ado moiety bound to Co and the orientation of the -$C_{4'}$-H plays a critical role in B_{12}-dependent enzymatic catalysis. Specifically, the ribose part of the Ado ligand can undergo various conformational changes including psuedorotation or rotation around the glycosyl angle. Typically, the conformational change of the ribose either favors radical recombination or promotes H-atom abstraction from the substrate. For the native enzymatic reaction of GLM, the psuedorotation of the ribose ring moves the $C_{5'}$ radical in the proper direction to generate a substrate radical instead of recombining with the Co(II). For diol dehydratase (DDH) or EAL, the $C_{5'}$ radical is moved over an approximately 7 Å distance towards the substrate via rotation around the N-glycosidic bond to facilitate H-atom abstraction (Masuda, Shibata, Morimoto, Toraya, & Yasuoka, 2000). Furthermore, the ribose portion of the Ado group can take on either the

2′-endo or 3′-endo conformation. For EAL, GLM, and CarH, the conformation of the ribose ring, prior to bond cleavage, was found to be 3′-endo (Gruber et al., 2001; Mamun et al., 2018). However, one of the key structural differences between CarH, EAL, and GLM, is that in the case of CarH-DS, the -$C_{4'}$-H in the equilibrium structure is oriented in such a way that would be suitable for H-atom abstraction (Fig. 10). On the other hand, in EAL and GLM, the -$C_{4'}$-H projects in a different orientation that is not favorable for H-atom abstraction from the $C_{4'}$ atom. The QM/MM results indicated that after photocleavage of the Co-$C_{5'}$ bond the -$C_{4'}$-H is in a perfect position to participate in β-H elimination and form a Co-hydride (HCbl). As discussed previously, having the Co-H bond, has been implicated to be a potential intermediate in proposed mechanisms for CarH-DS.

5.2 Photolytic properties of MeCbl

5.2.1 Photolysis of isolated MeCbl

To elucidate the photoreaction of MeCbl PESs associated with S_0 and S_1 states was constructed as a function of axial bond lengths (Fig. 11) (Lodowski, Jaworska, Andruniow, Garabato, & Kozlowski, 2014). The S_1 PES has been constructed by calculating the vertical excitation of each optimized point of S_0 PES. Like the AdoCbl the S_1 PES of MeCbl also carries two minima regions, the MLCT I (S_{1min}), which correspond to shorter axial bond distance, and the LF IIIB I (S_{1min}), which is observed along the elongated Co-N_{Im} distance and slightly stretched Co-C bond length. These regions in the S_1 PES are separated by a seam, which is a crossing between two excited state regions by MEP. The energy difference between MLCT and LF excited state in S_1 PES guides the photoreaction. Based on these topological observations two photodissociation pathways, namely Path A and Path B have been identified for the MeCbl. Path A initiates from the MLCT I (S_{1min}) and proceed along the elongated Co-C bond, followed by the displacement of the Co-N_{Im} bond. Path B is associated with the initial elongation of the Co-N_{Im} bond from the MLCT minima (I (S_{1min})) followed by the lengthening of the Co-C bond. While the intermediated involves in both pathways are different, the initial and final products are the same. These two pathways are associated with different excitation wavelengths, which can provide the wavelength sensitivity of MeCbl photochemistry. Excitation at the visible α/β band (~520 nm) can be represented by Path A, while Path B is associated with the excitation near the UV region (400 nm).

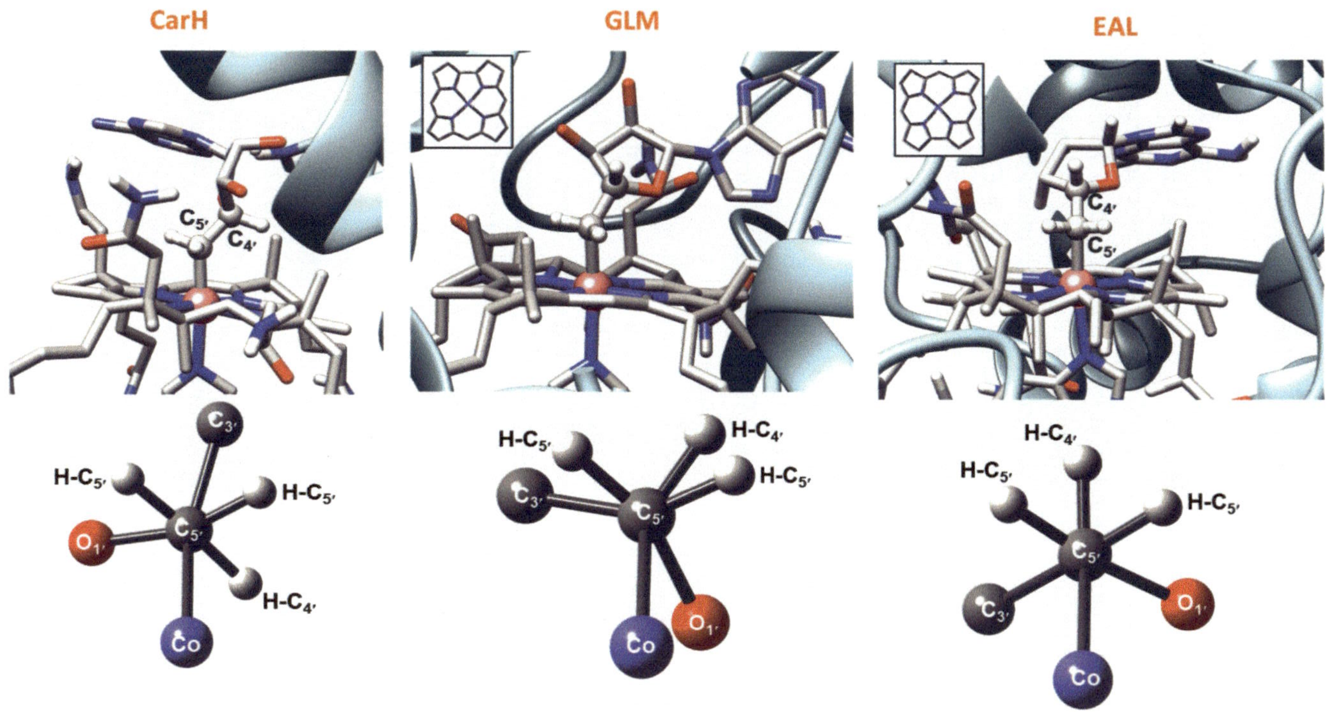

Fig. 10 Optimized ground state equilibrium structures of AdoCbl bound to (A) EAL, (B) GLM, and (C) CarH. $C_{5'}$ and $C_{4'}$ carbons and their corresponding hydrogens are represented in ball and stick form. The stereochemistry of the β-lactam system along the $C_{5'}$–$C_{4'}$ bond is presented in a gray circle. *Adapted from Toda, M. J., Mamun, A. A., Lodowski, P., & Kozlowski, P. M. (2020). Why is CarH photolytically active in comparison to other B_{12}-dependent enzymes?* Journal of Photochemistry & Photobiology, B: Biology, 209, *111919.*

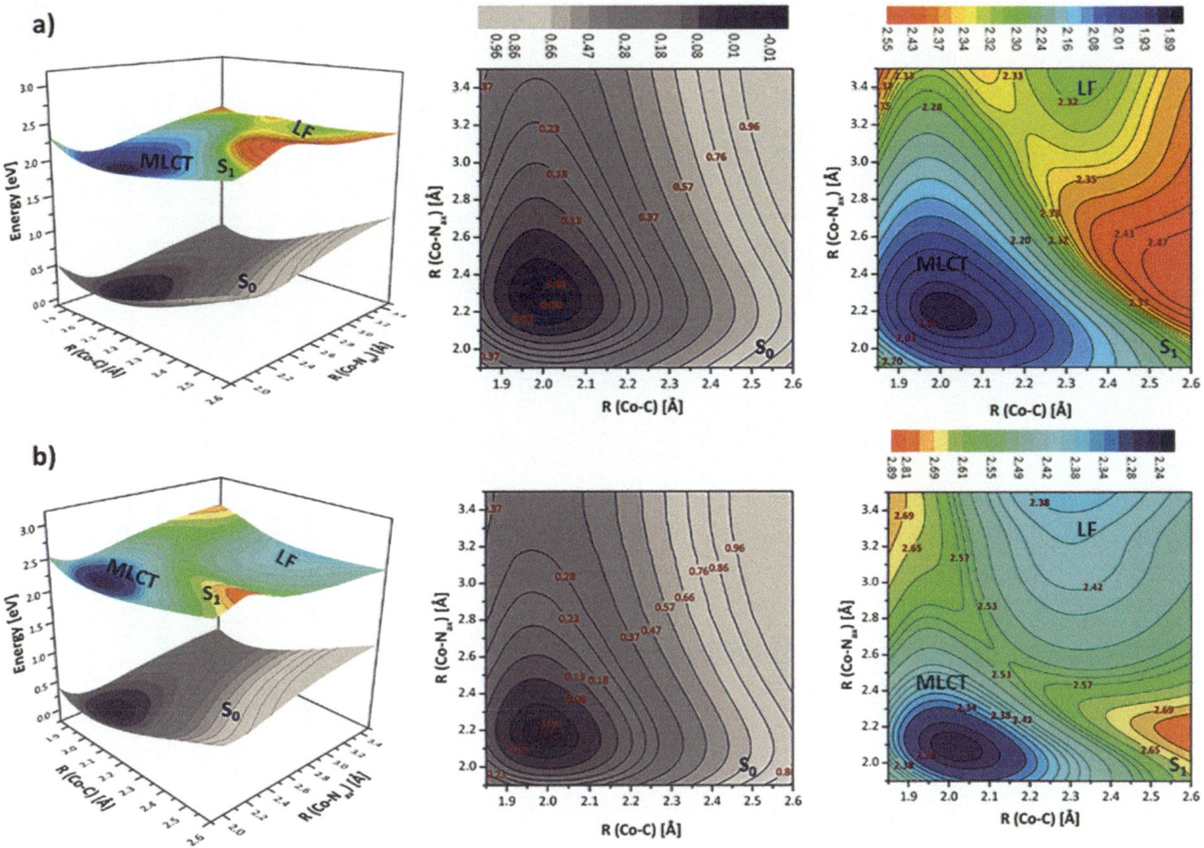

Fig. 11 Potential energy surfaces for the S_0 with vertical projections of the S_1 state plotted as functions of axial bond lengths along with the corresponding S_0 and S_1 contour PESs with color codes for (A) MeCbl inside methionine synthase (B) Im –[Co^{III}(corrin)]–Me$^+$ base-on model complex in water. *Adapted from Ghosh, A. P., Mamun, A. A., Lodowski, P., Jaworska, M., & Kozlowski, P. M. (2018). Mechanism of the photo-induced activation of Co-C bond in methylcobalamin-dependent methionine synthase.* Journal of Photochemistry and Photobiology B: Biology, 189, *306–317.*

5.2.2 Photolysis of MeCbl-dependent MetH

The S_1 PES of MeCbl-bound MetH has been analyzed and compared with the base-on MeCbl inside the solution to investigate the environmental effect on the photolytic properties of MeCbl (Ghosh et al., 2018). As usual, similar to the isolated MeCbl, the S_1 PES of MeCbl-bound MetH carries two minima regions: MLCT (I S_{1min}) and LF (IIIB S_{1min}) (Fig. 12). Although the topology S_1 PES of MeCbl and MeCbl-inside MetH both contain two energy minima, there is a significant difference in the energetics of the S_1 PESs. The energy difference between the MLCT and LF minima region in the S_1 PES MeCbl-inside MetH is almost 4 times higher compared to the isolated MeCbl. The LF region is energetically more stabilized in base-on MeCbl inside solution which indicates that the enzymatic environment is dis-stabilizing the LF region.

Analyzing the energetics and the topology of the S_1 PES it is evident that the photoreaction of MeCbl inside MetH proceeds through Path B, which the initial elongation of the Co-N_{Im} bond from the MLCT minimum (I S_{1min}) (Fig. 12). After the elongation of the Co-N_{Im} bond the photoreaction proceeds via the minimum energy crossing point (MECP), IIB, along the seam. After the photoreaction crosses the MECP (IIB) it reaches LF, where the photolysis of the Co-C bond occurs by generating the Co(II)/CH_3 RPs. The calculated energy barrier associated with Path B is 9.66 kcal/mol. On the other hand, for base-on MeCbl, the energy barrier to cross the MECP (IIB) along Path B is 7.05 kcal/mol. After the formation of RPs, the methyl radical can either diffuse through the "cage" formed by the protein residue or the RP can recombine through the deactivation to the ground state. On the other hand, Path A which is associated with the initial displacement of the Co-C bond prior to the elongation of the Co-N_{Im} bond is energetically unfavorable for the photoreaction. From Fig. 12 it is evident that in Path A the steep increase in energy associated with elongation of the Co-C bond makes the energy barrier for photoreaction very high, which makes it photolytically inactive. Moreover, the higher energetics of the LF state is the reason why we found only one pathway (Path B) for the photodissociation of MeCbl-bound MetH. This shows that the enzymatic environment induces dis-stabilization in the LF region and affecting the photolysis of MeCbl inside MetH.

Nevertheless, the photoreaction in enzyme-bound MeCbl initiates from the MLCT minima and proceeds along path B to reach the LF minima of S_1 PES at (IIIB S_{1min}) (Fig. 13). Once the reaction reaches the LF minima, the photolysis of the Co-C bond occurs generating the Co(II)/CH_3 RP.

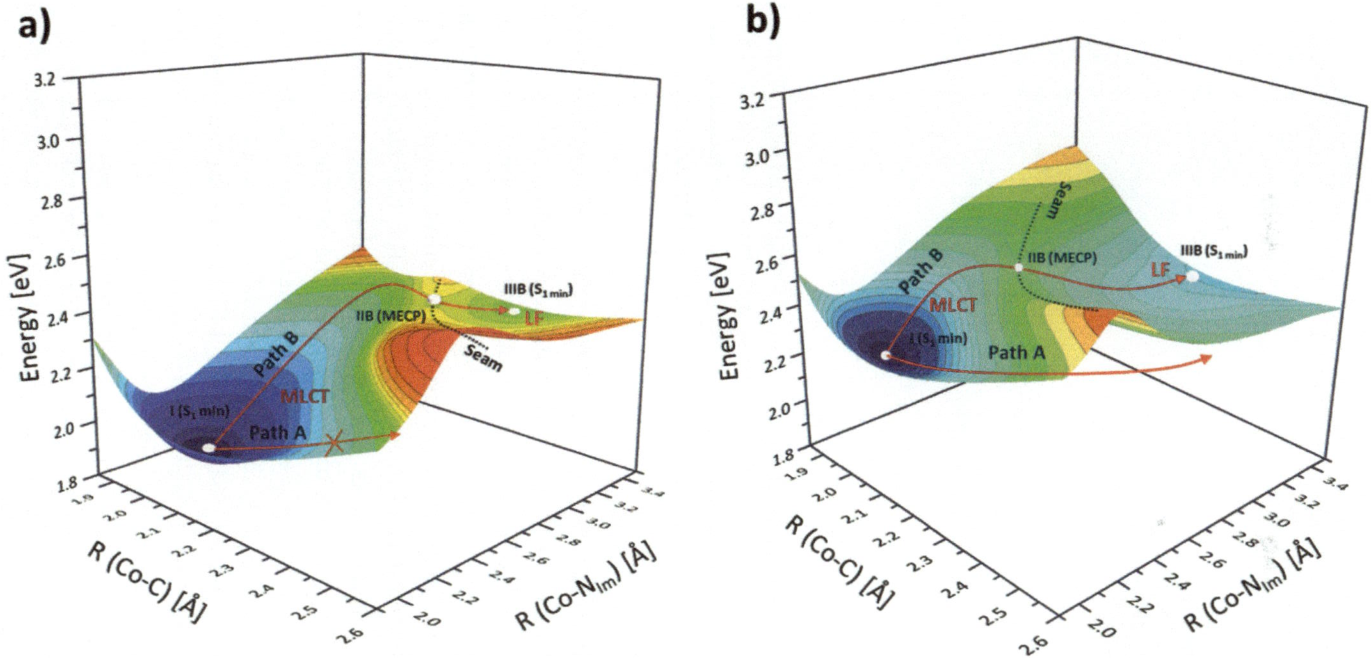

Fig. 12 Potential energy surfaces of S_1 state as a function of axial bond lengths with photoreaction pathway depicted by red arrows for (A) MeCbl inside methionine synthase (B) $Im-[Co^{III}(corrin)]-Me^+$ base-on model complex in water. *Adapted from Ghosh, A. P., Mamun, A. A., Lodowski, P., Jaworska, M., & Kozlowski, P. M. (2018). Mechanism of the photo-induced activation of Co-C bond in methylcobalamin-dependent methionine synthase.* Journal of Photochemistry and Photobiology B: Biology, 189, *306–317.*

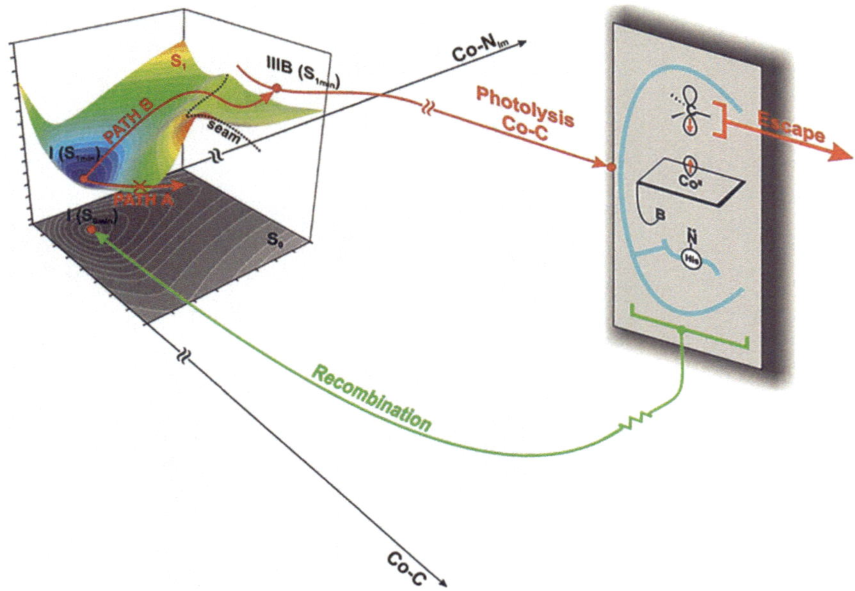

Fig. 13 Schematic representation of photoreaction for MeCbl-bound MetH. *Adapted from Ghosh, A. P., Mamun, A. A., Lodowski, P., Jaworska, M., & Kozlowski, P. M. (2018). Mechanism of the photo-induced activation of Co-C bond in methylcobalamin-dependent methionine synthase.* Journal of Photochemistry and Photobiology B: Biology, 189, *306–317.*

After the formation of RPs from the LF state, the methyl radical can either escape the protein cage formed by the amino acid residues of the cap domain of MetH or it can relax down to the ground state by recombining with the Co(II). The experimental studies have shown that the protein environment prevents the photolysis of MeCbl by caging the methyl radical and increasing the geminate recombination of photogenerated RP (Jarrett et al., 1996). Based on these calculations, it appears that several factors are responsible for preventing the photolysis of MeCbl inside the enzyme, (i) the protein environment is altering the photochemical yield of the Co(II)/CH_3 RP by increasing the energy difference between MLCT and LF minima. The energy difference between MLCT and LF minima for enzyme-bound MeCbl is almost three times higher than the isolated cofactor in the solution. (ii) it is increasing the energy barrier for photoreaction. These observations also complement the experimental finding which reported a 50-fold decrease in the rate of photoreaction for MetH-bound MeCbl (Jarrett et al., 1996). Therefore, considering all of these, we can predict that it is easier to form RP in the solution than inside the enzyme.

5.2.3 *Effects of mutation on the photolysis of MeCbl in MeCbl-dependent enzymes*

Considering all the existing photochemical data it is apparent that the protein environment protects the photolysis of MeCbl. To be precise, the amino acid residues (such as phenylalanine 708 (F708) and leucine 715 (L715)) of the cap domain which surround the upper face of the cofactor protect the photolysis of MeCbl by caging the photogenerated methyl radical. This cage formed by the hydrophobic residue of the cap domain limits the solvent accessibility of the methyl radical and prevents the radical escape from the protein cage. Matthews and co-workers also demonstrated that a mutation in the cap domain of the MetH-bound MeCbl, particularly at F708 positions significantly increases the rate of the photoreaction by almost 62-fold (Jarrett et al., 1996).

The excited-state properties of wild-type (WT)-MetH and F708A-MetH were computed and compared to investigate how does the mutation in the cap domain modulates the photolytic properties of MeCbl bound MetH (Fig. 14) (Ghosh, Mamun, & Kozlowski, 2019). The S_1 PES for WT-MetH and F708A-MetH carry two energy minima: MLCT and LF (Fig. 15). However, a noticeable difference can be seen in the topology and the energetics of the S_1 PESs. The energetic difference between the two minima region, i.e. MLCT (I (S_{1min})) and LF minima (IIIB (S_{1min})) (Figs. 15 and 16) for WT-MetH are almost 3-times higher while compared with the F708A-MetH. Thus, it appears that the LF electronic state of F708A-MetH is energetically stabilized than the WT-enzyme, which indicates that the mutation is playing a role in stabilizing the LF region. This stabilization of the LF region is also influencing the photodissociation pathway in the F708A-MetH. While the photoreaction in the WT-MetH only follows path B, for the F708A-MetH, both path A and path B is feasible. However, path B which involves the initial elongation of the Co-N_{Im} bond followed by the dissociation of the Co-C bond is energetically most favorable for the photolysis of MeCbl in the mutant enzyme. The calculated energy barrier associated with photoreaction along path B is 7.1 kcal/mol, which is lower than the energy barrier for the WT-MetH (Fig. 17). In addition to this, the mutation in the cap domain is also influencing the rate of geminate recombination. Based on the calculations, it can be seen that the rate of geminate recombination for WT-MetH is almost 20% higher than its mutant counterpart. The higher rate of geminate recombination suggests that the protein environment prevents the escape of methyl radical, which in turn decreases the rate of photolysis for the WT-MetH. Thus, the

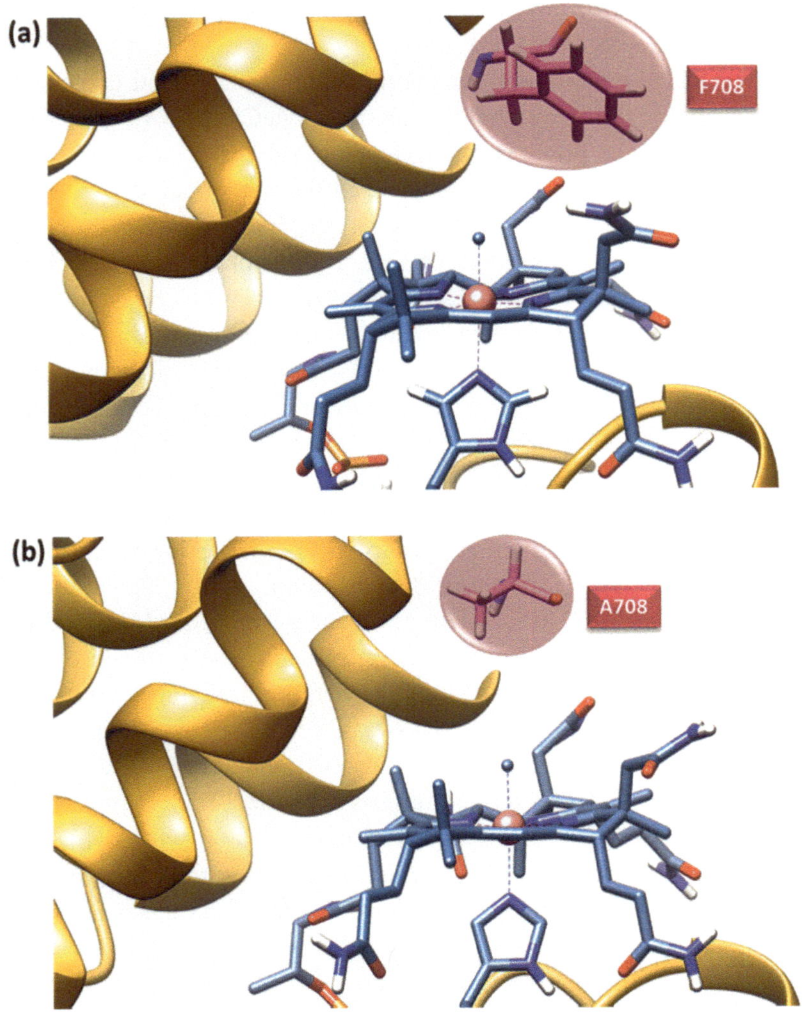

Fig. 14 The active site of MeCbl-dependent MetH. MeCbl cofactor, Phe, Ala amino group is depicted as a ball and stick model and protein residue using ribbon (A) WT-MetH (B) F708A-MetH. *Adapted from Ghosh, A. P., Mamun, A. A., & Kozlowski, P. M. (2019). How does the mutation in the cap domain of methylcobalamin-dependent methionine synthase influence the photoactivation of the Co–C bond?* Physical Chemistry Chemical Physics, 21*(37), 20628–20640.*

mutation in the cap domain impedes the rate of geminate recombination and increases the rate of cage escape, which presumably leads to an increase in the photolysis rate. The aforementioned observations complement the experimental findings which reported a significant rate enhancement of photolysis

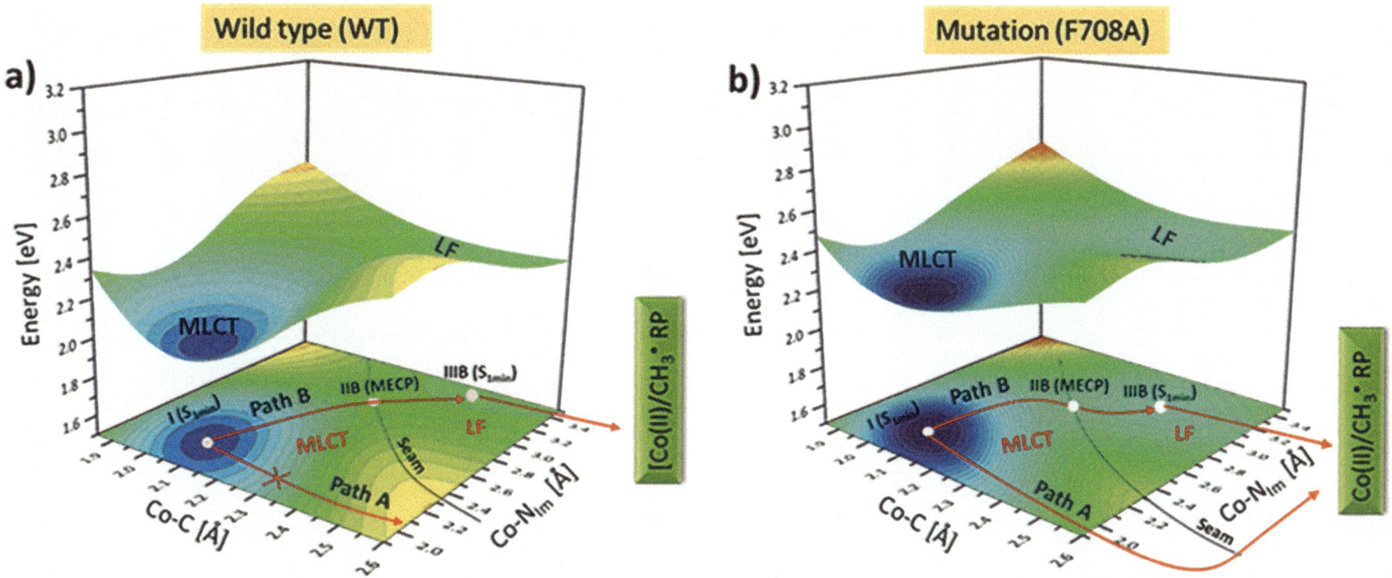

Fig. 15 Potential energy surfaces of S_1 state as a function of axial bond lengths with photoreaction pathway depicted by red arrows for (A) WT-MetH (B) F708A-MetH. *Adapted from Ghosh, A. P., Mamun, A. A., & Kozlowski, P. M. (2019). How does the mutation in the cap domain of methylcobalamin-dependent methionine synthase influence the photoactivation of the Co–C bond?* Physical Chemistry Chemical Physics, *21(37), 20628–20640.*

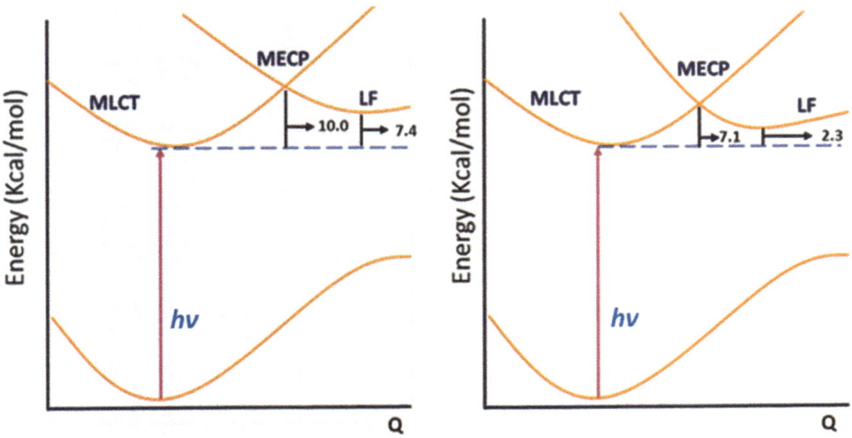

Fig. 16 Relative energetics of the MLCT and LF state for (A) WT-MetH (B) F708A-MetH. $Q = \sqrt{R^2_{Co-C} + R^2_{Co-N_{Im}}}$. *Adapted from Ghosh, A. P., Mamun, A. A., & Kozlowski, P. M. (2019). How does the mutation in the cap domain of methylcobalamin-dependent methionine synthase influence the photoactivation of the Co–C bond?* Physical Chemistry Chemical Physics, 21*(37), 20628–20640.*

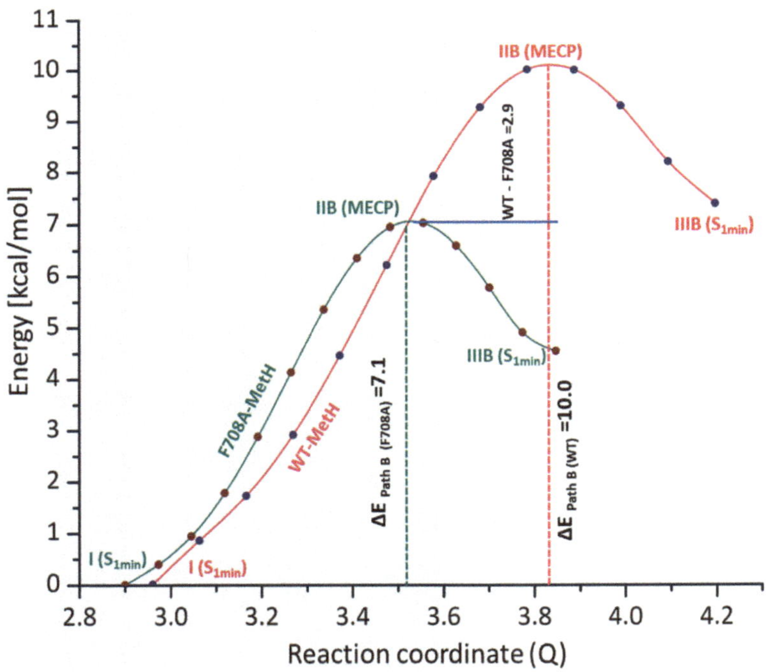

Fig. 17 Photoreaction along minimum energy paths (Path B) between MLCT and LF state plotted as a function of energy versus Q value. Pink represents WT-MetH and the teal curve is for the F708A-MetH. *Adapted from Ghosh, A. P., Mamun, A. A., & Kozlowski, P. M. (2019). How does the mutation in the cap domain of methylcobalamin-dependent methionine synthase influence the photoactivation of the Co–C bond?* Physical Chemistry Chemical Physics, 21*(37), 20628–20640.*

by 62-fold for F708A-MetH. Therefore, it is evident that the mutation in the cap domain influences the photolysis of MeCbl by modulating the low-lying excited states.

6. Summary and outlook

As indicated in the Introduction, the photochemistry of B_{12} derivatives, especially the biologically active forms, has applications across various disciplines. These include the generation of OH radicals, drug delivery, antivitamins B_{12}, and photoreceptors like CarH (Shell & Lawrence, 2011, 2015). Perhaps the most relevant example of this review is the CarH photoreceptor. CarH has been said to be the prototype for the B_{12}-dependent bacterial family of photoreceptors (Padmanabhan, Pérez-Castaño, & Elías-Arnanz, 2019). Not only is it representative of an entire group of enzymes that rely on light for their catalytic functions, but it has also proven to be a useful tool in the area of synthetic biology. Recently it has been shown that the Cbl binding domain can be replaced with other effector domains for applications in protein oligomerization. In addition, as an optogenetic tool, CarH can be used in light-regulated gene expression and receptor signaling. Applications are also emerging in the area of drug delivery and tissue engineering with the development of CarH inspired hydrogels and nanofilms (Wang, Yang, Luo, Hsing, & Sun, 2017). Clearly, the light applications for B_{12}-based systems will continue to emerge hand in hand with theoretical insights.

References

Andruniow, T., Kozlowski, P. M., & Zgierski, M. Z. (2001). Theoretical analysis of electronic absorption spectra of vitamin B_{12} models. *The Journal of Chemical Physics, 115*(16), 7522–7533. https://doi.org/10.1063/1.1405436.

Andruniow, T., Zgierski, M. Z., & Kozlowski, P. M. (2000). Density functional theory analysis of stereoelectronic properties of cobalamins. *The Journal of Physical Chemistry B, 104*(46), 10921–10927. https://doi.org/10.1021/jp000810x.

Banerjee, R. (2001). Radical peregrinations catalyzed by coenzyme B_{12}-dependent enzymes. *Biochemistry, 40*(21), 6191–6198. https://doi.org/10.1021/bi0104423.

Banerjee, R. (2003). Radical carbon skeleton rearrangements: Catalysis by coenzyme B_{12}-dependent mutases. *Chemical Reviews, 103*(6), 2083–2094. https://doi.org/10.1021/cr0204395.

Banerjee, R., & Padmakumar, R. (1995). Evidence from electron paramagnetic resonance spectroscopy of the participation of radical intermediates in the reaction catalyzed by methylmalonyl-coenzyme A mutase. *Journal of Biological Chemistry, 270*, 9295–9300.

Banerjee, R., & Ragsdale, S. W. (2003). The many faces of vitamin B_{12}: Catalysis by cobalamin-dependent enzymes. *Annual Review of Biochemistry, 72*, 209–247. https://doi.org/10.1146/annurev.biochem.72.121801.161828.

Barker, H. A., Smyth, R. D., Weissbach, H., Munch-Petersen, A., Toohey, J. I., Ladd, J. N., et al. (1960). Assay, purification, and properties of the adenylcobamide coenzyme. *Journal of Biological Chemistry*, *235*(1), 181–190.

Barker, H. A., Weissbach, H., & Smyth, R. D. (1958). A coenzyme containing pseudovitamin B_{12}. *Proceedings of the National Academy of Sciences of the United States of America*, *44*(11), 1093–1097.

Brooks, A. J., Vlasie, M., Banerjee, R., & Brunold, T. C. (2004). Spectroscopic and computational studies on the adenosylcobalamin-dependent methylmalonyl-CoA mutase: evaluation of enzymatic contributions to Co-C bond activation in the Co^{3+} ground state. *Journal of American Chemical Society*, *126*(26), 8167–8180. https://doi.org/10.1021/ja039114b.

Brown, K. L. (2005). Chemistry and enzymology of vitamin B_{12}. *Chemical Reviews*, *105*(6), 2075–2150. https://doi.org/10.1021/cr030720z.

Chowdhury, S., & Banerjee, R. (2000). Thermodynamic and kinetic characterization of Co − C bond homolysis catalyzed by coenzyme B_{12}-dependent methylmalonyl-CoA mutase. *Biochemistry*, *39*(27), 7998–8006. https://doi.org/10.1021/bi992535e.

Cole, A. G., Yoder, L. M., Shiang, J. J., Anderson, N. A., Walker, L. A., II, Banaszak Holl, M. M., et al. (2001). Time-resolved spectroscopic studies of B_{12} coenzymes: A comparison of the primary photolysis mechanism in methyl-, ethyl-, *n*-propyl-, and 5'-deoxyadenosylcobalamin. *Journal of American Chemical Society*, *124*(3), 434–441.

Drennan, C. L., Huang, S., Drummond, J. T., Matthews, R. G., & Lidwig, M. L. (1994). How a protein binds B_{12}: A 3.0 A X-ray structure of B_{12}-binding domains of methionine synthase. *Science*, *266*(5191), 1669–1674.

Finke, R. G., & Martin, B. D. (1990). Coenzyme $AdoB_{12}$ vs $AdoB_{12}$.-homolytic Co-C cleavage following electron transfer: A rate enhancement greater than or equal to 10^{12}. *Journal of Inorganic Biochemistry*, *40*(1), 19–22.

Firth, R. A., Hill, H. A. O., Mann, B. E., Pratt, J. M., Thorp, R. G., & Williams, R. J. P. (1968). The chemistry of vitamin B_{12}. Part IX. Evidence for five-co-ordinate cobalt(III) complexes. *Journal of the Chemical Society A: Inorganic, Physical, Theoretical*, 2419–2428. https://doi.org/10.1039/J19680002419.

Frey, P. A. (2001). Radical mechanisms of enzymatic catalysis. *Annual Review of Biochemistry*, *70*, 121–148.

Frey, P. A., & Reed, G. H. (2000). Radical mechanisms in adenosylmethionine- and adenosylcobalamin-dependent enzymatic reactions. *Archives of Biochemistry and Biophysics*, *382*(1), 6–14.

Garabato, B. D., Lodowski, P., Jaworska, M., & Kozlowski, P. M. (2016). Mechanism of Co-C photodissociation in adenosylcobalamin. *Physical Chemistry Chemical Physics*, *18*(28), 19070–19082. https://doi.org/10.1039/c6cp02136k.

Ghosh, A. P., Lodowski, P., Bazarganpour, A., Leks, M., & Kozlowski, P. M. (2020). Aerobic photolysis of methylcobalamin: Structural and electronic properties of Cbl-OO-CH_3 intermediate. *Dalton Transactions*, *49*, 4114–4124. https://doi.org/10.1039/C9DT03740C.

Ghosh, A. P., Mamun, A. A., & Kozlowski, P. M. (2019). How does the mutation in the cap domain of methylcobalamin-dependent methionine synthase influence the photoactivation of the Co–C bond? *Physical Chemistry Chemical Physics*, *21*(37), 20628–20640. https://doi.org/10.1039/C9CP01849B.

Ghosh, A. P., Mamun, A. A., Lodowski, P., Jaworska, M., & Kozlowski, P. M. (2018). Mechanism of the photo-induced activation of Co-C bond in methylcobalamin-dependent methionine synthase. *Journal of Photochemistry and Photobiology B: Biology*, *189*, 306–317. https://doi.org/10.1016/j.jphotobiol.2018.09.015.

Goulding, C. W., Postigo, D., & Matthews, R. G. (1997). Cobalamin-dependent methionine synthase is a modular protein with distinct regions for binding homocysteine, methyltetrahydrofolate, cobalamin, and adenosylmethionine. *Biochemistry, 36*(26), 8082–8091. https://doi.org/10.1021/bi9705164.

Gruber, K., Reitzer, R., & Kratky, C. (2001). Radical shuttling in a protein: Ribose pseudorotation controls alkyl-radical transfer in the coenzyme B_{12} dependent enzyme glutamate mutase. *Angewandte Chemie International Edition, 40*(18), 3377–3380. https://doi.org/10.1002/1521-3773(20010917)40:18.

Harris, D. A., Stickrath, A. B., Carroll, E. C., & Sension, R. J. (2007). Influence of environment on the electronic structure of Cob(III)alamins: Time-resolved absorption studies of the S_1 state spectrum and dynamics. *Journal of American Chemical Society, 129*(24), 7578–7585. https://doi.org/10.1021/ja066197y.

Hogenkamp, H. P. C. (1963). A cyclic nucleoside derived from coenzyme B_{12}. *Journal of Biological Chemistry, 238*(1), 477–480.

Hogenkamp, H. P. C., & Barker, H. A. (1961). The identification of a sugar derived from coenzyme B_{12}. *The Journal of Biological Chemistry, 236*(3097), 3101.

Hogenkamp, H. P. C., Ladd, J. N., & Barker, H. A. (1962). The identification of a nucleoside derived from a coenzyme B_{12}. *The Journal of Biological Chemistry, 237*(6), 1950–1952.

Jarrett, J. T., Drennan, C. L., Amaratunga, M., Scholten, J. D., Ludwig, M. L., & Matthews, R. G. (1996). A protein radical cage slows photolysis of methylcobalamin in methionine synthase from Escherichia coli. *Bioorganic & Medicinal Chemistry, 4*(8), 1237–1246.

Johnson, A. W., Mervyn, L., Shaw, N., & Smith, E. L. (1963). A partial synthesis of the vitamin B_{12} coenzyme and some of its analogues. *Journal of the Chemical Society (Resumed), 785*, 4146–4156. https://doi.org/10.1039/JR9630004146.

Jones, A. R. (2017). The photochemistry and photobiology of vitamin B_{12}. *Photochemical Photobiological Science, 16*(6), 820–834. https://doi.org/10.1039/c7pp00054e.

Jones, A. R., Hardman, S. J. O., Hay, S., & Scrutton, N. S. (2011). Is there a dynamic protein contribution to the substrate trigger in coenzyme B_{12}-dependent ethanolamine ammonia lyase? *Angewandte Chemie International Edition, 50*(46), 10843–10846. https://doi.org/10.1002/anie.201105132.

Jones, A. R., Hay, S., Woodward, J. R., & Scrutton, N. S. (2007). Magnetic field effect studies indicate reduced geminate recombination of the radical pair in substrate-bound adenosylcobalamin-dependent ethanolamine ammonia lyase. *Journal of American Chemical Society, 129*(50), 15718–15727. https://doi.org/10.1021/ja077124x.

Jones, A. R., Levy, C., Hay, S., & Scrutton, N. S. (2013). Relating localized protein motions to the reaction coordinate in coenzyme B_{12}-dependent enzymes. *FEBS Journal, 280*(13), 2997–3008. https://doi.org/10.1111/febs.12223.

Jones, A. R., Woodward, J. R., & Scrutton, N. S. (2009). Continuous wave photolysis magnetic field effect investigations with free and protein-bound alkylcobalamins. *Journal of American Chemical Society, 131*(47), 17246–17253. https://doi.org/10.1021/ja9059238.

Jost, M., Fernandez-Zapata, J., Polanco, M. C., Ortiz-Guerrero, J. M., Chen, P. Y., Kang, G., et al. (2015). Structural basis for gene regulation by a B_{12}-dependent photoreceptor. *Nature, 526*(7574), 536–541. https://doi.org/10.1038/nature14950.

Jost, M., Simpson, J. H., & Drennan, C. L. (2015). The transcription factor CarH safeguards use of adenosylcobalamin as a light sensor by altering the photolysis products. *Biochemistry, 54*(21), 3231–3234. https://doi.org/10.1021/acs.biochem.5b00416.

Kepp, K. P. (2014). Co–C dissociation of adenosylcobalamin (Coenzyme B_{12}): Role of dispersion, induction effects, solvent polarity, and relativistic and thermal corrections. *The Journal of Physical Chemistry A, 118*(34), 7104–7117. https://doi.org/10.1021/jp503607k.

Kornobis, K., Kumar, N., Lodowski, P., Jaworska, M., Piecuch, P., Lutz, J. J., et al. (2013). Electronic structure of the S_1 state in methylcobalamin: Insight from CASSCF/MC-XQDPT2, EOM-CCSD, and TD-DFT calculations. *Journal of Computational Chemistry, 34*(12), 987–1004. https://doi.org/10.1002/jcc.23204.

Kornobis, K., Kumar, N., Wong, B. M., Lodowski, P., Jaworska, M., Andruniow, T., et al. (2011). Electronically excited states of vitamin B_{12}: Benchmark calculations including time-dependent density functional theory and correlated ab initio methods. *The Journal of Physical Chemistry A*, *115*(7), 1280–1292. https://doi.org/10.1021/jp110914y.

Kozlowski, P. M., Garabato, B. D., Lodowski, P., & Jaworska, M. (2016). Photolytic properties of cobalamins: A theoretical perspective. *Dalton Transaction*, *45*(11), 4457–4470. https://doi.org/10.1039/c5dt04286k.

Kozlowski, P. M., Kumar, M., Piecuch, P., Li, W., Bauman, N. P., Hansen, J. A., et al. (2012). The cobalt-methyl bond dissociation in methylcobalamin: New benchmark analysis based on density functional theory and completely renormalized coupled-cluster calculations. *Journal of Chemical Theory and Computation*, *8*(6), 1870–1894. https://doi.org/10.1021/ct300170y.

Kutta, R. J., Hardman, S. J., Johannissen, L. O., Bellina, B., Messiha, H. L., Ortiz-Guerrero, J. M., et al. (2015). The photochemical mechanism of a B_{12}-dependent photoreceptor protein. *Nature Communication*, *6*, 7907. https://doi.org/10.1038/ncomms8907.

Lodowski, P., Jaworska, M., Andruniow, T., Garabato, B. D., & Kozlowski, P. M. (2014). Mechanism of Co-C bond photolysis in the base-on form of methylcobalamin. *The Journal of Physical Chemistry A*, *118*(50), 11718–11734. https://doi.org/10.1021/jp508513p.

Ludwig, M. L., & Matthews, R. G. (1997). Structure-based perspectives on B_{12}-dependent enzymes. *Annual Review of Biochemistry*, *66*(1), 269–313. https://doi.org/10.1146/annurev.biochem.66.1.269.

Mamun, A. A., Toda, M. J., Lodowski, P., Jaworska, M., & Kozlowski, P. M. (2018). Mechanism of light induced radical pair formation in coenzyme B_{12}-dependent ethanolamine ammonia-lyase. *ACS Catalysis*, *7164-7178*. https://doi.org/10.1021/acscatal.8b00120.

Mamun, A. A., Toda, M. J., Lodowski, P., & Kozlowski, P. M. (2019). Photolytic cleavage of Co–C bond in coenzyme B_{12}-dependent glutamate mutase. *The Journal of Physical Chemistry B*, *123*(12), 2585–2598. https://doi.org/10.1021/acs.jpcb.8b07547.

Masuda, J., Shibata, N., Morimoto, Y., Toraya, T., & Yasuoka, N. (2000). How a protein generates a catalytic radical from coenzyme B_{12}: X-ray structure of a diol-dehydratase-adeninylpentylcobalamin complex. *Structure (London, England: 1993)*, *8*(7), 775–788. https://doi.org/10.1016/s0969-2126(00)00164-7.

Matthews, R. G. (2001). Cobalamin-dependent methyltransferases. *Accounts of Chemical Research*, *34*(8), 681–689. https://doi.org/10.1021/ar0000051.

Matthews, R. G. (2009). Cobalamin- and corrinoid-dependent enzymes. *Metal Ions in Life Sciences*, *6*, 53–114. https://doi.org/10.1039/BK9781847559159-00053.

Matthews, R. G., Koutmos, M., & Datta, S. (2008). Cobalamin-dependent and cobamide-dependent methyltransferases. *Current Opinion in Structural Biology*, *18*(6), 658–666. https://doi.org/10.1016/j.sbi.2008.11.005.

Miller, N. A., Deb, A., Alonso-Mori, R., Garabato, B. D., Glownia, J. M., Kiefer, L. M., et al. (2017). Polarized XANES monitors femtosecond structural evolution of photoexcited vitamin B_{12}. *Journal of American Chemical Society*, *139*(5), 1894–1899. https://doi.org/10.1021/jacs.6b11295.

Miller, N. A., Kaneshiro, A. K., Konar, A., Alonso-Mori, R., Britz, A., Deb, A., et al. (2020). The photoactive excited state of the B_{12}-based photoreceptor CarH. *The Journal of Physical Chemistry B*, *124*(47), 10732–10738. https://doi.org/10.1021/acs.jpcb.0c09428.

Ortiz-Guerrero, J. M., Polanco, M. C., Murillo, F. J., Padmanabhan, S., & Elías-Arnanz, M. (2011). Light-dependent gene regulation by a coenzyme B_{12}-based photoreceptor. *Proceedings of the National Academy of Sciences of the United States of America*, *108*(18), 7565.

Padmakumar, R., Padmakumar, R., & Banerjee, R. (1997). Evidence that cobalt – carbon bond homolysis is coupled to hydrogen atom abstraction from substrate in methylmalonyl-CoA mutase. *Biochemistry*, *36*(12), 3713–3718. https://doi.org/10.1021/bi962503g.

Padmanabhan, S., Jost, M., Drennan, C. L., & Elías-Arnanz, M. (2017). A new facet of vitamin B_{12}: Gene regulation by cobalamin-based photoreceptors. *Annual Review of Biochemistry*, *86*(1), 485–514. https://doi.org/10.1146/annurev-biochem-061516-044500.

Padmanabhan, S., Pérez-Castaño, R., & Elías-Arnanz, M. (2019). B_{12}-based photoreceptors: From structure and function to applications in optogenetics and synthetic biology. *Current Opinion in Structural Biology*, *57*, 47–55. https://doi.org/10.1016/j.sbi.2019.01.020.

Pratt, J. M. (1964). The chemistry of vitamin B_{12}. Part II. Photochemical reactions. *Journal of the Chemical Society (Resumed)*, *988*, 5154–5160. https://doi.org/10.1039/JR9640005154.

Pratt, J. M., & Thorp, R. G. (1966). The chemistry of vitamin B_{12}. Part V. The class (b) character of the cobaltic ion. *Journal of the Chemical Society A: Inorganic, Physical, Theoretical*, *0*, 187–191. https://doi.org/10.1039/J19660000187.

Robertson, W. D., Wang, M., & Warncke, K. (2011). Characterization of protein contributions to cobalt-carbon bond cleavage catalysis in adenosylcobalamin-dependent ethanolamine ammonia-lyase by using photolysis in the ternary complex. *Journal of American Chemical Society*, *133*(18), 6968–6977. https://doi.org/10.1021/ja107052p.

Robertson, W. D., & Warncke, K. (2009). Photolysis of adenosylcobalamin and radical pair recombination in ethanolamine ammonia-lyase probed on the micro- to millisecond time scale by using time-resolved optical absorption spectroscopy. *Biochemistry*, *48*(1), 140–147. https://doi.org/10.1021/bi801659e.

Rury, A. S., Wiley, T. E., & Sension, R. J. (2015). Energy cascades, excited state dynamics, and photochemistry in cob(III)alamins and ferric porphyrins. *Accounts of Chemical Research*, *48*(3), 860–867. https://doi.org/10.1021/ar5004016.

Schrauzer, G. N., Lee, L. P., & Sibert, J. W. (1970). Alkylcobalamins and alkylcobaloximes. Electronic structure, spectra, and mechanism of photodealkylation. *Journal of American Chemical Society*, *92*(10), 2997–3005.

Schrauzer, G. N., Sibert, J. W., & Windgassen, R. J. (1968). Photochemical and thermal cobalt-carbon bond cleavage in alkylcobalamins and related organometallic compounds. Comparative study. *Journal of American Chemical Society*, *90*(24), 6681–6688. https://doi.org/10.1021/ja01026a021.

Sension, R. J., Cole, A. G., Harris, A. D., Fox, C. C., Woodbury, N. W., Lin, S., et al. (2004). Photolysis and recombination of adenosylcobalamin bound to glutamate mutase. *Journal of American Chemical Society*, *126*(6), 1598–1599. https://doi.org/10.1021/ja0396910.

Sension, R. J., Harris, D. A., Stickrath, A., Cole, A. G., Fox, C. C., & Marsh, E. N. (2005). Time-resolved measurements of the photolysis and recombination of adenosylcobalamin bound to glutamate mutase. *Journal of Physical Chemistry B*, *109*(38), 18146–18152. https://doi.org/10.1021/jp052492d.

Shell, T. A., & Lawrence, D. S. (2011). A new trick (hydroxyl radical generation) for an old vitamin (B_{12}). *Journal of American Chemical Society*, *133*(7), 2148–2150. https://doi.org/10.1021/ja111585c.

Shell, T. A., & Lawrence, D. S. (2015). Vitamin B_{12}: A tunable, long wavelength, light-responsive platform for launching therapeutic agents. *Accounts of Chemical Research*, *48*(11), 2866–2874. https://doi.org/10.1021/acs.accounts.5b00331.

Shiang, J. J., Walker, L. A., Anderson, N. A., Cole, A. G., & Sension, R. J. (1999). Time-resolved spectroscopic studies of B_{12} coenzymes: The photolysis of methylcobalamin is wavelength dependent. *The Journal of Physical Chemistry B*, *103*(47), 10532–10539. https://doi.org/10.1021/jp992358r.

Toda, M. J., Kozlowski, P. M., & Andruniów, T. (2019). Assessing electronically excited states of cobalamins via absorption spectroscopy and time-dependent density functional theory. In E. Broclawik, T. Borowski, & M. Radoń (Eds.), *Transition metals in coordination environments: Computational chemistry and catalysis viewpoints* (pp. 219–258). Cham: Springer International Publishing.

Toda, M. J., Lodowski, P., Mamun, A. A., Jaworska, M., & Kozlowski, P. M. (2019). Photolytic properties of the biologically active forms of vitamin B_{12}. *Coordination Chemistry Reviews*, *385*, 20–43. https://doi.org/10.1016/j.ccr.2018.12.017.

Toda, M. J., Mamun, A. A., Lodowski, P., & Kozlowski, P. M. (2020). Why is CarH photolytically active in comparison to other B_{12}-dependent enzymes? *Journal of Photochemistry and Photobiology B: Biology*, *209*, 111919. https://doi.org/10.1016/j.jphotobiol.2020.111919.

Walker, L. A., Jarrett, J. T., Anderson, N. A., Pullen, S. H., Matthews, R. G., & Sension, R. J. (1998). Time-resolved spectroscopic studies of B_{12} coenzymes: The identification of a metastable Cob(III)alamin photoproduct in the photolysis of methylcobalamin. *Journal of American Chemical Society*, *120*(15), 3597–3603. https://doi.org/10.1021/ja974024q.

Walker, L. A., Shiang, J. J., Anderson, N. A., Pullen, S. H., & Sension, R. J. (1998). Time-resolved spectroscopic studies of B_{12} coenzymes: The photolysis and geminate recombination of adenosylcobalamin. *Journal of American Chemical Society*, *120*(29), 7286–7292. https://doi.org/10.1021/ja981029u.

Wang, R., Yang, Z., Luo, J., Hsing, I.-M., & Sun, F. (2017). B_{12}-dependent photoresponsive protein hydrogels for controlled stem cell/protein release. *Proceedings of the National Academy of Sciences of the United States of America*, *114*, 5912–5917. https://doi.org/10.1073/pnas.1621350114.

Warncke, K., & Utada, A. S. (2001). Interaction of the substrate radical and the 5'-deoxyadenosine-5'-methyl group in vitamin B_{12} coenzyme-dependent ethanolamine deaminase. *Journal of American Chemical Society*, *123*(35), 8564–8572.

CHAPTER EIGHT

Antivitamins B_{12}

Bernhard Kräutler*
Institute of Organic Chemistry and Centre of Molecular Biosciences, University of Innsbruck, Innsbruck, Austria
*Corresponding author: e-mail address: bernhard.kraeutler@uibk.ac.at

Contents

Abstract

Antivitamins B_{12} have been designed as agents inducing B_{12}-deficiency by counteracting the metabolic effects of vitamin B_{12} and related cobalamins (Cbls). As expressed by our basic structure- and reactivity-based concept, antivitamins B_{12} are further specified as structural mimics of vitamin B_{12} carrying a chemical 'blockage' that prevents their metabolic conversion into the organometallic B_{12}-cofactors. Antivitamins B_{12} are either intact Cbls (that are) chemically inhibited from acting as precursors of the Cbl-cofactors, or isostructural metal analogues, in which the cobalt-center of the Cbls is replaced by another metal-ion. Hence, antivitamins B_{12} should be taken up by mammals via the natural pathway and be provided to their whole body, thus repressing the metabolic activity of the B_{12}-cofactors.

Antivitamins B_{12} may be specifically useful, in a diagnostic sense, in elucidating still largely puzzling pathophysiological phenomena associated with B_{12}-deficiency in mammals. They may help gain deep bio-structural insights and recognize cryptic physiological roles of B_{12}. Antivitamins B_{12} may, furthermore, display a unique potential in specific therapeutic applications, namely as novel antibiotics or inhibitors of B_{12}-dependent contributions to tumorigenic cellular growth.

Vitamins and Hormones, Volume 119
ISSN 0083-6729
https://doi.org/10.1016/bs.vh.2022.01.003

1. Introduction

In a qualitative sense, antivitamins are vitamin antagonists that oppose the metabolic roles of vitamins in humans and in other mammals (Friedrich, 1988; Kläui, 1979; Zelder, Sonnay, & Prieto, 2015). Antivitamins K may be mentioned here as specific examples, helpful as therapeutics counteracting the activity of vitamin K, in case the latter needs to be reduced in order to prevent excessive blood clotting and thrombosis (Hirsh, O'Donnell, & Eikelboom, 2007). Among the antivitamins K figure compounds unrelated to the vitamin by their structures, such as the coumarins "warfarin" and "acenocoumarol" (Freedman, 1996).

Antivitamins may also develop a beneficial effect as antimetabolites by interfering selectively with the metabolism of bacterial or fungal pathogens. Indeed, antivitamins may provide new approaches to antibiotics against drug-resistant infectious bacterial strains (Guzzo et al., 2016). Along these lines, "anti-folates," like methotrexate, inhibiting folate-dependent enzymatic processes, have become important antibiotics (Bermingham & Derrick, 2002).

Health problems due to (over)activity of cobalamins (Cbls) are unknown (Green & Miller, 2014). Hence, antivitamins B_{12} have not been taken into consideration in a therapeutic context, primarily, but for investigations of B_{12}-deficiency (Kräutler, 2015; Mutti, Ruetz, Birn, Kräutler, & Nexo, 2013; Ruetz et al., 2013; Zelder et al., 2015). Indeed, the study of B_{12}-deficiency in laboratory animals has helped to elaborate on the growing number of suspected contributions of B_{12}-deficiency to pathological phenomena, such as degenerative diseases of the central and peripheral nerve systems (Scalabrino, 2009). Until recently, experimental B_{12}-deficiency has, typically, been induced in laboratory mice by surgery, such as "total" gastrectomy (Scalabrino, 2001), with corresponding ethical implications and possible serious physiological interferences aside of the specific B_{12}-related questions. Proper antivitamins B_{12} would offer a "humane" means for inducing and studying (functional) B_{12}-deficiency in healthy and fully intact animals (Mutti et al., 2013; Ruetz, Gherasim, et al., 2013).

2. Vitamin B_{12} and organometallic B_{12}-cofactors

The compound vitamin B_{12}, also named cyanocobalamin (CNCbl, see Fig. 1), is a convenient chemical isolation form of the Cbls (Friedrich, 1988; Kräutler & Puffer, 2012), whose complex structure was characterized

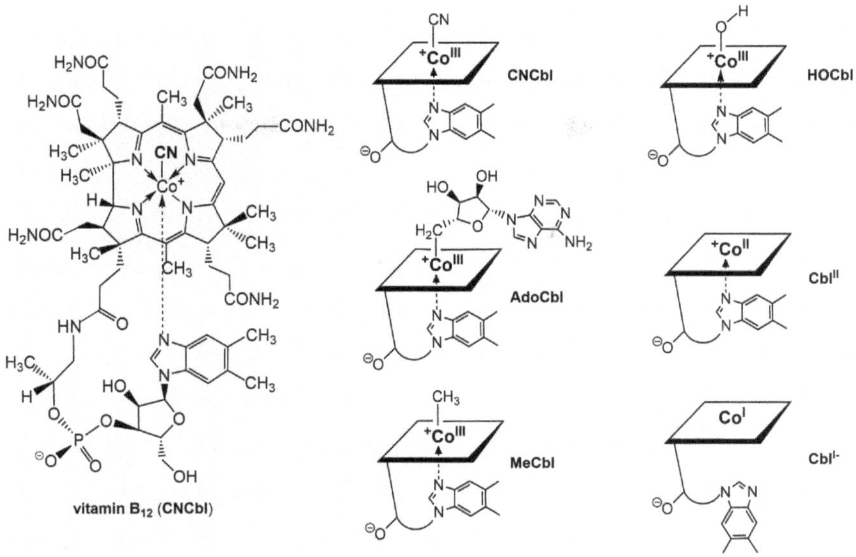

Fig. 1 Structural formula of vitamin B_{12} (CNCbl, *left*) and graphical symbols for representative cobalamins (Cbls, *middle* and *right*): vitamin B_{12} (CNCbl); coenzyme B_{12} (adenosylcobalamin, AdoCbl), methylcobalamin (MeCbl); hydroxocobalamin (HOCbl), cob(II)alamin (Cbl^{II}) and cob(I)alamin (Cbl^{I-}).

in the 1950s by Hodgkin (1965). Remarkably, CNCbl is not directly active, physiologically. The active B_{12}-cofactors are, rather, the organometallic B_{12}-analogues methylcobalamin (MeCbl) and adenosylcobalamin (AdoCbl, coenzyme B_{12}). In order to be useful in the mammalian metabolism, vitamin B_{12} and some other Cbls (B_{12}-vitamers, such as hydroxocobalamin (HOCbl) and its easily accessible protonation form aquocobalamin (aquoCbl^{+})) need to be converted into the B_{12}-cofactors as the proper co-catalysts of B_{12}-dependent enzymes (Banerjee & Ragsdale, 2003; Gruber, Puffer, & Kräutler, 2011; Matthews, 2009). Indeed, in humans (and in other mammals), CNCbl is an indirect biosynthetic precursor of its organometallic relatives MeCbl and AdoCbl (Banerjee, Gouda, & Pillay, 2021).

Clearly, the precise structural features of the Cbls are crucial, as are their specific redox-properties and their reactivity as organometallic cofactors (Kräutler & Puffer, 2012). Their biologically indispensable activity is a specific result of the unique interplay between the natural corrin ligand and the cobalt-ions (Kieninger, Deery, et al., 2019). In the B_{12}-cofactors MeCbl and AdoCbl this interaction provides the essential and characteristic capacity to generate, or cleave, their weak organometallic Co—C bonds under

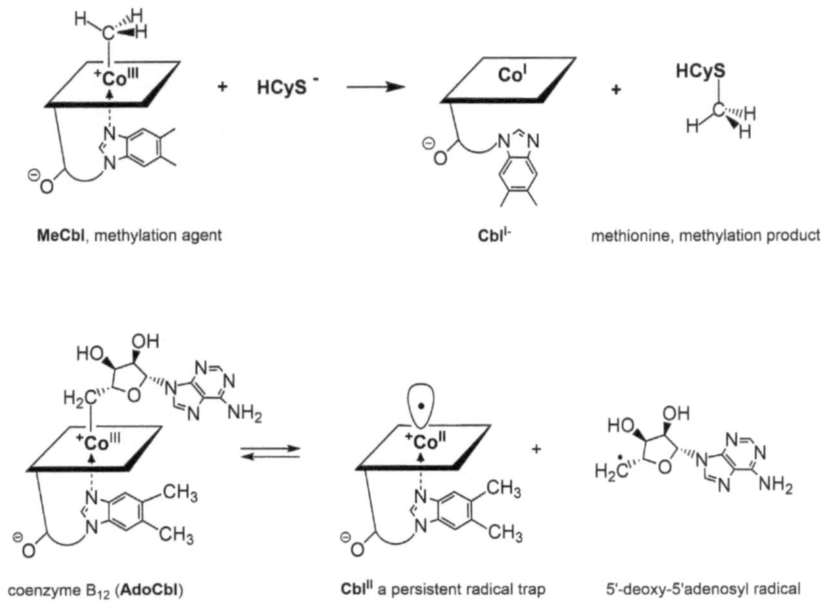

Fig. 2 *Top.* The Co—C bond of methylcobalamin (MeCbl) is cleaved by the attack of a nucleophile (e.g., the homocysteine anion $HCyS^-$ in methionine synthase) generating the methylation product (methionine). *Bottom.* The Co—C bond of coenzyme B_{12} cleaves by homolysis, reversibly furnishing the "radical trap" cob(II)alamin (Cbl^{II}) and the highly reactive (primary) 5′-desoxy-5′adenosyl radical ($Ado^{\bullet}$). In AdoCbl-dependent enzymes, such as methylmalonyl-CoA mutase (MCM), the protein-bound $Ado^{\bullet}$ initiates highly controlled radical reactions.

physiological conditions (Fig. 2) (Brown, 2005; Kräutler, 2019). On this basis, MeCbl acts as a methyl group transfer catalyst (Kräutler & Puffer, 2012; Matthews, Koutmos, & Datta, 2008), while AdoCbl operates as a reversibly functioning source of the highly reactive 5′-deoxy-5′-adenosyl radical (Buckel & Golding, 2006; Halpern, 1985; Kräutler & Puffer, 2012; Marsh & Drennan, 2001). Aside of humans and other mammals, a broad spectrum of living organisms depend upon the organometallic B_{12}-cofactors, which, perhaps, play the globally most broadly relevant role as bio-organometallic catalysts (Banerjee & Ragsdale, 2003; Bridwell-Rabb & Drennan, 2017; Bryant, Hunter, & Warren, 2020; Kräutler, 2019).

The known ("canonical") cofactor roles of Cbls ("B_{12}") in human metabolism are associated with two enzymes: B_{12}-dependend methionine-synthase (MetH) (Ludwig & Matthews, 2002) and AdoCbl-dependent methylmalonyl-CoA–mutase (MCM) (Banerjee, 2003; Gruber & Kratky, 2001), in which the B_{12}-derivatives MeCbl and AdoCbl act, respectively, as organometallic cofactors (Banerjee & Ragsdale, 2003; Brown, 2005; Buckel & Golding, 2012; Kräutler & Puffer, 2012). The enzyme MetH allows for the biosynthesis of methionine from homocysteine, coupled with the demethylation of *N*-methyl-tetrahydrofolate and liberation of tetrahydrofolate (Matthews et al., 2008). The AdoCbl-dependent MCM catalyzes the isomerization of (*R*)-methylmalonyl-CoA to succinyl CoA, as part of the complete metabolic degradation of odd-numbered fatty acids (Banerjee, 2003; Buckel & Golding, 2012; Frey & Hegeman, 2007). Both of these B_{12}-dependent enzymes bind their Cbl-cofactors at the interface of large domains and in the puzzling "base-off/His-on" form, in which a histidine residue of the protein displaces the 5,6-dimethylbenzimidazole (DMB) "base" of the pseudo-nucleotide appendage characteristic of the Cbls (Drennan, Huang, Drummond, Matthews, & Ludwig, 1994; Ludwig & Evans, 1999; Ludwig & Matthews, 1997; Mancia et al., 1996).

Cbls are fairly large "small molecules" that hardly diffuse passively through cell membranes. Hence, uptake and further metabolism of the very scarce Cbls is achieved nearly exclusively by a number of proteins and complementary receptors (Banerjee et al., 2021; Fedosov, 2012; Green et al., 2017; Nielsen, Rasmussen, Andersen, Nexo, & Moestrup, 2012). These must operate swiftly and interact precisely, in order to arrive at the proper supply of cells with the Cbls. In humans, the B_{12}-binding proteins intrinsic factor (IF), transcobalamin (TC) and haptocorrin (HC), manage uptake and cellular transport of the Cbls, in concert with the receptors mediating entry of Cbl-loaded B_{12}-transport proteins into intestinal and other cells of the human (mammalian) body (Fedosov, 2012).

When naturally occurring Cbls are taken up in the cells of healthy subjects, they are "tailored" by the enzyme CblC, a "Cbl-deligase," profoundly studied in the group of Ruma Banerjee (Michigan), which de-coordinates or removes the axial ligands of natural Cbls and generates cob(II)alamin (Cbl^{II}) in its "base-off" form (Fig. 3) (Banerjee et al., 2021; Banerjee, Gherasim, & Padovani, 2009). The enzyme CblC is remarkably versatile and crucial in removing the cobalt-bound ("upper") β-ligand of the known natural Cbls efficiently (Banerjee et al., 2021; Gherasim, Lofgren, & Banerjee, 2013). To achieve this, CblC makes use of the nucleophilic and the reducing activities of the thiol glutathione (GSH), which removes organic or inorganic ligands: the methyl group of MeCbl is detached by a nucleophilic displacement reaction, the cyanide ligand of vitamin B_{12} by a one-electron reduction (Banerjee et al., 2021; Hannibal et al., 2009). The eventual product of the de-ligation, "base-off" Cbl^{II}, is the corrinoid

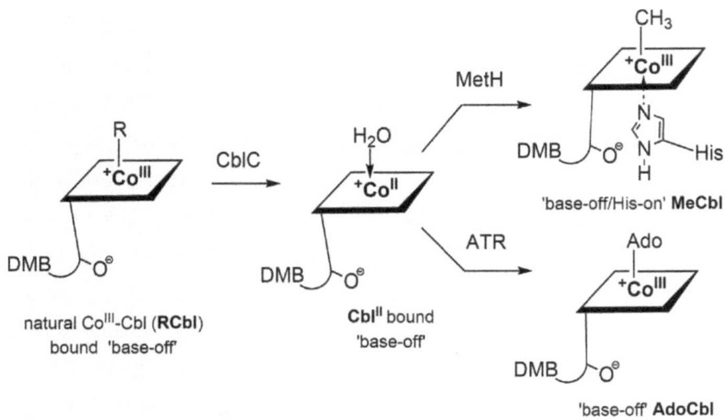

Fig. 3 The Cbl "de-ligase" CblC binds naturally occurring Cbls in their "base-off" form and converts them into "base-off" cob(II)alamin (Cbl^{II}). "Base-off" Cbl^{II} (center) is the direct precursor of methylcobalamin (MeCbl), assembled in the characteristic "base-off/His-on"-form by the cytosolic enzyme methionine synthase (MetH), or of coenzyme B_{12} (AdoCbl), assembled "base-off" by the mitochondrial adenosyltransferase (ATR).

substrate of the enzymes furnishing the proper B_{12}-cofactors, MeCbl and coenzyme B_{12} (AdoCbl). Thus the crucial and unique capacity of CblC paves an intracellular pathway to the two active organometallic B_{12}-cofactors. On the other hand, patients with Cbl-deficiency due to the impairment of CblC by critical mutations cannot be treated effectively by even large doses of CNCbl (Gherasim, Ruetz, Li, Hudolin, & Banerjee, 2015; Lerner-Ellis et al., 2006).

3. Antivitamins B_{12}

Antivitamins B_{12} have been conceived as "vitamin B_{12} dummies" that resist the "tailoring" by CblC, in spite of featuring molecular structures highly similar to CNCbl or AdoCbl (Kräutler, 2015; Ruetz, Gherasim, et al., 2013). According to this concept, antivitamins B_{12} should be taken up in mammals, like the natural Cbls, without becoming a metabolic source for the proper B_{12}-cofactors, thus leading to a rather complete inhibition of the B_{12}-based cellular metabolism.

3.1 Cobalamin-based antivitamins B_{12}

According to our original strategy, several antivitamins B_{12} were designed taking into account the structural aspects of cellular B_{12}-transport, of B_{12}-relevant structure-activity relationships and of the physiological modes of action of the natural Cbl-cofactors in humans, and were constructed rationally from Cbls, e.g., from vitamin B_{12} itself (Kräutler, 2015; Ruetz, Gherasim, et al., 2013). In fact, representatives of the previously unknown class of organometallic aryl-Cbls were posited to have the required properties as Cbl-based inert structural mimics of vitamin B_{12} (Ruetz, Gherasim, et al., 2013): their structural design, similar to CNCbl, would ensure their transport into the cells of the body; they were, furthermore, expected to have the required stability and chemical inertness against both (nucleophilic and reductive) mechanisms that CblC employs for "tailoring" incoming Cbls. 4-Ethyl-phenyl-cobalamin (EtPhCbl) was prepared as a first such Cbl-based potential antivitamin B_{12} (see Fig. 4) (Ruetz, Gherasim, et al., 2013). Indeed, EtPhCbl was bound well by proteins of the human B_{12}-transport system (IF, HC and TC). As expected, EtPhCbl was also bound well, yet not processed by the (human) Cbl-deligase CblC. Hence, the metabolic conversion of EtPhCbl into the proper B_{12}-cofactors MeCbl and AdoCbl would be blocked. Indeed, application of EtPhCbl to healthy laboratory mice induced the appearance of high levels of homocysteine and methyl-malonic acid, two common biomarkers of B_{12}-deficiency that indicate reduced activity of the B_{12}-dependent enzymes MetH and MCM (Mutti et al., 2013). Thus, the B_{12}-derivative EtPhCbl behaved as an antivitamin B_{12} and induced (functional) B_{12}-deficiency in these animals. The crucial locus of the physiological effect of EtPhCbl in inducing B_{12}-deficiency in

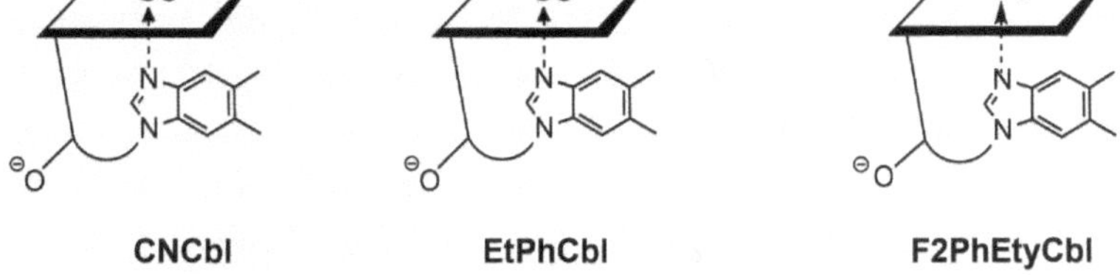

Fig. 4 Symbolic formulas of Cbls, known as vitamin B_{12} (*left*) or proposed to represent (potential) antivitamins B_{12}, such as the light-sensitive aryl-cobalamin 4-ethylphenyl-Cbl (EtPhCbl, *middle*) and the light-stable alkynyl-cobalamin 2-(2,4-difluoro-phenyl)-ethynyl-cobalamin (F2PhEtyCbl, *right*),

these animals has not yet been established, experimentally: it may, however, be associated with the direct inhibition and blockage of CblC by EtPhCbl and, hence, with a generally inhibited cellular net-import of Cbls.

A typical feature of aryl-cobalamins (and of most organometallic B_{12}-derivatives) is their light sensitivity, which may restrict their practical use: in an aerated aqueous solution, the antivitamin B_{12} EtPhCbl is converted by daylight into H_2OCbl^+, a B_{12} vitamer and metabolic precursor for the B_{12}-cofactors (Ruetz, Gherasim, et al., 2013). Aryl-cobalamins (such as EtPhCbl) are, thus, "conditional antivitamins B_{12}," in which the original inhibitory effect can be reversed by irradiation with light (an occasional asset in some animal experiments) (Miller et al., 2016; Ruetz, Gherasim, et al., 2013). Incidentally, in photo-physical studies with EtPhCbl using fsec to psec UV–visible transient absorption spectroscopy, a low quantum yield was determined for the readily observable light-induced Co—C bond cleavage (Miller et al., 2016).

The indicated limitation of the use of the light-sensitive aryl-cobalamins as antivitamins B_{12} was overcome by the organometallic alkynyl-Cbls 2-phenyl-ethynyl-cobalamin (PhEtyCbl) and 2-(2,4-difluorophenyl)-ethynyl-cobalamin (F2PhEtyCbl) (Ruetz et al., 2017; Ruetz, Salchner, Wurst, Fedosov, & Kräutler, 2013). These analogues of coenzyme B_{12} (see Fig. 4) proved to be exceptionally stable, both thermally and photochemically (Ruetz, Salchner, et al., 2013). PhEtyCbl bound to IF, TC and HC readily and also resisted the crucial partial degradation by the critical enzyme CblC (see Fig. 3) (Ruetz, Salchner, et al., 2013). Hence, light-stable alkynyl-Cbls (such as PhEtyCbl) represent another class of potential antivitamins B_{12} (Ruetz, Salchner, et al., 2013). However, as alkynyl-Cbls are prone to the acid-induced conversion into the B_{12}-vitamer aquocobalamin (by loss of the alkynyl ligand), the hydrolysis-resistant fluorinated alkynyl-Cbl 2-(2,4-difluorophenyl)-ethynyl-cobalamin (F2PhEtyCbl) was introduced as next generation version that was considerably more hydrolysis resistant (Ruetz et al., 2017). F2PhEtyCbl was bound unmodified to glutathione-loaded CblC, allowing for a first crystal structure analysis of a Cbl-binding holo-CblC (Ruetz et al., 2017).

The directly cobalt-bound C-atoms in EtPhCbl or in PhEtyCbl are unsaturated sp^2- or sp-carbons, respectively, inhibiting the "tailoring" of these antivitamins B_{12} by CblC, for geometric reasons mainly. In combination with a structural similarity to the natural Cbls, this metabolic resistance provides organometallic aryl- and alkynyl-Cbls with features of potential antivitamins-B_{12}. The (crystallographically deduced) strong bond from cobalt to the alkynyl sp-carbon of alkynyl-Cbls is a specific (second) factor contributing to their inertness (Chrominski, Lewalska, & Gryko, 2013; Ruetz, Salchner, et al., 2013). However, regarding the resistance of aryl-Cbls (such as the antivitamin B_{12} EtPhCbl) against their tailoring by the Cbl-deligase CblC, the aromatic nature of the directly cobalt-bound aryl-group appears to be the critical factor, rather than the strength of their organometallic bond. Indeed, the covalent organometallic Co-C_{sp2} bond in aryl-Cbls does not display the originally conjectured strength (Tsybizova, Brenig, Kieninger, Kräutler, & Chen, 2021).

3.2 Rhodium-based antivitamins B_{12}, the Rh-homologs of cobalamins

A fruitful alternative route to antivitamins B_{12} has been opened-up by the second-row transition metal rhodium, the group IX homolog of cobalt, also exhibiting most closely related chemical features (Kräutler, 2020). Indeed, as so far explored by crystallography (Widner et al., 2016; Widner et al., 2021) the basic coordination chemistry of corresponding Co- and Rh-corrins is closely related, and the most typical oxidation states of rhodium (Rh^{I}, Rh^{II} and Rh^{III}) correspond to the three biologically relevant oxidation states (Co^{I}, Co^{II} and Co^{III}) of the B_{12}-derivatives.

Besides having slightly longer bonds, rhodium-corrinoids are expected to be very similarly structured as the corresponding cobalt-corrinoids (Kräutler, 2020). Indeed, the now available crystallographic insights into the structures of adenosyl-rhodibalamin (AdoRhbl) (Widner et al., 2016), and chloro-rhodibalamin (ClRhbl) (Widner et al., 2021) (Fig. 5) indicate these Rhbls to be isostructural with the corresponding Cbls, AdoCbl (Lenhert & Hodgkin, 1961; Ouyang, Rulis, Ching, Nardin, & Randaccio, 2004), MeCbl (Randaccio et al., 2000) and ClCbl (Randaccio, Furlan, Geremia, & Slouf, 1998). However, the expected strong Rh—C bond of AdoRhbl (for the Rh—C bond of methyl-Rh(III)-porphyrins about 50 kcal/mol were determined (Fu & Wayland, 2005; Li et al., 2001)) contrasts with the weak Co—C bond (of about 30–36 kcal/mol) of the biological "radical starter" AdoCbl (Garr & Finke, 1993). Hence, rhodibalamins (Rhbls) are probably the closest structural transition-metal mimics of corresponding Cbls and represent particularly promising (potential) antivitamins B_{12} (Kräutler, 2020). So far, this suggestion has been most extensively tested with the B_{12}-cofactor mimic AdoRhbl, the Rh-analogue of AdoCbl.

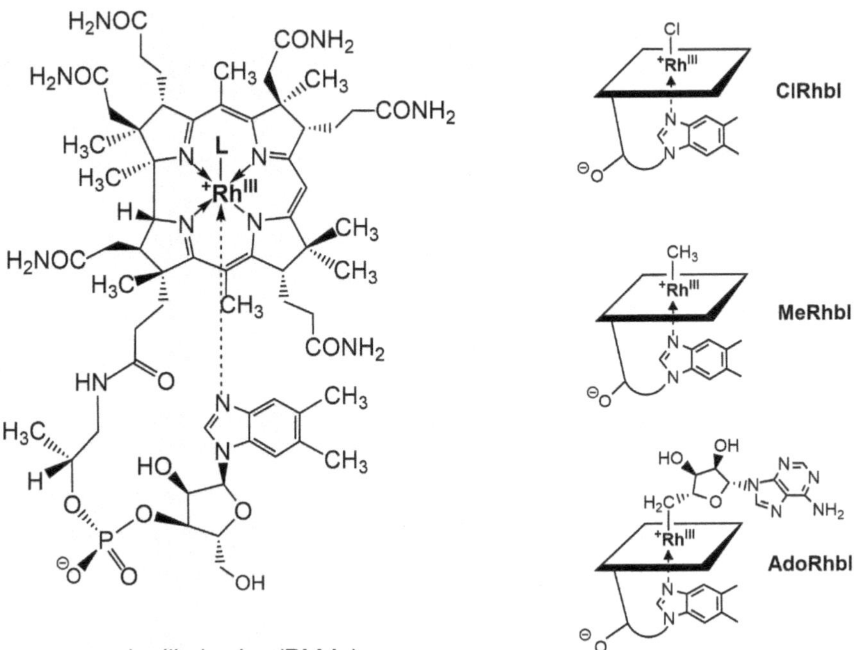

rhodibalamins (**Rhbls**)

Fig. 5 Generalized structural formula of Rh(III)-rhodibalamins (Rhbls) and graphical symbols of chlororhodibalamin (ClRhbl), methylrhodibalamin (MeRhbl) and deoxyadeosyl-rhodiblamin (AdoRhbl), analogues of ClCbl, MeCbl and AdoCbl, in which a Rh(III)-ion replaces the Co(III)-centre of the listed Cbls.

Indeed, exploratory studies with incompletely characterized Rhbls by Koppenhagen, Elsenhans, Wagner, and Pfiffner (1974) indicated "antivitamin B_{12}" effects in bacteria. AdoRhbl was also reported to "worsen the *in-vitro* behaviour" of bone marrow from vitamin B_{12}-deficient patients (whereas, strikingly, MeRhbl was able to correct the abnormality due to B_{12}-deficiency) (Carmel & Koppenhagen, 1977). In our studies, thoroughly characterized, pure AdoRhbl inhibited the catalysis by the AdoCbl-dependent 1.2-propanediol-dehydratase from *Citrobacter freundii*, as well as the growth of the bacterium *Salmonella typhimurum* (Widner et al., 2016). The light-stable AdoRhbl also bound to the photoreceptor/transcription factor CarH to enable its oligomerization and DNA binding, but not the AdoCbl-based light-dependent regulation in-vitro or in-vivo (Padmanabhan et al., n.d., publication in preparation). Furthermore, AdoRhbl associated with the AdoCbl-binding B_{12}-riboswitch btuB of *E.coli* with a similar high nM affinity as the "natural ligand" AdoCbl itself, thus causing the transcriptional inhibition of the outer membrane B_{12}-transporter protein BtuB as a result of the AdoCbl-induced conformational switch of this noncoding mRNA (Widner et al., n.d.. publication in preparation).

3.3 First-row transition metal-analogues of cobalamins

The direct, rational synthesis of some Rhbls and, likewise, of other genuine metal analogues of vitamin B_{12}, collectively named metbalamins (Metbls), has become feasible as a consequence of the preparative biotechnological production of the metal-free natural corrin hydrogenobyric acid (Hby) in the laboratory of Martin J. Warren (Canterbury) (Deery et al., 2012; Kieninger, Deery, et al., 2019) and of the partial synthesis (from Hby) of the complete metal-free B_{12}-ligand hydrogenobalamin (Hbl) (Kieninger et al., 2020; Kräutler, 2020). The Metbls, besides opening up a pathway to new biological and biomedical investigations, also invite to explore the largely unknown chemical territory of the transition metal complexes of the natural ring-contracted helical corrins, specifically in the direction of experimental studies on coordination chemistry and catalysis, in combination with computational theoretical work.

Among the Metbls synthesized from Hby and structurally characterized in detail (see Fig. 6), zincobalamin (Znbl) was prepared as a redox-inactive Zn(II)-analogue of vitamin B_{12} (Kieninger, Baker, et al., 2019). Znbl exists in a stable "base-on"form with a penta-coordinate Zn(II)-center (Kieninger, Baker, et al., 2019), constituting a luminescent structural mimic of the reduced

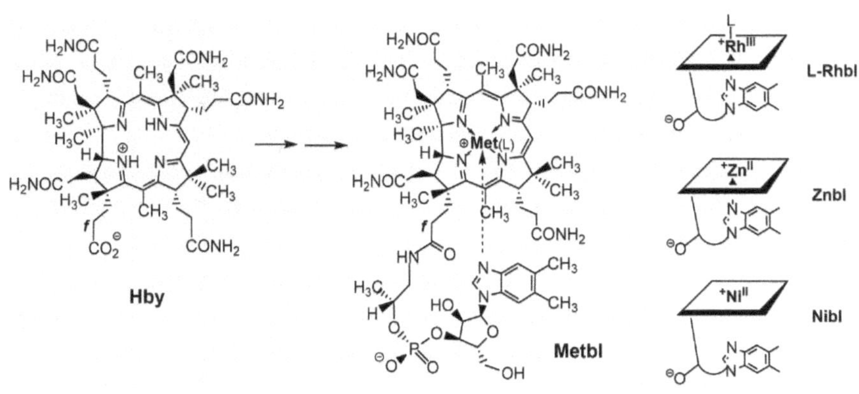

Fig. 6 Hydrogenobyric acid (Hby) has served as starting material for metbalamins (Metbls). The structures of rhodibalamins (L-Rhbl) resemble those of 6-coordinate Cbls, zincobalamin (Znbl) mimics the structure of 5-coordinate Cbl^{II} and nibalamin (Nibl) has been proposed to have a similar structure as 4-coordinate ("base-off") Cbl^{II} and Cbl^{I-}(Kräutler, 2020).

non-luminescent "base-on"Co(II)-cobalamin (Cbl^{II}), which features a corresponding penta-coordinate Co(II)-centre (Kräutler, Keller, & Kratky, 1989). Likewise, nibalamin (Nibl), the novel tetra-coordinate diamagnetic "base-off" Ni(II)-analogue of vitamin B_{12} (Kieninger et al., 2020), represents a largely redox-inactive structural mimic, not only of Co(I)-Cbl, but also of the highly activated tetra-coordinate "base-off" form of Co(II)-Cbl. These reduced Cbl-forms often occur as cryptic high-energy intermediates in Cbl-ependent enzymatic reactions, as well as in some essential B_{12}-biosynthetic organometallic transformations, as catalyzed, e.g., by adenosyl transferases (Banerjee et al., 2021; Moore, Newmister, Rayment, & Escalante-Semerena, 2012; Kräutler, 2019; Padovani, Labunska, Palfey, Ballou, & Banerjee, 2008).

The set of three types of Metbls (Fig. 6), comprising the newly available tetra-coordinate (i.e. "base-off") Nibl, penta-coordinate ("base-on") Znbl and hexa-coordinate ("base-on"). Rhbls, constitutes a complete suite of structural transition metal mimics of the biologically accessible redox states of the Cbls, i.e., of Co(I)-Cbls and tetra-coordinate "base-off" Co(II)-Cbls, of penta-coordinate "base-on" Co(II)-Cbl and of hexa-coordinate "base-on" Co(III)-Cbls. Thus, the three types of Metbls provide us with a structurally "complete" set of biochemically inactive B_{12}-antimetabolites that promise to have features of potent inhibitors of B_{12}-enzymes. As close ("nearly perfect") structural "dummies" of Cbls in (highly activated) intermediary stages of their participation in enzymatic reactions, the organometallic Rhbls (AdoRhbl and MeRhbl) and Znbl, may bind strongly to the apo-forms of some Cbl-dependent enzymes and stabilize them for bio-structural (spectroscopic or crystallographic) studies that may allow for unprecedented glimpses into the biochemical mechanism (and deduced transition state structures) of such enzyme processes.

Among the Metbls proposed to represent (potential) antivitamins B_{12} figure the Rhbls and Znbl, which are also presumed to interact, non-productively, with the human "B_{12}-deligase" CblC (Kräutler, 2020). As these ("base-on") Metbls are likely to be bound well by the mammalian B_{12}-transporter proteins that recognize Cbls in their characteristic "base-on" architecture, they appear to represent, indeed, promising antivitamins B_{12}. In contrast, the tetra-coordinate Ni(II)-analogue Nibl represents a "base-off" B_{12}-antimetabolite with the structural features of tetra-coordinate Cbl^{II}- and Cbl^{I}-forms (rather than a "base-on" structure), giving it characteristics of a potential inhibitor of some B_{12}-dependent enzymes relying on tetra-coordinate intermediary Cbls (Kräutler, 2020). In order to clarify the capacity of Nibl and specific other Metbls to serve as antivitamins B_{12} according to our concept (Kräutler, 2015), experimental studies of their ability to mimic the Cbls by high-affinity binding to the very structure-selective B_{12}-uptake and transport system in humans (and in other mammals) (Fedosov, 2012) are required, as well as deeper insights into their further B_{12}-like trafficking (Banerjee et al., 2021).

4. Summary and outlook

Antivitamins B_{12} feature common structural properties with vitamin B_{12} and coenzyme B_{12}, but they cannot engage in the known physiological functions of the B_{12}-vitamins. As observed in the case of studies with EtPhCbl and mice (Mutti et al., 2013), the application of antivitamins B_{12} is expected to cause (functional) B_{12}-deficiency in (otherwise) healthy animals. Such experiments with laboratory animals may, first of all, help to obtain a more complete picture of the consequences of impaired contributions of the two B_{12}-dependent enzymes and of the associated ("canonical") cofactor-roles of B_{12} in humans (Green et al., 2017). The resulting insights are important, as relevant neurological effects of Cbls (and of their deficiency) in mammals still appear particularly puzzling (Battaglia-Hsu et al., 2009; Kumar, 2014; Scalabrino, 2009) so that B_{12} has been classified as a "moonlighting" vitamin (Scalabrino & Peracchi, 2006). Hence, experiments with "antivitamins B_{12}" and laboratory animals may help to clarify the molecular basis for specific disastrous effects of B_{12}-deficiency, in particular in the context of a range of neuropathological defects (Kumar, 2014; McCaddon, 2013), in causing the development of neural tube defects in embryos (Moreno-Garcia, Rosenblatt, & Jerome-Majewska, 2013) and in being a factor relevant for the accelerated mental disorder in the elderly (Smith & Refsum, 2009).

The capacity of the Cbls (and other cobamides) to exist in their well-structured classical "base-on"-form, or to switch to a "base-off"-form, is a more broadly important structural asset of Cbls (and of other natural cobamides), discovered in the two B_{12}-enzymes, MetH (Drennan et al., 1994) and MCM (Ludwig and Evans, 1999), and particularly relevant in the intracellular human B_{12}-trafficking, as well (Banerjee et al., 2021; Kräutler & Puffer, 2012). The unique architecture of the "base-on"-forms of the ("complete") cobamides, in particular, is likely to be an essential basis for still largely elusive regulatory ("non-canonical") roles of Cbls (and other natural cobamides). Recently, AdoCbl was observed to act as an allosteric modulator of LRRK2-kinase activity, relevant in the development of Parkinson's disease (Schaffner et al., 2019). Indeed, the unique three-dimensional structures of ("base-on") Cbls have been exploited by Nature for regulatory roles of RNA-based B_{12}-riboswitches, in particular (however, such B_{12}-dependent regulation is not known in humans and in other mammals) (Serganov & Nudler, 2013). As observed in the B_{12}-riboswitches, the critical interaction between the mRNA and the bound B_{12}-ligand would be governed by the general structural features of the latter, and would be independent of the functional classification of the B_{12}-ligand as "B_{12}-vitamin" or as "antivitamin B_{12}." Thus, the structural role of AdoCbl as ligand in B_{12}-riboswitches could be mimicked by antivitamins B_{12}. Studies with antivitamins B_{12} could help to uncover such ("non-canonical" regulatory and other allosteric) effects of Cbls that are (merely) due to the binding of intact Cbls as ligands of the specific protein- or oligonucleotide-type bio-macromolecules involved.

B_{12}-antimetabolites, including the antivitamins B_{12}, have been considered to represent new types of antibiotics (Friedrich, 1988; Zelder, 2015). In addition to their possible direct role in impairing the metabolic formation of the (active) B_{12}-cofactors and, thus, in reducing their cellular availability (Kräutler, 2015; Mutti et al., 2013; Ruetz, Gherasim, et al., 2013), antivitamins B_{12} may also intercept the cobamide-uptake by B_{12}-dependent microorganisms due to their structure-based B_{12}-mimetic activity as ligands of B_{12}-regulatory elements, such as B_{12}-riboswitches (Widner et al., n.d.). Indeed, the response of B_{12}-regulatory elements to binding of a properly structured ligand is expected not to differentiate between the functional classification of the latter as "vitamin B_{12}" or as "antivitamin B_{12}." In consequence, antivitamins B_{12} are deduced to display a potentially very effective two-pronged bactericidal activity (both by the "canonical" bio-catalytic and the "non-canonical" B_{12}-regulatory roles played by the natural cobamides), as actually verified recently with AdoRhbl (Widner et al., 2016). Indeed, the capacity of the Cbl-based antivitamin B_{12} 4-ethylphenyl-cobalamin (EtPhCbl, see Fig. 4) has been successfully tested in promoting the effect of established drugs in causing bacterial thymineless death via the so called "methyl-tetradydrofolate trap." (Guzzo et al., 2016). Hence, antivitamins B_{12} may be particularly helpful in reducing the viability of bacterial pathogens that have developed resistance against classical antibiotics (Guzzo et al., 2016) and also depend on the ready uptake of cobamides and some of their biosynthetic precursors from the medium (Bradbeer, 1999), in order to drive their B_{12}-dependent metabolism. Natural purinyl-cobamides and some other non-cobalamin (nonCbl) cobamides are bound less well by the human B_{12}-uptake proteins (Fedosov, Fedosova, Kräutler, Nexo, & Petersen, 2007), and have a significant lesser affinity to human B_{12}-dependent enzymes (compared to the Cbls) (Bito et al., 2020; Sokolovskaya et al., 2021). Hence, their Cba-moiety may provide the basic part for the construction of corresponding nonCbl antivitamins B_{12} (e.g., EtPhCba analogous to EtPhCbl) that may display a decreased human uptake and metabolic use, inducing selectivity for some bacteria, thus increasing, correspondingly, their potential for the use of antibiotics.

As described elsewhere in this volume of "Vitamins and Hormones," a proper B_{12} status is a requirement for general health, and is crucial for normal neurological development, in particular (see also (Battaglia-Hsu et al., 2009; Green et al., 2017)). Cbl-deficiency also strongly affects epigenetic reactions, where lack of MeCbl and reduced activity of methionine synthase may specifically lead to reduced extent of methylation of DNA (Green et al., 2017). Among other consequences, Cbl-deficiency is known to reduce proliferation, while promoting differentiation of neuroblastoma cells (Battaglia-Hsu et al., 2009). Hence, the outlook of generating growth-inhibiting Cbl-deficiency in a controlled fashion by the application of antivitamins B_{12} (Kräutler, 2015), as well as the use of other Cbl-based drug carriers as related anti-proliferative and potential anti-cancerous agents (Tran et al., 2015; Zelder & Alberto, 2012), may raise further deep interest in experimental animal studies and cell-based biomedical investigations.

Antivitamins B_{12} differ by their designed, unnatural molecular inside from the B_{12}-vitamers and Cbl-cofactors, but display, like "a wolf in a sheep's clothing," a hardly distinguishable molecular outside. They are taken up naturally and may behave like Cbls (only) in regulatory and allosteric roles that depend upon the outside structure, but fail to fulfill any of the metabolically important functions as B_{12}-cofactors. Studies on antivitamins B_{12} and of other transition metal B_{12}-analogues are anticipated to have a profound basic impact on B_{12}-chemistry, and to lead to most important applications in biochemical, bio-structural, biological and biomedical areas. As suggested, antivitamins B_{12} are candidates for contributing novel lead compounds and small molecule structural motifs to the molecular toolboxes for antibiotics, anti-proliferative, and potential anti-cancerous agents, key areas of keen ongoing diagnostic and therapeutic interest.

Acknowledgments

For their essential and excellent contributions to the research on antivitamins B_{12}, I would like to thank my past and present co-workers, and a group of renowned research collaborators, all named in the text and in the references. I am grateful to the Austrian Science Fund (FWF) for continued financial support of our studies (current FWF project P-33059).

References

Banerjee, R. (2003). Radical carbon skeleton rearrangements: Catalysis by coenzyme B_{12}-dependent mutases. *Chemical Reviews*, *103*(6), 2083–2094.

Banerjee, R., Gherasim, C., & Padovani, D. (2009). The tinker, tailor, soldier in intracellular B_{12} trafficking. *Current Opinion in Chemical Biology*, *13*(4), 484–491.

Banerjee, R., Gouda, H., & Pillay, S. (2021). Redox-linked coordination chemistry directs vitamin B-12 trafficking. *Accounts of Chemical Research*, *54*(8), 2003–2013.

Banerjee, R., & Ragsdale, S. W. (2003). The many faces of vitamin B_{12}: Catalysis by cobalamin-dependent enzymes. *Annual Review of Biochemistry*, *72*, 209–247.

Battaglia-Hsu, S. F., Akchiche, N., Noel, N., Alberto, J. M., Jeannesson, E., Orozco-Barrios, C. E., et al. (2009). Vitamin B_{12} deficiency reduces proliferation and promotes differentiation of neuroblastoma cells and up-regulates PP2A, proNGF, and TACE. *Proceedings of the National Academy of Sciences of the United States of America*, *106*(51), 21930–21935.

Bermingham, A., & Derrick, J. P. (2002). The folic acid biosynthesis pathway in bacteria: Evaluation of potential for antibacterial drug discovery. *BioEssays*, *24*(7), 637–648.

Bito, T., Bito, M., Hirooka, T., Okamoto, N., Harada, N., Yamaji, R., et al. (2020). Biological activity of pseudovitamin B_{12} on cobalamin-dependent methylmalonyl-CoA mutase and methionine synthase in mammalian cultured COS-7 cells. *Molecules*, *25*(14), 3268.

Bradbeer, C. (1999). Cobalamin transport in bacteria. In R. Banerjee (Ed.), *Chemistry and biochemistry of B_{12}* (pp. 489–506). New York: John Wiley & Sons.

Bridwell-Rabb, J., & Drennan, C. L. (2017). Vitamin B_{12} in the spotlight again. *Current Opinion in Chemical Biology*, *37*, 63–70.

Brown, K. L. (2005). Chemistry and enzymology of vitamin B_{12}. *Chemical Reviews*, *105*(6), 2075–2149.

Bryant, D. A., Hunter, C. N., & Warren, M. J. (2020). Biosynthesis of the modified tetrapyrroles - the pigments of life. *The Journal of Biological Chemistry*, *295*(20), 6888–6925.

Buckel, W., & Golding, B. T. (2006). Radical enzymes in anaerobes. *Annual Review of Microbiology*, *60*, 27–49.

Buckel, W., & Golding, B. T. (2012). Radical Enzymes. In S. A. Chatgilialoglu (Ed.), *Encyclopedia of radicals in chemistry, biology and materials*, John Wiley & Sons.

Carmel, R., & Koppenhagen, V. B. (1977). Effect of rhodium and copper analogs of cobalamin on human cells in vitro. *Archives of Biochemistry and Biophysics*, *184*(1), 135–140.

Chrominski, M., Lewalska, A., & Gryko, D. (2013). Reduction-free synthesis of stable acetylide cobalamins. *Chemical Communications*, *49*, 11406–11408.

Deery, E., Schroeder, S., Lawrence, A. D., Taylor, S. L., Seyedarabi, A., Waterman, J., et al. (2012). An enzyme-trap approach allows isolation of intermediates in cobalamin biosynthesis. *Nature Chemical Biology*, *8*, 933–940.

Drennan, C. L., Huang, S., Drummond, J. T., Matthews, R. G., & Ludwig, M. L. (1994). How a protein binds B_{12}—a 3.0-angstrom X-ray structure of B_{12}-binding domains of methionine synthase. *Science*, *266*(5191), 1669–1674.

Fedosov, S. N. (2012) Physiological and molecular aspects of cobalamin transport. In J. R. Harris (Series Ed.), *Subcellular biochemistry: Vol. 56. Water soluble vitamins* (pp. 347-368): Springer.

Fedosov, S. N., Fedosova, N. U., Kräutler, B., Nexo, E., & Petersen, T. E. (2007). Mechanisms of discrimination between cobalamins and their natural analogues during their binding to the specific B_{12}-transporting proteins. *Biochemistry, 46*(21), 6446–6458.

Freedman, M. D. (1996). Warfarin and other "anti"-vitamin K anticoagulants: Pharmacodynamics and clinical use. *American Journal of Therapeutics, 3*, 771–783.

Frey, P. A., & Hegeman, A. D. (2007). *Enzymatic reaction mechanisms*. New York: Oxford University Press.

Friedrich, W. (1988). *Vitamins*. Berlin: Walter de Gruyter.

Fu, X. F., & Wayland, B. B. (2005). Thermodynamics of rhodium hydride reactions with CO, aldehydes, and olefins in water: Organo-rhodium porphyrin bond dissociation free energies. *Journal of the American Chemical Society, 127*(47), 16460–16467.

Garr, C. D., & Finke, R. G. (1993). Adocobalamin (Adocbl or coenzyme B-12) Co-C bond Homolysis radical-cage effects - product, kinetic, mechanistic, and cage efficiency factor (Fc) studies, plus the possibility that coenzyme B_{12}-dependent enzymes function as ultimate radical cages and ultimate radical traps. *Inorganic Chemistry, 32*(20), 4414–4421.

Gherasim, C., Lofgren, M., & Banerjee, R. (2013). Navigating the B_{12} road: Assimilation, delivery, and disorders of cobalamin. *The Journal of Biological Chemistry, 288*(19), 13186–13193.

Gherasim, C., Ruetz, M., Li, Z., Hudolin, S., & Banerjee, R. (2015). Pathogenic mutations differentially affect the catalytic activities of the human B_{12}-processing chaperone CblC and increase futile redox cycling. *The Journal of Biological Chemistry, 290*(18), 11393–11402.

Green, R., Allen, L. H., Bjorke-Monsen, A. L., Brito, A., Gueant, J. L., Miller, J. W., et al. (2017). Vitamin B_{12} deficiency. *Nature Reviews. Disease Primers, 3*, 17040.

Green, R., Miller, J. W., Gregory, J. F., & Stover, P. J. (2014). Vitamin B_{12}. In J. Zempleni, & J. W. Suttie (Eds.), *Handbook of vitamins* (5th ed., pp. 447–489). Boca Raton, USA: CRC Press.

Gruber, K., & Kratky, C. (2001). Methylmalonyl CoA mutase. In A. Messerschmidt, R. Huber, T. Poulos, & K. Wieghardt (Eds.), *Vol. II. Handbook of metalloproteins* (pp. 995–1009). Chichester: John Wiley & Sons.

Gruber, K., Puffer, B., & Kräutler, B. (2011). Vitamin B_{12}-derivatives—enzyme cofactors and ligands of proteins and nucleic acids. *Chemical Society Reviews, 40*, 4346–4363.

Guzzo, M. B., Nguyen, H. T., Pham, T. H., Wyszczelska-Rokiel, M., Jakubowski, H., Wolff, K. A., et al. (2016). Methylfolate trap promotes bacterial Thymineless death by sulfa drugs. *PLoS Pathogens, 12*(10), e1005949.

Halpern, J. (1985). Mechanisms of coenzyme B_{12}-dependent rearrangements. *Science, 227*(4689), 869–875.

Hannibal, L., Kim, J., Brasch, N. E., Wang, S. H., Rosenblatt, D. S., Banerjee, R., et al. (2009). Processing of alkylcobalamins in mammalian cells: A role for the MMACHC (cblC) gene product. *Molecular Genetics and Metabolism, 97*(4), 260–266.

Hirsh, J., O'Donnell, M., & Eikelboom, J. W. (2007). Beyond unfractionated heparin and warfarin—Current and future advances. *Circulation, 116*(5), 552–560.

Hodgkin, D. C. (1965). X-ray analysis of complicated molecules. *Science, 150*(3699), 979–988.

Kieninger, C., Baker, J. A., Podewitz, M., Wurst, K., Jockusch, S., Lawrence, A. D., et al. (2019). Zinc substitution of cobalt in vitamin B_{12}: Zincobyric acid and zincobalamin as luminescent structural B_{12}-mimics. *Angewandte Chemie, International Edition, 58*(41), 14568–14572.

Kieninger, C., Deery, E., Lawrence, A. D., Podewitz, M., Wurst, K., Nemoto-Smith, E., et al. (2019). The hydrogenobyric acid structure reveals the corrin ligand as an Entatic state module empowering B_{12}-cofactors for catalysis. *Angewandte Chemie, International Edition, 58*, 10756–10760.

Kieninger, C., Wurst, K., Podewitz, M., Stanley, M., Deery, E., Lawrence, A. D., et al. (2020). Replacement of the cobalt-center of vitamin B_{12} by Nickel–Nibalamin and Nibyric acid prepared from metal-free B_{12}-ligands hydrogenobalamin and hydrogenobyric acid. *Angewandte Chemie, International Edition, 59*, 20129.

Kläui, H. (1979). Inactivation of vitamins. *The Proceedings of the Nutrition Society, 38*, 135–141.

Koppenhagen, V. B., Elsenhans, B., Wagner, F., & Pfiffner, J. J. (1974). Methylrhodibalamin and 5'-deoxyadenosylrhodibalamin, rhodium analogs of methylcobalamin and cobalamin coenzyme. *The Journal of Biological Chemistry, 249*(20), 6532–6540.

Kräutler, B. (2015). Antivitamins B_{12}—A structure- and reactivity-based concept. *Chemistry - A European Journal, 21*, 11280–11287.

Kräutler, B. (2019). Biological organometallic chemistry of vitamin B_{12}-derivatives. In T. Hirao, & T. Moriuchi (Eds.), *Adv. Bioorganomet. Chem* (pp. 399–429). Cambridge, USA: Elsevier.

Kräutler, B. (2020). Antivitamins B_{12} - Some Inaugural Milestones. *Chemistry - A European Journal, 26*, 15438–15445.

Kräutler, B., Keller, W., & Kratky, C. (1989). Coenzyme B_{12}-chemistry: The crystal and molecular structure of cob(II)alamin. *Journal of the American Chemical Society, 111*, 8936–8938.

Kräutler, B., & Puffer, B. (2012). Vitamin B_{12}-derivatives: Organometallic catalysts, cofactors and ligands of bio-macromolecules. In K. M. Kadish, K. M. Smith, & R. Guilard (Eds.), *Vol. 25. Handbook of porphyrin science* (pp. 133–265). World Scientific.

Kumar, N. (2014). Neurologic aspects of cobalamin (B_{12}) deficiency. In J. Biller, & J. M. Ferro (Eds.), *Vol. 120. Handbook of clinical neurology* (pp. 915–926). Amsterdam: Elsevier.

Lenhert, P. G., & Hodgkin, D. C. (1961). Structure of 5,6-dimethylbenzimidazolylcobamide coenzyme. *Nature, 192*(480), 937.

Lerner-Ellis, J. P., Tirone, J. C., Pawelek, P. D., Dore, C., Atkinson, J. L., Watkins, D., et al. (2006). Identification of the gene responsible for methylmalonic aciduria and homocystinuria, cblC type. *Nature Genetics, 38*(1), 93–100.

Li, G., Zhang, F. F., Pi, N., Chen, H. L., Zhang, S. Y., & Chan, K. S. (2001). Determination of Rh–C bond dissociation energy in methyl(porphyrinato)rhodium(III) complexes: A new application of photoacoustic calorimetry. *Chemistry Letters*, 284–285.

Ludwig, M. L., & Evans, P. R. (1999). X-ray crystallography of B_{12} enzymes: Methylmalonyl-CoA mutase and methionine synthase. In R. Banerjee (Ed.), *Chemistry and biochemistry of B_{12}* (pp. 595–632). New York, Chichester: John Wiley & Sons.

Ludwig, M. L., & Matthews, R. G. (1997). Structure based perspectives on B_{12}-dependent enzymes. *Annual Review of Biochemistry, 66*, 269–313.

Ludwig, M. L., & Matthews, R. G. (2002). B_{12}-dependent methionine synthase: A structure that adapts to catalyze multiple methyl transfer reactions. *ACS Symposium Series: Structures and Mechanisms, from Ashes to enzymes, 827*(827), 186–201.

Mancia, F., Keep, N. H., Nakagawa, A., Leadlay, P. F., McSweeney, S., Rasmussen, B., et al. (1996). How coenzyme B_{12} radicals are generated: The crystal structure of methylmalonyl-coenzyme A mutase at 2 Å resolution. *Structure, 4*(3), 339–350.

Marsh, E. N. G., & Drennan, C. L. (2001). Adenosylcobalamin-dependent isomerases: New insights into structure and mechanism. *Current Opinion in Chemical Biology, 5*(5), 499–505.

Matthews, R. G. (2009). Metal ions in life sciences. In A. Sigel, H. Sigel, & R. K. O. Sigel (Eds.), *Vol. 6. Cobalamin- and corrinoid-dependent enzymes* (pp. 53–114). Cambridge (UK): RSC Publishing.

Matthews, R. G., Koutmos, M., & Datta, S. (2008). Cobalamin-dependent and cobamide-dependent methyltransferases. *Current Opinion in Structural Biology, 18*(6), 658–666.

McCaddon, A. (2013). Vitamin B_{12} in neurology and ageing; Clinical and genetic aspects. *Biochimie, 95*(5), 1066–1076.

Miller, N. A., Wiley, T. E., Spears, K. G., Ruetz, M., Kieninger, C., Kräutler, B., et al. (2016). Toward the design of Photoresponsive conditional antivitamins B_{12}: A transient absorption study of an Arylcobalamin and an Alkynylcobalamin. *Journal of the American Chemical Society, 138*(43), 14250–14256.

Moore, T. C., Newmister, S. A., Rayment, I., & Escalante-Semerena, J. C. (2012). Structural insights into the mechanism of four-coordinate cob(II)alamin formation in the active site of the Salmonella enterica ATP:Co(I)rrinoid adenosyltransferase enzyme: Critical role of residues Phe91 and Trp93. *The Biochemist, 51*(48), 9647–9657.

Moreno-Garcia, M. A., Rosenblatt, D. S., & Jerome-Majewska, L. A. (2013). Vitamin B_{12} metabolism during pregnancy and in embryonic mouse models. *Nutrients, 5*(9), 3531–3550.

Mutti, E., Ruetz, M., Birn, H., Kräutler, B., & Nexo, E. (2013). 4-Ethylphenyl-cobalamin impairs tissue uptake of vitamin B_{12} and causes vitamin B_{12} deficiency in mice. *PLoS One, 8*, e75312.

Nielsen, M. J., Rasmussen, M. R., Andersen, C. B. F., Nexo, E., & Moestrup, S. K. (2012). Vitamin B_{12} transport from food to the body's cells-a sophisticated, multistep pathway. *Nature Reviews. Gastroenterology & Hepatology, 9*(6), 345–354.

Ouyang, L., Rulis, P., Ching, W. Y., Nardin, G., & Randaccio, L. (2004). Accurate redetermination of the X-ray structure and electronic bonding in adenosylcobalamin. *Inorganic Chemistry, 43*(4), 1235–1241.

Padmanabhan, S., Pérez-Castaño, R., Widner, F. J., Kieninger, C., Deery, E., Warren, M. J., Elias-Arnanz, M., & Kräutler, B. (publication in preparation).

Padovani, D., Labunska, T., Palfey, B. A., Ballou, D. P., & Banerjee, R. (2008). Adenosylransferase tailors and delivers coenzyme B_{12}. *Nature Chemical Biology, 4*(3), 194–196.

Randaccio, L., Furlan, M., Geremia, S., & Slouf, M. (1998). Preparation and X-ray analysis of crystals of azido- and Chlorocobalamin containing LiCl: A structural model for the interactions of the corrin ring with ionic species. *Inorganic Chemistry, 37*(20), 5390–5393.

Randaccio, L., Furlan, M., Geremia, S., Slouf, M., Srnova, I., & Toffoli, D. (2000). Similarities and differences between cobalamins and cobaloximes. Accurate structural determination of methylcobalamin and of LiCl- and KCl-containing cyanocobalamins by synchrotron radiation. *Inorganic Chemistry, 39*(15), 3403–3413.

Ruetz, M., Gherasim, C., Fedosov, S. N., Gruber, K., Banerjee, R., & Kräutler, B. (2013). Radical synthesis opens access to organometallic aryl-Cobaltcorrins—4-Ethylphenyl-cobalamin, a potential "antivitamin B_{12}". *Angewandte Chemie, International Edition, 52*, 2606–2610.

Ruetz, M., Salchner, R., Wurst, K., Fedosov, S., & Kräutler, B. (2013). Phenylethynylcobalamin: A light-stable and Thermolysis-resistant organometallic vitamin B_{12} derivative prepared by radical synthesis. *Angewandte Chemie, International Edition, 52*, 11406–11409.

Ruetz, M., Shanmuganathan, A., Gherasim, C., Karasik, A., Salchner, R., Kieninger, C., et al. (2017). Antivitamin B_{12} inhibition of the human B_{12}-processing enzyme CblC: Crystal structure of an inactive ternary complex with glutathione as the Cosubstrate. *Angewandte Chemie, International Edition, 56*(26), 7387–7392.

Scalabrino, G. (2001). Subacute combined degeneration one century later. The neurotrophic action of cobalamin (vitamin B_{12}) revisited. *Journal of Neuropathology and Experimental Neurology, 60*, 109–120.

Scalabrino, G. (2009). The multi-faceted basis of vitamin B_{12} (cobalamin) Neurotrophism in adult central nervous system: Lessons learned from its deficiency. *Progress in Neurobiology*, *88*(3), 203–220.

Scalabrino, G., & Peracchi, M. (2006). New insights into the pathophysiology of cobalamin deficiency. *Trends in Molecular Medicine*, *12*(6), 247–254.

Schaffner, A., Li, X. T., Gomez-Llorente, Y., Leandrou, E., Memou, A., Clemente, N., et al. (2019). Vitamin B_{12} modulates Parkinson's disease LRRK2 kinase activity through allosteric regulation and confers neuroprotection. *Cell Research*, *29*(4), 313–329.

Serganov, A., & Nudler, E. (2013). A decade of riboswitches. *Cell*, *152*(1–2), 17–24.

Smith, A. D., & Refsum, H. (2009). Vitamin B_{12} and cognition in the elderly. *The American Journal of Clinical Nutrition*, *89*(2), 707s–711s.

Sokolovskaya, O. M., Plessl, T., Bailey, H., Mackinnon, S., Baumgartner, M. R., Yue, W. W., et al. (2021). Naturally occurring cobalamin (B_{12}) analogs can function as cofactors for human methylmalonyl-CoA mutase. *Biochimie*, *183*, 35–43.

Tran, M. T. Q., Stürup, S., Lambert, I. H., Gammelgaard, B., Furger, E., & Alberto, R. (2015). Cellular uptake of metallated cobalamins. *Metallomics*, *8*, 298–304.

Tsybizova, A., Brenig, C., Kieninger, C., Kräutler, B., & Chen, P. (2021). Surprising Homolytic gas phase Co-C bond dissociation energies of organometallic aryl-cobinamides reveal notable non-bonded intramolecular interactions. *Chemistry - A European Journal*, *27*(25), 7252–7264.

Widner, F. J., Khan, N. J., Deery, E., Warren, M.J., Taga, M. E., & Kräutler, B. (publication in preparation).

Widner, F. J., Kieninger, C., Wurst, K., Deery, E., Warren, M. J., & Kräutler, B. (2021). Synthesis, spectral characterization and crystal structure of chloro-rhodibalamin–A synthesis platform for rhodium analogues of vitamin B_{12} and for Rh-based antivitamins B_{12}. *Synthesis*, *53*, 332–337.

Widner, F. J., Lawrence, A. D., Deery, E., Heldt, D., Frank, S., Gruber, K., et al. (2016). Total synthesis, structure, and biological activity of adenosylrhodibalamin, the non-natural rhodium homologue of coenzyme B_{12}. *Angewandte Chemie, International Edition*, *55*(37), 11281–11286.

Zelder, F. (2015). Recent trends in the development of vitamin B_{12} derivatives for medicinal applications. *ChemComm*, *51*, 14004–14017.

Zelder, F., & Alberto, R. (2012). Vitamin B_{12} derivatives for spectroanalytical and medicinal applications. In K. M. Kadish, K. M. Smith, & R. Guilard (Eds.), *Vol. 25. Handbook of porphyrin science* (pp. 84–132). World Scientific.

Zelder, F., Sonnay, M., & Prieto, L. (2015). Antivitamins for medicinal applications. *ChemBioChem*, *16*, 1264–1278.

CHAPTER NINE

Vitamin B12 absorption and malabsorption

Jean-Louis Guéant[a,b,c,*], Rosa-Maria Guéant-Rodriguez[a,b], and David H. Alpers[d,*]

[a]University of Lorraine, INSERM UMR_S 1256, Nutrition, Genetics, and Environmental Risk Exposure (NGERE), Faculty of Medicine of Nancy, Nancy, France
[b]Department of Biochemistry, Molecular Biology, Nutrition, and Metabolism, University Hospital of Nancy, Nancy, France
[c]Department of Hepato-Gastroenterology, University Hospital of Nancy, Nancy, France
[d]Department of Internal Medicine, Division of Gastroenterology, Washington University School of Medicine, St Louis, MO, United States
*Corresponding authors: e-mail address: jean-louis.gueant@univ-lorraine.fr; dalpers@wustl.edu

Contents

Abstract

Vitamin B12 is assimilated and transported by complex mechanisms that involve three transport proteins, intrinsic factor (IF), haptocorrin (HC) and transcobalamin (TC) and their respective membrane receptors. Vitamin deficiency is mainly due to inadequate dietary intake in vegans, and B12 malabsorption is related to digestive diseases. This review explores the physiology of vitamin B12 absorption and the mechanisms and diseases that produce malabsorption. In the stomach, B12 is released from food carrier proteins and binds to HC. The degradation of HC by pancreatic proteases and

Vitamins and Hormones, Volume 119
ISSN 0083-6729
https://doi.org/10.1016/bs.vh.2022.01.016

the pH change trigger the transfer of B12 to IF in the duodenum. Cubilin and amnionless are the two components of the receptor that mediates the uptake of B12 in the distal ileum. Part of liver B12 is excreted in bile, and undergoes an enterohepatic circulation. The main causes of B12 malabsorption include inherited disorders (Intrinsic factor deficiency, Imerslund-Gräsbeck disease, Addison's pernicious anemia, obesity, bariatric surgery and gastrectomies. Other causes include pancreatic insufficiency, obstructive Jaundice, tropical sprue and celiac disease, bacterial overgrowth, parasitic infestations, Zollinger-Ellison syndrome, inflammatory bowel diseases, chronic radiation enteritis of the distal ileum and short bowel. The assessment of B12 deficit is recommended in the follow-up of subjects with bariatric surgery. The genetic causes of B12 malabsorption are probably underestimated in adult cases with B12 deficit. Despite its high prevalence in the general population and in the elderly, B12 malabsorption cannot be anymore assessed by the Schilling test, pointing out the urgent need for an equivalent reliable test.

1. Introduction

Vitamin B12 (B12, cobalamin (Cbl)) is a water-soluble vitamin with a highly complex structure, which anticipates the complexity of the mechanisms by which the vitamin is assimilated and transported. Three proteins, intrinsic factor (IF), haptocorrin (HC) and transcobalamin (TC) and their respective membrane receptors are involved in these mechanisms. B12 deficiency due to malabsorption may produce a wide range of clinical symptoms, which include hematological manifestations ranging from macrocytosis to megaloblastic anemia, central and peripheral neurological disorders, psychiatric disorders and thromboembolic outcomes (Green et al., 2017; Levy et al., 2021). Vitamin deficiency is mainly due (a) inadequate dietary intake in vegans, especially in subjects from countries with a high percentage of vegans (e.g., India) and (b) to B12 malabsorption related to digestive diseases. This review will explore the physiology of vitamin B12 absorption and the causes and pathophysiology of malabsorption.

2. Origin, food content and availability of vitamin B12

Metazoa (multicellular animals) do not have any gene coding B12 synthesis. In fact, B12 is synthesized only by microorganisms, i.e., bacteria, yeasts and possibly in some algae (Farquharson & Adams, 1976; Watanabe, Takenaka, Kittaka-Katsura, Ebara, & Miyamoto, 2002). Some of these bacteria are present in the ruminant herbivores and in the intestines of monogastric animals, which explains why these animals do not develop B12 deficiency despite a lack of intake through food. In addition, B12 synthesis by rumen or intestinal flora depends on cobalt intake. In some other mammals, like rodents, B12 intake is achieved by coprophagia, which leads to both absorption of vitamin and to subsequent assimilation of B12 synthesized by colonic flora. Sea fauna can accumulate B12 produced by environmental bacteria and archaea. These microorganisms are taken up by other B12-requiring bacteria and by eukaryotic phytoplankton, most of the latter being ingested by zooplankton, ending up in the bodies of fish (Watanabe & Bito, 2018). Phytoplankton are also siphoned by edible bi-valves.

Some plants touted to contain vitamin B12 include unwashed organic produce that contains B12-synthesizing microorganisms in soil, but the plants containing the most B12 include the dried purple laver alga (52 mcg/100 g) and dried shiitake mushrooms (5.6 mcg/100 g) (Watanabe, Yabuta, Bito, & Teng, 2014). Finally, yeast products provide a source for vitamin B12. Commercial nutritional yeast products provide in ¼ cup 8.3–24 mcg of B12, depending on the specific product (https://ods.od.nih.gov/factsheets/VitaminB12-HealthProfessional/).

Therefore, except for these plants and yeast products, B12-containing food are those of animal origin such as muscle meat, liver, kidneys, eggs, milk, fish and shellfish (Watanabe, 2007; Watanabe & Bito, 2018). For 100 g of each food, cooked mutton, cooked provides 2.6 μg of B12, cooked chicken 9.4 μg, cow's milk 0.4 μg, cooked eggs 1.3 μo, oysters 28.1 μg, cooked yellowfin tuna, 52.9 μg, and cooked rainbow trout, 4.9 μg (Heyssel, Bozian, Darby, & Bell, 1966; Watanabe, 2007). The U.S. Department of Agriculture's Food Data Central external link disclaimer provides a comprehensive list of foods containing vitamin B12 arranged by nutrient content and by food name (https://fdc.nal.usda.gov/). Although the vitamin B12 content of milk is not high, milk and dairy products are significant contributors of vitamin B12 intakes, considering their high intake in the general population, and a bioavailability of vitamin B12 comparable to other animal sources. About 20–60% of vitamin B12 that is originally presented in milk is recovered in cheese; this reduced content is the result of lactic acid fermentation (Arkbåge, Witthöft, Fondén, & Jägerstad, 2003).

In Western countries, a normal human diet provides 5–10 μg of vitamin B12 per day, whatever the age (Green et al., 2017) (Table 1). Vitamin B12 status varies among populations worldwide, with the highest

Table 1 Recommended Dietary Allowance of the US Food and Nutrition Board according to population groups.

Population group	Age	RDA (μg/day of vitamin B_{12})
Children	1–3 years	0.9
	4–8 years	1.2
Boys	9–13 years	1.8
	14–18 years	2.4
Girls	9–13 years	1.8
	14–18 years	2.4
Adult males	19–50 years	2.4
Adult females	19–50 years	2.4
Pregnancy	14–50 years	2.6
Lactation	14–18 years	2.8
Elderly males	51–70 years	2.4
	>70 years	2.4
Elderly females	51–70 years	2.4
	>70 years	2.4

Adapted from https://ods.od.nih.gov/factsheets/VitaminB12-HealthProfessional/).

rate of B12 dietary deficiency in the Indian subcontinent (Green et al., 2017). B12 deficiency is not uncommon in the vegetarian communities, in particular during pregnancy, and when diseases resulting in decreased absorption are present (Oussalah, Levy, Berthezène, Alpers, & Guéant, 2020).

The bioavailability of vitamin B12 varies according to food categories (Watanabe, 2007). It is estimated to be 56–89% in cooked mutton, to 61–66% in cooked chicken, 65% in Cow's milk, 9% in cooked eggs, and 42.0% in cooked rainbow trout. In bovine milk, vitamin B12 is bound to transcobalamin. Appreciable losses of vitamin B12 have been reported during the processing of milk; boiling for 2–5 min resulted in 30% loss of B12 (Ball, 1998), 5-min microwave cooking led to 50% loss and pasteurization to 5–10% loss (Ball, 1998; Watanabe et al., 1998).

Between 60% and 80% of ingested B12 is eliminated in stools or metabolized by gastrointestinal flora. Requirements correspond to daily losses of endogenous B12, and represent about 0.1% of the total body stores

(2–5 mg), e.g., 2–5 μg. B12 is eliminated mainly through the digestive tract (bile and mucous desquamation) and partly through the urinary tract (Heyssel et al., 1966).

3. Physiology of vitamin B12 absorption (Fig. 1)

3.1 Gastric phase

Two mechanisms occur in the gastric phase, the release of B12 from food proteins and its binding to endogenous carriers. Vitamin B12 contained in dietary samples is present in food proteins, largely in the two

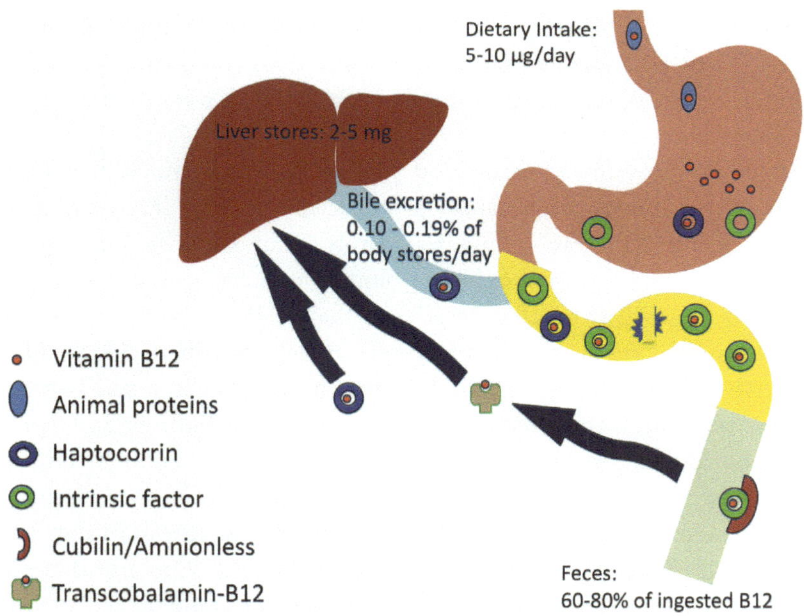

Fig. 1 The intraluminal transport, absorption and enterohepatic circulation of vitamin B12. In the stomach, vitamin B12 is released from food carrier proteins and binds to haptocorrin. The subsequent degradation of haptocorrin by pancreatic proteases and the pH change trigger the transfer of B12 to gastric intrinsic factor in the duodenum. Cubilin and amnionless are the two components of a receptor named cubam in the distal ileum. This receptor mediates the uptake of the vitamin B12-gastric intrinsic factor complex in presence of calcium at neutral pH. Haptocorrin in asialo form is involved in the liver uptake of vitamin B12. Part of liver B12 is excreted in bile, where it is mainly bound to haptocorrin. The subsequent degradation of haptocorrin, biliary B12 transfer to intrinsic factor and subsequent recycling by ileal uptake of B12-intrinsic factor are the other key steps of the enterohepatic circulation of B12. A significant amount of B12 is excreted in feces.

B12-dependent enzymes, methionine synthase and methylmalonic acid mutase (Kolhouse & Allen, 1977; Morrow, Brodie, Strimpler, & Barness, 1973). Cooking food helps to release B12 from these enzymes. However, it also alters B12 and may destroy it. Gastric juice acidity and pepsin are needed to efficiently produce soluble vitamin B12 from food proteins. This has been demonstrated in vitro in the presence of HCl and pepsin (Adams, Kennedy, Thomson, & Williamson, 1968). Clinical studies have corroborated these observations. The assimilation of food B12 has been quantified by a modified Schilling test, determining the urinary and/or fecal tracer excretion after ingestion of ^{57}Co-labeled B12 incorporated in various foods, such as egg, chicken, trout or chicken serum (Aimone-Gastin et al., 1997; Doscherholmen, McMahon, & Economon, 1981; Doscherholmen, McMahon, & Ripley, 1978; Doscherholmen & Swaim, 1973). When B12 was incorporated in food, the percentage excreted was lower than that observed with the free labeled B12 used in the standard Schilling test.

Once B12 is released from food proteins in gastric juice, it is bound tightly by two endogenous carrier proteins, haptocorrin (HC) and gastric intrinsic factor (IF). HC originates mainly from saliva in humans, and accounts for 10–40% of vitamin B_{12} binding capacity in the gastric juice. Its affinity for B12 is much greater than that of IF at the very acidic pH of the stomach (Allen, Seetharam, Allen, Podell, & Alpers, 1978). HC is the product of a single gene (TCN1) on chromosome 11 and is present in many body fluids and secreted by many cell types including glandular cells (Johnston, Bollekens, Allen, & Berliner, 1989). The 410 aa polypeptide has a molecular size of 45.5 kDa and the 30–40% glycosylation accounts for the 66 kDa size (Johnston et al., 1989). Pepsin has no observable effect on HC and IF (Allen, Seetharam, Podell, & Alpers, 1978) and therefore, B12 combines preferentially with HC under the intraluminal acidic conditions (Allen et al., 2013; Marcoullis, Parmentier, Nicolas, Jimenez, & Gerard, 1980). The degradation of HC by pancreatic enzymes is needed to ensure the transfer in B12 to IF. The B12–IF complex is required for >98% of the assimilation of vitamin B12 in the ileum by the process of receptor-mediated endocytosis. IF produced in parietal cells in humans and, is released along with hydrogen ions during stimulation of the parietal cell by cholinergic and hormonal mediators. The gene for IF is located at 11q13 and encodes a 417 amino acid protein (Hewitt, Gordon, Taggart, Mohandas, & Alpers, 1991). IF has a predicted molecular size of 45 kDa protein and its glycosylation accounts for the observed 58–60 kDa size of the secreted protein (Gueant et al., 1989). It binds about 30–50 pmol/mL B12 per mL of gastric juice, with a daily secreted capacity that greatly exceeds the daily intake of B12. IF binds B12 very specifically, with very poor recognition of B12 analogs produced by intestinal microbiota (Fedosov et al., 2005). The protein shows two distinct domains capable of binding B12, which appear to cooperatively interact in binding B12 (Fedosov et al., 2005). The specificity for B12 binding appears to reside in the beta 3 and beta 4 regions of the beta domains of the protein (Gueant et al., 1997; Mathews et al., 2007).

3.2 Duodenal phase and the enterohepatic circulation

The duodenum is the organ that collects fluids, gastric juice, bile and pancreatic juice. In the duodenum HC in the HC-B12 complex is partially degraded by pancreatic proteases (Allen, Seetharam, Allen, et al., 1978; Gueant, Champigneulle, et al., 1986; Gueant, Djalali, et al., 1986; Marcoullis et al., 1980). The partially degraded HC has an apparent molecular mass only half that of the non-degraded form and an affinity for B12, which is reduced by about 150 times. In contrast, IF remains intact in the presence of pancreatic proteases, both in vitro and in vivo (Allen, Seetharam, Podell, & Alpers, 1978; Gueant, Champigneulle, et al., 1986; Gueant, Djalali, et al., 1986). Its affinity for B12 is not modified and becomes much greater than that of partially degraded HC. This allows the transfer to IF of B12 (Allen, Seetharam, Podell, & Alpers, 1978; Gueant, Champigneulle, et al., 1986; Gueant, Djalali, et al., 1986; Marcoullis et al., 1980). The respective behavior of HC and IF in the presence of proteases may be explained by the role of the carbohydrate core in protecting IF, compared to HC (Gordon, Hu, Chokshi, Hewitt, & Alpers, 1991; Gueant, Monin, et al., 1988). The sequential treatment of IF and HC by exoglycosidases and proteases causes their complete degradation (Gueant, Monin, et al., 1988).

An enterohepatic circulation of B12 has been demonstrated in the human, hog, rat, baboon and rabbit (Abdel Nour, Ringot, Gueant, & Chango, 2007; Grasbeck, Nyberg, & Reizenstein, 1958; Green, Jacobsen, Van Tonder, Kew, & Metz, 1982; Gueant, Monin, Boissel, Gaucher, & Nicolas, 1984; Kolhouse & Allen, 1977; Okuda, Grasbeck, & Chow, 1958). B12 excreted in the bile each day appears to represent about 0.10–0.19% of the total amount stored in the body. Biliary excretion of B12 has been demonstrated after parenteral injection of labeled B12 (el Kholty et al., 1991; Gueant, Monin, et al., 1984; Okuda et al., 1958). HC in bile is degraded by pancreatic enzymes, causing a transfer of "endogenous" B12 to IF (el Kholty et al., 1991). Bile plays a prominent role in the clearance of inactive B12 analogs (Gueant, Monin, et al., 1984; Kanazawa, Herbert, Herzlich, Drivas, & Manusselis, 1983). B12 analogs, including cobinamide, are excreted in bile, combined with HC at a concentration of about 1 nmol/L. There is a negative correlation between their serum and biliary concentrations (el Kholty et al., 1991; Gueant, Champigneulle, et al., 1986). The amount of B12 stored in the liver decreases significantly in prolonged B12 deficiency. The body thus adapts to modify the distribution of B12 by decreasing turnover and encouraging tissue storage (Linnell, Hoffbrand, Hussein, Wise, & Matthews, 1974). This would minimize the importance of the enterohepatic cycle in further depleting tissue stores during prolonged B12 deficiency.

The presence of unsaturated and biologically active IF has been detected in the intestinal fluid throughout the intestine as far as the distal ileum (Gueant, Champigneulle, et al., 1986; Kapadia, Mathan, & Baker, 1976). A significant proportion of labeled B12 excreted in the bile can be absorbed in the absence of IF in gastrectomized baboons (Green et al., 1982). It is estimated that ~1% of doses of vitamin B12 ranging from 100 mcg to 5 mg are absorbed in the absence of intrinsic factor (Chan, Low, & Lee, 2016). This absorption has been attributed to passive diffusion, although the size of B12 (MW 1354) is much larger than those water soluble molecules known to be absorbed by passive paracellular diffusion (e.g., monosaccharides, urea). Although this passive absorption is said to occur in the ileum, the surface area of the jejunum is about twice that of the ileum, and it seems likely that passive absorption occurs throughout the small intestine (Dahlgren & Lennernäs, 2019). The effectiveness of oral vitamin B12 therapy in patients with pernicious anemia has been repeatedly demonstrated in randomized controlled trials (Andrès, Zulfiqar, Serraj, Vogel, & Kaltenbach, 2018; Chan et al., 2016).

3.3 Ileo-colonic intraluminal stage and microbiota

There is a gradient in digestive flora in the oral-aboral direction, with about $10–10^5$ microorganisms per gram and $10^5–10^7$ microorganisms per gram of intestinal fluid, in the jejunum and the ileum, respectively. These include aerobic and anaerobic microorganisms. The biosynthesis of B12 is restricted to certain bacteria and requires approximately 30 enzymatic steps for its complete de novo synthesis, which undertakes two distinct biosynthetic pathways, namely, aerobic and anaerobic routes (Moore & Warren, 2012). The major intestinal bacteria synthesizing B12 include *Klebsiella* and *Pseudomonas*, the synthesizing activity of which was demonstrated using intestinal liquid taken from vegetarians. Other bacteria also synthesize B12 and analogs, such as *Clostridium*, *Veillonella* and *Fusobacterium* (Albert, Mathan, & Baker, 1980). In the colon, anaerobic flora becomes predominant, with the total number of microorganisms being 10^8–10^{11} per gram. The bacteria behave either autotrophically with respect to B12, or heterotrophically, i.e., they use the B12 synthesized by other bacteria. HC and IF may have a protective role with respect to the uptake of B12 by bacteria. The intestinal flora can also behave as a B12 provider for the host, as evidenced in apparently healthy southern Indian subjects (Albert et al., 1980). Indeed, at least two groups of organisms in the small bowel, *Pseudomonas* and *Klebsiella*, may synthesize significant amounts of the vitamin (Albert et al., 1980). Stools contain a considerable quantity of vitamin B12 analogs (Gueant, Vidailhet, Pasquet, Djalali, & Nicolas, 1984), a large proportion of which are represented by cobinamide (Djalali et al., 1990). Stool also contains partially degraded haptocorrin, which retains a certain capacity to bind vitamin B12 (Gueant, Vidailhet, et al., 1984; Gullberg, 1974).

Digestive flora can transform B12 into analogs, namely, adenine cyanocobamide (pseudo-vitamin B12), 2,6 methyladenine-cyanocobamide (factor A) and dicyanocobinamide (factor B) (Brandt, Bernstein, & Wagle, 1977). The relatively high specificity of IF for B12 limits the IF-dependent assimilation of other corrinoids. However, there is a higher concentration of analogs in the portal circulation than in the peripheral circulation, indicative of a digestive origin of plasma analogs (el Kholty et al., 1991). The treatment of cystic fibrosis patients with pancreatic extracts increase their concentration, and could reflect the influence of intraluminal HC to sequester analogs, at least in subjects with pancreatic insufficiency (Gueant, Vidailhet, et al., 1984).

The current understanding of the roles of cobamides in microbial communities is the result of multilayered approaches using laboratory cultures of single species, biochemical characterization of cobamide-dependent enzymes, and comparative genomic analyses of thousands of microbial species, as reviewed recently (Sokolovskaya, Shelton, & Taga, 2020). Genomic studies of bacteria estimate that 58–69% of bacteria are unable to produce cobamides de novo, while 86% use cobamides in their metabolism (Sokolovskaya et al., 2020; Zhang, Rodionov, Gelfand, & Gladyshev, 2009). Cobamides regulate bacterial genes by riboswitches, located within the 5′ untranslated region of mRNAs that directly bind cobamides to affect the expression of downstream genes (Sokolovskaya et al., 2020). Cobamides can also influence the metabolism of microbes for which they are not essential, e.g., for succinate production. It has been thought for decades that B12 analogs could behave as antagonists of B12 through potential competitive inhibition of target enzymes, but this view has been challenged recently. Although cobalamin is the most effective cofactor for MMUT, one of the two cobalamin-dependent enzymes in humans, diverse cobamides can also support this function (Sokolovskaya et al., 2021). Moreover, disease-associated MMUT variants that cause cobalamin-responsive impairments in enzymatic activity also respond to other cobamides. These data challenge the current view regarding the exclusive need for cobalamin in human physiology, and suggest various cobamides can support the function of human cobalamin-dependent enzymes (Sokolovskaya et al., 2021).

3.4 Uptake and fate of the IF–B12 complex in distal ileal enterocytes

Two mechanisms are involved in the absorption of B12 in the intestine, namely, passive diffusion most likely along the entire small intestine for pharmacological doses, and uptake via receptor mediated endocytosis of the IF–B12 complex for physiological doses in the ileum. Passive diffusion occurs only with supraphysiological doses of 100–1000 μg and rises according to the dose ingested. Uptake of the IF–B12 complex via the receptor is the preferred mechanism for vitamin B12 absorption provided by food or supplementation. The maximum dose absorbed by this pathway is 1.5 μg. B12 absorption has been studied mainly using animal models such as rats, mice, guinea pigs, hamsters, pigs and dogs.

As early as 1954, Glass et al. suspected the existence of an ileal receptor for vitamin B12 (Glass, Boyd, & Stephanson, 1954). Several groups subsequently demonstrated that IF was involved in the uptake of B12 by the mucosa (Booth, Chanarin, Anderson, & Mollin, 1957; Citrin, Derosa, & Halsted, 1957; Donaldson, Mackenzie, & Trier, 1967; Wolff & Vuillemin-Weis, 1958) and that this occurred selectively in the ileum (Booth et al., 1957). In adult humans, the receptor expression is present in significant amounts in the entire distal three-fifths of the intestine (Hagedorn & Alpers, 1977), while it is distributed in the entire intestine in the embryo (Schohn et al., 1992). The presence of the receptor is also detected in the colon of human fetuses before 25 weeks of gestation (Schohn et al., 1992). Cells derived from human colon cancer such as Caco-2 and HT 29, which have the characteristics of fetal epithelial cells with brush border express the receptor for IF-B12 on their surfaces (Dix, Hassan, Obray, Shah, & Wilson, 1990; Gueant et al., 1992; Seetharam, Ramanujam, Seetharam, Li, & Fyfe, 1991). The IF-B12 receptor was initially reported with a high molecular size and to be a hydrophobic protein (Seetharam, Alpers, & Allen, 1981; Guéant et al., 1991). It was later identified as cubilin, a multiligand binding protein (Aminoff et al., 1999; Kozyraki et al., 1998). The cubilin protein is a 460 kDa glycoprotein of 3623 aminoacids containing an amino-terminal domain required for oligomerization, 8 EGF-like repeats and 27 CUB domains including the ligand binding site. Cubilin possesses a high affinity for the IF–B12 complex but is devoid of any transmembrane or cytosolic domains. In addition to the intrinsic factor-vitamin B12 complex, cubilin can simultaneously bind and internalize various ligands, including albumin, transferrin or vitamin D-binding protein. Because cubilin is a peripheral membrane protein without marked hydrophobicity, the existence of another protein involved in its membrane attachment was suspected and evidence for its presence demonstrated (Gueant, Chery, & Namour, 2001; He et al., 2005; Tanner et al., 2003). This second protein component, amnionless (AMN), is a 50-kD transmembrane protein of 453 aminoacids that is expressed at the apical membrane of enterocytes (Jensen, Andersen, Hager, & Madsen, 2014). AMN is needed for cubilin membrane expression and internalization (He et al., 2005; Namour et al., 2011) through interaction of three cubilin subunits with a corresponding three-faced β-helix domain in AMN (Larsen et al., 2018). Megalin is another multiligand transmembrane protein (approximately 600 kD) that belongs to the LDL receptor family (LRP2) (Kozyraki & Gofflot, 2007). Megalin binds cubilin with high affinity and may contribute to the internalization of cubilin-ligand complexes in the fetus and adult kidney, but is not present in adult human intestine (Jensen et al., 2014; Kozyraki & Gofflot, 2007).

The endocytosis of IF–B12 complex has been studied in vivo in animal models and different types of polarized cells (Dan & Cutler, 1994; Gueant, Gerard, et al., 1988; Gueant et al., 1992; Pons, Guy, Lambert, Hatier, & Gueant, 2000; Ramanujam, Seetharam, Ramasamy, & Seetharam, 1991; Robertson & Gallagher, 1985). Internalization of the complex is rapid, as it beginning as soon as 5–10 min after the IF-B12 is attached to the receptor

(Gueant et al., 1992). The fate of the IF–B12 complex in the cell has been studied in vivo under the action of drugs such as chloroquine or cycloheximide (Dan & Cutler, 1994; Gueant et al., 1992). These studies have established the role of the intracellular acidic compartments in the liberation of B12 from IF. Separation of the complex from its receptor takes place in pre-lysosomal vesicles. Proteases sensitive to leupeptin and glycosidases are responsible for the degradation of IF in vitro and in cell models (Dan & Cutler, 1994; Gueant, Monin, et al., 1988). B12 is converted in part into the two co-enzyme forms, Ado- and Me-Cbl (Pons et al., 2000) released from the basolateral membrane (Ramanujam et al., 1991). TC synthesis is needed intracellulary for the transcytosis of B12, which is released in the basolateral side in a form complexed to TC (Cho, Jung, & Choi, 2012; Pons et al., 2000). These in vitro data confirmed previous experiments performed in healthy volunteers with Schilling tests. The subjects received an oral dose of 0.05 mcg of ^{57}Co-B12 and an injected dose of 800 mcg of B12 and ^{58}Co-B12. Free B12 saturated HC and TC and ^{58}Co-B12 was mainly in free form from in plasma, with a minor fraction bound to TC that remained stable during the whole experiment. Oral ^{57}Co-B12 absorbed from the gut appeared in plasma attached to TC, with a peak observed 8 h after B12 ingestion. A second peak of ^{57}Co-B12 in free form was observed at a much lower magnitude after 10 h. This experiment showed that ^{57}Co-B12 was prominently exported from gut mucosa in a form complexed to TC and that a minor route could lead to export of free B12 (Chanarin, Muir, Hughes, & Hoffbrand, 1978). The involvement of intestinal TC in B12 export from enterocytes may explain why inherited TC deficiency can produce abnormal absorption of cobalamin (Barshop et al., 1990). The multidrug resistance protein 1 (MRP1/ABCC1) serves as molecular gate for the cellular export of B12 in free form in the basolateral side of enterocytes and its release into the circulating blood (Beedholm-Ebsen et al., 2010).

3.5 Absorption of B12 in the foetus

TC and IF are present in amniotic fluid and have mainly a foetal origin (Gueant, Hambaba, et al., 1989). This could mediate the delivery of B12 to the foetus in utero, since amniotic fluid is swallowed by the foetus. In order to study the ability of B12 to be transported and metabolized across the embryo-fetal digestive tract, human embryonic stomachs and intestines were transplanted into nude mice. IF was secreted by the non-stimulated embryonic stomach and lacked sialic acid (Aimone-Gastin et al., 1999). B12 bound to IF was subsequently transported across the xenografted embryo-fetal intestine, suggesting that amniotic fluid may contribute to B12 delivery to the embryo-fetus (Aimone-Gastin et al., 1999).

4. Causes of vitamin B12 malabsorption (Table 2)

4.1 Inherited diseases of Cbl malabsorption

Cbl malabsorption results mainly from mutations in the gene encoding gastric IF involved in the intraluminal transport of B12 (*CBLGIF*) and/or the two genes needed for the intestinal uptake of Cbl, *Amnion Associated Transmembrane Protein* (*AMN*) and *Cubilin* (*CUBN*).

4.1.1 Intrinsic factor deficiency (IFD, OMIM 261000)

Most of the cases of IF deficiency reported in the literature have presented with a homozygous mutation in *CBLGIF* (Gordon et al., 2004; Tanner, Sturm, Baack, Liyanarachchi, & de la Chapelle, 2012; Yassin et al., 2004). The first reported genetic defect was a 68A > G variant, which predicted a Q5R change in the mature protein (Gordon et al., 2004; Yassin et al., 2004). However, the recombinant IF bound Cbl comparably to native IF. Despite absence of functional consequence, the *68AG* heterozygous genotype of *CBLGIF* was a predictor of low vitamin B12, with a higher allele frequency in cases with pernicious anemia and other cases with low plasma vitamin B12 (3.25% and 4.35%) than in control subjects (1.1%) (Yassin et al., 2004). This association between *CBLGIF 68A > G* variant and plasma vitamin B12 was not confirmed, however, in two populations from West Africa and South Europe (Chery et al., 2013). Tanner et al. made a large genetic screening study of 154 families or patients with suspected hereditary vitamin B12 malabsorption, with systematic genetic testing of the three genes *CUBN*, *AMN*, and *CBLGIF* (Tanner et al., 2012). Among patients with hereditary cobalaminl (cbl) malabsorption 22% of the identified mutations were in *CBLGIF*. Among the IF deficiency cases, they found one family with a c.79 + 1G > A and c.290 T > C compound mutation and three cases with heterozygous *CBLGIF* mutations, including one case with a *290 T > C* mutation. Chery et al. reported two families with heterozygous *CBLGIF* mutation, 290 T > C, M97T, with decreased binding affinity of IF for vitamin B12 and a K145_N146delinsN deletion that led to complete absence of vitamin B12 binding capacity in the mutated recombinant IF (Chery et al., 2013). A *FUT2* secretor variant

Table 2 The causes of B12 malabsorption.[a]

Cause	Mechanism	Diagnosis
Intrinsic factor deficiency	Pathogenic mutations in CBLGIF	No AIG, family context Gene sequencing
Imerslund-Gräsbeck disease	Pathogenic mutations in CUBN and/or AMN	No AIG, family context Gene sequencing
Biermer-Addison's disease	AIG with impaired IF and gastric acid secretion	Endoscopy with fundic and antral biopsies IF and parietal cell auto-immunity
Non-Biermer-Addison's gastritis	Impaired gastric acid secretion Malabsorption of protein bound B12	No AIG
Bariatric surgery and gastrectomies	Impaired IF and gastric acid secretion Bacterial overgrowth	
Pancreatic insufficiency	Impaired degradation of HC	Pancreas imaging, lipasemia, steatorrhea
Obstructive Jaundice	Impaired B12 binding to IF and uptake	Imaging and biology
Tropical sprue and celiac disease	Villous atrophy with impaired B12-IF uptake	Anti-tranglutaminase auto-immunity
Blind loop syndrome	Bacterial overgrowth	Breath test
Parasitic infestations	Villous atrophy and B12 consumption	Stool analyses
Zollinger-Ellison syndrome	Acid intestinal pH and Impaired B12-IF uptake	Imaging and gastrinemia
Inflammatory bowel diseases	Parietal injury with impaired IF-B12 uptake	Endoscopy, imaging
Chronic radiation enteritis of the distal ileum	Parietal injury with impaired IF-B12 uptake	Imaging
Iatrogenic effects of drugs	Impaired IF-B12 uptake	

[a]AIG, auto-immune gastritis.

was found to worsen B12 status in cases with heterozygous *CBLGIF* mutations by impairing IF secretion, independently of its association with *Helicobacter pylori*-related gastritis.

4.1.2 Imerslund-Gräsbeck disease (IGD, OMIM 261100)

Imerslund-Gräsbeck syndrome is a rare recessive disorder of intestinal vitamin B12 (cobalamin) malabsorption resulting in megaloblastic anemia, and deficient renal protein reabsorption, occasionally accompanied by mental disorders. Consequently, proteinuria is frequently observed in IGD but not in IFD. In contrast to IFD, the cbl absorption Schilling test does not improve when cbl is administered with IF in IGD patients. However, this test is not anymore available, due to the lack of radio-labeled cobalamin (Green et al., 2017). Mutations in *CUBN* or *AMN* are responsible for IGD (Namour et al., 2011; Tanner et al., 2003, 2005). The disease mutations produce a variety of phenotypes, including decreased activity of the cbl-intrinsic factor receptor (Gueant et al., 1995), expression of an unstable receptor (Eaton, Livingston, Seetharam, & Puntis, 1998), and decreased binding affinity for IF-cbl (Gueant, Chery, Namour, Aimone-Gastin, & Wustinger, 1999).

4.2 Autoimmune gastritis (AIG) and Biermer-Addison's pernicious anemia

This disease, referred to as Addison's pernicious anemia in the United States and United Kingdom and Biermer's anemia in many European countries, is caused by the interruption of IF secretion associated with an auto-immune fundic atrophic gastritis. The autoimmune gastritis (AIG) is an increasingly prevalent, organ-specific, autoimmune disorder characterized by the destruction of gastric parietal cells, leading to the loss of IF and acid secretion (Lenti et al., 2020). The more or less complete absence of IF causes B12 deficiency as a consequence of the malabsorption of both dietary and biliary B12 and progressive loss of body reserves. Type 1 anti-IF autoantibodies prevent B12-IF binding and type II autoantibodies may inhibit the binding of Cbl-IF to the ileal receptor. The epitope of type 1 anti-intrinsic factor autoantibodies is located in the 251–265 amino acid sequence of the protein, which shares mimicry with peptides from herpesvirus Saimiri and from pathogenic *Escherichia coli*, suggesting a possible role of the microbiota in the pathophysiology of anti-IF auto-immunity (Gueant et al., 1997).

Anemia is only part of the AIG clinical spectrum, and the confusion between the gastric disease and Biermer-Addison's pernicious anemia is frequent in the literature. This is the reason why we propose to refer the AIG clinical spectrum to Biermer-Addison's disease than to pernicious anemia. The clinical symptoms with B12 deficiency include neurological outcomes, megaloblastic anemia and thromboembolic manifestations. Anemia is found in ~50% of patients at the time of AIG diagnosis. Iron deficiency anemia is more common in cases <50 years of age than in older subjects. Many patients present with combined B12 and iron deficiencies, leading to normocytic instead of macrocytic anemia. The prevalence of diabetes type 1 and Hashimoto thyroiditis is three- to fivefold and three- to eightfold higher in patients with AIG than in the general population (Lenti et al., 2020). The early step of fundic mucosal injury is characterized by inflammatory erosion and oxyntic gland inflammation with lymphocytes, plasma cells, mast cells and eosinophils, and CD4+ T cells are felt to be important in cellular destruction (Hall & Appelman, 2019). In a later stages, mucosal atrophy and metaplasia develop.

Gastric biopsies of the fundus and antrum are mandatory to diagnose AIG. Gastritis is graded according to the severity of inflammation, activity (the degree of polymorphonuclear neutrophil infiltration), atrophy, and intestinal metaplasia on a scale from 0 to 3. Patients with either severe atrophic gastritis or moderate–severe mucosal dysplasia, whether due to AIG or chronic *H. pylori* infection, should be followed endoscopically for the risk of gastric cancer. The upper digestive tract endoscopy should also involve duodenum/proximal jejunum examination with biopsies for diagnosing villous atrophy. Although there is increased risk of gastric cancer in patients with intestinal metaplasia, the risk in patients with atrophic gastritis is still insufficiently documented. Current recommendations for patients with autoimmune gastritis include performing a gastrointestinal endoscopy every 3–5 years.

4.3 Non-Addison-Biermer gastritis with protein-bound vitamin B12 malabsorption

Some forms of gastritis are accompanied by B12 deficit or deficiency with low B12, mild increase of homocysteine and/or methylmalonic acid and/or macrocytosis, despite normal schilling test. The patients are sometimes difficult to distinguish from the genuine AIG of Biermer-Addison's disease (Aimone-Gastin et al., 1997; Doscherholmen et al., 1981). Similar to AIG, they are frequent in people over 60 years of age. No clear fundic atrophy is found in the examination of fundic biopsies. A decreased gastric acid secretion and intermediate-to-normal level of IF secretion are observed after pentagastrin stimulation, despite absence or low Sydney score of fundic lesions. The deficient absorption of B12 occurs despite sufficient IF secretion and is related to a lack of cobalamin release from food proteins, as a consequence of decreased acid and peptic secretion. This explains that unlike the standard Schilling test, absorption tests for B12 combined with a protein medium are almost always disturbed (Gastin et al., 1997). A study assessed vitamin B12 deficiency in 394 elderly hospitalized patients elderly patients by performing upper endoscopy and gastric biopsies in antrum and fundus, anti-IF and anti-parietal cell auto-antibodies as markers of AIG, pepsinogen as marker of gastric atrophy and standard and modified Schilling test performed with trout flesh (Gastin et al., 1997). Among these subjects, 14 subjects had low B12 with anemia and/or macrocytosis and increased homocysteine and/or methylmalonate serum level. The protein bound modified Schilling test showed malabsorption in five of these patients, including three AIG cases who had a Schilling test corrected by IF and two subjects bearing a non-Addison-Biermer disease with normal standard Schilling test, normal/subnormal Sydney score of fundic gastritis, absence of anti-IF and anti-parietal cell auto-antibodies and normal pepsinogen concentration. A significant negative correlation of the modified Schilling test was found with the Sydney score for fundic mucosa but not with the score for antral mucosa nor with pepsinogen blood levels. The deficient gastric acid secretion was therefore the predominant factor responsible for the protein bound vitamin B12 malabsorption (Gastin et al., 1997). It is therefore possible that this gastritis entity with protein bound vitamin B12 malabsorption represents the first stage of AIG of Biermer-Addison disease. More recently another study evaluated the vitamin B-12 status according to serum concentration of pepsinogens I and II as markers of gastritis, in 3299 elderly volunteers from the TUDA Irish cohort, including 1216 users of proton pump inhibitors (PPI) and 2083 PPI non-users (Porter et al., 2021). They found a 2.5-fold higher prevalence of vitamin B12 deficiency in the subjects with abnormal pepsinogens I and II ratio (38%), compared with PPIs non-users with normal values of pepsinogens (15%; $P<0.001$). Even if anti-IF and anti-parietal cell auto-antibodies were not considered in this observational study, the results suggest that gastritis is a common cause of B12 deficiency in the elderly.

4.4 Obesity, bariatric surgery and gastrectomies

Patients with morbid obesity who have a BMI $> 40\,kg/m^2$ have an increased risk of nutritional deficiencies after bariatric surgery such as Roux-en-Y gastric bypass and sleeve gastrectomy, with a relatively high frequency of B12 deficiency (Dogan et al., 2018; Kornerup, Hvas, Abild, Richelsen, & Nexo, 2019). The malabsorption is produced through diverse mechanisms that include gastric achlorhydria, decreased secretion of IF and bacterial overgrowth (Green et al., 2017). These mechanisms lead also to malabsorption of iron and folate. B12 deficiency occurred with a prevalence of 13.2% in a cohort of 399 bariatric surgery cases, and the prevalence increased with time from surgery (Antoine et al., 2021). Vitamin B12 concentration at surgery was a predictor of post-surgery B12 deficit, along with age and nonalcoholic fatty liver disease (NAFLD) (Antoine et al., 2021). Similar mechanisms produce malabsorption and deficiency of cbl deficiency in cases with partial gastrectomies. Total gastrectomies are inevitably accompanied by cbl deficiency as IF secretion is abolished (Green et al., 2017).

4.5 Other causes of B12 malabsorption

4.5.1 Pancreatic insufficiency

Allen et al. demonstrated that trypsin, chymotrypsin degrades HC, which results in B12 being transferred to IF. Consequently, the malabsorption of B12 in cases of exocrine pancreatic insufficiency was found to result from inadequate degradation of HC (Allen, Seetharam, Allen, et al., 1978). However, chronic pancreatitis in general is not a common cause of B12 malabsorption (Guéant, Champigneulle, Gaucher, & Nicolas, 1990).

4.5.2 Obstructive jaundice

Bile has an enhancing effect on B12-IF ileal uptake and the Schilling test reveals malabsorption of crystalline cbl in about 40% of cases of obstructive jaundice and about 50% of patients with external bile drainage (Teo, Scott, Neale, & Weir, 1980). This report has not been repeated, and the mechanism by which the presence of bile stimulates B12 absorption is obscure. It is possible that increased water flux through the ileum secondary to sodium-linked bile acid absorption plays a role. Whatever the cause and similar to chronic pancreatitis, biliary obstruction by lithiasis is not a common cause of B12 deficiency. In some other cases, the mechanisms of malabsorption are complex and may result from the association of pernicious anemia with primary biliary cirrhosis through autoimmune mechanisms shared by both diseases (Jazia et al., 2010).

4.5.3 Blind loop syndrome

An imbalance of the intestinal flora, known as blind loop syndrome, occurs in various pathological circumstances, such as gastrectomies, segmental intestinal resection with end-to-side anastomosis, ileocolic resection, diverticulosis or gastric achlorhydria (Tabaqchali, 1970). Thanks to ready access to hydrogen breath testing, the possibility of small intestinal bacterial overgrowth (SIBO) can now be easily considered. The definition criteria, however, remain controversial and consequently the prevalence is undefined (Quigley, Murray, & Pimentel, 2020). Laboratory findings can include elevated folate related to microbiota overproduction and vitamin B_{12} deficiency related to malabsorption (Quigley et al., 2020). The physiopathology of this malabsorption reflects the disruption of microbial homeostasis between uptake and use of B12 by intestinal flora (Rodionov et al., 2019; Tabaqchali, 1970). B12 is produced de novo by some gut bacteria (prototrophs) but must be provided for many other bacterial species by prototrophs and dietary B12 supply (auxotrophs). As a consequence any dysbiosis with excess of auxotrophs with produce a bacterial consumption, which divert the B12 from its physiological needs by the host. These auxotrophs are mostly anaerobic bacteria as the malabsorption can be corrected by oral administration of metronidazole (Welkos, Toskes, & Baer, 1981). In addition, Brandt et al. showed that anerobic bacteria predominated in the intestinal liquid of three patients suffering from blind loop syndrome, particularly *Bacteroides fragilis* and *Clostridium per-fringens* (Brandt et al., 1977). Other bacteria could be implicated, such as *Proteus mirabilis*, *Klebsiella pneumoniae*, *E. coli*, *Lactobacillus*, *Streptococcus A* and *Bacillus subtilis*. All these bacteria possess proteins needed for B12 uptake. B12 uptake could occur through competition with the B12 acceptors present in the intestinal fluid (Giannella, Broitman, & Zamcheck, 1971, 1972). B12 could also be removed from receptor-mediated absorption after dissociation of the IF–B12 complex by the action of proteases and glycosidases of bacterial origin (Gueant, Monin, et al., 1988).

4.5.4 Tropical sprue and celiac disease

Tropical sprue and celiac disease are two diseases characterized by progressive villus atrophy in the small intestine. The loss of intestinal villi (i.e., surface area and increased immature crypt-like cells along with decreased mature epithelial cells) affects intestinal absorptive function, and patients present with nutrient malabsorption, including folate, iron, vitamin D and vitamin B12. Patients with tropical sprue typically may present with macrocytic anemia due to the malabsorption of folate and/or vitamin B_{12}. Zinc and iron are the most frequent micronutrients with deficit, followed by order by vitamin D, vitamin B12 and folate, in celiac disease (Bledsoe et al., 2019). The low prevalence of B12 malabsorption is related to the predominance of celiac pathology in the proximal intestine of most patients, whereas nearly all IF–B12 complex absorption occurs in the ileum. The etiology of celiac disease is related to environmental, genetic, and immune mechanisms, including the deamidation of gluten peptides by the enzyme transglutaminase 2 (TG2) generating peptides that bind strongly to the predisposing HLA-DQ2 or HLA-DQ8 molecules, with subsequent T cell altered immunity (Voisine & Abadie, 2021). The treatment of celiac disease includes dietary restriction of gluten and supplementations for correction of micronutrient deficits. In contrast, the etiology of tropical sprue remains obscure. Considering the benefit of antibiotic therapy, the favored hypothesis is that it is initiated or maintained by a still-undefined infection (Westergaard, 2004). Treatment of tropical sprue with the combination of folic acid supplementation and antiobiotic therapy with tetracycline for 6 months has become the standard recommended treatment. Vitamin B12 supplementation is added if there is evidence of B12 deficiency and/or malabsorption. B12 malabsorption occurs in most cases and can be reversed by antibiotic treatment over a period of time (Westergaard, 2004).

4.5.5 Parasitic infestations

The two main parasites involved are *Diphyllobothrium latum* (diphyllobothriasis) and *Giardia lamblia* (lambliasis), but B12 deficiency is seen only in rare cases. Fish tapeworm infestation, which used to be common in the Scandinavian countries, is now anecdotal. It is accompanied by a decrease in IF secretion and atrophic gastritis. The parasite traps B12 bound with IF as well as free B12 and also inhibits the binding of B12 with IF (Gräsbeck, 1984; Nyberg, 1963). B12 malabsorption in cases of lambliasis is the result of B12 being trapped by the parasite but also of villous atrophy (Notis, 1972). The odd feature of the rare cases of giardiasis reported with B12 deficiency is that the organism grows only in the very proximal intestine, as it requires phospholipids for growth, supplied by the delivery of bile into the duodenum. Thus, the proximal villous atrophy probably plays little role, and the associated diarrhea may be a factor in B12 malabsorption, perhaps in patients with already low B12 body stores.

4.5.6 Zollinger-Ellison syndrome

Hypergastrinemia is accompanied by hyperacidity of the duodenal chyme. However, the pH in the ileum is above 6 and therefore cannot be responsible for malabsorption. The acidic pH found in the duodenum may irreversibly inhibit pancreatic proteases, causing inadequate HC degradation, leading to lack of transfer of B12 from HC to IF. The treatment with PPI does not reverse the malabsorption probably because of insufficient decrease of acidic secretion (Hirschowitz, Worthington, & Mohnen, 2008).

4.5.7 Inflammatory bowel diseases

Crohn's disease (CD) with lesions in the terminal ileum or ileal resection may produce vitamin B12 malabsorption. In a single-center evaluation of 381 patients with CD, the prevalence of B12 deficiency among patients with CD was 33% compared with 16% in ulcerative colitis ($P < 0.0001$). Holotranscobalamin (holoTC) and/or methylmalonic acid (MMA) are more accurate than serum B12 in identifying B12 deficit. Serum B12 identified B12 deficiency in 5% of patients with CD, and holoTC/MMA in 32% (Ward et al., 2015). Resection of the terminal ileum is an important factor in loss of B12 body stores (Battat et al., 2014). It is accompanied usually by loss of the ileo-cecal valve, which decreases the residence time of intestinal fluid in the remaining ileum.

4.5.8 Short bowel syndrome (SBS)

Vitamin B12 deficit is common in patients with short bowel, in particular when a long section of the distal ileum is resected (defined as <180–200 cm of residual small bowel) (González et al., 2005). Most cases of B12 malabsorption in patients with IBD (see above) fall into this category. Many, but not all patients with SBS have lost the ileo-cecal valve as well. Many patients with ileal resection receive parenteral B12, and so have elevated serum B12 levels. A paradoxical elevation of methylmalonic acid (MMA) despite normal or elevated serum B12 is common in children, which is reversed after treatment with B12 injections (Davis, Strogach, Carobene, Shaul, & Thompson, 2020; Jimenez, Stamm, Depaula, & Duggan, 2018). The suggested explanation for this discrepancy is an overgrowth in intestinal bacteria that produce propionate, a precursor for MMA. This indicates that caution should be used when interpreting serum B12 and MMA values in children with SBS. A similar discrepancy has not been reported in adult patients with SBS.

4.5.9 Chronic radiation enteritis of the distal ileum

Chronic radiation enteritis is most commonly seen after radiotherapy for pelvic and abdominal malignancies and reduces the absorptive capacity of micronutrients, which is most commonly seen after radiotherapy for pelvic and abdominal malignancies. Malnutrition is a common complication of chronic radiation enteritis (CRE). The malabsorption of B12 is observed when the CRE encompasses the pelvis, where most of the ileum resides, and is frequently associated with bile acid malabsorption (Webb, Brooke, & De Silva, 2013). Chronic radiation enteritis often occurs following treatment for gynecological tumors, but the risk factors to predict its occurrence are not known (Kuku, Fragkos, McCormack, & Forbes, 2013).

4.5.10 Iatrogenic effects of drugs

Proton-pump inhibitors (PPIs), H2-receptor antagonists (H2RAs), used in the treatment of oeso-gastric diseases and Zollinger-Ellison syndrome and metformin, used in the treatment of type 2 diabetes, prediabetes, and polycystic ovary syndrome are widely prescribed drugs that can reduce serum vitamin B_{12} level (Miller, 2018). Proton pump inhibitors (PPIs) and H2-receptor antagonists (H2RAs) reduce the gastric acid secretion through direct effect on the gastric H+K+-ATPase and inhibition of histamine binding to parietal cell histamine H2 receptor, respectively. Consequently, the reduced gastric acid secretion leads to impaired protein-bound vitamin B12 absorption, as reported in Sections 2.1 and 3.2. B12 impaired absorption may be also related to the bacterial overgrowth produced by decreased gastric acid secretion (Guéant, Djalali, Perrin, Duc, & Nicolas, 1983). An association of long-term use of these drugs with low B12 has been reported in a systematic review of several observational and case control studies (Jung, Nagaraja, Kapur, & Eslick, 2015). It was concluded that there is limited evidence that these drugs leads to B_{12} deficiency with clinical manifestations. However, several case reports suggest that long-term treatment increases the risk of B12 deficiency. For example, B12 deficiency with megaloblastic anemia has been reported in one patient under long-term omeprazole therapy at a daily dose of 40–60 mg for 4 years as treatment for a gastro-oesophagal reflux complicated by peptic oesophagitis (Bellou et al., 1996). Serum vitamin B12 was dramatically decreased at 80 pmol. The standard Schilling test was normal and it was at 0% with cobalamin-labeled trout meat. All other assimilation tests were normal except an expiratory hydrogen breath test performed with lactulose. The hematological status was restored after intramuscular treatment with cobalamin, showing that prolonged omeprazole therapy can be responsible for a cobalamin deficiency due to protein-bound cobalamin malabsorption (Bellou et al., 1996). In conclusion, it is recommended to provide vitamin B_{12} supplements if altered blood biomarkers of B12 deficit or clinical signs consistent with B12 deficiency develop (Miller, 2018).

Metformin can also reduce vitamin B_{12} concentrations after prolonged use (Miller, 2018). Metformin is thought to impair vitamin B_{12} absorption by inhibiting the calcium-dependent binding of IF–B_{12} to the ileum IF receptor. Several studies evaluated the effect of metformin on biomarkers of B12 deficit/deficiency, including holoTC, homocysteine and methylmalonic acid (Greibe, Trolle, Bor, Lauszus, & Nexo, 2013; Leung et al., 2010; Obeid et al., 2013). Their results suggest that metformin has no significant effect on vitamin B_{12} status. Other drugs have been occasionally associated with B12 malabsorption in case reports, including cholestyramine, colchicine and neomycin. Cholestyramine is a resin which chelates IF, while colchicine and neomycin act as inhibitors of IF-Cbl endocytosis (Davis, 1985).

5. Measurements of vitamin B12 malabsorption in humans

About 99% of luminal B12 absorbed from diet or the enterohepatic circulation occurs via receptor-mediated endocytosis of the IF–B12 complex. Thus, to measure B12 absorption one must estimate the production of gastric IF and the capacity of the ileal absorption. One mole of IF (MW 55,000) binds 1 mol of B12 (1.34 mg). A unit of IF has been defined as the quantity needed to bind 1 ng of B12, and 400–500 units of IF are required for optimal absorption of 1 mcg of B12 (Chanarin, 1971). Normal daily output of IF in the human exceeds 70,000 units, and 1 mL of gastric contents after secretory stimulation contains about 55 units of IF. To absorb the RDA for B12 of 2.4 mcg would require only about 1400 IF units, <2% of the estimated daily production of IF. These estimates make many assumptions, but they confirm that it is only severe gastric atrophy with loss of most parietal cells that leads to IF production that is inadequate to support normal absorption of B12. Therefore, the first step in assessing B12 absorption is to determine the degree of gastric mucosal atrophy in the body of the stomach. This evaluation requires biopsy with histological evaluation, but does not require quantitation of IF.

The dual-isotope Schilling test was the gold standard method to evaluate the functional capacity of the ileal IF-B12 receptor, but it required the use of cobalt-labeled B12, and this compound has not been available worldwide since the late 1990s. The major advantage of the test was that it directly measured B12 absorption. The major disadvantage was that the sensitivity was only 83% for patients with pernicious anemia, and even lower (67%) for cases of malabsorption syndrome (Domstad, Choy, Kim, & DeLand, 1981). The second stage (ingestion of cobalt-57 labeled cyanocobalamin-IF) was started 4 days after the first stage (ingestion of cyano-^{58}Co-cobalamin). Urine was collected for 48 h at each stage. The limit of normal values was established at 10% and 11% of urinary excretion of the tracer for the first and second stages, respectively, although the range of 7–10% excretion was found in many patients with no other evidence of B12 deficiency. To ensure urinary excretion of absorbed labeled B12, before its ingestion 1000 μg of non-labeled vitamin B12 were injected intramuscularly to restore body stores, and urine was collected for 48 h. The determination of excreted radioactivity was estimated by 10 min counting of five 10 mL urine samples (Aimone-Gastin et al., 1997; Brito, Habeych, Silva-Zolezzi, Galaffu, & Allen, 2018).

The CobaSorb test, based on an assessment of the plasma holoTC concentration following an oral dose of B12 has been used as an indirect measurement of B12 absorption. HoloTC is determined before (day 0) and after intake of 9 μg of B12 as cyanocobalamin three times over 24 h (day 1) and serum is collected over day 2. The patients are classified as absorbers if holoTC increases more than 10 pmol/L and more than 22% on day 2 compared with day 0. Those with a holoTC above 65 pmol/L at day 0 and not fulfilling the criteria are classified as uncertain, while those with a holoTC below 65 pmol/L at day 0 not fulfilling the criteria were classified as non-absorbers (Hvas, Morkbak, & Nexo, 2007). The test is helpful to identify patients who respond to supplemental B12 but cannot be a substitute of the Schilling test, because the results reflect relative absorption only. It is somewhat limited in addition since it needs 3 days to complete, and cannot be used after B12 supplementation.

Two methods with isotopes are in development but need further evaluation and validation before their possible use in the clinic. The first used carbon-14 (^{14}C)-labeled cobalamin administered orally at a dose compatible with physiological conditions (1.5 μg, 2.2 kBq/59 nCi). Accelerator mass spectrometry is used for its detection and study of absorption and kinetics. This method uses a very low amount of radioactivity, and can measure levels of ^{14}C in microliter volumes of biological samples with negligible exposure of subjects to radioactivity (Carkeet et al., 2006). The plasma appearance and clearance curves are consistent with those of the pure vitamin. The tracer has been also used to determine the bioavailability and absorbed B12 from chicken eggs endogenously labeled (Garrod et al., 2019). A second method uses a stable isotope, (^{13}C)-cyanocobalamin to measure the bioavailability of B12. (^{13}C)-cyanocobalamin is rapidly converted into (^{13}C)-methylcobalamin and plasma (^{13}C)-methylcobalamin reached a peak between 5 and 8 h, similar to that reported in the study with (^{14}C)-labeled vitamin B12. This method is safe but requires serial blood sampling to provide an accurate index of B12 absorption and bioavailability using plasma concentrations between 5 and 7 h after the dose. It needs further testing and validation in patients of different age with different causes of B12 malabsorption (Devi et al., 2020).

In conclusion, the metabolic importance of vitamin B12, as well as its low content in food can partly explain the remarkable complexity of B12 transport and cellular metabolism in all mammals, and in particular in humans. Such complexity could also reflect mechanisms adapted to the relative fragility of the molecule, as well as the need to protect it from the avidity of the digestive flora. B_{12} deficiency due to malabsorption may produce a wide range of clinical symptoms, which include neurological and cardiovascular outcomes, beside or independently from hematological manifestations. Pernicious anemia is in fact a subset of the clinical presentation of AIG, which is considered the most frequent cause of B12 malabsorption. A particular attention to B12 deficit is recommended in the follow-up of subjects with bariatric surgery. The genetic causes of B12 malabsorption are classically diagnosed in infancy and childhood, but they are probably underestimated in adult cases with B12 deficit. Despite its high prevalence in the general population and in the elderly, B12 malabsorption cannot be anymore assessed by Schilling test, pointing out the urgent need for an equivalent reliable test.

References

Abdel Nour, A. M., Ringot, D., Gueant, J. L., & Chango, A. (2007). Folate receptor and human reduced folate carrier expression in HepG2 cell line exposed to fumonisin B1 and folate deficiency. *Carcinogenesis*, *28*(11), 2291–2297. https://doi.org/10.1093/carcin/bgm149.

Adams, J. F., Kennedy, E. H., Thomson, J., & Williamson, J. (1968). The effect of acid peptic digestion on free and tissue-bound colbalamins. *The British Journal of Nutrition*, *22*(1), 111–114. https://doi.org/10.1079/bjn19680013.

Aimone-Gastin, I., Gueant, J. L., Plenat, F., Muhale, F., Maury, F., Djalali, M., et al. (1999). Assimilation of [57Co]-labeled cobalamin in human fetal gastrointestinal xenografts into nude mice. *Pediatric Research*, *45*(6), 860–866. https://doi.org/10.1203/00006450-199906000-00014.

Aimone-Gastin, I., Pierson, H., Jeandel, C., Bronowicki, J. P., Plenat, F., Lambert, D., et al. (1997). Prospective evaluation of protein bound vitamin B12 (cobalamin) malabsorption in the elderly using trout flesh labelled in vivo with 57Co-cobalamin. *Gut*, *41*(4), 475–479.

Albert, M. J., Mathan, V. I., & Baker, S. J. (1980). Vitamin B12 synthesis by human small intestinal bacteria. *Nature*, *283*(5749), 781–782. https://doi.org/10.1038/283781a0.

Allen, A. S., Berkovic, S. F., Cossette, P., Delanty, N., Dlugos, D., & Eichler, E. E. (2013). De novo mutations in epileptic encephalopathies. *Nature*, *501*. https://doi.org/10.1038/nature12439.

Allen, R. H., Seetharam, B., Allen, N. C., Podell, E. R., & Alpers, D. H. (1978). Correction of cobalamin malabsorption in pancreatic insufficiency with a cobalamin analogue that binds with high affinity to R protein but not to intrinsic factor. In vivo evidence that a failure to partially degrade R protein is responsible for cobalamin malabsorption in pancreatic insufficiency. *The Journal of Clinical Investigation*, *61*(6), 1628–1634. https://doi.org/10.1172/JCI109083.

Allen, R. H., Seetharam, B., Podell, E., & Alpers, D. H. (1978). Effect of proteolytic enzymes on the binding of cobalamin to R protein and intrinsic factor. In vitro evidence that a failure to partially degrade R protein is responsible for cobalamin malabsorption in pancreatic insufficiency. *The Journal of Clinical Investigation*, *61*(1), 47–54. https://doi.org/10.1172/JCI108924.

Aminoff, M., Carter, J. E., Chadwick, R. B., Johnson, C., Grasbeck, R., Abdelaal, M. A., et al. (1999). Mutations in CUBN, encoding the intrinsic factor-vitamin B12 receptor, cubilin, cause hereditary megaloblastic anaemia 1. *Nature Genetics*, *21*(3), 309–313. https://doi.org/10.1038/6831.

Andrès, E., Zulfiqar, A. A., Serraj, K., Vogel, T., & Kaltenbach, G. (2018). Systematic review and pragmatic clinical approach to Oral and nasal vitamin B12 (cobalamin) treatment in patients with vitamin B12 deficiency related to gastrointestinal disorders. *Journal of Clinical Medicine*, 7(10), 304. https://doi.org/10.3390/jcm7100304.

Antoine, D., Li, Z., Quilliot, D., Sirveaux, M. A., Meyre, D., Mangeon, A., et al. (2021). Medium term post-bariatric surgery deficit of vitamin B12 is predicted by deficit at time of surgery. *Clinical Nutrition*, *40*(1), 87–93. https://doi.org/10.1016/j.clnu.2020.04.029.

Arkbåge, K., Witthöft, C., Fondén, R., & Jägerstad, M. (2003). Retention of vitamin B12 during manufacture of six fermented dairy products using a validated radio protein-binding assay. *International Dairy Journal*, *13*(2), 101–109. https://doi.org/10.1016/S0958-6946(02)00146-2.

Ball, G. F. M. (1998). *Bioavailability and analysis of vitamins in foods* (pp. 497–515). London: Chapman & Hall.

Barshop, B. A., Wolff, J., Nyhan, W. L., Yu, A., Prodanos, C., Jones, G., et al. (1990). Transcobalamin II deficiency presenting with methylmalonic aciduria and homocystinuria and abnormal absorption of cobalamin. *American Journal of Medical Genetics*, *35*(2), 222–228. https://doi.org/10.1002/ajmg.1320350216.

Battat, R., Kopylov, U., Szilagyi, A., Saxena, A., Rosenblatt, D. S., Warner, M., et al. (2014). Vitamin B12 deficiency in inflammatory bowel disease: Prevalence, risk factors, evaluation, and management. *Inflammatory Bowel Diseases*, *20*(6), 1120–1128. https://doi.org/10.1097/mib.0000000000000024.

Beedholm-Ebsen, R., van de Wetering, K., Hardlei, T., Nexø, E., Borst, P., & Moestrup, S. K. (2010). Identification of multidrug resistance protein 1 (MRP1/ABCC1) as a molecular gate for cellular export of cobalamin. *Blood*, *115*(8), 1632–1639. https://doi.org/10.1182/blood-2009-07-232587.

Bellou, A., Aimone-Gastin, I., De Korwin, J. D., Bronowicki, J. P., Moneret-Vautrin, A., Nicolas, J. P., et al. (1996 Sep). 1996. Cobalamin deficiency with megaloblastic anaemia in one patient under long-term omeprazole therapy. *Journal of Internal Medicine*, *240*(3), 161–164. https://doi.org/10.1046/j.1365-2796.1996.20846000.x.

Bledsoe, A. C., King, K. S., Larson, J. J., Snyder, M., Absah, I., Choung, R. S., et al. (2019). Micronutrient deficiencies are common in contemporary celiac disease despite lack of overt malabsorption symptoms. *Mayo Clinic Proceedings*, *94*(7), 1253–1260. https://doi.org/10.1016/j.mayocp.2018.11.036.

Booth, C. C., Chanarin, I., Anderson, B. B., & Mollin, D. L. (1957). The site of absorption and tissue distribution of orally administered 56Co-labelled vitamin B12 in the rat. *British Journal of Haematology*, *3*(3), 253–261. https://doi.org/10.1111/j.1365-2141.1957.tb05794.x.

Brandt, L. J., Bernstein, L. H., & Wagle, A. (1977). Production of vitamin B 12 analogues in patients with small-bowel bacterial overgrowth. *Annals of Internal Medicine*, *87*(5), 546–551.

Brito, A., Habeych, E., Silva-Zolezzi, I., Galaffu, N., & Allen, L. H. (2018). Methods to assess vitamin B12 bioavailability and technologies to enhance its absorption. *Nutrition Reviews*, *76*(10), 778–792. https://doi.org/10.1093/nutrit/nuy026.

Carkeet, C., Dueker, S. R., Lango, J., Buchholz, B. A., Miller, J. W., Green, R., et al. (2006). Human vitamin B12 absorption measurement by accelerator mass spectrometry using specifically labeled (14)C-cobalamin. *Proceedings of the National Academy of Sciences of the United States of America*, *103*(15), 5694–5699. https://doi.org/10.1073/pnas.0601251103.

Chan, C. Q., Low, L. L., & Lee, K. H. (2016). Oral vitamin B12 replacement for the treatment of pernicious anemia. *Frontiers in Medicine (Lausanne)*, *3*, 38. https://doi.org/10.3389/fmed.2016.00038.

Chanarin, I. (1971). Absorption of cobalamins. *Journal of Clinical Pathology*, *24*(Suppl 5), 60–65.

Chanarin, I., Muir, M., Hughes, A., & Hoffbrand, A. V. (1978). Evidence for intestinal origin of transcobalamin II during vitamin B12 absorption. *British Medical Journal*, *1*(6125), 1453–1455. https://doi.org/10.1136/bmj.1.6125.1453.

Chery, C., Hehn, A., Mrabet, N., Oussalah, A., Jeannesson, E., Besseau, C., et al. (2013). Gastric intrinsic factor deficiency with combined GIF heterozygous mutations and FUT2 secretor variant. *Biochimie*, *95*(5), 995–1001. https://doi.org/10.1016/j.biochi.2013.01.022.

Cho, S. J., Jung, U. J., & Choi, M. S. (2012). Differential effects of low-dose resveratrol on adiposity and hepatic steatosis in diet-induced obese mice. *The British Journal of Nutrition*, *108*(12), 2166–2175. https://doi.org/10.1017/S0007114512000347.

Citrin, Y., Derosa, C., & Halsted, J. A. (1957). Sites of absorption of vitamin B12. *The Journal of Laboratory and Clinical Medicine*, *50*(5), 667–672.

Dahlgren, D., & Lennernäs, H. (2019). Intestinal permeability and drug absorption: Predictive experimental, computational and in vivo approaches. *Pharmaceutics*, *11*(8). https://doi.org/10.3390/pharmaceutics11080411.

Dan, N., & Cutler, D. F. (1994). Transcytosis and processing of intrinsic factor-cobalamin in Caco-2 cells. *The Journal of Biological Chemistry*, *269*(29), 18849–18855.

Davis, R. E. (1985). Clinical chemistry of vitamin B12. *Advances in Clinical Chemistry*, *24*, 163–216. https://doi.org/10.1016/s0065-2423(08)60273-5.

Davis, E. T., Strogach, I., Carobene, M., Shaul, E., & Thompson, J. (2020). Paradoxical elevation of both serum B12 and methylmalonic acid levels in assessing B12 status in children with short-bowel syndrome. *Journal of Parenteral and Enteral Nutrition*, *44*(7), 1257–1262. https://doi.org/10.1002/jpen.1764.

Devi, S., Pasanna, R. M., Shamshuddin, Z., Bhat, K., Sivadas, A., Mandal, A. K., et al. (2020). Measuring vitamin B-12 bioavailability with [13C]-cyanocobalamin in humans. *The American Journal of Clinical Nutrition*, *112*(6), 1504–1515. https://doi.org/10.1093/ajcn/nqaa221.

Dix, C. J., Hassan, I. F., Obray, H. Y., Shah, R., & Wilson, G. (1990). The transport of vitamin B12 through polarized monolayers of Caco-2 cells. *Gastroenterology*, *98*(5 Pt 1), 1272–1279. https://doi.org/10.1016/0016-5085(90)90344-z.

Djalali, M., Gueant, J. L., Lambert, D., el Kholty, S., Saunier, M., & Nicolas, J. P. (1990). High-performance liquid chromatographic separation and dual competitive binding assay of corrinoids in biological material. *Journal of Chromatography*, *529*(1), 81–91. https://doi.org/10.1016/s0378-4347(00)83809-8.

Dogan, K., Homan, J., Aarts, E. O., de Boer, H., van Laarhoven, C., & Berends, F. J. (2018). Long-term nutritional status in patients following Roux-en-Y gastric bypass surgery. *Clinical Nutrition*, *37*(2), 612–617. https://doi.org/10.1016/j.clnu.2017.01.022.

Domstad, P. A., Choy, Y. C., Kim, E. E., & DeLand, F. H. (1981). Reliability of the dual-isotope schilling test for the diagnosis of pernicious anemia or malabsorption syndrome. *American Journal of Clinical Pathology*, *75*(5), 723–726. https://doi.org/10.1093/ajcp/75.5.723.

Donaldson, R. M., Jr., Mackenzie, I. L., & Trier, J. S. (1967). Intrinsic factor-mediated attachment of vitamin B12 to brush borders and microvillous membranes of hamster intestine. *The Journal of Clinical Investigation*, *46*(7), 1215–1228. https://doi.org/10.1172/JCI105615.

Doscherholmen, A., McMahon, J., & Economon, P. (1981). Vitamin B12 absorption from fish. *Proceedings of the Society for Experimental Biology and Medicine*, *167*(4), 480–484. https://doi.org/10.3181/00379727-167-41201.

Doscherholmen, A., McMahon, J., & Ripley, D. (1978). Vitamin B12 assimilation from chicken meat. *The American Journal of Clinical Nutrition*, *31*(5), 825–830.

Doscherholmen, A., & Swaim, W. R. (1973). Impaired assimilation of egg co 57 vitamin B 12 in patients with hypochlorhydria and achlorhydria and after gastric resection. *Gastroenterology*, *64*(5), 913–919.

Eaton, D. M., Livingston, J. H., Seetharam, B., & Puntis, J. W. (1998). Overexpression of an unstable intrinsic factor-cobalamin receptor in Imerslund-Grasbeck syndrome. *Gastroenterology*, *115*(1), 173–176. https://doi.org/10.1016/s0016-5085(98)70379-8.

el Kholty, S., Gueant, J. L., Bressler, L., Djalali, M., Boissel, P., Gerard, P., et al. (1991). Portal and biliary phases of enterohepatic circulation of corrinoids in humans. *Gastroenterology*, *101*(5), 1399–1408.

Farquharson, J., & Adams, J. F. (1976). The forms of vitamin B12 in foods. *The British Journal of Nutrition*, *36*(1), 127–136. https://doi.org/10.1079/bjn19760063.

Fedosov, S. N., Fedosova, N. U., Berglund, L., Moestrup, S. K., Nexo, E., & Petersen, T. E. (2005). Composite organization of the cobalamin binding and cubilin recognition sites of intrinsic factor. *Biochemistry*, *44*(9), 3604–3614. https://doi.org/10.1021/bi047936v.

Garrod, M. G., Rossow, H. A., Calvert, C. C., Miller, J. W., Green, R., Buchholz, B. A., et al. (2019). (14)C-cobalamin absorption from endogenously labeled chicken eggs assessed in humans using accelerator mass spectrometry. *Nutrients*, *11*(9). https://doi.org/10.3390/nu11092148.

Giannella, R. A., Broitman, S. A., & Zamcheck, N. (1971). Vitamin B12 uptake by intestinal microorganisms: Mechanism and relevance to syndromes of intestinal bacterial overgrowth. *The Journal of Clinical Investigation*, *50*(5), 1100–1107. https://doi.org/10.1172/jci106581.

Giannella, R. A., Broitman, S. A., & Zamcheck, N. (1972). Competition between bacteria and intrinsic factor for vitamin B 12: Implications for vitamin B 12 malabsorption in intestinal bacterial overgrowth. *Gastroenterology*, *62*(2), 255–260.

Glass, G. B., Boyd, L. J., & Stephanson, L. (1954). Intestinal absorption of vitamin B12 in man. *Science*, *120*(3106), 74–75. https://doi.org/10.1126/science.120.3106.74-a.

González, H. F., Pérez, N. B., Malpeli, A., Martínez, M. I., Del Buono, B., & Viteri, F. E. (2005). Nutrition and immunological status in long-term follow up of children with short bowel syndrome. *Journal of Parenteral and Enteral Nutrition*, *29*(3), 186–191. https://doi.org/10.1177/0148607105029003186.

Gordon, M. M., Brada, N., Remacha, A., Badell, I., del Rio, E., Baiget, M., et al. (2004). A genetic polymorphism in the coding region of the gastric intrinsic factor gene (GIF) is associated with congenital intrinsic factor deficiency. *Human Mutation*, *23*(1), 85–91. https://doi.org/10.1002/humu.10297.

Gordon, M. M., Hu, C., Chokshi, H., Hewitt, J. E., & Alpers, D. H. (1991). Glycosylation is not required for ligand or receptor binding by expressed rat intrinsic factor. *The American Journal of Physiology*, *260*(5 Pt 1), G736–G742. https://doi.org/10.1152/ajpgi.1991.260.5.G736.

Gräsbeck, R. (1984). Biochemistry and clinical chemistry of vitamin B12 transport and the related diseases. *Clinical Biochemistry*, *17*(2), 99–107. https://doi.org/10.1016/s0009-9120(84)90279-0.

Grasbeck, R., Nyberg, W., & Reizenstein, P. (1958). Biliary and fecal vit. B12 excretion in man: An isotope study. *Proceedings of the Society for Experimental Biology and Medicine*, *97*(4), 780–784. https://doi.org/10.3181/00379727-97-23879.

Green, R., Allen, L. H., Bjorke-Monsen, A. L., Brito, A., Gueant, J. L., Miller, J. W., et al. (2017). Vitamin B12 deficiency. *Nature Reviews. Disease Primers*, *3*, 17040. https://doi.org/10.1038/nrdp.2017.40.

Green, R., Jacobsen, D. W., Van Tonder, S. V., Kew, M. C., & Metz, J. (1982). Absorption of biliary cobalamin in baboons following total gastrectomy. *The Journal of Laboratory and Clinical Medicine*, *100*(5), 771–777.

Greibe, E., Trolle, B., Bor, M. V., Lauszus, F. F., & Nexo, E. (2013). Metformin lowers serum cobalamin without changing other markers of cobalamin status: A study on women with polycystic ovary syndrome. *Nutrients*, *5*(7), 2475–2482. https://doi.org/10.1016/j.biochi.2013.02.002.

Gueant, J. L., Champigneulle, B., Djalali, M., Bigard, M. A., Gaucher, P., Hassouni, A., et al. (1986). In-vitro test of haptocorrin degradation for biological diagnosis of exocrine pancreatic dysfunction using duodenal juice collected during endoscopy. *Lancet*, *2*(8509), 709–712. https://doi.org/10.1016/s0140-6736(86)90230-8.

Guéant, J. L., Champigneulle, B., Gaucher, P., & Nicolas, J. P. (1990). Malabsorption of vitamin B12 in pancreatic insufficiency of the adult and of the child. *Pancreas*, *5*(5), 559–567. https://doi.org/10.1097/00006676-199009000-00011.

Gueant, J. L., Chery, C., & Namour, F. (2001). Cublin and the hydrophobic intrinsic factor receptor are distinct molecules. *Blood*, *97*(10), 3316–3317. https://doi.org/10.1182/blood.v97.10.3316.

Gueant, J. L., Chery, C., Namour, F., Aimone-Gastin, I., & Wustinger, M. (1999). Decreased affinity of urinary intrinsic factor-cobalamin receptor in a case of Grasbeck-Imerslund syndrome. *Gastroenterology*, *116*(5), 1274–1276. https://doi.org/10.1016/s0016-5085(99)70041-7.

Gueant, J. L., Djalali, M., Aouadj, R., Gaucher, P., Monin, B., & Nicolas, J. P. (1986). In vitro and in vivo evidences that the malabsorption of cobalamin is related to its binding on haptocorrin (R binder) in chronic pancreatitis. *The American Journal of Clinical Nutrition*, *44*(2), 265–277.

Guéant, J. L., Djalali, M., Perrin, M. O., Duc, M., & Nicolas, J. P. (1983). May an anomaly of the intestinal flora be a factor aggravating vitamin B12 deficiency in achlorhydric gastritis? *Gastroentérologie Clinique et Biologique*, 7(2), 225–226.

Gueant, J. L., Gerard, A., Monin, B., Champigneulle, B., Gerard, H., & Nicolas, J. P. (1988). Radioautographic localisation of iodinated human intrinsic factor in the Guinea pig ileum using electron microscopy. *Gut*, *29*(10), 1370–1378. https://doi.org/10.1136/gut.29.10.1370.

Gueant, J. L., Hambaba, L., Vidailhet, M., Schaefer, C., Wahlstedt, V., & Nicolas, J. P. (1989). Concentration and physicochemical characterisation of unsaturated cobalamin binding proteins in amniotic fluid. *Clinica Chimica Acta*, *181*(2), 151–161.

Gueant, J. L., Jokinen, O., Schohn, H., Monin, B., Nicolas, J. P., & Grasbeck, R. (1989). Purification of intrinsic factor receptor from pig ileum using as affinity medium human intrinsic factor covalently bound to Sepharose. *Biochimica et Biophysica Acta*, *992*(3), 281–288. https://doi.org/10.1016/0304-4165(89)90086-x.

Gueant, J. L., Masson, C., Schohn, H., Girr, M., Saunier, M., & Nicolas, J. P. (1992). Receptor-mediated endocytosis of the intrinsic factor-cobalamin complex in HT 29, a human colon carcinoma cell line. *FEBS Letters*, *297*(3), 229–232. https://doi.org/10.1016/0014-5793(92)80544-q.

Gueant, J. L., Monin, B., Boissel, P., Gaucher, P., & Nicolas, J. P. (1984). Biliary excretion of cobalamin and cobalamin analogues in man. *Digestion*, *30*(3), 151–157.10.1159/000199099.

Gueant, J. L., Monin, B., Djalali, M., Wahlstedt, V., Bois, F., & Nicolas, J. P. (1988). Effect of glycosidases and proteinases on cobalamin binding and physicochemical properties of purified saturated haptocorrin and intrinsic factor. *Biochimica et Biophysica Acta*, *957*(3), 390–398. https://doi.org/10.1016/0167-4838(88)90230-0.

Gueant, J. L., Safi, A., Aimone-Gastin, I., Rabesona, H., Bronowicki, J. P., Plenat, F., et al. (1997). Autoantibodies in pernicious anemia type I patients recognize sequence 251-256 in human intrinsic factor. *Proceedings of the Association of American Physicians*, *109*(5), 462–469.

Gueant, J. L., Saunier, M., Gastin, I., Safi, A., Lamireau, T., Duclos, B., et al. (1995). Decreased activity of intestinal and urinary intrinsic factor receptor in Grasbeck-Imerslund disease [corrected]. *Gastroenterology*, *108*(6), 1622–1628.

Guéant, J. L., Schohn, H., Brulé, H., Saunier, M., Gräsbeck, R., & Nicolas, J. P. (1991). Binding assay and physicochemical characteristics of solubilized intrinsic factor receptor in ileal mucosal homogenates using phenyl-Sepharose to separate the saturated receptor from free intrinsic factor. *Biochimica et Biophysica Acta*, *1073*(3), 614–618. https://doi.org/10.1016/0304-4165(91)90238-c.

Gueant, J. L., Vidailhet, M., Pasquet, C., Djalali, M., & Nicolas, J. P. (1984). Effect of pancreatic extracts on the faecal excretion and on the serum concentration of cobalamin and cobalamin analogues in cystic fibrosis. *Clinica Chimica Acta*, *137*(1), 33–41. https://doi.org/10.1016/0009-8981(84)90309-7.

Gullberg, R. (1974). Possible influence of vitamin B 12-binding protein in milk on the intestinal flora in breast-fed infants. II. Contents of unsaturated B 12-binding protein in meconium and faeces from breast-fed and bottle-fed infants. *Scandinavian Journal of Gastroenterology*, *9*(3), 287–292.

Hagedorn, C. H., & Alpers, D. H. (1977). Distribution of intrinsic factor-vitamin B12 receptors in human intestine. *Gastroenterology*, *73*(5), 1019–1022.

Hall, S. N., & Appelman, H. D. (2019). Autoimmune Gastritis. *Archives of Pathology & Laboratory Medicine*, *143*(11), 1327–1331. https://doi.org/10.5858/arpa.2019-0345-RA.

He, Q., Madsen, M., Kilkenney, A., Gregory, B., Christensen, E. I., Vorum, H., et al. (2005). Amnionless function is required for cubilin brush-border expression and intrinsic factor-cobalamin (vitamin B12) absorption in vivo. *Blood*, *106*(4), 1447–1453. https://doi.org/10.1182/blood-2005-03-1197.

Hewitt, J. E., Gordon, M. M., Taggart, R. T., Mohandas, T. K., & Alpers, D. H. (1991). Human gastric intrinsic factor: Characterization of cDNA and genomic clones and localization to human chromosome 11. *Genomics, 10*(2), 432–440. https://doi.org/10.1016/0888-7543(91)90329-d.

Heyssel, R. M., Bozian, R. C., Darby, W. J., & Bell, M. C. (1966). Vitamin B12 turnover in man. The assimilation of vitamin B12 from natural foodstuff by man and estimates of minimal daily dietary requirements. *The American Journal of Clinical Nutrition, 18*(3), 176–184. https://doi.org/10.1093/ajcn/18.3.176.

Hirschowitz, B. I., Worthington, J., & Mohnen, J. (2008). Vitamin B12 deficiency in hypersecretors during long-term acid suppression with proton pump inhibitors. *Alimentary Pharmacology & Therapeutics, 27*(11), 1110–1121. https://doi.org/10.1111/j.1365-2036.2008.03658.x.

Hvas, A. M., Morkbak, A. L., & Nexo, E. (2007). Plasma holotranscobalamin compared with plasma cobalamins for assessment of vitamin B12 absorption; optimisation of a non-radioactive vitamin B12 absorption test (CobaSorb). *Clinica Chimica Acta, 376*(1–2), 150–154. https://doi.org/10.1016/j.cca.2006.08.009.

Jazia, E. B., Khalifa, M., Abdelkader, A. B., Kaabia, N., Ghannouchi, N., Braham, A., et al. (2010). A case of primary biliary cirrhosis associated with pernicious anemia: A case report. *Cases Journal, 3*, 11. https://doi.org/10.1186/1757-1626-3-11.

Jensen, L. L., Andersen, R. K., Hager, H., & Madsen, M. (2014). Lack of megalin expression in adult human terminal ileum suggests megalin-independent cubilin/amnionless activity during vitamin B12 absorption. *Physiological Reports, 2*(7). https://doi.org/10.14814/phy2.12086.

Jimenez, L., Stamm, D. A., Depaula, B., & Duggan, C. P. (2018). Is serum methylmalonic acid a reliable biomarker of vitamin B12 status in children with short bowel syndrome: A case series. *The Journal of Pediatrics, 192*, 259–261. https://doi.org/10.1016/j.jpeds.2017.09.024.

Johnston, J., Bollekens, J., Allen, R. H., & Berliner, N. (1989). Structure of the cDNA encoding transcobalamin I, a neutrophil granule protein. *The Journal of Biological Chemistry, 264*(27), 15754–15757.

Jung, S. B., Nagaraja, V., Kapur, A., & Eslick, G. D. (2015). Association between vitamin B12 deficiency and long-term use of acid-lowering agents: A systematic review and meta-analysis. *Internal Medicine Journal, 45*(4), 409–416. https://doi.org/10.1111/imj.12697.

Kanazawa, S., Herbert, V., Herzlich, B., Drivas, G., & Manusselis, C. (1983). Removal of cobalamin analogue in bile by enterohepatic circulation of vitamin B12. *Lancet, 1*(8326 Pt 1), 707–708. https://doi.org/10.1016/s0140-6736(83)91996-7.

Kapadia, C. R., Mathan, V. I., & Baker, S. J. (1976). Free intrinsic factor in the small intestine in man. *Gastroenterology, 70*(5 PT.1), 704–706.

Kolhouse, J. F., & Allen, R. H. (1977). Absorption, plasma transport, and cellular retention of cobalamin analogues in the rabbit. Evidence for the existence of multiple mechanisms that prevent the absorption and tissue dissemination of naturally occurring cobalamin analogues. *The Journal of Clinical Investigation, 60*(6), 1381–1392. https://doi.org/10.1172/JCI108899.

Kornerup, L. S., Hvas, C. L., Abild, C. B., Richelsen, B., & Nexo, E. (2019). Early changes in vitamin B12 uptake and biomarker status following roux-en-Y gastric bypass and sleeve gastrectomy. *Clinical Nutrition, 38*(2), 906–911. https://doi.org/10.1016/j.clnu.2018.02.007.

Kozyraki, R., & Gofflot, F. (2007). Multiligand endocytosis and congenital defects: Roles of cubilin, megalin and amnionless. *Current Pharmaceutical Design, 13*(29), 3038–3046. https://doi.org/10.2174/138161207782110507.

Kozyraki, R., Kristiansen, M., Silahtaroglu, A., Hansen, C., Jacobsen, C., Tommerup, N., et al. (1998). The human intrinsic factor-vitamin B12 receptor, cubilin: Molecular characterization and chromosomal mapping of the gene to 10p within the autosomal recessive megaloblastic anemia (MGA1) region. *Blood*, *91*(10), 3593–3600.

Kuku, S., Fragkos, C., McCormack, M., & Forbes, A. (2013). Radiation-induced bowel injury: The impact of radiotherapy on survivorship after treatment for gynaecological cancers. *British Journal of Cancer*, *109*(6), 1504–1512. https://doi.org/10.1038/bjc.2013.491.

Larsen, C., Etzerodt, A., Madsen, M., Skjødt, K., Moestrup, S. K., & Andersen, C. B. F. (2018). Structural assembly of the megadalton-sized receptor for intestinal vitamin B(12) uptake and kidney protein reabsorption. *Nature Communications*, *9*(1), 5204. https://doi.org/10.1038/s41467-018-07468-4.

Lenti, M. V., Rugge, M., Lahner, E., Miceli, E., Toh, B. H., Genta, R. M., et al. (2020). Autoimmune gastritis. *Nature Reviews. Disease Primers*, *6*(1), 56. https://doi.org/10.1038/s41572-020-0187-8.

Leung, S., Mattman, A., Snyder, F., Kassam, R., Meneilly, G., & Nexo, E. (2010). Metformin induces reductions in plasma cobalamin and haptocorrin bound cobalamin levels in elderly diabetic patients. *Clinical Biochemistry*, *43*(9), 759–760. https://doi.org/10.1016/j.biochi.2013.02.002.

Levy, J., Rodriguez-Guéant, R. M., Oussalah, A., Jeannesson, E., Wahl, D., Ziuly, S., et al. (2021). Cardiovascular manifestations of intermediate and major hyperhomocysteinemia due to vitamin B12 and folate deficiency and/or inherited disorders of one-carbon metabolism: A 3.5-year retrospective cross-sectional study of consecutive patients. *The American Journal of Clinical Nutrition*, *113*(5), 1157–1167. https://doi.org/10.1093/ajcn/nqaa432.

Linnell, J. C., Hoffbrand, A. V., Hussein, H. A., Wise, I. J., & Matthews, D. M. (1974). Tissue distribution of coenzyme and other forms of vitamin B12 in control subjects and patients with pernicious anaemia. *Clinical Science and Molecular Medicine*, *46*(2), 163–172.

Marcoullis, G., Parmentier, Y., Nicolas, J. P., Jimenez, M., & Gerard, P. (1980). Cobalamin malabsorption due to nondegradation of R proteins in the human intestine. Inhibited cobalamin absorption in exocrine pancreatic dysfunction. *The Journal of Clinical Investigation*, *66*(3), 430–440. https://doi.org/10.1172/JCI109873.

Mathews, F. S., Gordon, M. M., Chen, Z., Rajashankar, K. R., Ealick, S. E., Alpers, D. H., et al. (2007). Crystal structure of human intrinsic factor: Cobalamin complex at 2.6-a resolution. *Proceedings of the National Academy of Sciences of the United States of America*, *104*(44), 17311–17316. https://doi.org/10.1073/pnas.0703228104.

Miller, J. W. (2018). Proton pump inhibitors, H2-receptor antagonists, metformin, and vitamin B-12 deficiency: Clinical implications. *Advances in Nutrition*, *9*(4), 511S–518S. https://doi.org/10.1093/advances/nmy023.

Moore, S. J., & Warren, M. J. (2012). The anaerobic biosynthesis of vitamin B12. *Biochemical Society Transactions*, *40*(3), 581–586. https://doi.org/10.1042/bst20120066.

Morrow, G., 3rd, Brodie, J. D., Strimpler, A., & Barness, L. A. (1973). Subcellular distribution of methylmalonyl CoA carbonylmutase in human liver extracts. *Biochemical Medicine*, *8*(2), 240–248. https://doi.org/10.1016/0006-2944(73)90028-8.

Namour, F., Dobrovoljski, G., Chery, C., Audonnet, S., Feillet, F., Sperl, W., et al. (2011). Luminal expression of cubilin is impaired in Imerslund-Grasbeck syndrome with compound AMN mutations in intron 3 and exon 7. *Haematologica*, *96*(11), 1715–1719. https://doi.org/10.3324/haematol.2011.043984.

Notis, W. M. (1972). Giardiasis and vitamin B 12 malabsorption. *Gastroenterology*, *63*(6), 1085.

Nyberg, W. (1963). Diphyllobothrium latum and human nutrition, with particular reference to vitamin B12 deficiency. *The Proceedings of the Nutrition Society*, *22*, 8–14. https://doi.org/10.1079/pns19630004.

Obeid, R., Jung, J., Falk, J., Herrmann, W., Geisel, J., Friesenhahn-Ochs, B., et al. (2013). Serum vitamin B12 not reflecting vitamin B12 status in patients with type 2 diabetes. *Biochimie*, *95*(5), 1056–1061. https://doi.org/10.1016/j.biochi.2012.10.028.

Okuda, K., Grasbeck, R., & Chow, B. F. (1958). Bile and vitamin B12 absorption. *The Journal of Laboratory and Clinical Medicine*, *51*(1), 17–23.

Oussalah, A., Levy, J., Berthezène, C., Alpers, D. H., & Guéant, J. L. (2020). Health outcomes associated with vegetarian diets: An umbrella review of systematic reviews and meta-analyses. *Clinical Nutrition*, *39*(11), 3283–3307. https://doi.org/10.1016/j.clnu.2020.02.037.

Pons, L., Guy, M., Lambert, D., Hatier, R., & Gueant, J. (2000). Transcytosis and coenzymatic conversion of [(57)co]cobalamin bound to either endogenous transcobalamin II or exogenous intrinsic factor in caco-2 cells. *Cellular Physiology and Biochemistry*, *10*(3), 135–148 (16344).

Porter, K. M., Hoey, L., Hughes, C. F., Ward, M., Clements, M., Strain, J., et al. (2021). Associations of atrophic gastritis and proton-pump inhibitor drug use with vitamin B-12 status, and the impact of fortified foods, in older adults. *American Journal of Clinical Nutrition*, *114*(4), 1286–1294. https://doi.org/10.1093/ajcn/nqab193.

Quigley, E. M. M., Murray, J. A., & Pimentel, M. (2020). AGA clinical practice update on small intestinal bacterial overgrowth: Expert review. *Gastroenterology*, *159*(4), 1526–1532. https://doi.org/10.1053/j.gastro.2020.06.090.

Ramanujam, K. S., Seetharam, S., Ramasamy, M., & Seetharam, B. (1991). Expression of cobalamin transport proteins and cobalamin transcytosis by colon adenocarcinoma cells. *The American Journal of Physiology*, *260*(3 Pt 1), G416–G422. https://doi.org/10.1152/ajpgi.1991.260.3.G416.

Robertson, J. A., & Gallagher, N. D. (1985). In vivo evidence that cobalamin is absorbed by receptor-mediated endocytosis in the mouse. *Gastroenterology*, *88*(4), 908–912. https://doi.org/10.1016/s0016-5085(85)80007-x.

Rodionov, D. A., Arzamasov, A. A., Khoroshkin, M. S., Iablokov, S. N., Leyn, S. A., Peterson, S. N., et al. (2019). Micronutrient requirements and sharing capabilities of the human gut microbiome. *Frontiers in Microbiology*, *10*, 1316. https://doi.org/10.3389/fmicb.2019.01316.

Schohn, H., Gueant, J. L., Leheup, B., Saunier, M., Grignon, G., & Nicolas, J. P. (1992). Intrinsic factor receptor during fetal development of the human intestine. *The Biochemical Journal*, *286*(Pt 1), 153–156. https://doi.org/10.1042/bj2860153.

Seetharam, B., Alpers, D. H., & Allen, R. H. (1981). Isolation and characterization of the ileal receptor for intrinsic factor-cobalamin. *The Journal of Biological Chemistry*, *256*(8), 3785–3790.

Seetharam, B., Ramanujam, K. S., Seetharam, S., Li, N., & Fyfe, J. C. (1991). Normal and abnormal physiology of intrinsic factor mediated absorption of cobalamin (vitamin B12). *Indian Journal of Biochemistry & Biophysics*, *28*(5–6), 324–330.

Sokolovskaya, O. M., Plessl, T., Bailey, H., Mackinnon, S., Baumgartner, M. R., Yue, W. W., et al. (2021). Naturally occurring cobalamin (B(12)) analogs can function as cofactors for human methylmalonyl-CoA mutase. *Biochimie*, *183*, 35–43. https://doi.org/10.1016/j.biochi.2020.06.014.

Sokolovskaya, O. M., Shelton, A. N., & Taga, M. E. (2020). Sharing vitamins: Cobamides unveil microbial interactions. *Science*, *369*(6499). https://doi.org/10.1126/science.aba0165.

Tabaqchali, S. (1970). The pathophysiological role of small intestinal bacterial flora. *Scandinavian Journal of Gastroenterology. Supplement*, *6*, 139–163.

Tanner, S. M., Aminoff, M., Wright, F. A., Liyanarachchi, S., Kuronen, M., Saarinen, A., et al. (2003). Amnionless, essential for mouse gastrulation, is mutated in recessive hereditary megaloblastic anemia. *Nature Genetics*, *33*(3), 426–429. https://doi.org/10.1038/ng1098.

Tanner, S. M., Li, Z., Perko, J. D., Oner, C., Cetin, M., Altay, C., et al. (2005). Hereditary juvenile cobalamin deficiency caused by mutations in the intrinsic factor gene. *Proceedings of the National Academy of Sciences of the United States of America*, *102*(11), 4130–4133. https://doi.org/10.1073/pnas.0500517102.

Tanner, S. M., Sturm, A. C., Baack, E. C., Liyanarachchi, S., & de la Chapelle, A. (2012). Inherited cobalamin malabsorption. Mutations in three genes reveal functional and ethnic patterns. *Orphanet Journal of Rare Diseases*, *7*, 56. https://doi.org/10.1186/1750-1172-7-56.

Teo, N. H., Scott, J. M., Neale, G., & Weir, D. G. (1980). Effect of bile on vitamin B12 absorption. *British Medical Journal*, *281*(6244), 831–833. https://doi.org/10.1136/bmj.281.6244.831.

Voisine, J., & Abadie, V. (2021). Interplay between gluten, HLA, innate and adaptive immunity orchestrates the development of coeliac disease. *Frontiers in Immunology*, *12*, 674313. https://doi.org/10.3389/fimmu.2021.674313.

Ward, M. G., Kariyawasam, V. C., Mogan, S. B., Patel, K. V., Pantelidou, M., Sobczyńska-Malefora, A., et al. (2015). Prevalence and risk factors for functional vitamin B12 deficiency in patients with Crohn's disease. *Inflammatory Bowel Diseases*, *21*(12), 2839–2847. https://doi.org/10.1097/mib.0000000000000559.

Watanabe, F. (2007). Vitamin B12 sources and bioavailability. *Experimental Biology and Medicine (Maywood, N.J.)*, *232*(10), 1266–1274. https://doi.org/10.3181/0703-MR-67.

Watanabe, F., Abe, K., Fujita, T., Goto, M., Hiemori, M., & Nakano, Y. (1998). Effects of microwave heating on the loss of vitamin B(12) in foods. *Journal of Agricultural and Food Chemistry*, *46*(1), 206–210. https://doi.org/10.1021/jf970670x.

Watanabe, F., & Bito, T. (2018). Vitamin B(12) sources and microbial interaction. *Experimental Biology and Medicine (Maywood, N.J.)*, *243*(2), 148–158.10.1177/1535370217746612.

Watanabe, F., Takenaka, S., Kittaka-Katsura, H., Ebara, S., & Miyamoto, E. (2002). Characterization and bioavailability of vitamin B12-compounds from edible algae. *Journal of Nutritional Science and Vitaminology (Tokyo)*, *48*(5), 325–331. https://doi.org/10.3177/jnsv.48.325.

Watanabe, F., Yabuta, Y., Bito, T., & Teng, F. (2014). Vitamin B_{12}-containing plant food sources for vegetarians. *Nutrients*, *6*(5), 1861–1873. https://doi.org/10.3390/nu6051861.

Webb, G. J., Brooke, R., & De Silva, A. N. (2013). Chronic radiation enteritis and malnutrition. *Journal of Digestive Diseases*, *14*(7), 350–357. https://doi.org/10.1111/1751-2980.12061.

Welkos, S. L., Toskes, P. P., & Baer, H. (1981). Importance of anaerobic bacteria in the cobalamin malabsorption of the experimental rat blind loop syndrome. *Gastroenterology*, *80*(2), 313–320.

Westergaard, H. (2004). Tropical Sprue. *Current Treatment Options in Gastroenterology*, 7(1), 7–11. https://doi.org/10.1007/s11938-004-0020-6.

Wolff, R., & Vuillemin-Weis, J. (1958). Research on the intestinal absorption of vitamin B12 in rats. *Bulletin de la Société de Chimie Biologiquel (Paris)*, *40*(11), 1539–1558.

Yassin, F., Rothenberg, S. P., Rao, S., Gordon, M. M., Alpers, D. H., & Quadros, E. V. (2004). Identification of a 4-base deletion in the gene in inherited intrinsic factor deficiency. *Blood*, *103*(4), 1515–1517. https://doi.org/10.1182/blood-2003-07-2239.

Zhang, Y., Rodionov, D. A., Gelfand, M. S., & Gladyshev, V. N. (2009). Comparative genomic analyses of nickel, cobalt and vitamin B12 utilization. *BMC Genomics*, *10*, 78. https://doi.org/10.1186/1471-2164-10-78.

CHAPTER TEN

Intracellular processing of vitamin B_{12} by MMACHC (CblC)

Luciana Hannibal[a,*] and Donald W. Jacobsen[b]
[a]Laboratory of Clinical Biochemistry and Metabolism, Department of General Pediatrics, Adolescent Medicine and Neonatology, Faculty of Medicine, Medical Center—University of Freiburg, Freiburg, Germany
[b]Department of Cardiovascular and Metabolic Sciences, Lerner Research Institute, Cleveland Clinic, Cleveland, OH, United States
*Corresponding author: e-mail address: luciana.hannibal@uniklinik-freiburg.de

Contents

Abstract

Vitamin B_{12} (cobalamin, Cbl, B_{12}) is a water-soluble micronutrient synthesized exclusively by a group of microorganisms. Human beings are unable to make B_{12} and thus obtain the vitamin *via* intake of animal products, fermented plant-based foods or supplements. Vitamin B_{12} obtained from the diet comprises three major chemical forms, namely hydroxocobalamin (HOCbl), methylcobalamin (MeCbl) and adenosylcobalamin (AdoCbl). The most common form of B_{12} present in supplements is cyanocobalamin (CNCbl). Yet, these chemical forms cannot be utilized directly as they come, but instead, they undergo chemical processing by the MMACHC protein, also known as CblC. Processing of dietary B_{12} by CblC involves removal of the upper-axial ligand (beta-ligand) yielding the one-electron reduced intermediate cob(*II*)alamin. Newly formed cob(*II*)alamin undergoes trafficking and delivery to the two B_{12}-dependent enzymes, cytosolic methionine synthase (MS) and mitochondrial methylmalonyl-CoA mutase (MUT). The catalytic cycles of MS and MUT incorporate cob(*II*)alamin as a precursor to regenerate the coenzyme forms MeCbl and AdoCbl, respectively. Mutations and epimutations in the *MMACHC* gene result in cblC disease, the most common inborn error of B_{12} metabolism, which manifests with

Vitamins and Hormones, Volume 119
ISSN 0083-6729
https://doi.org/10.1016/bs.vh.2022.02.001

combined homocystinuria and methylmalonic aciduria. Elevation of metabolites homocysteine and methylmalonic acid occurs because the lack of an active CblC blocks formation of the indispensable precursor cob(*II*)alamin that is necessary to activate MS and MUT. Thus, in patients with cblC disease, vitamin B_{12} is absorbed and present in circulation in normal to high concentrations, yet, cells are unable to make use of it. Mutations in seemingly unrelated genes that modify MMACHC gene expression also result in clinical phenotypes that resemble cblC disease. We review current knowledge on structural and functional aspects of intracellular processing of vitamin B_{12} by the versatile protein CblC, its partners and possible regulators.

Abbreviations

B_{12}	vitamin B_{12}
Hcy	homocysteine
MMA	methylmalonic acid
Cbl	cobalamin
MS	methionine synthase
MCM	methylmalonyl-CoA mutase
MMACHC	methylmalonic aciduria combined with homocystinuria type C
MMADHC	methylmalonic aciduria combined with homocystinuria type D
TC	transcobalamin
apoTC	apo-transcobalamin
holoTC	holo-transcobalamin
FMN	flavin mononucleotide
FAD	flavin adenine dinucleotide
GSH	reduced glutathione
GSSG	oxidized glutathione
MeCbl	methylcobalamin
AdoCbl	adenosylcobalamin
HOCbl/H_2OCbl	hydroxo/aquacobalamin
CNCbl	cyanocobalamin

1. Overview of cellular metabolism of cobalamins

The pathways for the biosynthesis of cobalamin (Cbl, B_{12}) are exclusive to relatively few members of the Archaea and Bacteria domains (Martens, Barg, Warren, & Jahn, 2002). Comprising complex aerobic and anaerobic reactions, the elucidation of cobalamin synthesis by microorganisms has been matter of intense research (Croft, Lawrence, Raux-Deery, Warren, & Smith, 2005; Martens et al., 2002; Moore et al., 2013; Raux, Lanois, Rambach, Warren, & Thermes, 1998; Roth, Lawrence, & Bobik, 1996; Warren, 2006; Warren, Raux, Schubert, & Escalante-Semerena, 2002). The important discoveries have led experts in the field to propose cobalamin as one of the primordial biomolecules enabling evolution from chemistry to biology (Eschenmoser, 2011). Unlike plants and fungi, most Eukaryota have an absolute requirement for B_{12} metabolism. Humans are no exception and thus rely on dietary intake of the micronutrient. While required in minute quantities of 2.4–4.0 μg daily (Stabler & Allen, 2004; Strohle et al., 2019), insufficiency and deficiency of B_{12} result in hematological and neurological impairments. The first chemical structure of cobalamin was published by Hodgkin et al. (1955). This landmark in the field along the subsequent identification of monogenic disorders of B_{12} metabolism stimulated research to elucidate the mechanisms of transport, cellular uptake and intracellular handling of B_{12}.

1.1 Transport, cellular uptake and intracellular handling of B_{12}

After intake and gastrointestinal absorption (Nielsen, Rasmussen, Andersen, Nexo, & Moestrup, 2012), B_{12} is distributed to every cell in the body bound it transport protein transcobalamin (TC, gene *TCN2*). Autosomal recessive mutations in genes encoding proteins for transport, cellular uptake and intracellular handling of B_{12} are known as Cbl complementation groups (*cblA-cblH, cblX, epiCblC* and *mut*) (Rosenblatt & Whitehead, 1999; Watkins & Rosenblatt, 2011) (Fig. 1). The name "complementation" derives from the method employed to identify the loss of function in cells, which involves fusion of new patient fibroblasts with fibroblasts carrying known defects in B_{12} metabolism (cell fusion with 40% polyethylene glycol to form heterokaryons (Watkins, Matiaszuk, & Rosenblatt, 2000)).

Cobalamin complementation groups are a group of inborn errors of metabolism that result in either isolated or combined homocystinuria and methylmalonic aciduria. This phenotype emerges from functional inactivation of the two B_{12}-dependent enzymes that reside in every human cell, namely, cytosolic methionine synthase (MS, gene *MTR*) and mitochondrial methylmalonyl-CoA mutase (MUT, gene *MUT*). The first step in cellular cobalamin utilization is the import of B_{12}-transcobalamin complex, holoTC, by absorptive endocytosis *via* the transcobalamin receptor (TCR, gene *CD320*) (Fig. 1) (Quadros, Nakayama, & Sequeira, 2009; Quadros & Sequeira, 2013). At this stage, the bound cobalamin could be one of four chemical forms found in dietary sources and supplements, methylcobalamin (MeCbl), adenosylcobalamin (AdoCbl), hydroxocobalamin (HOCbl) or cyanocobalamin (CNCbl), depending on the nature of the β-axial ligand ("X" in Fig. 1) of the micronutrient, methyl-, adenosyl-, hydroxo- and cyano,

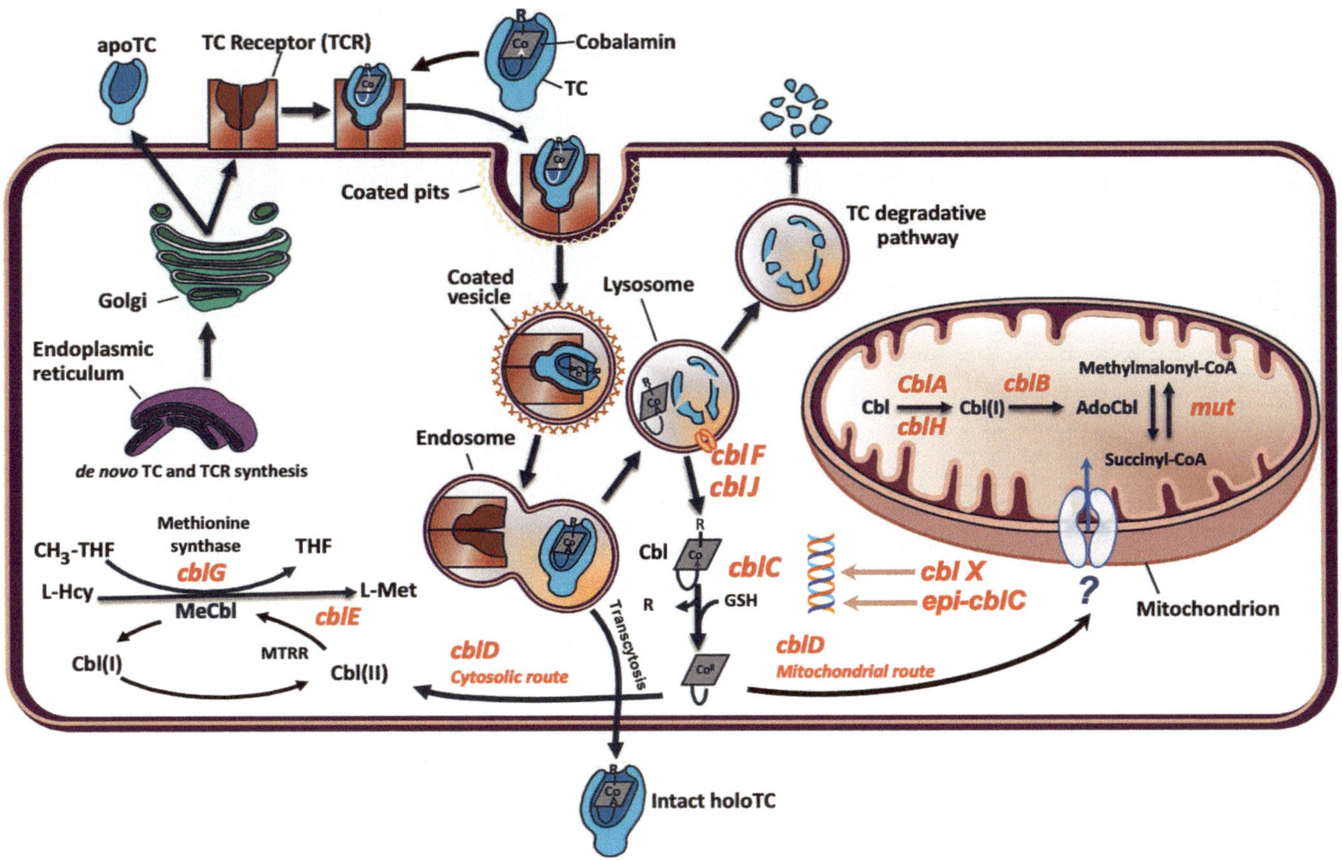

Fig. 1 See figure legend on opposite page.

respectively. After vesicular internalization of holoTC, early endosomes fuse with acidic lysosomes, wherein the B_{12} molecule is freed from TC and exported into the cytosol with the aid of lysosomal transporters cblF (gene *LMBRD1*) and cblJ (gene *ABCD4*) (Fig. 1) (Coelho et al., 2012; Gailus, Hohne, et al., 2010; Gailus, Suormala, et al., 2010; Rutsch, Gailus, Suormala, & Fowler, 2011). The CD320 (TCR) receptor previously thought to recycle back to the cell surface (Amagasaki, Green, & Jacobsen, 1990) has been recently shown to undergo proteolytic degradation in the cytosol (Gick et al., 2020). Transfer of B_{12} from the lysosome into the cytosol appears to be mediated by interactions between the lysosomal transporters CblF/CblJ and the cytosolic processing enzyme and chaperone MMACHC (Deme et al., 2014) (for *methylmalonic aciduria type C with homocystinuria*; hereafter referred to as CblC), which is the product of the *MMACHC* gene. The CblC protein is a unique enzyme catalyzing the removal of the β-axial ligand "X" with reduction of the cobalt center from cob(*III*)alamin to cob(II)- and cob(*I*)alamins (Hannibal, Kim, et al., 2009; Kim, Gherasim, & Banerjee, 2008; Kim, Hannibal, Gherasim, Jacobsen, & Banerjee, 2009). Subsequent delivery of newly processed B_{12} to recipient enzymes MS and MUT is mediated by CblC interactions with adaptor protein MMADHC (Coelho et al., 2008; Suormala et al., 2004) (for *methylmalonic aciduria type D and homocystinuria*; hereafter referred to CblD) (Gherasim, Hannibal, Rajagopalan, Jacobsen, & Banerjee, 2013; Jusufi et al., 2014). Previously thought to not bind to B_{12}, recent studies demonstrated that CblD forms a cob(II)-thiolatocobalamin complex *via* its amino acid residue Cys261 (Li et al., 2020). The redox state of Cys261 is critical for cobalamin binding and important for orchestrating downstream steps in B_{12} partition into cytosolic and mitochondrial pathways

Fig. 1 Cellular metabolism of B_{12} in humans. After absorption in the gastrointestinal tract, cobalamin is taken up by every cell in the body *via* receptor mediated endocytosis of its protein-bound form, holoTC. The TCR receptor takes up holoTC by absorptive endocytosis. Fusion of early endosomes with acidic lysosomes permits the liberation of the B_{12} molecule from its protein binder TC. Free B_{12} carrying naturally occurring β-axial ligands (X = methyl, adenosyl, hydroxo-, cyano-) is shuttled into the cytosol *via* lysosomal transporters cblF and cblJ. After exit from the lysosome, B_{12} is taken up by cblC for chemical processing of the its β-axial ligand X and reduction of its cobalt center. Newly processed B_{12} bound to cblC interacts with adaptor protein cblD, and the cob(*II*)alamin-cblC-cblD complex is partitioned into the cytosolic route toward MS or the mitochondrial route to supply MUT. *Figure adapted from Hannibal, L., Lysne, V., Bjorke-Monsen, A. L., Behringer, S., Grunert, S. C., Spiekerkoetter, U., . . . Blom, H. J. (2016). Biomarkers and algorithms for the diagnosis of vitamin B12 deficiency.* Frontiers in Molecular Biosciences, 3, *27. doi:10.3389/fmolb.2016.00027.*

(Banerjee, Gouda, & Pillay, 2021). It has also been reported that interactions between CblC/CblD and MS modulate B_{12} processing (Bassila et al., 2017; Fofou-Caillierez et al., 2013).

1.2 Defective intracellular processing of B_{12}: Mutations in CblC

A paradoxical fact in human B_{12} metabolism is that dietary MeCbl and AdoCbl must undergo processing of their upper axial ligand prior to their incorporation into B_{12}-dependent methionine synthase (MS) and B_{12}-dependent methylmalonyl-CoA mutase (MCM), wherein they undergo redox cycling between reduced forms and respective cofactor forms MeCbl and AdoCbl that are made *de novo* (Chu, Begley, Colligan, & Hall, 1993).

The first patient with cblC disease was identified by Harvey Mudd et al. over 5 decades ago (Mudd et al., 1969). The case was confirmed by complementation analysis and by combined elevation of homocysteine and methylmalonic acid (Mudd, Levy, Abeles, & Jennedy, 1969). Cultured fibroblasts from the cblC patient exhibited impaired biosynthesis of both MeCbl and AdoCbl, slightly reduced uptake of and export of B_{12} from cells at long incubation times (Mudd et al., 1969). This characterization represents the first line of evidence that the cblC gene was required for the biosynthesis of both cofactor forms of B_{12}, MeCbl and AdoCbl. Studies performed with cultured fibroblast from cblC patients and metabolic labeling with radioactive [^{57}Co]-HOCbl and [^{57}Co]-CNCbl by Mellman et al. provided convincing evidence for a "processing activity" of B_{12} being required prior to the biosynthesis of MeCbl and AdoCbl (Mellman, Willard, Youngdahl-Turner, & Rosenberg, 1979). This study also provided for the first time, a rationale for the better response of cblC patients to HOCbl compared to CNCbl (Mellman et al., 1979).

Attempts to elucidate the type of enzymatic activity associated to the cblC gene followed thereafter, and revealed that cblC fibroblasts exhibited diminished B_{12} β-transferase activity and/or Cbl reductase activity (Pezacka, 1993; Pezacka, Denison, Green, & Jacobsen, 1988; Pezacka, Green, & Jacobsen, 1990; Pezacka & Rosenblatt, 1994). These *in vitro* studies also showed that reduced glutathione (GSH), the most abundant low molecular weight intracellular thiol, was required in the reactions that occurred prior to MeCbl and AdoCbl biosynthesis (Pezacka, 1993; Pezacka et al., 1988; Pezacka et al., 1990; Pezacka & Rosenblatt, 1994). Altogether, these early works proposed CblC as the protein responsible for processing of dietary Cbls, a finding that was fully confirmed by molecular genetics in 2006 (Lerner-Ellis et al., 2006) and by several subsequent studies described in the sections below.

2. Structure and stability of CblC

The identification of the MMACHC gene underlying the cblC disease (Lerner-Ellis et al., 2006) and the first studies on genotype–phenotype associations (Lerner-Ellis et al., 2009), facilitated the expression and characterization of human CblC in recombinant form by four independent research groups (Deme et al., 2012; Froese, Zhang, Healy, & Gravel, 2009; Kim et al., 2008; Plesa et al., 2011; Wingert et al., 2021). Additionally, bovine (Jeong, Ha, & Kim, 2011) and *C. elegans* (Li, Gherasim, Lesniak, & Banerjee, 2014; Park & Kim, 2015) orthologues of human CblC have been expressed and characterized. Human recombinant CblC is a monomer with a predicted molecular weight of 29 kDa and an experimental molecular weight of 31.7 kDa. Evidence also exists that human CblC can form dimers, both in purified form (Froese et al., 2012) and in phage display assays that suggest self-assembly regions in the protein (Deme et al., 2012). The relevance of oligomerization in CblC remains to be investigated. Binding of B_{12} to CblC (human as well as from other species) leads to the base-off configuration of the B_{12} molecule, wherein the dimethylbenzimidazole moiety is detached from the cobalt center (Fig. 2).

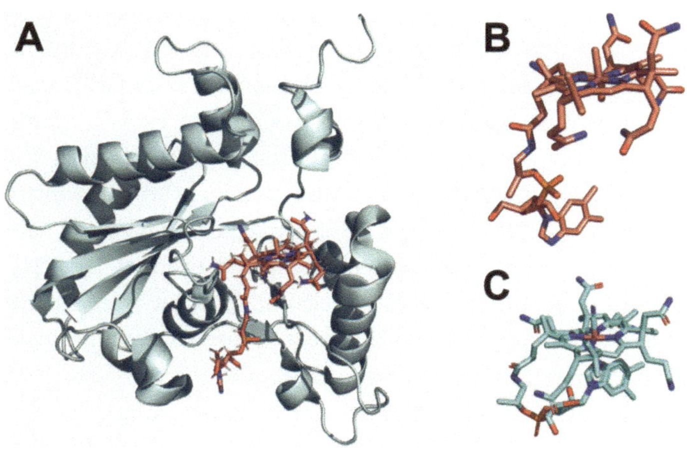

Fig. 2 Structure of human CblC bound to MeCbl. (A) Structure of human CblC with MeCbl bound in its base-off conformation (PDB 3SCO). MeCbl is shown as red sticks. (B) View of the structure of MeCbl base-off from structure shown in A (PDB 3SCO). (C) Structure of CNCbl base-on as seen bound to a fragment of human transcobalamin (PDB 5NP4).

Cobalamins bind to CblC with affinities in the low micromolar range, with reported K_d values between 1 and 15 μM (Froese et al., 2010; Jeong & Kim, 2011; Kim et al., 2008; Park & Kim, 2015; Plesa et al., 2011). One study showed that binding of GSH to bovine CblC increases its affinity for B_{12} (Jeong & Kim, 2011), and protects the bound cofactor from redox cycling with GSH to form GSSG (oxidized glutathione) (Jeong et al., 2011).

2.1 Crystal structure of human CblC and its orthologues

High-resolution X-ray crystal structures of human CblC have been published by Koutmos, Gherasim, Smith, and Banerjee (2011) and Froese et al. (2012). CblC possesses an N-terminal flavodoxin nitroreductase domain that can employ FMN or FAD for the reductive decyanation of CNCbl (Koutmos et al., 2011). The B_{12} molecule is accommodated within a large cavity in the CblC protein (Fig. 2, panel A) that enables binding in the base-off configuration (Fig. 2, panels A-B). In contrast to transcobalamin that binds B_{12} in the base-on configuration (Fig. 2, panel C, base-on CNCbl as seen bound to a human TC fragment (Bloch, Ruetz, Krautler, & Locher, 2017)) and MS and MUT that bind B_{12} in the base-off/His-on mode (Hannibal & Jacobsen, 2017), the coordination position of the lower-axial ligand DMBI remains unoccupied in CblC-bound B_{12} (Koutmos et al., 2011). Binding of MeCbl to CblC induces measurable conformational changes in three different loop-structured domains around the B_{12} cavity compared to apo-CblC. The structure of CblC bound to AdoCbl has also been elucidated (Froese et al., 2012). The overall fold of CblC is conserved in comparison to that of MeCbl bound to CblC. A novel finding was the presence of a highly conserved dimerization cap for the β-axial 5′-adenosyl moiety, and an arginine-rich pocket that could anchor co-substrate GSH above the β-axial ligand of the B_{12} molecule (Froese et al., 2012). Importantly, the arginine-rich pocket comprises residues Arg161, Arg206 and Arg230 (Fig. 3, panel A, cyan sticks) all of which are sites for point mutations that occur in humans, leading to *cblC* disease. Recombinant expression of variant Arg206Gln was insoluble, and variants Arg161Gln and Arg161Gly had a much reduced dealkylation activity compared to wild type human CblC (Froese et al., 2012). The study also revealed that FMN, and to a significantly lesser extent Cbl, induce dimerization of human recombinant CblC (Froese et al., 2012). Independent studies elucidated the mechanism of reduced activity of pathogenic variants Arg161Gln and Arg161Gly (Gherasim, Ruetz, Li, Hudolin, & Banerjee, 2015) as well as the possible repair of activity loss with thiolatocobalamins (Wingert et al., 2021).

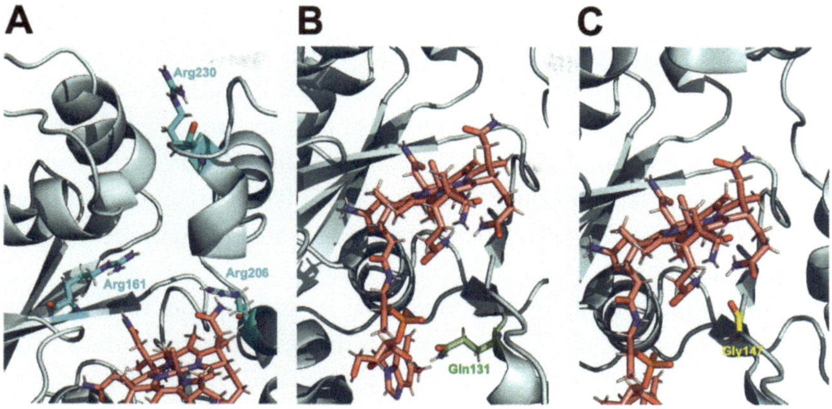

Fig. 3 Amino acid residues in CblC with essential roles in GSH and B_{12} binding known to be sites for pathogenic mutations in humans. (A) The Arg-triad required for GSH binding in human CblC (PDB 3SC0) comprises residues Arg161, Arg206 and Arg230 (shown as cyan sticks). Mutations Arg161Gly and Arg161Gln are the cause of early and late onset cblC disease in humans, respectively. (B) Amino acid residue Gln131 (green sticks) establishes direct H-bonding contacts with the B_{12} molecule. Deletion of Gln131 is pathogenic in humans with nearly zero residual enzymatic activity. (C) Amino acid residue Gly147 (yellow sticks) is required for B_{12} binding. Mutation Gly147Asp leads to early onset cblC disease that is irresponsive to B_{12} therapy. MeCbl is shown as red sticks.

An interesting finding is that amino acid residues Arg161 and Arg206 are required for the interaction of CblC with MS (Fofou-Caillierez et al., 2013). Results from molecular modeling and docking studies support the existence of a direct interaction between Arg161/Arg206 of CblC with a loop of MS (Fofou-Caillierez et al., 2013). This is supported by substantial solvent exposure of these two residues in CblC (Fig. 3, panel A), which could enable protein–protein interactions.

Crystallization of human CblC with enzymatically inactive B_{12} forms, the so called "anti-vitamin B_{12}" shed light on additional structural details that may be important for function (Ruetz et al., 2017). Molecular details of the B_{12} binding site identified Tyr215, Asp77 and Asp80 as additional residues for H-bonding with GSH. The study revealed that the S-atom of GSH is too far away from the β-axial ligand of B_{12}, which in principle is not consistent with the SN_2 mechanism proposed for GSH-driven dealkylation of Co—C bonded B_{12}. The authors proposed that the different β-axial ligands of B_{12} could modify Co-β-ligand-S proximity and electronics required for catalysis (Ruetz et al., 2017).

Besides impairments in binding to co-substrate GSH, pathogenic variants that disrupt B_{12} binding have also been described. For example, deletion of Gln131 was shown to disrupt cobalamin binding, specifically, the base-on to base-off transition (Backe et al., 2013). Deletion of residue Gln131 left the protein with nearly zero residual enzymatic activity (Backe et al., 2013). Residue Gln131 establishes direct hydrogen bond contacts with the B_{12} molecule (Fig. 3B). Another study identified mutation Gly147Asp to completely abolish binding of CblC to CNCbl and HOCbl, thus rendering the protein completely inactive (Froese et al., 2009). This explains the early onset cblC disease and the lack of response to B_{12} treatment in these patients (Froese et al., 2009).

Examination of CblC protein expression in other species revealed that a truncated variant of CblC lacking the C-terminus is the predominant form present in murine tissues (Koutmos et al., 2011). This is in contrast to the case of human CblC, wherein studies in HEK293 cells and fibroblasts demonstrated the presence of a full-length protein (Deme et al., 2012). The function of the C-terminus of human CblC comprising residues 236–282, if any, is unknown. The crystal structure revealed that the C-terminus of human CblC is highly disordered (Koutmos et al., 2011), which could serve as a recognition motif for cellular chaperones that recognize partially folded motifs, such as heat-shock proteins, or to support protein–protein interactions that require structural flexibility. In conclusion, the structural requirements that enable CblC function appear to vary across species.

2.2 Thermolability of CblC

The human CblC protein is labile to high temperature. Studies by Froese et al. showed that full-length human CblC has a melting temperature of 39°C, and that the most frequent pathogenic mutations decrease thermal stability further and also impact binding to B_{12} (Froese et al., 2010). Among the B_{12} forms tested *in vitro*, AdoCbl and MeCbl were the species that afforded maximal stability, as measured by augmented polydispersity and hydrodynamic radii (Froese et al., 2010). It is possible that the seemingly modest conformational changes that occur upon binding of B_{12} to CblC are important to increase its stability, as has been seen in bacterial B_{12}-binding protein such as apo-BtuF vs cyanocobalamin (CNCbl)-BtuF (James, Hancock, Gagnon, & Coulton, 2009). Another stabilizing element in CblC is binding to GSH (Jeong et al., 2011; Jeong & Kim, 2011; Park, Jeong, & Kim, 2012). *In vitro* characterization of bovine CblC showed that GSH stabilizes CblC, hence, intracellular availability of GSH and more generally, redox control, could play a role in the regulation of the protein's lifetime.

3. Enzymatic activity of CblC

Studies by Chu et al. indicated that dietary B_{12}, regardless of the nature of their β-axial-ligands, must undergo intracellular processing to produce MeCbl and AdoCbl *de novo* (Chu et al., 1993). This is consistent with clinical studies showing lack of phenotype correction in cblC (Ribes et al., 1990; Wang, Yan, Wen, & Zhao, 2019) and cblE (Fowler, Schutgens, Rosenblatt, Smit, & Lindemans, 1997) patients treated with MeCbl, and cblB patients treated with AdoCbl (Batshaw, Thomas, Cohen, Matalon, & Mahoney, 1984; Bhatt, Linnell, & Barltrop, 1986; Chalmers, Bain, Mistry, Tracey, & Weaver, 1991). Guidelines continue to recommend HOCbl as the most suitable chemical form of B_{12} to treat patients with cblC disease and other genetic disorders of the B_{12} pathway (Huemer et al., 2005, 2017). The following section describes catalysis by CblC, illustrating the versatility of the enzyme toward a variety of natural and synthetic of β-axial ligands.

3.1 Catalysis by CblC

The CblC protein catalyzes the decyanation of CNCbl in the presence of a flavoprotein reductase and NADPH (Kim et al., 2008) or with GSH (Li, Gherasim, et al., 2014). Decyanation of CNCbl was demonstrated in a multi-protein system *in vitro* with electron transfer from recombinant MS reductase (MSR) and novel reductase 1 (NR1) as well as with GSH under anaerobic conditions (Kim et al., 2008; Li, Gherasim, et al., 2014). Initially thought to not process Co—C bonded MeCbl and AdoCbl (Kim et al., 2008), the two major dietary forms, it was later discovered that the *cblC* gene is indeed indispensable for the dealkylation of natural (MeCbl and AdoCbl) and synthetic (MeCbl analogues with increased length of the hydrocarbon chain in the β-axial ligand position) alkylcobalamins (Hannibal, Kim, et al., 2009). Findings in cultured fibroblasts isolated from cblC patients were later confirmed by *in vitro* studies carried out with human recombinant CblC (Kim et al., 2009). These studies demonstrated the following: (i) CblC catalyzes dealkylation of MeCbl and AdoCbl in the presence of GSH, with demethylation of MeCbl being orders of magnitude faster than deadenosylation of AdoCbl; (ii) CblC could dealkylate a series of synthetic MeCbl analogues (Hannibal, Kim, et al., 2009; Hannibal, Smith, et al., 2009) with increased hydrocarbon chain (ethylcobalamin to hexylcobalamin) albeit, the reaction was progressively slower with increased β-axial ligand chain length; (iii) the reaction products were the S-alkylated variant of GSH, i.e., *S*-methylglutathione, S-ethylglutathione, S-adenosylglutathione, and so forth. In conclusion, CblC presents features reminiscent of the GSH S-transferase family (removal of alkyl groups) as well as of MS (transfer of methyl groups to a sulfur acceptor; i.e., as seen in the methylation of homocysteine to form methionine). Mechanistic studies revealed that dealkylation of alkylcobalamins forms the super-nucleophile cob(*I*)alamin as an intermediate, which undergoes one-electron oxidation to form the more stable cob(*II*)alamin (Kim et al., 2009). Studies with the CblC orthologue present in *C. elegans* showed an unusual stabilization of cob(*I*)alamin under aerobic conditions (Li, Lesniak, & Banerjee, 2014), which is thought to be unique to the structure of CblC in this organism.

Besides decyanation of CNCbl and dealkylation of MeCbl, AdoCbl and synthetic alkylcobalamins, CblC has been remarkable versatile toward other forms of B_{12} with β-axial ligands that include Co—S (thiolato-cobalamins) (Pezacka et al., 1990) and Co—N bonding (nitrocobalamin) (Mascarenhas, Li, Gherasim, Ruetz, & Banerjee, 2020).

Thiolatocobalamin dethiolation was first demonstrated in early studies with cell extracts by Pezacka et al. (1990) and in more recent work with bovine recombinant CblC (Jeong, Park, Park, & Kim, 2014) and human recombinant CblC (Wingert et al., 2021). *In vitro* studies with the recombinant CblC protein showed that dethiolation is orders of magnitude faster than the dealkylation of alkylcobalamins, and requires lower concentration of co-substrate GSH (Jeong et al., 2014; Wingert et al., 2021). These findings have generated renewed interest in the use of thiolatocobalamins as possible drug candidates in diseases involving a defective cobalamin metabolism (Wingert et al., 2021). An unexpected finding was the occurrence of spontaneous dethiolation of synthetic thiolatocobalamins cysteaminylcobalamin and mercaptopropionylglycylcobalamin upon binding to human CblC and in the absence of GSH (Wingert et al., 2021). This finding may prove promising for rescuing residual activity of pathogenic CblC variants characterized by impaired GSH binding. A most exciting finding is the identification of a protein-bound thiolatocobalamin as part of the CblC-CblD protein–protein interaction driving partition of the processed B_{12} into cytosolic and mitochondrial routes (Li et al., 2020). The study performed with human recombinant CblC and CblD showed that adaptor protein CblD, with its cysteine residues in the reduced form, could bind to cob(*II*)alamin to form a transient interprotein thiolatocob(*II*)alamin

(Li et al., 2020). Amino acid residue Cys261 in the CblD protein serves as the ligand for protein-bound thiolatocobalamin formation. This suggests that redox modulation of thiol groups in proteins of the B_{12} pathway may be important for controlling trafficking of B_{12} in cells.

Denitration of nitrocobalamin (NO_2Cbl) was also demonstrated as a favorable reaction catalyzed by human CblC (Mascarenhas et al., 2020). The reaction catalyzed by CblC in the presence of GSH converts NO_2Cbl to cob(*II*)alamin, which under aerobic conditions oxidizes to form aquacobalamin (Mascarenhas et al., 2020). In the same study it was demonstrated that cob(*I*)alamin bound to CblC catalyzes the reduction of nitrite (NO_2^-) to nitric oxide ($^{\bullet}NO$) with formation of cob(*II*)alamin. Rapid reaction of cob(*II*)alamin with $^{\bullet}NO$ follows to form nitrosylcobalamin, a well-characterized, air-sensitive B_{12} form (Hannibal, Smith, Jacobsen, & Brasch, 2007; Hassanin, Hannibal, Jacobsen, Brown, et al., 2009; Hassanin, Hannibal, Jacobsen, El-Shahat, et al., 2009; Selcuki, van Eldik, & Clark, 2004; Wolak, Stochel, Hamza, & van Eldik, 2000; Zheng & Birke, 2001).

In sum, the CblC protein exhibits remarkable versatility in terms of substrate use and reaction mechanisms. A summary of reactions catalyzed by CblC and its interactions with adaptor protein CblD is provided in Fig. 4.

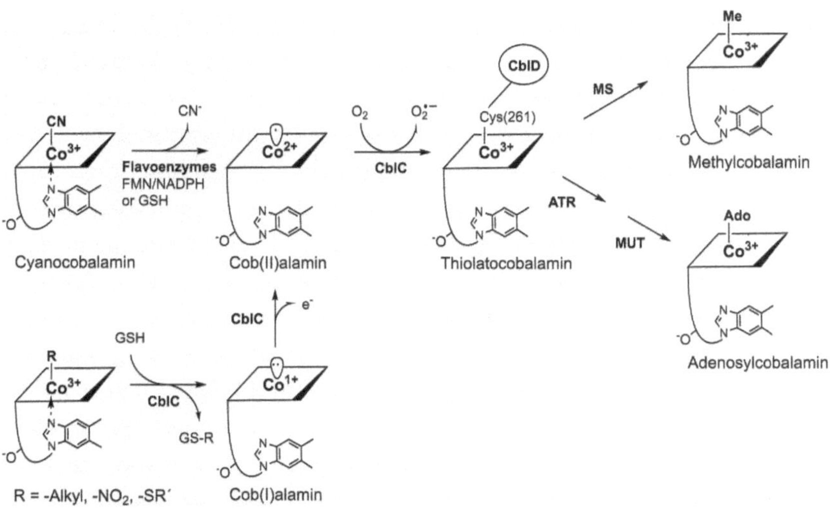

Fig. 4 Reactions catalyzed by CblC. A variety of B_{12} forms including CNCbl, alkylcobalamins (Co—S), thiolatocobalamins (Co—S) and NO_2Cbl undergo β-axial ligand removal and reduction of their cobalt center to form cob(*II*)alamin. Cob(*II*)alamin bound to CblC is recognized by CblD and forms an inter-protein thiolatocobalamin intermediate that involves coordination in the β-ligand position of Cys261 of the CblD protein.

3.2 Production of reactive species by pathogenic variants of CblC

While the CblC protein exhibits remarkable versatility with respect of its B_{12} substrates, certain pathogenic mutations can uncouple electron transfer from product formation and instead, lead to deleterious redox cycling that generates superoxide and hydrogen peroxide, contributing to oxidative stress (Gherasim et al., 2015). It has been noted that patients with *cblC* disease have an abnormal GSH/GSSG pool (Pastore et al., 2014) and increased oxidative stress (Jorge-Finnigan, Gamez, Perez, Ugarte, & Richard, 2010; Mc Guire, Parikh, & Diaz, 2009; Richard et al., 2009). A recent study showed that certain thiolatocobalamins support CblC activity in these uncoupled pathogenic variants without the abnormal production of reactive oxidants. Thus, besides certain amino acids in the protein structure, it appears that the nature of the β-axial ligand of the B_{12} molecule can influence the degree of productive electron transfer for catalysis by CblC.

4. Regulation of CblC expression and clinical phenotypes mimicking cblC disease

Advances in molecular genetics have facilitated the fast and accurate diagnosis of patients with inborn errors of metabolism (Pupavac, Tian, et al., 2016). Routine genetic screening of patients with combined homocystinuria and methylmalonic aciduria has uncovered additional cases of cblC disease, and unexpectedly, mutations in genes seemingly unrelated to *MMACHC* but presenting with similar clinical manifestations. These cblC disease-mimics include mutations in *PRDX1* (epimutation, also known as epi-CblC), *HCFC1*, *THAP11* and *ZNF143* (Cavicchi et al., 2021; Gueant et al., 2018; Pupavac, Watkins, et al., 2016; Quintana et al., 2017; Yu et al., 2013; Zhang et al., 2021). Studies by Yu et al. revealed that the expression of CblC depends on the activity of transcription factor THAP11 that forms a complex with co-regulator HCFC1 (Yu et al., 2013). Mutations in *HCFC1*, designated as cblX, results in a phenotype reminiscent of cblC disease while in the presence of an intact MMACHC gene (Yu et al., 2013). A patient with an unclear presentation of combined homocystinuria and methylmalonic aciduria was found to carry a mutation in *THAP11* and was designated cblX-like (Quintana et al., 2017). The presence of mutations in the highly conserved Kelch domain of HCFC1 or within the DNA-binding THAP domain of *THAP11* downregulate the expression of *MMACHC* (Quintana et al., 2017; Yu et al., 2013). A recent study showed that mutations located outside the Kelch domain of *HCFC1* also lead to cblX-like disease, but with dissociation of the biochemical and clinical phenotypes; patients exhibited elevated serum vitamin B_{12} without abnormalities in Hcy and MMA, yet presented with developmental delay, intellectual disability and dysmorphic features, reminiscent of moderate to severe presentations of the cblC disease (Wongkittichote, Wegner, & Shinawi, 2021). Mutations in transcriptional activator ZNF143, another interactor of the HCFC1-THAP11 dyad, also yields a cblC disease-like phenotype (Pupavac, Watkins, et al., 2016).

A most interesting discovery was the identification of the so-called CblC epimutations, or epi-CblC first described by Gueant et al. (2018). This cblC disease-like phenotype is brought about by mutations in the *PRDX1* gene, which causes antisense transcription of *MMACHC* and hypermethylation of the promoter region with silencing of *MMACHC* expression (Gueant et al., 2018). According to two subsequent reports, epi-CblC mutations are a rather common cause of combined homocystinuria and methylmalonic aciduria (Cavicchi et al., 2021; Zhang et al., 2021).

5. Future directions

Our mechanistic understanding of intracellular B_{12} processing by CblC has deepened steadily since the identification of its encoding gene in 2006 (Lerner-Ellis et al., 2006). However, significantly less is known about the consequences of a mutated or absent CblC protein in other pathways of metabolism and how this impacts the heterogeneous clinical manifestations of cblC disease. Proteomic studies performed in fibroblasts (Hannibal, DiBello, & Jacobsen, 2013; Hannibal et al., 2011) and lymphocytes (Caterino et al., 2015) isolated from *cblC* patients revealed abnormal expression of proteins in pathways outside the inner sphere of B_{12} metabolism, such as cellular detoxification, glutathione and energy metabolism, protein folding and cytoskeleton organization and assembly (Caterino et al., 2015; Hannibal et al., 2013, 2011). Notably, several of the proteins identified as differentially expressed in cblC mutant cells are pathological hallmarks of muscular dystrophy, Alzheimer's and Parkinson's diseases. Symptoms commonly seen in these disorders are also characteristic of the cblC clinical phenotype, such as cognitive decline, muscular dystonia and atrophy (Carrillo-Carrasco, Chandler, & Venditti, 2012). There is thus much to investigate concerning possible additional roles of CblC beyond that of its catalytic B_{12} processing activity, especially in the most affected organs in patients, such as the brain, eyes, kidney and skeletal muscle. The role of CblC in the cytosolic branch of B_{12} handling comprises interactions with adaptor protein CblD (Gherasim et al., 2013; Jusufi et al., 2014; Li et al., 2020; Mah et al., 2013; Stucki et al., 2012; Yamada, Gherasim, Banerjee, & Koutmos, 2015) and with MS (Fofou-Caillierez et al., 2013). The function of CblC in the mitochondrial route is comparably less well understood, including the long-standing mystery of how is B_{12} imported into the organelle.

Besides inborn errors of metabolism, other studies have identified that abnormal expression of the MMACHC gene is implicated in dysregulation of methionine metabolism and cancer (Gotoh et al., 2014; Gueant et al., 2020; Loewy et al., 2009; Sorin, Watkins, Gilfix, & Rosenblatt, 2021). Further evidence of MMACHC being involved in chronic illnesses emerged from GWAS performed in diverse populations, which revealed that polymorphic variants of *MMACHC* are associated with vascular and cardiovascular outcomes (Finsterer, 2009; Raffield et al., 2018; van Meurs et al., 2013). Based on these findings, it is possible to speculate that the CblC protein may have moonlighting roles in the life of cells that are yet to be discovered. Multi-omic approaches along with the investigation of CblC expression and function in organ-relevant cell types, obtained for example, *via* CRISPR-Cas editing of the *MMACHC* gene, will bring us closer to a comprehension of the non-enzymatic functions of CblC.

References

Amagasaki, T., Green, R., & Jacobsen, D. W. (1990). Expression of transcobalamin II receptors by human leukemia K562 and HL-60 cells. *Blood*, *76*(7), 1380–1386. Retrieved from http://www.ncbi.nlm.nih.gov/pubmed/2169922.

Backe, P. H., Ytre-Arne, M., Rohr, A. K., Brodtkorb, E., Fowler, B., Rootwelt, H., et al. (2013). Novel deletion mutation identified in a patient with late-onset combined Methylmalonic Acidemia and Homocystinuria, cblC type. *JIMD Reports*, *11*, 79–85. https://doi.org/10.1007/8904_2013_225.

Banerjee, R., Gouda, H., & Pillay, S. (2021). Redox-linked coordination chemistry directs vitamin B12 trafficking. *Accounts of Chemical Research*, *54*(8), 2003–2013. https://doi.org/10.1021/acs.accounts.1c00083.

Bassila, C., Ghemrawi, R., Flayac, J., Froese, D. S., Baumgartner, M. R., Gueant, J. L., et al. (2017). Methionine synthase and methionine synthase reductase interact with MMACHC and with MMADHC. *Biochimica et Biophysica Acta - Molecular Basis of Disease*, *1863*(1), 103–112. https://doi.org/10.1016/j.bbadis.2016.10.016.

Batshaw, M. L., Thomas, G. H., Cohen, S. R., Matalon, R., & Mahoney, M. J. (1984). Treatment of the cbl B form of methylmalonic acidaemia with adenosylcobalamin. *Journal of Inherited Metabolic Disease*, 7(2), 65–68. https://doi.org/10.1007/BF01805805.

Bhatt, H. R., Linnell, J. C., & Barltrop, D. (1986). Treatment of hydroxocobalamin-resistant methylmalonic acidaemia with adenosylcobalamin. *Lancet*, *2*(8504), 465. https://doi.org/10.1016/s0140-6736(86)92176-8.

Bloch, J. S., Ruetz, M., Krautler, B., & Locher, K. P. (2017). Structure of the human transcobalamin beta domain in four distinct states. *PLoS One, 12*(9), e0184932. https://doi.org/10.1371/journal.pone.0184932.

Carrillo-Carrasco, N., Chandler, R. J., & Venditti, C. P. (2012). Combined methylmalonic acidemia and homocystinuria, cblC type. I. Clinical presentations, diagnosis and management. *Journal of Inherited Metabolic Disease, 35*(1), 91–102. https://doi.org/10.1007/s10545-011-9364-y.

Caterino, M., Pastore, A., Strozziero, M. G., Di Giovamberardino, G., Imperlini, E., Scolamiero, E., et al. (2015). The proteome of cblC defect: In vivo elucidation of altered cellular pathways in humans. *Journal of Inherited Metabolic Disease, 38*(5), 969–979. https://doi.org/10.1007/s10545-014-9806-4.

Cavicchi, C., Oussalah, A., Falliano, S., Ferri, L., Gozzini, A., Gasperini, S., et al. (2021). PRDX1 gene-related epi-cblC disease is a common type of inborn error of cobalamin metabolism with mono- or bi-allelic MMACHC epimutations. *Clinical Epigenetics, 13*(1), 137. https://doi.org/10.1186/s13148-021-01117-2.

Chalmers, R. A., Bain, M. D., Mistry, J., Tracey, B. M., & Weaver, C. (1991). Enzymologic studies on patients with methylmalonic aciduria: Basis for a clinical trial of deoxyadenosylcobalamin in a hydroxocobalamin-unresponsive patient. *Pediatric Research, 30*(6), 560–563. https://doi.org/10.1203/00006450-199112000-00014.

Chu, R. C., Begley, J. A., Colligan, P. D., & Hall, C. A. (1993). The methylcobalamin metabolism of cultured human fibroblasts. *Metabolism, 42*(3), 315–319. Retrieved from http://www.ncbi.nlm.nih.gov/pubmed/8487649.

Coelho, D., Kim, J. C., Miousse, I. R., Fung, S., du Moulin, M., Buers, I., et al. (2012). Mutations in ABCD4 cause a new inborn error of vitamin B12 metabolism. *Nature Genetics, 44*(10), 1152–1155. https://doi.org/10.1038/ng.2386.

Coelho, D., Suormala, T., Stucki, M., Lerner-Ellis, J. P., Rosenblatt, D. S., Newbold, R. F., et al. (2008). Gene identification for the cblD defect of vitamin B12 metabolism. *The New England Journal of Medicine, 358*(14), 1454–1464. https://doi.org/10.1056/NEJMoa072200.

Croft, M. T., Lawrence, A. D., Raux-Deery, E., Warren, M. J., & Smith, A. G. (2005). Algae acquire vitamin B12 through a symbiotic relationship with bacteria. *Nature, 438*(7064), 90–93. https://doi.org/10.1038/nature04056.

Deme, J. C., Hancock, M. A., Xia, X., Shintre, C. A., Plesa, M., Kim, J. C., et al. (2014). Purification and interaction analyses of two human lysosomal vitamin B12 transporters: LMBD1 and ABCD4. *Molecular Membrane Biology, 31*(7–8), 250–261. https://doi.org/10.3109/09687688.2014.990998.

Deme, J. C., Miousse, I. R., Plesa, M., Kim, J. C., Hancock, M. A., Mah, W., et al. (2012). Structural features of recombinant MMADHC isoforms and their interactions with MMACHC, proteins of mammalian vitamin B12 metabolism. *Molecular Genetics and Metabolism, 107*(3), 352–362. https://doi.org/10.1016/j.ymgme.2012.07.001.

Eschenmoser, A. (2011). Etiology of potentially primordial biomolecular structures: From vitamin B12 to the nucleic acids and an inquiry into the chemistry of life's origin: A retrospective. *Angewandte Chemie (International Ed. in English), 50*(52), 12412–12472. https://doi.org/10.1002/anie.201103672.

Finsterer, J. (2009). Cardiogenetics, neurogenetics, and pathogenetics of left ventricular hypertrabeculation/noncompaction. *Pediatric Cardiology, 30*(5), 659–681. https://doi.org/10.1007/s00246-008-9359-0.

Fofou-Caillierez, M. B., Mrabet, N. T., Chery, C., Dreumont, N., Flayac, J., Pupavac, M., et al. (2013). Interaction between methionine synthase isoforms and MMACHC: Characterization in cblG-variant, cblG and cblC inherited causes of megaloblastic anaemia. *Human Molecular Genetics, 22*(22), 4591–4601. https://doi.org/10.1093/hmg/ddt308.

Fowler, B., Schutgens, R. B., Rosenblatt, D. S., Smit, G. P., & Lindemans, J. (1997). Folate-responsive homocystinuria and megaloblastic anaemia in a female patient with functional methionine synthase deficiency (cblE disease). *Journal of Inherited Metabolic Disease*, *20*(6), 731–741. https://doi.org/10.1023/a:1005372730310.

Froese, D. S., Healy, S., McDonald, M., Kochan, G., Oppermann, U., Niesen, F. H., et al. (2010). Thermolability of mutant MMACHC protein in the vitamin B12-responsive cblC disorder. *Molecular Genetics and Metabolism*, *100*(1), 29–36. https://doi.org/10.1016/j.ymgme.2010.02.005.

Froese, D. S., Krojer, T., Wu, X., Shrestha, R., Kiyani, W., von Delft, F., et al. (2012). Structure of MMACHC reveals an arginine-rich pocket and a domain-swapped dimer for its B12 processing function. *Biochemistry*, *51*(25), 5083–5090. https://doi.org/10.1021/bi300150y.

Froese, D. S., Zhang, J., Healy, S., & Gravel, R. A. (2009). Mechanism of vitamin B12-responsiveness in cblC methylmalonic aciduria with homocystinuria. *Molecular Genetics and Metabolism*, *98*(4), 338–343. https://doi.org/10.1016/j.ymgme.2009.07.014.

Gailus, S., Hohne, W., Gasnier, B., Nurnberg, P., Fowler, B., & Rutsch, F. (2010). Insights into lysosomal cobalamin trafficking: Lessons learned from cblF disease. *Journal of Molecular Medicine (Berlin, Germany)*, *88*(5), 459–466. https://doi.org/10.1007/s00109-010-0601-x.

Gailus, S., Suormala, T., Malerczyk-Aktas, A. G., Toliat, M. R., Wittkampf, T., Stucki, M., et al. (2010). A novel mutation in LMBRD1 causes the cblF defect of vitamin B(12) metabolism in a Turkish patient. *Journal of Inherited Metabolic Disease*, *33*(1), 17–24. https://doi.org/10.1007/s10545-009-9032-7.

Gherasim, C., Hannibal, L., Rajagopalan, D., Jacobsen, D. W., & Banerjee, R. (2013). The C-terminal domain of CblD interacts with CblC and influences intracellular cobalamin partitioning. *Biochimie*, *95*(5), 1023–1032. https://doi.org/10.1016/j.biochi.2013.02.003.

Gherasim, C., Ruetz, M., Li, Z., Hudolin, S., & Banerjee, R. (2015). Pathogenic mutations differentially affect the catalytic activities of the human B12-processing chaperone CblC and increase futile redox cycling. *The Journal of Biological Chemistry*, *290*(18), 11393–11402. https://doi.org/10.1074/jbc.M115.637132.

Gick, G. G., Arora, K., Sequeira, J. M., Nakayama, Y., Lai, S. C., & Quadros, E. V. (2020). Cellular uptake of vitamin B12: Role and fate of TCblR/CD320, the transcobalamin receptor. *Experimental Cell Research*, *396*(1), 112256. https://doi.org/10.1016/j.yexcr.2020.112256.

Gotoh, M., Ichikawa, H., Arai, E., Chiku, S., Sakamoto, H., Fujimoto, H., et al. (2014). Comprehensive exploration of novel chimeric transcripts in clear cell renal cell carcinomas using whole transcriptome analysis. *Genes, Chromosomes & Cancer*, *53*(12), 1018–1032. https://doi.org/10.1002/gcc.22211.

Gueant, J. L., Chery, C., Oussalah, A., Nadaf, J., Coelho, D., Josse, T., et al. (2018). APRDX1 mutant allele causes a MMACHC secondary epimutation in cblC patients. *Nature Communications*, *9*(1), 67. https://doi.org/10.1038/s41467-017-02306-5.

Gueant, J. L., Oussalah, A., Zgheib, R., Siblini, Y., Hsu, S. B., & Namour, F. (2020). Genetic, epigenetic and genomic mechanisms of methionine dependency of cancer and tumor-initiating cells: What could we learn from folate and methionine cycles. *Biochimie*, *173*, 123–128. https://doi.org/10.1016/j.biochi.2020.03.015.

Hannibal, L., DiBello, P. M., & Jacobsen, D. W. (2013). Proteomics of vitamin B12 processing. *Clinical Chemistry and Laboratory Medicine*, *51*(3), 477–488. https://doi.org/10.1515/cclm-2012-0568.

Hannibal, L., DiBello, P. M., Yu, M., Miller, A., Wang, S., Willard, B., et al. (2011). The MMACHC proteome: Hallmarks of functional cobalamin deficiency in humans. *Molecular Genetics and Metabolism*, *103*(3), 226–239. https://doi.org/10.1016/j.ymgme.2011.03.008.

Hannibal, L., & Jacobsen, D. W. (2017). Intracellular processing and utilization of cobalamins. In R. Obeid (Ed.), *Vitamin B12: Advances and insights* (p. 366). Boca Raton: CRC Press/Taylor & Francis Group.

Hannibal, L., Kim, J., Brasch, N. E., Wang, S., Rosenblatt, D. S., Banerjee, R., et al. (2009). Processing of alkylcobalamins in mammalian cells: A role for the MMACHC (cblC) gene product. *Molecular Genetics and Metabolism, 97*(4), 260–266. https://doi.org/10.1016/j.ymgme.2009.04.005.

Hannibal, L., Smith, C. A., Jacobsen, D. W., & Brasch, N. E. (2007). Nitroxylcob(III)alamin: Synthesis and X-ray structural characterization. *Angewandte Chemie (International Ed. in English), 46*(27), 5140–5143. https://doi.org/10.1002/anie.200701131.

Hannibal, L., Smith, C. A., Smith, J. A., Axhemi, A., Miller, A., Wang, S., et al. (2009). High resolution crystal structure of the methylcobalamin analogues ethylcobalamin and butylcobalamin by X-ray synchrotron diffraction. *Inorganic Chemistry, 48*(14), 6615–6622. https://doi.org/10.1021/ic900590p.

Hassanin, H. A., Hannibal, L., Jacobsen, D. W., Brown, K. L., Marques, H. M., & Brasch, N. E. (2009). NMR spectroscopy and molecular modelling studies of nitrosylcobalamin: Further evidence that the deprotonated, base-off form is important for nitrosylcobalamin in solution. *Dalton Transactions, 3*, 424–433. https://doi.org/10.1039/b810895a.

Hassanin, H. A., Hannibal, L., Jacobsen, D. W., El-Shahat, M. F., Hamza, M. S., & Brasch, N. E. (2009). Mechanistic studies on the reaction between R2N-NONOates and aquacobalamin: Evidence for direct transfer of a nitroxyl group from R2N-NONOates to cobalt(III) centers. *Angewandte Chemie (International Ed. in English), 48*(47), 8909–8913. https://doi.org/10.1002/anie.200904360.

Hodgkin, D. G., Pickworth, J., Robertson, J. H., Trueblood, K. N., Prosen, R. J., & White, J. G. (1955). The crystal structure of the hexacarboxylic acid derived from B12 and the molecular structure of the vitamin. *Nature, 176*(4477), 325–328. Retrieved from http://www.ncbi.nlm.nih.gov/pubmed/13253565.

Huemer, M., Diodato, D., Schwahn, B., Schiff, M., Bandeira, A., Benoist, J. F., et al. (2017). Guidelines for diagnosis and management of the cobalamin-related remethylation disorders cblC, cblD, cblE, cblF, cblG, cblJ and MTHFR deficiency. *Journal of Inherited Metabolic Disease, 40*(1), 21–48. https://doi.org/10.1007/s10545-016-9991-4.

Huemer, M., Simma, B., Fowler, B., Suormala, T., Bodamer, O. A., & Sass, J. O. (2005). Prenatal and postnatal treatment in cobalamin C defect. *The Journal of Pediatrics, 147*(4), 469–472. https://doi.org/10.1016/j.jpeds.2005.04.040.

James, K. J., Hancock, M. A., Gagnon, J. N., & Coulton, J. W. (2009). TonB interacts with BtuF, the Escherichia coli periplasmic binding protein for cyanocobalamin. *Biochemistry, 48*(39), 9212–9220. https://doi.org/10.1021/bi900722p.

Jeong, J., Ha, T. S., & Kim, J. (2011). Protection of aquo/hydroxocobalamin from reduced glutathione by a B12 trafficking chaperone. *BMB Reports, 44*(3), 170–175. https://doi.org/10.5483/BMBRep.2011.44.3.170.

Jeong, J., & Kim, J. (2011). Glutathione increases the binding affinity of a bovine B(1)(2) trafficking chaperone bCblC for vitamin B(1)(2). *Biochemical and Biophysical Research Communications, 412*(2), 360–365. https://doi.org/10.1016/j.bbrc.2011.07.103.

Jeong, J., Park, J., Park, J., & Kim, J. (2014). Processing of glutathionylcobalamin by a bovine B12 trafficking chaperone bCblC involved in intracellular B12 metabolism. *Biochemical and Biophysical Research Communications, 443*(1), 173–178. https://doi.org/10.1016/j.bbrc.2013.11.075.

Jorge-Finnigan, A., Gamez, A., Perez, B., Ugarte, M., & Richard, E. (2010). Different altered pattern expression of genes related to apoptosis in isolated methylmalonic aciduria cblB type and combined with homocystinuria cblC type. *Biochimica et Biophysica Acta, 1802*(11), 959–967. https://doi.org/10.1016/j.bbadis.2010.08.002.

Jusufi, J., Suormala, T., Burda, P., Fowler, B., Froese, D. S., & Baumgartner, M. R. (2014). Characterization of functional domains of the cblD (MMADHC) gene product. *Journal of Inherited Metabolic Disease*, *37*(5), 841–849. https://doi.org/10.1007/s10545-014-9709-4.

Kim, J., Gherasim, C., & Banerjee, R. (2008). Decyanation of vitamin B12 by a trafficking chaperone. *Proceedings of the National Academy of Sciences of the United States of America*, *105*(38), 14551–14554. https://doi.org/10.1073/pnas.0805989105.

Kim, J., Hannibal, L., Gherasim, C., Jacobsen, D. W., & Banerjee, R. (2009). A human vitamin B12 trafficking protein uses glutathione transferase activity for processing alkylcobalamins. *The Journal of Biological Chemistry*, *284*(48), 33418–33424. https://doi.org/10.1074/jbc.M109.057877.

Koutmos, M., Gherasim, C., Smith, J. L., & Banerjee, R. (2011). Structural basis of multifunctionality in a vitamin B12-processing enzyme. *The Journal of Biological Chemistry*, *286*(34), 29780–29787. https://doi.org/10.1074/jbc.M111.261370.

Lerner-Ellis, J. P., Anastasio, N., Liu, J., Coelho, D., Suormala, T., Stucki, M., et al. (2009). Spectrum of mutations in MMACHC, allelic expression, and evidence for genotype-phenotype correlations. *Human Mutation*, *30*(7), 1072–1081. https://doi.org/10.1002/humu.21001.

Lerner-Ellis, J. P., Tirone, J. C., Pawelek, P. D., Dore, C., Atkinson, J. L., Watkins, D., et al. (2006). Identification of the gene responsible for methylmalonic aciduria and homocystinuria, cblC type. *Nature Genetics*, *38*(1), 93–100. https://doi.org/10.1038/ng1683.

Li, Z., Gherasim, C., Lesniak, N. A., & Banerjee, R. (2014). Glutathione-dependent one-electron transfer reactions catalyzed by a B(1)(2) trafficking protein. *The Journal of Biological Chemistry*, *289*(23), 16487–16497. https://doi.org/10.1074/jbc.M114.567339.

Li, Z., Lesniak, N. A., & Banerjee, R. (2014). Unusual aerobic stabilization of Cob(I)alamin by a B12-trafficking protein allows chemoenzymatic synthesis of organocobalamins. *Journal of the American Chemical Society*, *136*(46), 16108–16111. https://doi.org/10.1021/ja5077316.

Li, Z., Mascarenhas, R., Twahir, U. T., Kallon, A., Deb, A., Yaw, M., et al. (2020). An Interprotein Co-S coordination complex in the B12-trafficking pathway. *Journal of the American Chemical Society*, *142*(38), 16334–16345. https://doi.org/10.1021/jacs.0c06590.

Loewy, A. D., Niles, K. M., Anastasio, N., Watkins, D., Lavoie, J., Lerner-Ellis, J. P., et al. (2009). Epigenetic modification of the gene for the vitamin B(12) chaperone MMACHC can result in increased tumorigenicity and methionine dependence. *Molecular Genetics and Metabolism*, *96*(4), 261–267. https://doi.org/10.1016/j.ymgme.2008.12.011.

Mah, W., Deme, J. C., Watkins, D., Fung, S., Janer, A., Shoubridge, E. A., et al. (2013). Subcellular location of MMACHC and MMADHC, two human proteins central to intracellular vitamin B(12) metabolism. *Molecular Genetics and Metabolism*, *108*(2), 112–118. https://doi.org/10.1016/j.ymgme.2012.11.284.

Martens, J. H., Barg, H., Warren, M. J., & Jahn, D. (2002). Microbial production of vitamin B12. *Applied Microbiology and Biotechnology*, *58*(3), 275–285. https://doi.org/10.1007/s00253-001-0902-7.

Mascarenhas, R., Li, Z., Gherasim, C., Ruetz, M., & Banerjee, R. (2020). The human B12 trafficking protein CblC processes nitrocobalamin. *The Journal of Biological Chemistry*, *295*(28), 9630–9640. https://doi.org/10.1074/jbc.RA120.014094.

Mc Guire, P. J., Parikh, A., & Diaz, G. A. (2009). Profiling of oxidative stress in patients with inborn errors of metabolism. *Molecular Genetics and Metabolism*, *98*(1–2), 173–180. https://doi.org/10.1016/j.ymgme.2009.06.007.

Mellman, I., Willard, H. F., Youngdahl-Turner, P., & Rosenberg, L. E. (1979). Cobalamin coenzyme synthesis in normal and mutant human fibroblasts. Evidence for a processing enzyme activity deficient in cblC cells. *The Journal of Biological Chemistry, 254*(23), 11847–11853. Retrieved from http://www.ncbi.nlm.nih.gov/pubmed/500677.

Moore, S. J., Lawrence, A. D., Biedendieck, R., Deery, E., Frank, S., Howard, M. J., et al. (2013). Elucidation of the anaerobic pathway for the corrin component of cobalamin (vitamin B12). *Proceedings of the National Academy of Sciences of the United States of America, 110*(37), 14906–14911. https://doi.org/10.1073/pnas.1308098110.

Mudd, S. H., Levy, H. L., Abeles, R. H., & Jennedy, J. P., Jr. (1969). A derangement in B 12 metabolism leading to homocystinemia, cystathioninemia and methylmalonic aciduria. *Biochemical and Biophysical Research Communications, 35*(1), 121–126. Retrieved from http://www.ncbi.nlm.nih.gov/entrez/query.fcgi?cmd=Retrieve&db=PubMed&dopt=Citation&list_uids=5779140.

Nielsen, M. J., Rasmussen, M. R., Andersen, C. B., Nexo, E., & Moestrup, S. K. (2012). Vitamin B12 transport from food to the body's cells- -a sophisticated, multistep pathway. *Nature Reviews. Gastroenterology & Hepatology, 9*(6), 345–354. https://doi.org/10.1038/nrgastro.2012.76.

Park, J., Jeong, J., & Kim, J. (2012). Destabilization of a bovine B(1)(2) trafficking chaperone protein by oxidized form of glutathione. *Biochemical and Biophysical Research Communications, 420*(3), 547–551. https://doi.org/10.1016/j.bbrc.2012.03.031.

Park, J., & Kim, J. (2015). Characterization of a B(1)(2)trafficking chaperone protein from caenorhabditis elegans. *Protein and Peptide Letters, 22*(1), 31–38. Retrieved from http://www.ncbi.nlm.nih.gov/pubmed/25256267.

Pastore, A., Martinelli, D., Piemonte, F., Tozzi, G., Boenzi, S., Di Giovamberardino, G., et al. (2014). Glutathione metabolism in cobalamin deficiency type C (cblC). *Journal of Inherited Metabolic Disease, 37*(1), 125–129. https://doi.org/10.1007/s10545-013-9605-3.

Pezacka, E. H. (1993). Identification and characterization of two enzymes involved in the intracellular metabolism of cobalamin. Cyanocobalamin beta-ligand transferase and microsomal cob(III)alamin reductase. *Biochimica et Biophysica Acta, 1157*(2), 167–177. Retrieved from http://www.ncbi.nlm.nih.gov/pubmed/8507652.

Pezacka, E., Denison, C., Green, R., & Jacobsen, D. (1988). Biosynthesis of methylcobalamin: Chemical model studies with thiol-cobalamin adducts and S-adenosylmethionine. *Journal of Cellular Physiology, 107*, 860a.

Pezacka, E., Green, R., & Jacobsen, D. W. (1990). Glutathionylcobalamin as an intermediate in the formation of cobalamin coenzymes. *Biochemical and Biophysical Research Communications, 169*(2), 443–450. Retrieved from http://www.ncbi.nlm.nih.gov/pubmed/2357215.

Pezacka, E. H., & Rosenblatt, D. S. (1994). Intracellular metabolism of cobalamin. Altered activities of 13-axial-ligand transferase and microsomal cob(III)alamin reducatase in cblC and cblD fibroblasts. In H. R. Bhatt, V. H. T. James, G. M. Besser, G. F. Bottazzo, & H. Keen (Eds.), *Advances in Thomas Addison's Diseases (Journal of Endocrinology)* (pp. 315–323). Bristol, London.

Plesa, M., Kim, J., Paquette, S. G., Gagnon, H., Ng-Thow-Hing, C., Gibbs, B. F., et al. (2011). Interaction between MMACHC and MMADHC, two human proteins participating in intracellular vitamin B(1)(2) metabolism. *Molecular Genetics and Metabolism, 102*(2), 139–148. https://doi.org/10.1016/j.ymgme.2010.10.011.

Pupavac, M., Tian, X., Chu, J., Wang, G., Feng, Y., Chen, S., et al. (2016). Added value of next generation gene panel analysis for patients with elevated methylmalonic acid and no clinical diagnosis following functional studies of vitamin B12 metabolism. *Molecular Genetics and Metabolism, 117*(3), 363–368. https://doi.org/10.1016/j.ymgme.2016.01.008.

Pupavac, M., Watkins, D., Petrella, F., Fahiminiya, S., Janer, A., Cheung, W., et al. (2016). Inborn error of cobalamin metabolism associated with the intracellular accumulation of transcobalamin-bound cobalamin and mutations in ZNF143, which codes for a transcriptional activator. *Human Mutation*, *37*(9), 976–982. https://doi.org/10.1002/humu.23037.

Quadros, E. V., Nakayama, Y., & Sequeira, J. M. (2009). The protein and the gene encoding the receptor for the cellular uptake of transcobalamin-bound cobalamin. *Blood*, *113*(1), 186–192. https://doi.org/10.1182/blood-2008-05-158949.

Quadros, E. V., & Sequeira, J. M. (2013). Cellular uptake of cobalamin: Transcobalamin and the TCblR/CD320 receptor. *Biochimie*, *95*(5), 1008–1018. https://doi.org/10.1016/j.biochi.2013.02.004.

Quintana, A. M., Yu, H. C., Brebner, A., Pupavac, M., Geiger, E. A., Watson, A., et al. (2017). Mutations in THAP11 cause an inborn error of cobalamin metabolism and developmental abnormalities. *Human Molecular Genetics*, *26*(15), 2838–2849. https://doi.org/10.1093/hmg/ddx157.

Raffield, L. M., Ellis, J., Olson, N. C., Duan, Q., Li, J., Durda, P., et al. (2018). Genome-wide association study of homocysteine in African Americans from the Jackson heart study, the multi-ethnic study of atherosclerosis, and the coronary artery risk in young adults study. *Journal of Human Genetics*, *63*(3), 327–337. https://doi.org/10.1038/s10038-017-0384-9.

Raux, E., Lanois, A., Rambach, A., Warren, M. J., & Thermes, C. (1998). Cobalamin (vitamin B12) biosynthesis: Functional characterization of the *Bacillus megaterium* cbi genes required to convert uroporphyrinogen III into cobyrinic acid a,c-diamide. *The Biochemical Journal*, *335*(Pt. 1), 167–173. Retrieved from http://www.ncbi.nlm.nih.gov/pubmed/9742226.

Ribes, A., Briones, P., Vilaseca, M. A., Lluch, M., Rodes, M., Maya, A., et al. (1990). Methylmalonic aciduria with homocystinuria: Biochemical studies, treatment, and clinical course of a Cbl-C patient. *European Journal of Pediatrics*, *149*(6), 412–415. https://doi.org/10.1007/BF02009662.

Richard, E., Jorge-Finnigan, A., Garcia-Villoria, J., Merinero, B., Desviat, L. R., Gort, L., et al. (2009). Genetic and cellular studies of oxidative stress in methylmalonic aciduria (MMA) cobalamin deficiency type C (cblC) with homocystinuria (MMACHC). *Human Mutation*, *30*(11), 1558–1566. https://doi.org/10.1002/humu.21107.

Rosenblatt, D. S., & Whitehead, V. M. (1999). Cobalamin and folate deficiency: Acquired and hereditary disorders in children. *Seminars in Hematology*, *36*(1), 19–34. Retrieved from http://www.ncbi.nlm.nih.gov/pubmed/9930566.

Roth, J. R., Lawrence, J. G., & Bobik, T. A. (1996). Cobalamin (coenzyme B12): Synthesis and biological significance. *Annual Review of Microbiology*, *50*, 137–181. https://doi.org/10.1146/annurev.micro.50.1.137.

Ruetz, M., Shanmuganathan, A., Gherasim, C., Karasik, A., Salchner, R., Kieninger, C., et al. (2017). Antivitamin B12 inhibition of the human B12 -processing enzyme CblC: Crystal structure of an inactive ternary complex with glutathione as the Cosubstrate. *Angewandte Chemie (International Ed. in English)*, *56*(26), 7387–7392. https://doi.org/10.1002/anie.201701583.

Rutsch, F., Gailus, S., Suormala, T., & Fowler, B. (2011). LMBRD1: The gene for the cblF defect of vitamin B(1)(2) metabolism. *Journal of Inherited Metabolic Disease*, *34*(1), 121–126. https://doi.org/10.1007/s10545-010-9083-9.

Selcuki, C., van Eldik, R., & Clark, T. (2004). NO binding to cobalamin: Influence of the metal oxidation state. *Inorganic Chemistry*, *43*(9), 2828–2833. https://doi.org/10.1021/ic0347945.

Sorin, M., Watkins, D., Gilfix, B. M., & Rosenblatt, D. S. (2021). Methionine dependence in tumor cells: The potential role of cobalamin and MMACHC. *Molecular Genetics and Metabolism*, *132*(3), 155–161. https://doi.org/10.1016/j.ymgme.2021.01.006.

Stabler, S. P., & Allen, R. H. (2004). Vitamin B12 deficiency as a worldwide problem. *Annual Review of Nutrition*, *24*, 299–326. https://doi.org/10.1146/annurev.nutr.24.012003.132440.

Strohle, A., Richter, M., Gonzalez-Gross, M., Neuhauser-Berthold, M., Wagner, K. H., Leschik-Bonnet, E., et al. (2019). The revised D-A-CH-reference values for the intake of vitamin B12: Prevention of deficiency and beyond. *Molecular Nutrition & Food Research*, *63*(6), e1801178. https://doi.org/10.1002/mnfr.201801178.

Stucki, M., Coelho, D., Suormala, T., Burda, P., Fowler, B., & Baumgartner, M. R. (2012). Molecular mechanisms leading to three different phenotypes in the cblD defect of intracellular cobalamin metabolism. *Human Molecular Genetics*, *21*(6), 1410–1418. https://doi.org/10.1093/hmg/ddr579.

Suormala, T., Baumgartner, M. R., Coelho, D., Zavadakova, P., Kozich, V., Koch, H. G., et al. (2004). The cblD defect causes either isolated or combined deficiency of methylcobalamin and adenosylcobalamin synthesis. *The Journal of Biological Chemistry*, *279*(41), 42742–42749. https://doi.org/10.1074/jbc.M407733200.

van Meurs, J. B., Pare, G., Schwartz, S. M., Hazra, A., Tanaka, T., Vermeulen, S. H., et al. (2013). Common genetic loci influencing plasma homocysteine concentrations and their effect on risk of coronary artery disease. *The American Journal of Clinical Nutrition*, *98*(3), 668–676. https://doi.org/10.3945/ajcn.112.044545.

Wang, S. J., Yan, C. Z., Wen, B., & Zhao, Y. Y. (2019). Clinical feature and outcome of late-onset cobalamin C disease patients with neuropsychiatric presentations: A Chinese case series. *Neuropsychiatric Disease and Treatment*, *15*, 549–555. https://doi.org/10.2147/NDT.S196924.

Warren, M. J. (2006). Finding the final pieces of the vitamin B12 biosynthetic jigsaw. *Proceedings of the National Academy of Sciences of the United States of America*, *103*(13), 4799–4800. https://doi.org/10.1073/pnas.0601030103.

Warren, M. J., Raux, E., Schubert, H. L., & Escalante-Semerena, J. C. (2002). The biosynthesis of adenosylcobalamin (vitamin B12). *Natural Product Reports*, *19*(4), 390–412. Retrieved from http://www.ncbi.nlm.nih.gov/pubmed/12195810.

Watkins, D., Matiaszuk, N., & Rosenblatt, D. S. (2000). Complementation studies in the cblA class of inborn error of cobalamin metabolism: Evidence for interallelic complementation and for a new complementation class (cblH). *Journal of Medical Genetics*, *37*(7), 510–513. Retrieved from http://www.ncbi.nlm.nih.gov/pubmed/10882753.

Watkins, D., & Rosenblatt, D. S. (2011). Inborn errors of cobalamin absorption and metabolism. *American Journal of Medical Genetics. Part C, Seminars in Medical Genetics*, *157C*(1), 33–44. https://doi.org/10.1002/ajmg.c.30288.

Wingert, V., Mukherjee, S., Esser, A. J., Behringer, S., Tanimowo, S., Klenzendorf, M., et al. (2021). Thiolatocobalamins repair the activity of pathogenic variants of the human cobalamin processing enzyme CblC. *Biochimie*, *183*, 108–125. https://doi.org/10.1016/j.biochi.2020.10.006.

Wolak, M., Stochel, G., Hamza, M., & van Eldik, R. (2000). Aquacobalamin (vitamin B12a) does not bind NO in aqueous solution. Nitrite impurities account for observed reaction. *Inorganic Chemistry*, *39*(10), 2018–2019. Retrieved from http://www.ncbi.nlm.nih.gov/entrez/query.fcgi?cmd=Retrieve&db=PubMed&dopt=Citation&list_uids=12526506.

Wongkittichote, P., Wegner, D. J., & Shinawi, M. S. (2021). Novel exon-skipping variant disrupting the basic domain of HCFC1 causes intellectual disability without metabolic abnormalities in both male and female patients. *Journal of Human Genetics*, *66*(7), 717–724. https://doi.org/10.1038/s10038-020-00892-9.

Yamada, K., Gherasim, C., Banerjee, R., & Koutmos, M. (2015). Structure of human B12 trafficking protein CblD reveals molecular mimicry and identifies a new subfamily of nitro-FMN reductases. *The Journal of Biological Chemistry*, *290*(49), 29155–29166. https://doi.org/10.1074/jbc.M115.682435.

Yu, H. C., Sloan, J. L., Scharer, G., Brebner, A., Quintana, A. M., Achilly, N. P., et al. (2013). An X-linked cobalamin disorder caused by mutations in transcriptional coregulator HCFC1. *American Journal of Human Genetics*, *93*(3), 506–514. https://doi.org/10.1016/j.ajhg.2013.07.022.

Zhang, X., Chen, Q., Song, Y., Guo, P., Wang, Y., Luo, S., et al. (2021). Epimutation of MMACHC compound to a genetic mutation in cblC cases. *Molecular Genetics & Genomic Medicine*, *9*(6), e1625. https://doi.org/10.1002/mgg3.1625.

Zheng, D., & Birke, R. L. (2001). Spectroscopic evidence for nitric oxide binding with cob(II)alamin. *Journal of the American Chemical Society*, *123*(19), 4637–4638. Retrieved from http://www.ncbi.nlm.nih.gov/entrez/query.fcgi?cmd=Retrieve&db=PubMed&dopt=Citation&list_uids=11457265.

CHAPTER ELEVEN

Telomere length and vitamin B12

Guruvaiah Praveen, Mudili Sivaprasad, and G. Bhanuprakash Reddy*

Department of Biochemistry, ICMR-National Institute of Nutrition, Hyderabad, India
*Corresponding author: e-mail addresses: reddyg.bp@icmr.gov.in; geereddy@yahoo.com

Contents

Abstract

Telomeres are non-coding nucleoprotein structures consisting of a highly conserved tandem repeat DNA sequence that caps the ends of chromosomes in eukaryotes. Telomeres confer chromosomal stability, protect the genome from nucleolytic degradation, avoid aberrant recombination and improper repair, and prevent random fusion of chromosomes. The end-replication problem results in telomere shortening with every cell division, eventually leading to cellular senescence and aging. Telomere length (TL) is thereby an ideal candidate for "biological aging." Telomeres possess guanine-rich repeats, which are highly susceptible to oxidative stress. Epidemiological studies have indicated the association of telomere attrition with mortality and various age-related diseases. Micronutrients comprising vitamins and minerals act as potential modulators of stress and can influence TL. Research has indicated that vitamin B12 (B12) regulates oxidative stress and maintains genomic stability, thereby influencing telomere integrity and cellular aging. The deficiency of B12 leads to elevated levels of homocysteine, which reduces the methylation potential and increases oxidative stress, thereby compromising the TL. Telomere shortening and mitochondrial dysfunction are independently linked to

Vitamins and Hormones, Volume 119
ISSN 0083-6729
https://doi.org/10.1016/bs.vh.2022.01.014

aging. However, they are connected through telomerase reverse transcriptase activity, which regulates mitochondrial biogenesis. Further, experimental evidence indicated the positive association of B12 with relative TL and mitochondrial DNA copy number, an indirect index of mitochondrial biogenesis. The present chapter provides some insights into the role of B12 in influencing TL. Exploring their association might open new avenues to understand the pathophysiology of aging and age-related diseases.

1. Telomeres

Telomeres are specialized non-coding nucleoprotein structures that cap the ends of eukaryotic chromosomes. In mammals, telomeres are formed of a highly conserved six base-pair *5′-TTAGGG-3′* tandem repeat DNA sequence. Telomeres safeguard the chromosomes from detrimental degradation and random recombination. Since eukaryotic chromosomes are linear, they need to deal with the end replication problem, which causes telomere shortening in each round of cell division, eventually leading to cellular senescence and apoptosis. Telomerase activity acts as a counterbalance mechanism that lengthens the telomeres and protects from telomere attrition elicited DNA-damage response at chromosomal ends. However, telomerase activity is prominently observed in germ cells and stem cells and negligible in somatic cells.

Hermann Muller and Barbara McClintock (Zakian, 2012) first predicted telomere structure in Drosophila and Maize. Muller coined the term telomere and predicted that telomeres are required to maintain genomic integrity and stability. Later in 1941, McClintock concluded that telomeres could stabilize the ends of the eukaryotic chromosomes and protect them from being recognized as double-strand DNA breaks. Elizabeth Blackburn, Carol Greider, and Jack Szostak were jointly awarded the Nobel Prize in Physiology or Medicine in 2009 for the discovery of the protection of chromosomes by the telomeres and enzyme telomerase.

Telomerase helps in solving the “end replication problem” in eukaryotes. The telomerase contains an RNA component (TERC) and a reverse transcriptase (TERT) subunit. Human TERC is expressed in almost all normal cells, but TERT is limited to stem cells, cancer cells, T and B cells (Hiyama et al., 1995; Kim et al., 1994). Further, TERC acts as a template for telomere biosynthesis and helps in regulating the telomerase activity (Zheng, Huang, & Wang, 2019). Likewise, TERT is involved in maintaining the length of telomeric DNA and chromosomal stability, thereby plays a critical role in cellular immortalization, cancer development, and progression (Terracciano, Piscuoglio, & Ng, 2019). The telomere ends are rich in guanine (G) nucleotides and resemble a DNA double-stranded break with an overhang at the 3′ end measuring 35–600 nucleotides (Samassekou, Gadji, Drouin, & Yan, 2010). The G-rich strand of the telomere pairs with TERC, adds telomeric sequences and elongates the overhang (Masutomi et al., 2003). The G-overhang invades the double-stranded telomeric helix and forms a telomere loop/T-loop, which prevents telomere ends from being recognized as breakpoints by the DNA damage repair machinery (de Lange, 2002) (Fig. 1). Other than telomerase, the telomere function is maintained by six telomere-specific proteins called telosome or "Shelterin complex" (Xin, Liu, & Songyang, 2008). These telomeric proteins include POT1 (protection of telomere-1), Rap1 (the human ortholog of the yeast Repressor/Activator Protein 1), TIN2 (TRF1-interacting protein 2), TPP1/TINT1/PTOP/PIP1 (tripeptidyl peptidase 1), TRF1 and TRF2 (telomere repeat factor-1 and -2). The functions and characteristics of Shelterin complex proteins are given in Table 1.

Recent studies have revealed that apart from providing structural stability to chromosomal ends, mammalian telomeres are transcribed into telomeric repeat-containing RNA (TERRA). The TERRA plays a role

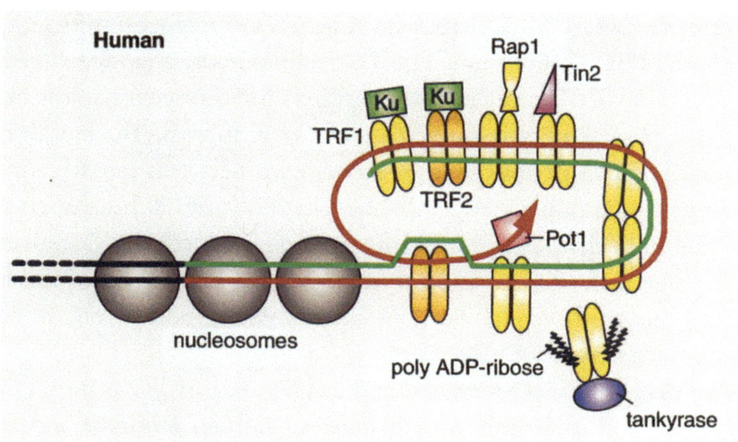

Fig. 1 Structure of telomere loop/T-loop (Blackburn, 2001). Red line represents the G-rich telomeric repeat strand synthesized by telomerase which interacts with and extends the 3′ single-stranded overhang indicated by the arrow; green line represents complementary C-rich strand of the telomeric repeats; heavy black lines represent subtelomeric DNA.

Table 1 Shelterin complex proteins (Turner, Vasu, & Griffin, 2019).

Protein	Interactions	Functions
Protection of Telomeres 1 (POT1)	Interact with the single-stranded overhang	Prevent DNA damage and maintain telomere length
RAP1/TERF2IP	Interact with TERF2	Maintain telomere length
TIN2/TINF2	Associate with TERF1, TERF2 and POT1	Stabilization of TERF2 with telomere
TPP1/ACD	Associate with TINF2 and POT1	Enhances the POT1 binding to the telomere
TRF1/TERF1	bind to double-stranded TTAGGG repeats	Maintain telomere length
TRF2/TERF2	bind to double-stranded TTAGGG repeats	Stabilizes the T-Loop and regulate telomere length

in the regulation of telomere length, telomerase activity, heterochromatin formation, and the DNA damage response (Moslehi, DePinho, & Sahin, 2012).

1.1 Telomere homeostasis (telomere shortening)

During fertilization, male and female gametes fuse to form a zygote. In the zygote, the length of telomeres must be reset to maintain sufficient telomere length (TL) in off-springs. The TL declines in the cleavage stage of the embryo, and it will continue until the morula stage. However, in the blastula stage, telomerase activity is elevated, which will increase the length of the telomeres. The telomere extension slows down between the 8th and 11th weeks of gestation (Cheng et al., 2013). The TL varies in human embryos, fetuses, and both the mother and father influence its length. Moreover, the rate of telomere shortening is pronounced in the early years of the life of birth, and the attrition rate declines during young adulthood. Its length reduces in older stages of life.

Earlier investigations revealed that TL is highly variable at birth (Turner et al., 2019), and its length also diverges in human embryos and fetuses (Holmes et al., 2009). Furthermore, in new-borns, TL is associated with the parental telomere lengths. Interestingly, TL appears to maintain synchrony between tissues at the time of birth and, TL appears similar between male and female babies. This observation contrasts with the finding in adults, which indicates that while telomere lengths in different tissues may be correlated, they are highly variable. Telomere length is longer in women than in men (Daniali et al., 2013; Praveen, Shalini, Sivaprasad, & Reddy, 2020), and two theories explain how TL is safeguarded in women. Telomeres are sensitive to oxidative stress. Studies have reported that women produce less reactive oxygen species (ROS) and possess higher stress handling capacity than men (Nawrot, Staessen, Gardner, & Aviv, 2004). Secondly, there is an estrogen-responsive element in telomerase that promotes telomerase activity (Mayer et al., 2006).

Since the telomeres are dynamic, few telomeric sequences will be lost in each cell cycle, and consequently, telomere shortening occurs, which is associated with cellular proliferation. An imbalance between TL maintenance and cellular proliferation leads to premature cellular aging and the development of cancer. Deterioration of the telomere structure generates the end-to-end chromosome fusion and chromosomal aberrations, which promotes genomic instability. Cellular genetic instability forces the cells to enter senescence or apoptosis by induction of p53 and p16. The cells that escape senescence/apoptosis divide to generate unbalanced chromosomes, leading to cellular transformation.

1.2 Telomere length and biological aging

Aging is an irreversible process that results from the interaction between molecular and genetic events over the lifespan. In somatic cells, each cell division results in the shortening of the telomere sequences by 50–200 bp. Beyond a critical threshold, it leads to cellular senescence and apoptosis (Kim et al., 1994).

The significance of adequate telomerase activity and TL maintenance stems from the pivotal studies carried out in human fibroblasts. The study demonstrated that continuous passage of fibroblasts ended up in telomere shortening and entered the senescence phase after a finite number of cell divisions, a barrier known as "Hayflick limit" (Hayflick & Moorhead, 1961). Further, another study re-established the importance of TL in imparting stability and the capacity to overcome replicative senescence, using TERC positive cell lines (Bodnar et al., 1998). Thus, TL is considered as an ideal marker of "biological aging" (von Zglinicki & Martin-Ruiz, 2005), and several studies have linked TL with premature aging (Gasser & le Coutre, 2013; Kong, Lee, & Wang, 2013). Furthermore, cellular factors such as mitochondrial dysfunction and chronic inflammation and external factors like a high-fat diet, radiation, and chemotherapy accelerate the cellular aging process by damaging G-rich telomere repeat sequences. These factors are strongly associated with ROS generation, which is known to induce cellular senescence, and the G-rich telomere repeat sequence is particularly susceptible to oxidative damage (Finkel & Holbrook, 2000). Since telomere attrition is an initiator of cell senescence in vitro, numerous studies have investigated the relationship between telomere shortening and age-related diseases (Bar & Blasco, 2016).

Telomere shortening during cell division is a physiological process. The human fibroblast cells lose 50–100 base pairs of telomeres during each cell division. After 50–70 cell divisions, telomeres reach a critical length and lose the capping function, and this loss is recognized as DNA damage and thereby induces DNA damage checkpoints. The first checkpoint is called senescence (first mortality stage), which is induced by p53. The cells with mutated p53 bypass the senescence checkpoint and carry out cell division with dysfunctional telomeres. This step induces a second checkpoint called crisis checkpoint (second mortality stage), p53 independent, distinguished by immense chromosomal instability and cell death. Telomere shortening, senescence, and crisis limit the cells' replicative capacity, reducing the life span of the human cells. Most human tissues and organs are subjected to telomere shortening in the aging process (Djojosubroto, Choi, Lee, & Rudolph, 2003).

1.3 Telomere-mitochondria-aging axis

Though the functions of telomeres and mitochondria were studied separately, recent studies have discovered an intimate connection between them. Both telomere dysfunction and mitochondrial dysfunction, which are regarded as the hallmarks of aging, contribute to genomic instability (Zheng et al., 2019). Mitochondrial dysfunctions lead to telomere attrition, while telomere damage results in the reprogramming of mitochondrial biology. Under oxidative stress, TERT, the catalytic component of telomerase, shuttles between the nucleus and mitochondria, indicating the connection between mitochondria and telomeres (Haendeler et al., 2004). Further, a classical experiment carried out by Sahin et al. in mice lacking TERT established the direct link between telomere and mitochondrial biogenesis. They observed that telomere dysfunction via p53 suppressed the promoters of peroxisome proliferator-activated receptor-gamma, co-activator 1 alpha, and beta (PGC-1α and PGC-1β), master regulators of mitochondrial physiology and metabolism, respectively (Sahin et al., 2011).

Additionally, studies on SIRT1 knock-out mice provided strong evidence for the telomere and mitochondrial axis role in the aging process. SIRT1 is a NAD-dependent deacetylase that acts as a metabolic sensor. Ameliorated levels of SIRT1 in mice diminished the rate of telomere erosion in the liver and kidney. The increased expression of SIRT1 also deacetylates the p53, thereby reduces its transcriptional activity and inhibits p53-dependent apoptosis, which in turn leads to cell survival. SIRT1 regulates mitochondrial biogenesis by deacetylation and activation of PGC-1α, Nuclear respiratory factor (NRF) 1, and Nuclear factor erythroid 2-related factor 2 (NRF2). Furthermore, NRF1 and NRF2 regulate several genes involved in mitochondrial biogenesis and its function. Also, SIRT1 inhibits the mitochondrial transcription factor A (TFAM), a vital regulator of the mitochondrial DNA replication via the SIRT1-HIF-1α-Myc-TFAM pathway (Gomes et al., 2013).

1.4 Telomere length and human diseases

The progressive shortening of telomeres with each cell cycle acts as a "molecular clock" leading to aging (Blackburn, 2001). In addition to their role in aging, telomere shortening facilitates a tumor-suppressive mechanism by preventing cell proliferation with telomere dysfunction, which might also carry defective chromosomes. Loss of critical length causes loss of telomere protection leading to chromosomal instability. Defective TL is associated with various age-related diseases (Blasco, 2005; Calado & Young, 2009).

Reduced TL is observed in lymphocytes of HIV and Alzheimer's subjects, hepatocytes of hepatitis, intestinal epithelium of chronic inflammatory bowel disease patients, and endothelial cells of atherosclerotic plaques. Telomere attrition is also associated with the progression of liver cirrhosis and myelodysplastic syndrome (MDS). In liver cirrhosis, progressive chronic hepatocyte damage is observed, which later evolves into end-stage chronic liver disease. Rapid hepatocyte cell division accelerates telomere shortening, which induces hepatocyte cell senescence, leading to reduced liver regenerative reserve, fibrotic scarring, and liver failure. Similarly, MDS is an age-associated hematopoietic stem cell disorder that promotes bone marrow failure due to intensive telomere shortening. Studies have reported that short TL is associated with various age-related diseases including cardiovascular diseases, acute myocardial infarction, atherosclerosis, hypertension, and type 2 diabetes mellitus (Herrmann & Herrmann, 2020). Hence, TL may act as a biomarker for future risk prediction health outcomes.

1.5 Telomere length—Demographic and lifestyle factors

Many studies revealed that TL is associated with age. Alongside chronological aging, emerging pieces of evidence showed that genetic and environmental factors might modulate the TL. Several genetic loci are reported to be associated with TL (Andrew et al., 2006; Codd et al., 2013). Ethnicity also affects TL as white individuals have slightly longer telomeres than Black and Hispanic individuals (Diez Roux et al., 2009). Psychosocial stress elevates oxidative stress, which reduces the telomerase enzyme activity, thereby TL (Wolkowitz et al., 2011). Besides, TL is highly variable among individuals of similar age groups.

Furthermore, smoking, high body mass index, low socioeconomic status, less physical activity reduces the length of telomeres (Cherkas et al., 2008). Available evidence shows that obesity causes telomere shortening by elevation of ROS, chronic inflammation, while weight loss can improve the TL (Valdes et al., 2005). Diet, physical activity, and other lifestyle factors have been implicated in the pathophysiology of several chronic diseases and lower total mortality and longer life expectancy. One of the mechanisms in which diet and nutrition can reduce disease risk is its impact on telomeres (Galie et al., 2020). Diet and nutrition may influence TL by several mechanisms, such as regulating oxidative stress and inflammation or modulating epigenetic reactions.

2. Nutrition and telomere length

There is an increasing interest in exploring whether age-related diseases can be prevented by dietary means through nutrients. Because of the slow nature of the aging process, biomarkers of aging, such as TL, are highly useful in investigating the effect of nutrition on aging. Epidemiologic studies provide support for the putative effects of diet components on TL and the aging process. Furthermore, variability in telomere shortening independent of chronological age signifies its flexible nature, which may be explained partly by lifestyle factors, precisely by nutrition (Freitas-Simoes, Ros, & Sala-Vila, 2016). There is increasing evidence that nutritional factors are associated with telomere length. The results from epidemiologic and clinical trials conducted in humans evaluating the role of nutrients, food groups, and dietary patterns on TL suggest that the consumption of fruits and vegetables, some antioxidant nutrients, and the Mediterranean diet are mainly associated with longer telomeres

(Balan, Decottignies, & Deldicque, 2018; Galie et al., 2020). However, a Finnish study in the elderly population found that the different dietary approaches had little impact on telomere length and attrition (Meinila et al., 2019).

Various nutrients influence TL possibly through mechanisms that include oxidative stress, inflammation, DNA integrity and methylation, and telomerase activity (Paul, 2011). A longitudinal study reported an inverse relationship between marine omega-3 fatty acid levels in the blood and the rate of telomere attrition (Farzaneh-Far et al., 2010). In the context of nutritional factors influencing telomere length, the role of micronutrients, in particular, is very relevant as they are known to influence various molecular mechanisms of the aging process.

2.1 Micronutrients and telomere length

Evidence from epidemiologic studies suggests the association of shorter TL with higher mortality and age-related chronic diseases (Xu et al., 2009). Experimental studies have stressed the contribution of oxidative stress, chronic inflammation, and lifestyle variables toward telomere attrition (Mundstock et al., 2015). Micronutrients comprising vitamins and minerals with anti-oxidative properties can mitigate the stress and inflammation and may influence TL (Furumoto, Inoue, Nagao, Hiyama, & Miwa, 1998; Richards et al., 2007; Richards et al., 2008; Tanaka, Moritoh, & Miwa, 2007).

A study found that multivitamin usage in women resulted in longer TL (Xu et al., 2009). Furthermore, minerals such as zinc and selenium, which are involved in the antioxidant response, may influence TL and telomere base damage, but the shape of the dose-response relationship within the physiological range has yet to be determined (Ferguson, Karunasinghe, Zhu, & Wang, 2012; Sharif, Thomas, Zalewski, & Fenech, 2012). A cross-sectional study that investigated the blood micronutrient levels and TL found that TL was inversely associated with plasma zinc (Milne et al., 2015). Folate and vitamin B12 (B12) are critical factors for cellular metabolic functions, and studies indicate that subjects with low levels of these micronutrients have shown increased oxidative stress (Lee, Wang, Lin, & Lin, 2016; van de Lagemaat, de Groot, & van den Heuvel, 2019; Weber et al., 2014). Investigations have revealed that folate and B12 influence telomere integrity and cellular aging by regulating cellular oxidative stress, nucleotide synthesis, and DNA stability (Pusceddu et al., 2019; van de Lagemaat et al., 2019). Furthermore, epidemiological studies have shown an association of plasma B12 levels with TL (Xu et al., 2009).

2.2 Vitamin B12 and telomere length

Telomere shortening is usually a by-product of either oxidative stress or inflammation. Micronutrients with antioxidant properties can protect the telomeres from this damage. B12 is crucial in maintain DNA integrity through nucleotide biosynthesis and methylation. Moreover, B12 is known to possess potent antioxidant properties, which can help in combating oxidative stress in the body (Ling & Chow, 1953). Usage of B12 has been linked to longer telomeres in women than non-users (Xu et al., 2009). Further, studies have shown that low B12 levels via homocysteine (Hcys) accumulation and high B12 levels via inflammation might influence telomere shortening (Pusceddu et al., 2019). Therefore, experimental evidence links the role of B12 in preventing telomere shortening.

2.3 Vitamin B12

Vitamin B12 is a generic name for all corrin nuclei having compounds that exhibit the biological activity of cyanocobalamin. The Nobel Prize in Medicine was awarded to Minot, Murphy, and Whipple for isolating and identifying B12 and Hodgkin for mapping the chemical structure of B12 (Fig. 2). When R-group in B12 is occupied by the methyl or 5'-deoxyadenosyl group (yielding enzymatically active cofactors), it is utilized by cytosolic methionine synthase and mitochondrial methylmalonyl CoA (MMA), respectively. The binding of hydroxyl or cyano group yields hydroxocobalamin and cyanocobalamin (synthetic inactive forms), respectively, which need to be metabolized to active forms in humans. Under physiological conditions, cobalt exists in three possible oxidation states in B12 derivatives: Co(III), Co(II), and Co(I), which determine the function of B12 as enzyme cofactors. Humans and other mammals, fungi, and plants are incapable of synthesizing B12, unlike prokaryotes (mostly anaerobic). The gastrointestinal environment of the animals (primarily ruminant) favors the growth of B12 synthesizing microbial flora. B12, thus produced is absorbed and stored in the tissues like the liver, muscles, and milk of the animals, which, when consumed by humans, enter the human food chain.

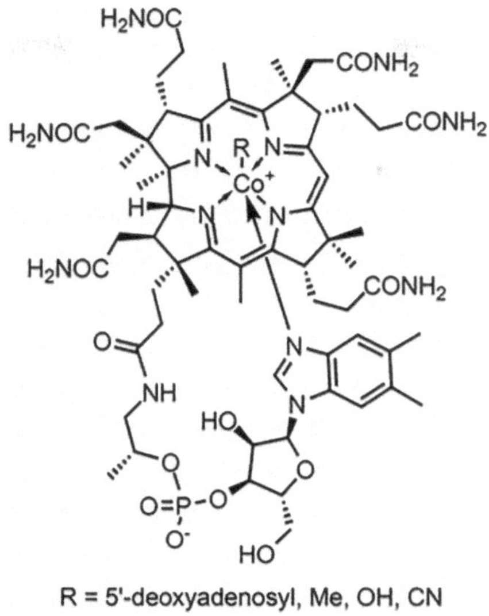

Fig. 2 Structure of B12 (Lakshmiah, Lakshmi, & Reddy, 2016).

B12 is widely distributed in animal foods such as liver, eggs, milk, meat, and fish. Spirulina (blue-green algae) and some seaweed contain appreciable amounts of B12.

Primary reasons for B12 insufficiency are dietary intake or poor absorption; however, factors such as age, gender, genetic factors, ethnicity, socio-cultural factors might affect B12 status (Hinds, Johnson, Webb, & Graham, 2011; Villamor, Mora-Plazas, Forero, Lopez-Arana, & Baylin, 2008). Defects in the gastric mucosa, chronic atrophic gastritis, gastrectomy, malabsorption in the ileum, intestinal stasis, drugs, and intestinal parasites can impair B12 absorption. Other causes include old age and pernicious anemia, characterized by autoimmune destruction of gastric parietal cells resulting in a lack of intrinsic factor (IF), which facilitates the absorption of B12. Subclinical cobalamin deficiency (SCCD) is epidemiologically more prevalent than clinical cobalamin deficiency in the elderly and vegetarian populations. It differs from cobalamin deficiency in many aspects: SCCD is milder yet more widespread. The underlying causes are less severe; hence it is less likely to progress (Carmel, 2012). It lacks clinical manifestations, so it is challenging to diagnose and may require more than one biomarker for diagnosis (Hinds et al., 2011). Prevalence of sub-clinical B12 deficiency (around 30–35%) among the apparently healthy individuals has been a consistent finding in some of our recent studies (Satyanarayana et al., 2011; Shalini et al., 2019; Sivaprasad et al., 2016; Sivaprasad et al., 2019).

2.4 Biological functions affected by vitamin B12

Vitamin B12 is a biologically active corrinoid, existing in different forms—cyanocobalamin, methylcobalamin, hydroxycobalamin, and deoxyadenosyl cobalamin. It is involved in various cellular biochemical reactions, either as a cofactor or a coenzyme (Sivaprasad, Shalini, Das, Vijayalakshmi, & Reddy, 2016). B12 dependent enzymes fall into four classes, which catalyze the methylation, reductive dehalogenation, isomerization, and radical S-adenosylmethionine (SAM) reactions.

B12 is essential for synthesizing red blood cells, DNA, and the myelin sheath, essential for healthy growth and development. The crucial roles of B12 in the healthy maintenance of skin, mucous membrane, hematological, and central nervous system make them noteworthy. The deficiency of B12 can lead to a broad spectrum of neuropsychiatric disorders that affect all age group individuals, and that can often be reversed by early diagnosis and prompt treatment (Sahoo, Avasthi, & Singh, 2011) (Bottiglieri, 1996). Deficiency of B12 increases the plasma concentration of homocysteine (Hcys), which is a risk factor for cardiovascular diseases (CVD), megaloblastic anemia, macrocytosis, and cognitive impairment in the elderly (Mahalle, Kulkarni, Garg, & Naik, 2013). B12 is an essential cofactor that regulates the expression of various genes by methylation and metabolism. Adenosylcobalamin (in mitochondria) and methylcobalamin (in the cytoplasm) are two essential coenzyme forms of the B12 that regulate cellular methylmalonic acid and Hcys homeostasis. In the absence of B12, conversion of MMA to succinyl CoA is inhibited, which leads to disruption of odd-chain fatty acid metabolism in neuronal cells. The patients with dementia, depression, reduced cognition, and Alzheimer's disease have reduced B12 levels and severely damaged the brain. Cognitive defects can be reversed by B12 supplementation (Health Quality, 2013; Jatoi et al., 2020; Moore et al., 2012; Smith & Refsum, 2009). The risk of CVD shows higher levels of Hcys and reduced B12. Lower concentrations of B12 contribute to CVD development via increased levels of Hcys; and stroke by increasing blood pressure and atrial fibrillation (Spence, 2017).

Cobalamins act as effective antioxidants invitro exhibiting tough protection against pro-oxidants Hcys and hydrogen peroxide. The novel forms of thiolatocobalamins including thiolatocobalamins *N*-acetyl-L-cysteinylcobalamin and glutathionylcobalamin, reduced apoptotic and necrotic cell death by inhibiting intracellular peroxide generation and maintaining intracellular glutathione levels. Further, they are non-toxic even at supraphysiological concentrations. Since cobalamin variants are effective against inflammatory oxidative stress which plays a major role in telomere shortening, their potential for treating such problems warrants further research (Birch, Brasch, McCaddon, & Williams, 2009).

Vitamin B12 along with B6 interfere with immune function by participating in the nucleic acid and protein biosynthesis. Vitamin B12 may act as an immunomodulator for cellular immunity by suppressing the NK cell activity. Further, it reduces the number of lymphocytes and CD8+ cells, as well as the proportion of CD4+ cells, resulting in an excessively high CD4+/CD8+ ratio (Haryanto, Suksmasari, Wintergerst, & Maggini, 2015). The ability of B12 to modulate the host immune response makes it an important target in immunocompromised conditions such as Covid-19 pandemic.

A recent study by Tan et al. in older COVID-19 patients reported that the administration of B12 in combination with vitamin D and magnesium was linked with a substantial reduction in the percentage of patients with clinical deterioration requiring oxygen support, intensive care support, or both (Tan et al., 2020). However, in a study carried out in South Korean adults with covid 19, there were no substantial increase in the prevalence of deficiency in vitamins B12 and other B vitamins (Im et al., 2020). In another study with patients admitted for COVID Pneumonia, poor outcome was significantly associated with old age, high P/F ratio and higher plasma B12 levels (Dalbeni et al., 2021).

3. Role of vitamin B12 in telomere homeostasis

Studies have indicated the association of plasma B12 levels with telomere length, and it is argued that deficiency B12 may affect genomic stability and the normal methylation state of genes (Pusceddu et al., 2019; Xin et al., 2008). Synthesis of DNA and epigenetic methylation reactions require B12. Studies revealed a direct relationship between the TL and micronutrients such as B12 and folic acid. Telomere length is epigenetically regulated by DNA methylation; hence anything that can modulate DNA methylation would, in turn, influence TL (Crider, Yang, Berry, & Bailey, 2012). Folate is a crucial methyl group donor that allows the synthesis of methionine from Hcys in the presence of methionine synthase, a B12 dependent enzyme. In invertebrates, methyl groups are added to cytosine residues, particularly at cytosine phosphate-guanine (CpG) islands in the presence of DNA-methyltransferase (Elgin & Reuter, 2013). The hexameric telomere repeats do not have CpG islands in their region, so the methylation occurs at the CpG islands in subtelomeres. Mouse embryonic stem cells that lack the DNA-methyltransferase showed a decreased subtelomere methylation and increased TL, possibly because of elevated telomeric recombination (Gonzalo et al., 2006). Human tumor cell lines with higher telomerase expression showed increased methylation in the subtelomeric region (Choudhury, Cui, Milton, Li, & Irudayaraj, 2015; Zhang et al., 2014). Therefore, the correlation between TL and DNA methylation depends on the cell type and other underlying circumstances.

Deficiency of B12 lowers the methionine synthase activity, which diminishes the SAM levels, thereby reducing DNA methylation and causing folate unavailability required for the conversion of dUMP to dTMP. Folate is essential for DNA hypomethylation and the prevention of uracil incorporation into DNA. Folate and B12 deficiency result in a chromosome-breaking effect (Blount et al., 1997). In-vivo studies observed that increased Hcys levels lead to increased micronucleus formation and reduced TL. In yeast cells, decreased synthesis of dTMP promotes the telomere shortening by destabilizing replication forks during DNA replication (Toussaint, Dionne, & Wellinger, 2005).

B12 acts as a cofactor for the MMA mutase, which converts MMA to succinyl CoA. The deficiency of B12 causes elevated levels of MMA and reduced succinyl CoA, a TCA cycle intermediate of mitochondria (Takahashi-Iniguez, Garcia-Hernandez, Arreguin-Espinosa, & Flores, 2012). Abnormal levels of these two substrates disrupt the mitochondrial functions, increase ROS, and reduce mitochondrial ATP production. B12 deficiency-induced ROS stimulates the release of proinflammatory cytokines such as tumor necrosis factor-alpha (TNF-α), whose levels are further exaggerated by the increased Hcys levels. Telomeres are highly susceptible to oxidative stress induced by ROS and elevated ROS attack on the telomeres' guanine-rich repeats, converting it to 8-oxo-7,-8-dihydro-2′-deoxyguanine (8-oxoG) (Ko, Seo, & Jung, 2018). The 8-oxoG is one of the most abundant DNA lesions. Its chronic formation and persistence in telomeres hazardously cause telomere damage, leading to the formation of chromatin bridges, micronuclei, and fusion of chromosome ends. 8-oxoG causes telomere shortening by triggering the replication stress at telomeres, which further drives the genomic instability (Fouquerel et al., 2019).

Furthermore, the presence of 8-oxoG in the 3′ overhangs of telomere disrupts the G-quadruplex structures, (Fouquerel et al., 2016) prevents binding of TRF1 and TRF2, and promotes guanine to thymine transitional mutation. Altogether, these events diminish telomere stability. When the 8-oxoG is present in the cellular nucleotide pool, telomerase uses them to elongate the telomeres, resulting in the abrupt termination of telomere extension (Fouquerel et al., 2016). Excessive ROS in cells also acts on the deoxyribonucleoside triphosphates (dNTPs), inhibiting the human TERT mediated telomere elongation.

Studies have demonstrated the link between telomere shortening and mitochondrial biogenesis (Sahin et al., 2011; Tyrka et al., 2015). Measuring the mitochondrial DNA copy number (mtCN) molecules per cell serves as an indirect index of mitochondrial biogenesis (Sahin & DePinho, 2012). In this context, recently, we have shown a positive correlation between the relative telomere length (rTL) and mtCN in a community-based cross-sectional study with 428 subjects (Fig. 3) (Praveen et al., 2020). Most importantly, both the rTL and mtCN were positively associated with plasma folate and B12 levels (Fig. 4). The study

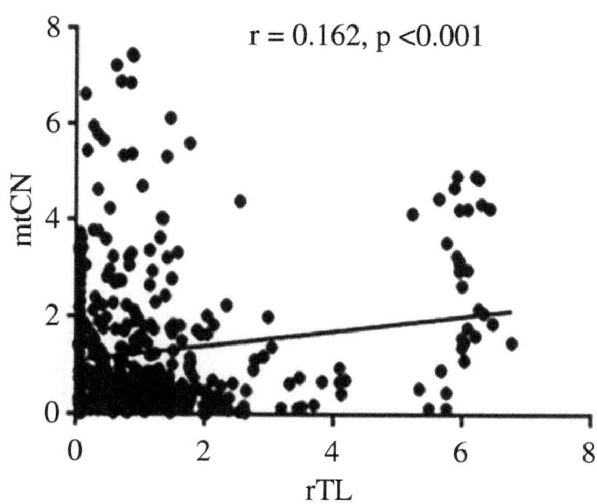

Fig. 3 Correlation between human leucocyte relative TL (rTL) and mitochondrial DNA copy number (mtCN). Spearman rank correlation indicates a positive correlation between the variables (Praveen et al., 2020).

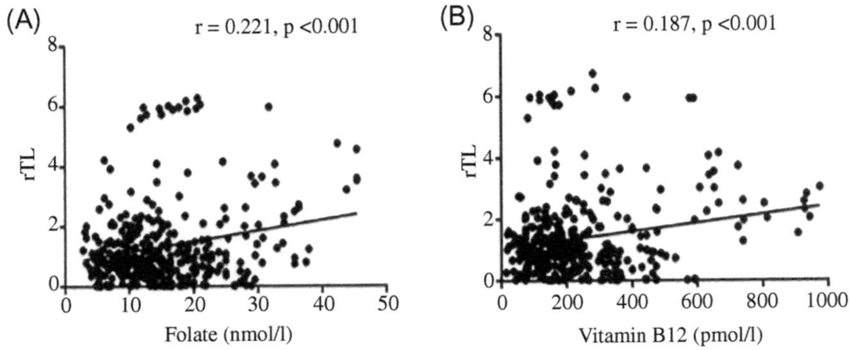

Fig. 4 Correlation of (A) folate and (B) B12 with relative TL (rTL). Spearman rank correlation indicates a positive correlation between the variables (Praveen et al., 2020).

revealed a decline of rTL and mtCN with age in the Indian population, and their association suggests that they may co-regulate each other with age. Further, folate and B12 may delay aging by preventing the reduction in rTL length and mtCN. A study by Audrey et al. reported a positive correlation between mtDNA and TL in younger adults (Tyrka et al., 2015). Another cross-sectional investigation conducted in 5581 US individuals from the National Health and Nutrition Examination Survey (NHANES) revealed that in women, the dietary B12 was linearly related to TL but not in men ($F=4.3$, $P=0.0468$) (Tucker, 2019). Population studies conducted in the United Kingdom and Italy also followed a similar trend (Paul et al., 2009; Richards et al., 2008). However, a cross-sectional study with 798 men and women aged 55–79 years found no association between leukocyte TL and serum folate and B12 in multiple models adjusted for age, sex, serum high sensitive C-reactive protein levels. Nevertheless, there was a significant inverse association between serum Hcys levels and leucocyte TL (Shin & Baik, 2016).

Intake of B12 deficient diet and methionine-rich proteins are known to increase the circulating Hcys levels (Chambers, Obeid, & Kooner, 1999; Yakub, Iqbal, & Iqbal, 2010). Recent studies revealed that Hcys could potentially induce oxidative stress and inflammation (Shin & Baik, 2016). In the presence of oxygen, two Hcys molecules form a disulfide bond and superoxide radical. Studies conducted in rats and humans showed that a modest increase in Hcys levels could elevate oxidative stress, which triggers the pro-inflammatory responses within the cell. The Hcys can directly affect the length of telomeres by involving pathways that affect DNA methylation and nucleotide synthesis (Wierzbicki, 2007). An elevated level of Hcys reduces the methylation potential and increases oxidative stress, which further compromises the cell TL. Human telomerase reverse transcriptase is involved in telomere extension by its RNA and reverse transcriptase activity. Homocysteine levels maintain the TL by transcriptional regulation of human TERT, which is strictly controlled by DNA methylation. In cells, hypermethylation of CTCF repressor binding site in the human TERT promoter region leads to reduced human TERT expression, and the use of demethylating agents such as 5-aza-CdR can elevate the human TERT expression by removing methyl groups (demethylation of the promoter) (Meeran, Patel, & Tollefsbol, 2010; Renaud, Loukinov, Bosman, Lobanenkov, & Benhattar, 2005). Studies in mice with high plasma Hcys levels decreased the methylation of the TERT promoter region, thereby promoting telomere attrition (Zhang, Wen, Zhang, & Cui, 2014). Previously Ritz et al. revealed that humans and mice have similar TERT expression patterns (Ritz et al., 2005). Mice leukocytes showed a significant decrease in DNA methylation of CTCF binding sites on the promoter of mice TERT and mice TERT mRNA levels in 4-weeks after high levels of Hcys induction (Horikawa et al., 2005; Ritz et al., 2005). All these studies indicate that Hcys induces telomere attrition by epigenetic hypermethylation of CTCF binding sites of the human TERT promoter region. Besides telomere shortening, Hcys can also reduce DNA stability by forming micronuclei and DNA interstrand cross-links (Fenech, 2012).

Several epidemiological reports have shown inconsistent results with plasma Hcys and telomere length. Bekaert et al. first demonstrated the association of Hcys with a length of telomeres in a Belgian cohort study (Bekaert et al., 2007). A study involving the Singapore Chinese individuals with age between 45 and 74 years showed an inverse relationship between the plasma Hcys and TL (Rane et al., 2015), and a similar observation was reported in the United Kingdom population, which is conducted in the 1319 healthy subjects (Richards et al., 2008). In a South Australian population study, the older males showed a negative correlation between the plasma Hcys and telomere length. This study consisted of 43 younger (18 males and 25 females) and 47 older (24 males and 23 females) adults. The younger cohort had 11.5% greater TL than the older ones ($P < 0.015$) whereas, in the older cohort, females showed 12.5% greater TL compared to the males ($P < 0.028$). A significant inverse correlation was observed between the TL and Hcys ($r = -0.57$, $P < 0.004$) in males. The individuals in the lowest tertile for plasma Hcys (6.0 ± 0.85 μmol/L) had the longest TL, but this outcome was not observed in older females and younger subjects

(Bull, O'Callaghan, Mayrhofer, & Fenech, 2009). A total of 1715 subjects from the Nurses' Health Study aged 30–55 displayed a neutral association between Hcys and TL (Liu et al., 2013). Such inconsistencies in these studies may be attributed to the different conditions and characteristics of the subjects, ethnicity, age, nutritional status, and B12.

The association between TL and Hcys is widely studied in various human diseases. Previously, subjects with mild cognitive impairment and Alzheimer's disease showed a lower TL with a higher concentration of Hcys in the plasma (Ma et al., 2019). The TL may be an intermediary in the biological mechanism by which higher serum Hcys concentration leads to Alzheimer's disease. In both in vivo and in vitro environments, higher levels of Hcys induced the endothelial cell senescence by hypomethylation of the human TERT promoter region and attrition of telomeres (Zhang, Wang, et al., 2014). Endothelial cell senescence can accelerate the development of atherosclerosis and cardiovascular diseases. Furthermore, in the Northern Chinese Han population, the negative correlation between the TL and Hcys ($r=-0.357$, $P=0.024$) increased the progression of ischemic stroke.

4. Conclusion

This chapter has focussed on the role of B12 in influencing telomere length. Telomeres are specialized structures that confer protection to the eukaryotic chromosomes. Telomere attrition majorly stems from oxidative stress and is attributed to both aging and various age-related disorders. Micronutrients with antioxidant potential can modulate oxidative stress. Vitamin B12, known for its role in nucleotide biosynthesis and methylation reactions, is also a potent antioxidant that reduces oxidative stress. The deficiency of B12 increases the Hcys, which can compromise TL via increased oxidative stress.

Further, B12 is also associated with TL and mitochondrial biology. Altogether, B12 seems to be a potential factor that can influence telomere homeostasis. Further research is warranted to explore the role of B12 in modulating TL and age-related diseases.

Acknowledgments

GBR is supported by grants from the Department of Biotechnology, Science and Engineering Research Board and Indian Council of Medical Research.

References

Andrew, T., Aviv, A., Falchi, M., Surdulescu, G. L., Gardner, J. P., Lu, X., et al. (2006). Mapping genetic loci that determine leukocyte telomere length in a large sample of unselected female sibling pairs. *American Journal of Human Genetics*, *78*(3), 480–486. https://doi.org/10.1086/500052.

Balan, E., Decottignies, A., & Deldicque, L. (2018). Physical activity and nutrition: Two promising strategies for telomere maintenance? *Nutrients*, *10*(12). https://doi.org/10.3390/nu10121942.

Bar, C., & Blasco, M. A. (2016). Telomeres and telomerase as therapeutic targets to prevent and treat age-related diseases. *F1000Res*, *5*. https://doi.org/10.12688/f1000research.7020.1.

Bekaert, S., De Meyer, T., Rietzschel, E. R., De Buyzere, M. L., De Bacquer, D., Langlois, M., et al. (2007). Telomere length and cardiovascular risk factors in a middle-aged population free of overt cardiovascular disease. *Aging Cell*, *6*(5), 639–647. https://doi.org/10.1111/j.1474-9726.2007.00321.x.

Birch, C. S., Brasch, N. E., McCaddon, A., & Williams, J. H. (2009). A novel role for vitamin B(12): Cobalamins are intracellular antioxidants in vitro. *Free Radical Biology & Medicine*, *47*(2), 184–188. https://doi.org/10.1016/j.freeradbiomed.2009.04.023.

Blackburn, E. H. (2001). Switching and signaling at the telomere. *Cells*, *106*(6), 661–673. https://doi.org/10.1016/s0092-8674(01)00492-5.

Blasco, M. A. (2005). Telomeres and human disease: Ageing, cancer and beyond. *Nature Reviews. Genetics*, *6*(8), 611–622. https://doi.org/10.1038/nrg1656.

Blount, B. C., Mack, M. M., Wehr, C. M., MacGregor, J. T., Hiatt, R. A., Wang, G., et al. (1997). Folate deficiency causes uracil misincorporation into human DNA and chromosome breakage: Implications for cancer and neuronal damage. *Proceedings of the National Academy of Sciences of the United States of America*, *94*(7), 3290–3295. https://doi.org/10.1073/pnas.94.7.3290.

Bodnar, A. G., Ouellette, M., Frolkis, M., Holt, S. E., Chiu, C. P., Morin, G. B., et al. (1998). Extension of life-span by introduction of telomerase into normal human cells. *Science*, *279*(5349), 349–352. https://doi.org/10.1126/science.279.5349.349.

Bottiglieri, T. (1996). Folate, vitamin B12, and neuropsychiatric disorders. *Nutrition Reviews*, *54*(12), 382–390. https://doi.org/10.1111/j.1753-4887.1996.tb03851.x.

Bull, C. F., O'Callaghan, N. J., Mayrhofer, G., & Fenech, M. F. (2009). Telomere length in lymphocytes of older south Australian men may be inversely associated with plasma homocysteine. *Rejuvenation Research*, *12*(5), 341–349. https://doi.org/10.1089/rej.2009.0868.

Calado, R. T., & Young, N. S. (2009). Telomere diseases. *The New England Journal of Medicine*, *361*(24), 2353–2365. https://doi.org/10.1056/NEJMra0903373.

Carmel, R. (2012). Subclinical cobalamin deficiency. *Current Opinion in Gastroenterology*, *28*(2), 151–158. https://doi.org/10.1097/MOG.0b013e3283505852.

Chambers, J. C., Obeid, O. A., & Kooner, J. S. (1999). Physiological increments in plasma homocysteine induce vascular endothelial dysfunction in normal human subjects. *Arteriosclerosis, Thrombosis, and Vascular Biology*, *19*(12), 2922–2927. https://doi.org/10.1161/01.atv.19.12.2922.

Cheng, G., Kong, F., Luan, Y., Sun, C., Wang, J., Zhang, L., . . . Xu, D. (2013). Differential shortening rate of telomere length in the development of human fetus. Biochemical and Biophysical Research Communications, 442(1–2), 112–115. doi:https://doi.org/10.1016/j.bbrc.2013.11.022.

Cherkas, L. F., Hunkin, J. L., Kato, B. S., Richards, J. B., Gardner, J. P., Surdulescu, G. L., et al. (2008). The association between physical activity in leisure time and leukocyte telomere length. *Archives of Internal Medicine*, *168*(2), 154–158. https://doi.org/10.1001/archinternmed.2007.39.

Choudhury, S. R., Cui, Y., Milton, J. R., Li, J., & Irudayaraj, J. (2015). Selective increase in subtelomeric DNA methylation: An epigenetic biomarker for malignant glioma. *Clinical Epigenetics*, 7, 107. https://doi.org/10.1186/s13148-015-0140-y.

Codd, V., Nelson, C. P., Albrecht, E., Mangino, M., Deelen, J., Buxton, J. L., et al. (2013). Identification of seven loci affecting mean telomere length and their association with disease. *Nature Genetics*, *45*(4), 422–427. 427e421-422 https://doi.org/10.1038/ng.2528.

Crider, K. S., Yang, T. P., Berry, R. J., & Bailey, L. B. (2012). Folate and DNA methylation: A review of molecular mechanisms and the evidence for folate's role. *Advances in Nutrition*, *3*(1), 21–38. https://doi.org/10.3945/an.111.000992.

Dalbeni, A., Bevilacqua, M., Teani, I., Normelli, I., Mazzaferri, F., & Chiarioni, G. (2021). Excessive vitamin B12 and poor outcome in COVID-19 pneumonia. *Nutrition, Metabolism, and Cardiovascular Diseases*, *31*(3), 774–775. https://doi.org/10.1016/j.numecd.2020.12.005.

Daniali, L., Benetos, A., Susser, E., Kark, J. D., Labat, C., Kimura, M., et al. (2013). Telomeres shorten at equivalent rates in somatic tissues of adults. *Nature Communications*, *4*, 1597. https://doi.org/10.1038/ncomms2602.

de Lange, T. (2002). Protection of mammalian telomeres. *Oncogene*, *21*(4), 532–540. https://doi.org/10.1038/sj.onc.1205080.

Diez Roux, A. V., Ranjit, N., Jenny, N. S., Shea, S., Cushman, M., Fitzpatrick, A., et al. (2009). Race/ethnicity and telomere length in the multi-ethnic study of atherosclerosis. *Aging Cell*, *8*(3), 251–257. https://doi.org/10.1111/j.1474-9726.2009.00470.x.

Djojosubroto, M. W., Choi, Y. S., Lee, H. W., & Rudolph, K. L. (2003). Telomeres and telomerase in aging, regeneration and cancer. Molecules and Cells, 15(2), 164–175. Retrieved from http://www.ncbi.nlm.nih.gov/pubmed/12803478.

Elgin, S. C., & Reuter, G. (2013). Position-effect variegation, heterochromatin formation, and gene silencing in Drosophila. *Cold Spring Harbor Perspectives in Biology*, *5*(8), a017780. https://doi.org/10.1101/cshperspect.a017780.

Farzaneh-Far, R., Lin, J., Epel, E. S., Harris, W. S., Blackburn, E. H., & Whooley, M. A. (2010). Association of marine omega-3 fatty acid levels with telomeric aging in patients with coronary heart disease. *JAMA*, *303*(3), 250–257. https://doi.org/10.1001/jama.2009.2008.

Fenech, M. (2012). Folate (vitamin B9) and vitamin B12 and their function in the maintenance of nuclear and mitochondrial genome integrity. *Mutation Research*, *733*(1–2), 21–33. https://doi.org/10.1016/j.mrfmmm.2011.11.003.

Ferguson, L. R., Karunasinghe, N., Zhu, S., & Wang, A. H. (2012). Selenium and its' role in the maintenance of genomic stability. *Mutation Research*, *733*(1–2), 100–110. https://doi.org/10.1016/j.mrfmmm.2011.12.011.

Finkel, T., & Holbrook, N. J. (2000). Oxidants, oxidative stress and the biology of ageing. *Nature*, *408*(6809), 239–247. https://doi.org/10.1038/35041687.

Fouquerel, E., Barnes, R. P., Uttam, S., Watkins, S. C., Bruchez, M. P., & Opresko, P. L. (2019). Targeted and persistent 8-oxoguanine base damage at telomeres promotes telomere loss and crisis. *Molecular Cell*, *75*(1), 117–130 e116. https://doi.org/10.1016/j.molcel.2019.04.024.

Fouquerel, E., Lormand, J., Bose, A., Lee, H. T., Kim, G. S., Li, J., et al. (2016). Oxidative guanine base damage regulates human telomerase activity. *Nature Structural & Molecular Biology*, *23*(12), 1092–1100. https://doi.org/10.1038/nsmb.3319.

Freitas-Simoes, T. M., Ros, E., & Sala-Vila, A. (2016). Nutrients, foods, dietary patterns and telomere length: Update of epidemiological studies and randomized trials. *Metabolism*, *65*(4), 406–415. https://doi.org/10.1016/j.metabol.2015.11.004.

Furumoto, K., Inoue, E., Nagao, N., Hiyama, E., & Miwa, N. (1998). Age-dependent telomere shortening is slowed down by enrichment of intracellular vitamin C via suppression of oxidative stress. *Life Sciences*, *63*(11), 935–948. https://doi.org/10.1016/s0024-3205(98)00351-8.

Galie, S., Canudas, S., Muralidharan, J., Garcia-Gavilan, J., Bullo, M., & Salas-Salvado, J. (2020). Impact of nutrition on telomere Health: Systematic review of observational cohort studies and randomized clinical trials. *Advances in Nutrition*, *11*(3), 576–601. https://doi.org/10.1093/advances/nmz107.

Gasser, S. M., & le Coutre, J. (2013). Nutrition and the biology of human ageing: Molecular mechanisms underlying ageing. *The Journal of Nutrition, Health & Aging*, *17*(8), 710–711. https://doi.org/10.1007/s12603-013-0373-4.

Gomes, A. P., Price, N. L., Ling, A. J., Moslehi, J. J., Montgomery, M. K., Rajman, L., et al. (2013). Declining NAD(+) induces a pseudohypoxic state disrupting nuclear-mitochondrial communication during aging. *Cells*, *155*(7), 1624–1638. https://doi.org/10.1016/j.cell.2013.11.037.

Gonzalo, S., Jaco, I., Fraga, M. F., Chen, T., Li, E., Esteller, M., et al. (2006). DNA methyltransferases control telomere length and telomere recombination in mammalian cells. *Nature Cell Biology*, *8*(4), 416–424. https://doi.org/10.1038/ncb1386.

Haendeler, J., Hoffmann, J., Diehl, J. F., Vasa, M., Spyridopoulos, I., Zeiher, A. M., et al. (2004). Antioxidants inhibit nuclear export of telomerase reverse transcriptase and delay replicative senescence of endothelial cells. *Circulation Research*, *94*(6), 768–775. https://doi.org/10.1161/01.RES.0000121104.05977.F3.

Haryanto, B., Suksmasari, T., Wintergerst, E., & Maggini, S. (2015). Multivitamin supplementation supports immune function and ameliorates conditions triggered by reduced air quality. *Vitamins and Minerals*, *4*, 1–15.

Hayflick, L., & Moorhead, P. S. (1961). The serial cultivation of human diploid cell strains. *Experimental Cell Research*, *25*, 585–621. https://doi.org/10.1016/0014-4827(61)90192-6.

Health Quality, O. (2013). Vitamin B12 and cognitive function: An evidence-based analysis. Ontario Health Technology Assessment Series, 13(23), 1–45. Retrieved from http://www.ncbi.nlm.nih.gov/pubmed/24379897.

Herrmann, W., & Herrmann, M. (2020). The importance of telomere shortening for atherosclerosis and mortality. *Journal of Cardiovascular Development and Disease*, 7-(3). https://doi.org/10.3390/jcdd7030029.

Hinds, H. E., Johnson, A. A., Webb, M. C., & Graham, A. P. (2011). Iron, folate, and vitamin B12 status in the elderly by gender and ethnicity. *Journal of the National Medical Association*, *103*(9–10), 870–877. https://doi.org/10.1016/s0027-9684(15)30442-9.

Hiyama, K., Hirai, Y., Kyoizumi, S., Akiyama, M., Hiyama, E., Piatyszek, M. A., . . . Yamakido, M. (1995). Activation of telomerase in human lymphocytes and hematopoietic progenitor cells. Journal of Immunology, 155(8), 3711–3715. Retrieved from http://www.ncbi.nlm.nih.gov/pubmed/7561072.

Holmes, D. K., Bellantuono, I., Walkinshaw, S. A., Alfirevic, Z., Johnston, T. A., Subhedar, N. V., et al. (2009). Telomere length dynamics differ in foetal and early post-natal human leukocytes in a longitudinal study. *Biogerontology*, *10*(3), 279–284. https://doi.org/10.1007/s10522-008-9194-y.

Horikawa, I., Chiang, Y. J., Patterson, T., Feigenbaum, L., Leem, S. H., Michishita, E., et al. (2005). Differential cis-regulation of human versus mouse TERT gene expression in vivo: Identification of a human-specific repressive element. *Proceedings of the National Academy of Sciences of the United States of America*, *102*(51), 18437–18442. https://doi.org/10.1073/pnas.0508964102.

Im, J. H., Je, Y. S., Baek, J., Chung, M. H., Kwon, H. Y., & Lee, J. S. (2020). Nutritional status of patients with COVID-19. *International Journal of Infectious Diseases*, *100*, 390–393. https://doi.org/10.1016/j.ijid.2020.08.018.

Jatoi, S., Hafeez, A., Riaz, S. U., Ali, A., Ghauri, M. I., & Zehra, M. (2020). Low vitamin B12 levels: An underestimated cause of minimal cognitive impairment and dementia. *Cureus*, *12*(2), e6976. https://doi.org/10.7759/cureus.6976.

Kim, N. W., Piatyszek, M. A., Prowse, K. R., Harley, C. B., West, M. D., Ho, P. L., et al. (1994). Specific association of human telomerase activity with immortal cells and cancer. *Science*, *266*(5193), 2011–2015. https://doi.org/10.1126/science.7605428.

Ko, E., Seo, H. W., & Jung, G. (2018). Telomere length and reactive oxygen species levels are positively associated with a high risk of mortality and recurrence in hepatocellular carcinoma. *Hepatology*, *67*(4), 1378–1391. https://doi.org/10.1002/hep.29604.

Kong, C. M., Lee, X. W., & Wang, X. (2013). Telomere shortening in human diseases. *The FEBS Journal*, *280*(14), 3180–3193. https://doi.org/10.1111/febs.12326.

Lakshmiah, N., Lakshmi, A. V., & Reddy, G. B. (2016). Water solublevitamins. In M. S. Bamji, K. Krishnaswamy, & G. N. V. Brahmam (Eds.), *Text book of human nutrition* (4th ed.). New Delhi: Oxford & IBH Publishing Co Pvt Ltd.

Lee, Y. J., Wang, M. Y., Lin, M. C., & Lin, P. T. (2016). Associations between vitamin B-12 status and oxidative stress and inflammation in diabetic vegetarians and omnivores. *Nutrients*, *8*(3), 118. https://doi.org/10.3390/nu8030118.

Ling, C. T., & Chow, B. F. (1953). Effect of vitamin B12 on the levels of soluble sulfhydryl compounds in blood. The Journal of Biological Chemistry, 202(1), 445–456. Retrieved from http://www.ncbi.nlm.nih.gov/pubmed/13061470.

Liu, J. J., Prescott, J., Giovannucci, E., Hankinson, S. E., Rosner, B., & De Vivo, I. (2013). One-carbon metabolism factors and leukocyte telomere length. *The American Journal of Clinical Nutrition*, *97*(4), 794–799. https://doi.org/10.3945/ajcn.112.051557.

Ma, F., Lv, X., Du, Y., Chen, H., Liu, S., Zhao, J., et al. (2019). Association of Leukocyte Telomere Length with mild cognitive impairment and Alzheimer's disease: Role of folate and homocysteine. *Dementia and Geriatric Cognitive Disorders*, *48*(1–2), 56–67. https://doi.org/10.1159/000501958.

Mahalle, N., Kulkarni, M. V., Garg, M. K., & Naik, S. S. (2013). Vitamin B12 deficiency and hyperhomocysteinemia as correlates of cardiovascular risk factors in Indian subjects with coronary artery disease. *Journal of Cardiology*, *61*(4), 289–294. https://doi.org/10.1016/j.jjcc.2012.11.009.

Masutomi, K., Yu, E. Y., Khurts, S., Ben-Porath, I., Currier, J. L., Metz, G. B., et al. (2003). Telomerase maintains telomere structure in normal human cells. *Cells*, *114*(2), 241–253. https://doi.org/10.1016/s0092-8674(03)00550-6.

Mayer, S., Bruderlein, S., Perner, S., Waibel, I., Holdenried, A., Ciloglu, N., et al. (2006). Sex-specific telomere length profiles and age-dependent erosion dynamics of individual chromosome arms in humans. *Cytogenetic and Genome Research*, *112*(3–4), 194–201. https://doi.org/10.1159/000089870.

Meeran, S. M., Patel, S. N., & Tollefsbol, T. O. (2010). Sulforaphane causes epigenetic repression of hTERT expression in human breast cancer cell lines. *PLoS One*, *5*(7), e11457. https://doi.org/10.1371/journal.pone.0011457.

Meinila, J., Perala, M. M., Kautiainen, H., Mannisto, S., Kanerva, N., Shivappa, N., et al. (2019). Healthy diets and telomere length and attrition during a 10-year follow-up. *European Journal of Clinical Nutrition*, *73*(10), 1352–1360. https://doi.org/10.1038/s41430-018-0387-4.

Milne, E., O'Callaghan, N., Ramankutty, P., de Klerk, N. H., Greenop, K. R., Armstrong, B. K., et al. (2015). Plasma micronutrient levels and telomere length in children. *Nutrition*, *31*(2), 331–336. https://doi.org/10.1016/j.nut.2014.08.005.

Moore, E., Mander, A., Ames, D., Carne, R., Sanders, K., & Watters, D. (2012). Cognitive impairment and vitamin B12: A review. *International Psychogeriatrics*, *24*(4), 541–556. https://doi.org/10.1017/S1041610211002511.

Moslehi, J., DePinho, R. A., & Sahin, E. (2012). Telomeres and mitochondria in the aging heart. *Circulation Research*, *110*(9), 1226–1237. https://doi.org/10.1161/CIRCRESAHA.111.246868.

Mundstock, E., Sarria, E. E., Zatti, H., Mattos Louzada, F., Kich Grun, L., Herbert Jones, M., et al. (2015). Effect of obesity on telomere length: Systematic review and meta-analysis. *Obesity (Silver Spring)*, *23*(11), 2165–2174. https://doi.org/10.1002/oby.21183.

Nawrot, T. S., Staessen, J. A., Gardner, J. P., & Aviv, A. (2004). Telomere length and possible link to X chromosome. *Lancet*, *363*(9408), 507–510. https://doi.org/10.1016/S0140-6736(04)15535-9.

Paul, L. (2011). Diet, nutrition and telomere length. *The Journal of Nutritional Biochemistry*, *22*(10), 895–901. https://doi.org/10.1016/j.jnutbio.2010.12.001.

Paul, L., Cattaneo, M., D'Angelo, A., Sampietro, F., Fermo, I., Razzari, C., et al. (2009). Telomere length in peripheral blood mononuclear cells is associated with folate status in men. *The Journal of Nutrition*, *139*(7), 1273–1278. https://doi.org/10.3945/jn.109.104984.

Praveen, G., Shalini, T., Sivaprasad, M., & Reddy, G. B. (2020). Relative telomere length and mitochondrial DNA copy number variation with age: Association with plasma folate and vitamin B12. *Mitochondrion*, *51*, 79–87. https://doi.org/10.1016/j.mito.2020.01.007.

Pusceddu, I., Herrmann, W., Kleber, M. E., Scharnagl, H., Marz, W., & Herrmann, M. (2019). Telomere length, vitamin B12 and mortality in persons undergoing coronary angiography: The Ludwigshafen risk and cardiovascular health study. *Aging (Albany NY)*, *11*(17), 7083–7097. https://doi.org/10.18632/aging.102238.

Rane, G., Koh, W. P., Kanchi, M. M., Wang, R., Yuan, J. M., & Wang, X. (2015). Association between leukocyte telomere length and plasma homocysteine in a Singapore Chinese population. *Rejuvenation Research*, *18*(3), 203–210. https://doi.org/10.1089/rej.2014.1617.

Renaud, S., Loukinov, D., Bosman, F. T., Lobanenkov, V., & Benhattar, J. (2005). CTCF binds the proximal exonic region of hTERT and inhibits its transcription. *Nucleic Acids Research*, *33*(21), 6850–6860. https://doi.org/10.1093/nar/gki989.

Richards, J. B., Valdes, A. M., Gardner, J. P., Kato, B. S., Siva, A., Kimura, M., et al. (2008). Homocysteine levels and leukocyte telomere length. *Atherosclerosis*, *200*(2), 271–277. https://doi.org/10.1016/j.atherosclerosis.2007.12.035.

Richards, J. B., Valdes, A. M., Gardner, J. P., Paximadas, D., Kimura, M., Nessa, A., et al. (2007). Higher serum vitamin D concentrations are associated with longer leukocyte telomere length in women. *The American Journal of Clinical Nutrition*, *86*(5), 1420–1425. https://doi.org/10.1093/ajcn/86.5.1420.

Ritz, J. M., Kuhle, O., Riethdorf, S., Sipos, B., Deppert, W., Englert, C., et al. (2005). A novel transgenic mouse model reveals humanlike regulation of an 8-kbp human TERT gene promoter fragment in normal and tumor tissues. *Cancer Research*, *65*(4), 1187–1196. https://doi.org/10.1158/0008-5472.CAN-04-3046.

Sahin, E., Colla, S., Liesa, M., Moslehi, J., Muller, F. L., Guo, M., et al. (2011). Telomere dysfunction induces metabolic and mitochondrial compromise. *Nature*, *470*(7334), 359–365. https://doi.org/10.1038/nature09787.

Sahin, E., & DePinho, R. A. (2012). Axis of ageing: Telomeres, p53 and mitochondria. *Nature Reviews. Molecular Cell Biology*, *13*(6), 397–404. https://doi.org/10.1038/nrm3352.

Sahoo, M. K., Avasthi, A., & Singh, P. (2011). Negative symptoms presenting as neuropsychiatric manifestation of vitamin B12 deficiency. *Indian Journal of Psychiatry*, *53*(4), 370–371. https://doi.org/10.4103/0019-5545.91914.

Samassekou, O., Gadji, M., Drouin, R., & Yan, J. (2010). Sizing the ends: Normal length of human telomeres. *Annals of Anatomy*, *192*(5), 284–291. https://doi.org/10.1016/j.aanat.2010.07.005.

Satyanarayana, A., Balakrishna, N., Pitla, S., Reddy, P. Y., Mudili, S., Lopamudra, P., et al. (2011). Status of B-vitamins and homocysteine in diabetic retinopathy: Association with vitamin-B12 deficiency and hyperhomocysteinemia. *PLoS One*, *6*(11), e26747. https://doi.org/10.1371/journal.pone.0026747.

Shalini, T., Sivaprasad, M., Balakrishna, N., Madhavi, G., Radhika, M. S., Kumar, B. N., et al. (2019). Micronutrient intakes and status assessed by probability approach among the urban adult population of Hyderabad city in South India. *European Journal of Nutrition*, *58*(8), 3147–3159. https://doi.org/10.1007/s00394-018-1859-y.

Sharif, R., Thomas, P., Zalewski, P., & Fenech, M. (2012). The role of zinc in genomic stability. *Mutation Research*, *733*(1–2), 111–121. https://doi.org/10.1016/j.mrfmmm.2011.08.009.

Shin, C., & Baik, I. (2016). Leukocyte telomere length is associated with serum vitamin B12 and homocysteine levels in older adults with the presence of systemic inflammation. *Clinical Nutrition Research*, *5*(1), 7–14. https://doi.org/10.7762/cnr.2016.5.1.7.

Sivaprasad, M., Shalini, T., Balakrishna, N., Sudarshan, M., Lopamudra, P., Suryanarayana, P., et al. (2016). Status of vitamin B12 and folate among the urban adult population in South India. *Annals of Nutrition & Metabolism*, *68*(2), 94–102. https://doi.org/10.1159/000442677.

Sivaprasad, M., Shalini, T., Das, S., Vijayalakshmi, V., & Reddy, G. B. (2016). Vitamins B6, folic acid, and B12: Molecular functions and prevalence of deficiency in India. *Proceedings of Indian National Science Acadamy*, *82*(5), 1395–1412.

Sivaprasad, M., Shalini, T., Reddy, P. Y., Seshacharyulu, M., Madhavi, G., Kumar, B. N., et al. (2019). Prevalence of vitamin deficiencies in an apparently healthy urban adult population: Assessed by subclinical status and dietary intakes. *Nutrition*, *63-64*, 106–113. https://doi.org/10.1016/j.nut.2019.01.017.

Smith, A. D., & Refsum, H. (2009). Vitamin B-12 and cognition in the elderly. *The American Journal of Clinical Nutrition*, *89*(2), 707S–711S. https://doi.org/10.3945/ajcn.2008.26947D.

Spence, J. D. (2017). Increased coagulation with aging: Importance of homocysteine and vitamin B12. *Circulation Journal*, *81*(2), 268. https://doi.org/10.1253/circj.CJ-16-1011.

Takahashi-Iniguez, T., Garcia-Hernandez, E., Arreguin-Espinosa, R., & Flores, M. E. (2012). Role of vitamin B12 on methylmalonyl-CoA mutase activity. *Journal of Zhejiang University. Science. B*, *13*(6), 423–437. https://doi.org/10.1631/jzus.B1100329.

Tan, C. W., Ho, L. P., Kalimuddin, S., Cherng, B. P. Z., Teh, Y. E., Thien, S. Y., et al. (2020). Cohort study to evaluate the effect of vitamin D, magnesium, and vitamin B12 in combination on progression to severe outcomes in older patients with coronavirus (COVID-19). *Nutrition*, *79-80*, 111017. https://doi.org/10.1016/j.nut.2020.111017.

Tanaka, Y., Moritoh, Y., & Miwa, N. (2007). Age-dependent telomere-shortening is repressed by phosphorylated alpha-tocopherol together with cellular longevity and intracellular oxidative-stress reduction in human brain microvascular endotheliocytes. *Journal of Cellular Biochemistry*, *102*(3), 689–703. https://doi.org/10.1002/jcb.21322.

Terracciano, L. M., Piscuoglio, S., & Ng, C. K. Y. (2019). Hepatocellular carcinoma: Pathology and genetics. In P. Boffetta, & P. Hainaut (Eds.), *Encyclopedia of cancer* (3rd ed., pp. 198–210). Oxford: Academic Press.

Toussaint, M., Dionne, I., & Wellinger, R. J. (2005). Limited TTP supply affects telomere length regulation in a telomerase-independent fashion. *Nucleic Acids Research, 33*(2), 704–713. https://doi.org/10.1093/nar/gki219.

Tucker, L. A. (2019). Serum and dietary folate and vitamin B12 levels account for differences in cellular aging: Evidence based on telomere findings in 5581 U.S. adults. *Oxidative Medicine and Cellular Longevity, 2019*, 4358717. https://doi.org/10.1155/2019/4358717.

Turner, K. J., Vasu, V., & Griffin, D. K. (2019). Telomere biology and human phenotype. *Cell, 8*(1). https://doi.org/10.3390/cells8010073.

Tyrka, A. R., Carpenter, L. L., Kao, H. T., Porton, B., Philip, N. S., Ridout, S. J., et al. (2015). Association of telomere length and mitochondrial DNA copy number in a community sample of healthy adults. *Experimental Gerontology, 66*, 17–20. https://doi.org/10.1016/j.exger.2015.04.002.

Valdes, A. M., Andrew, T., Gardner, J. P., Kimura, M., Oelsner, E., Cherkas, L. F., et al. (2005). Obesity, cigarette smoking, and telomere length in women. *Lancet, 366*(9486), 662–664. https://doi.org/10.1016/S0140-6736(05)66630-5.

van de Lagemaat, E. E., de Groot, L., & van den Heuvel, E. (2019). Vitamin B12 in relation to oxidative stress: A systematic review. *Nutrients, 11*(2). https://doi.org/10.3390/nu11020482.

Villamor, E., Mora-Plazas, M., Forero, Y., Lopez-Arana, S., & Baylin, A. (2008). Vitamin B-12 status is associated with socioeconomic level and adherence to an animal food dietary pattern in Colombian school children. *The Journal of Nutrition, 138*(7), 1391–1398. https://doi.org/10.1093/jn/138.7.1391.

von Zglinicki, T., & Martin-Ruiz, C. M. (2005). Telomeres as biomarkers for ageing and age-related diseases. *Current Molecular Medicine, 5*(2), 197–203. https://doi.org/10.2174/1566524053586545.

Weber, D., Stuetz, W., Bernhard, W., Franz, A., Raith, M., Grune, T., et al. (2014). Oxidative stress markers and micronutrients in maternal and cord blood in relation to neonatal outcome. *European Journal of Clinical Nutrition, 68*(2), 215–222. https://doi.org/10.1038/ejcn.2013.263.

Wierzbicki, A. S. (2007). Homocysteine and cardiovascular disease: A review of the evidence. *Diabetes & Vascular Disease Research, 4*(2), 143–150. https://doi.org/10.3132/dvdr.2007.033.

Wolkowitz, O. M., Mellon, S. H., Epel, E. S., Lin, J., Dhabhar, F. S., Su, Y., et al. (2011). Leukocyte telomere length in major depression: correlations with chronicity, inflammation and oxidative stress- -preliminary findings. *PLoS One, 6*(3), e17837. https://doi.org/10.1371/journal.pone.0017837.

Xin, H., Liu, D., & Songyang, Z. (2008). The telosome/shelterin complex and its functions. *Genome Biology, 9*(9), 232. https://doi.org/10.1186/gb-2008-9-9-232.

Xu, Q., Parks, C. G., DeRoo, L. A., Cawthon, R. M., Sandler, D. P., & Chen, H. (2009). Multivitamin use and telomere length in women. *The American Journal of Clinical Nutrition, 89*(6), 1857–1863. https://doi.org/10.3945/ajcn.2008.26986.

Yakub, M., Iqbal, M. P., & Iqbal, R. (2010). Dietary patterns are associated with hyperhomocysteinemia in an urban Pakistani population. *The Journal of Nutrition, 140*(7), 1261–1266. https://doi.org/10.3945/jn.109.120477.

Zakian, V. A. (2012). Telomeres: The beginnings and ends of eukaryotic chromosomes. *Experimental Cell Research*, *318*(12), 1456–1460. https://doi.org/10.1016/j.yexcr.2012.02.015.

Zhang, Z. X., Wang, Y., Tao, Z. Z., Chen, S. M., Xiao, B. K., & Zhou, T. (2014). Subtelomeric demethylation deregulated hTERT expression, telomerase activity, and telomere length in four nasopharyngeal carcinoma cell lines. *Cancer Biotherapy & Radiopharmaceuticals*, *29*(7), 289–294. https://doi.org/10.1089/cbr.2013.1581.

Zhang, D. H., Wen, X. M., Zhang, L., & Cui, W. (2014). DNA methylation of human telomerase reverse transcriptase associated with leukocyte telomere length shortening in hyperhomocysteinemia-type hypertension in humans and in a rat model. *Circulation Journal*, *78*(8), 1915–1923. https://doi.org/10.1253/circj.cj-14-0233.

Zheng, Q., Huang, J., & Wang, G. (2019). Mitochondria, telomeres and telomerase subunits. *Frontiers in Cell and Development Biology*, 7, 274. https://doi.org/10.3389/fcell.2019.00274.

CHAPTER TWELVE

Vitamin B12 and chronic kidney disease

Henry H.L. Wu[a,b,*] **and Angela Yee-Moon Wang**[c]

[a]Department of Renal Medicine, Lancashire Teaching Hospitals NHS Foundation Trust, Preston, United Kingdom

[b]Faculty of Medical and Human Sciences, University of Manchester, Manchester, United Kingdom

[c]Department of Medicine, Queen Mary Hospital, The University of Hong Kong, Pokfulam, Hong Kong, China

*Corresponding author: e-mail address: honlinhenry.wu@postgrad.manchester.ac.uk

Contents

Abstract

Chronic Kidney Disease (CKD) is an emerging public health issue with a fast-growing global prevalence. Impairment in vitamin B12 metabolism is considered a nontraditional risk factor of poor outcomes associated with CKD, and there is greater interest from the scientific community than ever before to explore the role and influence of vitamin B12 in CKD. Homocysteine metabolism forms an important component of the vitamin B12 metabolic pathway. Hyperhomocysteinemia is frequently observed in CKD and End-Stage Kidney Disease (ESKD), but its representation as a prognostic marker for CKD outcomes is still not fully clear. This chapter reviews the vitamin B12 and homocysteine metabolic pathways and their dysfunction in CKD states. Biochemical factors and the MTHFR genetic polymorphisms which disrupt vitamin B12 and homocysteine metabolism are explored. The mechanisms of homocysteine-mediated and

Vitamins and Hormones, Volume 119
ISSN 0083-6729
https://doi.org/10.1016/bs.vh.2022.01.011

vitamin B12-mediated tissue damage in CKD are discussed. This chapter reviews current perspective on definition and measurement of plasma vitamin B12 levels in the CKD population. Updated evidence investigating the prognostic role of vitamin B12 for CKD outcomes is presented. Findings from major clinical trials conducted relating to outcomes from multivitamin (including folic acid and vitamin B12) supplementation in nondialysis and dialysis-dependent CKD are highlighted. The prognostic value of vitamin B12 and effects of vitamin B12 supplementation in the context of kidney transplantation and acute kidney injury is also reviewed. Future research considerations are summarized based on evidence gaps in our knowledge base of this topic. Greater abundance of high-level evidence to guide an approach toward vitamin B12 measurement, monitoring and supplementation in CKD may contribute to improved clinical outcomes.

1. Vitamin B12 and chronic kidney disease: An overview

Chronic Kidney Disease (CKD) is defined by Kidney Disease: Improving Global Outcomes (KDIGO) working group as reduction of the estimated glomerular filtration rate (eGFR) below $60\,mL/min/1.73\,m^2$, increase of the urinary albumin:creatinine ratio above 30 mg/g or an identified structural renal impairment which persists more than 3 months. CKD is divided into five main stages, with further subcategories reflecting chronic progression of disease (Fig. 1) (Inker et al., 2014; Levin et al., 2013). End-Stage Kidney Disease (ESKD) is the most severe stage of CKD. ESKD patients are in kidney failure and mortality is inevitable if they are not duly treated with renal replacement therapy in the form of kidney transplantation or dialysis (United States Renal Data System, 2020).

The global impact of CKD is well-known, with a disease prevalence of between 10% and 15% worldwide currently (Levin et al., 2013; United States Renal Data System, 2020). By 2040, CKD is estimated to become the fifth leading cause of death globally and one of the largest projected increases of any major cause of death (Foreman et al., 2018). This projection is likely explained by an increased aging general population and associated health issues such as protein energy wasting (PEW), malnutrition and sarcopenia, cognitive impairment, psychosocial problems and poverty (Clegg, Young, Iliffe, Rikkert, & Rockwood, 2013; Kalantar-Zadeh, Jafar, Nitsch, Neuen, & Perkovic, 2021; Kim, Kalantar-Zadeh, & Kopple, 2013; Nixon et al., 2018). PEW, malnutrition and sarcopenia in the context of renal impairment is related to reduced energy intake from uremia, superimposed infections, chronic inflammation, medication side-effects, low sunlight exposure, iron deficiency and under-utilization,

CURRENT CHRONIC KIDNEY DISEASE (CKD) NOMENCLATURE USED BY KDIGO

CKD is defined as abnormalities of kidney structure or function, present for >3 months, with implications for health and CKD is classified based on cause, GFR category, and albuminuria category (CGA).

Prognosis of CKD by GFR and albuminuria category

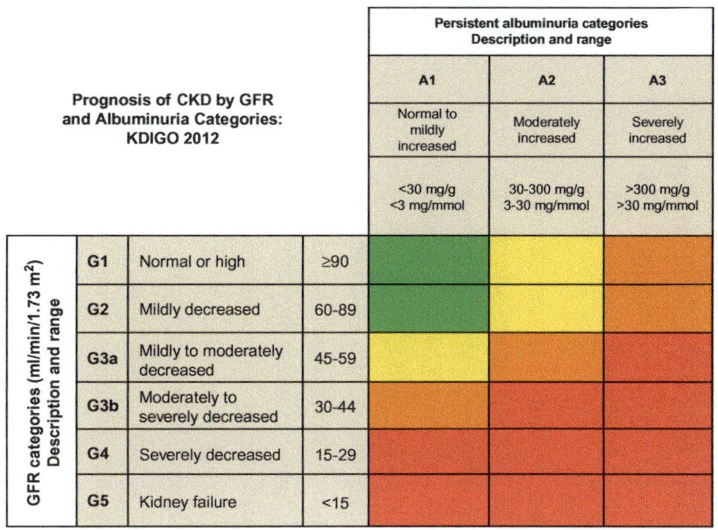

Green: low risk (if no other markers of kidney disease, no CKD); Yellow: moderately increased risk; Orange: high risk; Red, very high risk.

Fig. 1 Chronic kidney disease nomenclature from the KDIGO 2012 clinical practice guideline for the evaluation and management of chronic kidney disease. *From Levin, A., Stevens, P. E., Bilous, R. W., Coresh, J., De Francisco, A. L., De Jong, P. E., et al. (2013). Kidney Disease: Improving Global Outcomes (KDIGO) CKD Work Group. KDIGO 2012 clinical practice guideline for the evaluation and management of chronic kidney disease.* Kidney International Supplements, 3, *1–150. Permission obtained from Elsevier.*

impaired pathways of vitamin absorption and metabolism, and dialysis-related nutritional losses among other factors (Carrero et al., 2013; Walston et al., 2006). Cognitive impairment and psychosocial problems may lead to reduced ability to have an adequate dietary intake. Poverty limits access to dietary resources (Crews et al., 2015; Kim, Kim, Jang, Kim, & Chang, 2014). These factors contribute to the build-up of oxidative stress, cardiovascular risks, muscle weakness, worsened frailty status and mortality (Kovesdy & Kalantar-Zadeh, 2009). In advanced CKD, poor vitamin dietary intake is common irrespective of treatment received (Fouque, Pelletier, Mafra, & Chauveau, 2011). The myriad of unspecific symptoms related to CKD makes it difficult to distinguish whether symptoms are ascribed to severe uremia or nutritional inadequacy (Guarnieri, Antonione, & Biolo, 2003). Misjudgment of true nutritional status is frequently observed in clinical practice.

As the largest and most structurally complex vitamin in the human body, vitamin B12 is a water-soluble vitamin which plays a major role in the synthesis of myelin and maturation phase of erythropoiesis in the bone marrow (Brown, 2005; Green & Miller, 2014). The intricate associations between vitamin B12 and CKD have not been frequently explored in detail historically, though vitamin B12 deficiency in the background of CKD and ESKD is gradually being emphasized (Li, Spence, et al., 2020; Supakul et al., 2020). The "homocysteine hypothesis" concept in CKD is an intriguing topic. Elevated serum homocysteine levels (normal plasma homocysteine levels ranges from 4 to 10 mmol/L) in kidney impairment is commonly observed, kidneys being the primary organ of homocysteine metabolism in the human body. Significant associations exist between vitamin B12 and homocysteine in mammalian cells (Dhonukshe-Rutten et al., 2009; Shen & Ji, 2015). Vitamin B12 deficiency may lead to mild, moderate or severe impairments in homocysteine metabolism, dependent on the severity of vitamin B12 deficiency and other coexisting factors such as genetic polymorphisms and comorbidities such as thyroid and kidney disease. Hyperhomocysteinemia has strong correlations with atherosclerosis, vascular endothelial damage and cardiovascular mortality (Cattaneo, 1999; Cesari et al., 2005; De Koning, Werstuck, Zhou, & Austin, 2003). The role played by folic acid and vitamin B12, both water-soluble vitamins in the homocysteine metabolic pathway is acknowledged. Both are not merely essential cofactors for homocysteine metabolism, but the disruption to the homeostasis of folic acid and vitamin B12 is also directly detrimental to patient outcomes related to cardiovascular risk and CKD progression (Strain, Dowey, Ward, Pentieva, & McNulty, 2004; Verhoef et al., 1996).

In the first section, we will recap the chemical structure of homocysteine and explore its synthesis pathway in normal physiological conditions. Details of homocysteine metabolism and elimination with well-functioning kidneys will be summarized, and the processes of remethylation and transsulfuration, important in the homocysteine synthesis pathway are to be explored. The mechanisms contributing to vitamin B12 deficiency in renal impairment will be described and the impact of vitamin B12 deficiency on raised serum homocysteine levels will be discussed. The role of methylenetetrahydrofolate reductase (MTHFR) gene polymorphisms will be reviewed in the context of a CKD patient. The pathophysiology of homocysteine-mediated and vitamin B12-mediated tissue damage leading to endothelial stress and renal tissue damage will also be reviewed.

The latter section explores on the clinical outcomes of vitamin B12 deficiency and the effects of vitamin B12 supplementation in CKD. We will discuss current definition and monitoring approaches of vitamin B12 and homocysteine levels and recap the updated guidance on vitamin B12 dietary recommendations for CKD patients. The role of vitamin B12 and homocysteine as prognostic markers to predict cardiovascular, renal function and mortality outcomes will be discussed. We will discuss evidence highlighting the clinical impact of vitamin B12 supplementation on mortality risk, cardiovascular outcomes and CKD progression for nondialysis and dialysis dependent individuals. A global review of the current stance on vitamin B12 supplementation in CKD will be conducted. There will also be a brief discussion on issues relating to potassium and phosphate dietary restriction, drug interaction and polypharmacy with vitamin B12 supplementation. Current evidence relating to vitamin B12 prognostication and supplementation within the context of kidney transplantation and acute kidney injury (AKI) will be reviewed.

To conclude, we will highlight the gaps from our knowledge base in this topic and potential avenues of future research. Going forward, advancement of our knowledge on the associations between vitamin B12 and CKD, and to improve delivery of vitamin B12 supplementation for CKD patients should be a major aim within nephrology.

2. Vitamin B12 and the homocysteine metabolic pathway in chronic kidney disease

2.1 Homocysteine chemical structure and forms

Homocysteine, alongside its related structures of methionine and cysteine, are primary sulfur-containing amino acids (chemical structure of

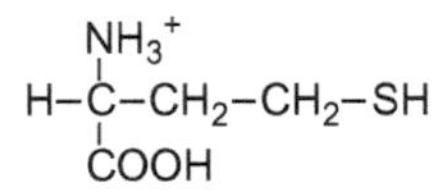

Fig. 2 Chemical structure of homocysteine. *From Miller, J. W. (2013). Homocysteine. In B. Caballero, L. Allen, & A. Prentice (Eds.),* Encyclopedia of human nutrition *(pp. 424–430). New York: Elsevier. Permission obtained from Elsevier.*

homocysteine shown in Fig. 2) which play a major role in cellular metabolism and protein synthesis (Miller, 2013). Free sulfhydryl groups are present at the end of the side chains for homocysteine and cysteine, and are highly susceptible to oxidation and formation of disulfide linkages with other sulfhydryl compounds. Methionine does not have a free sulfhydryl group and does not form disulfide compounds. Serum homocysteine can form disulfide linkages with cysteine residues within the primary sequences of albumin and other plasma proteins in protein-bound form, mixed disulfides (free cysteines or cysteine-containing peptides) or other homocysteine molecules. Approximately 1% of serum homocysteine is in free-reduced form (Kumar et al., 2017).

2.2 Homocysteine metabolism and elimination in normal functioning kidneys

Dietary methionine is the ultimate source of homocysteine, methionine being the initiating amino acid within the protein synthesis pathway. Homocysteine exists as an intermediate product during the conversion of methionine to cysteine in this biochemical cycle, a process called transsulfuration. A diagram overview of the methionine cycle and transsulfuration pathway in homocysteine metabolism is shown (Fig. 3) (Miller, 2013).

The cycle begins with the activation of methionine from addition of an adenosyl group such as adenosine triphosphate to form S-adenosylmethionine (SAM). SAM is the methylating agent in a variety of essential reactions involving DNA, RNA, proteins, membrane phospholipids, neurotransmitters, and creatine synthesis, and the end product of these reactions is S-adenosylhomocysteine (SAH). In turn, SAH is hydrolyzed to eventually form adenosine and homocysteine. Homocysteine then faces a crossroad in its metabolic journey, where it could either be remethylated to form methionine or undergo transsulfuration (or catabolism) through cystathionine synthesis. Homocysteine reacquires a methyl group during remethylation, supported by catalyzation from methionine synthase (5-methyltetrahydrofolate–homocysteine methyltransferase), a zinc-dependent enzyme. In this reaction,

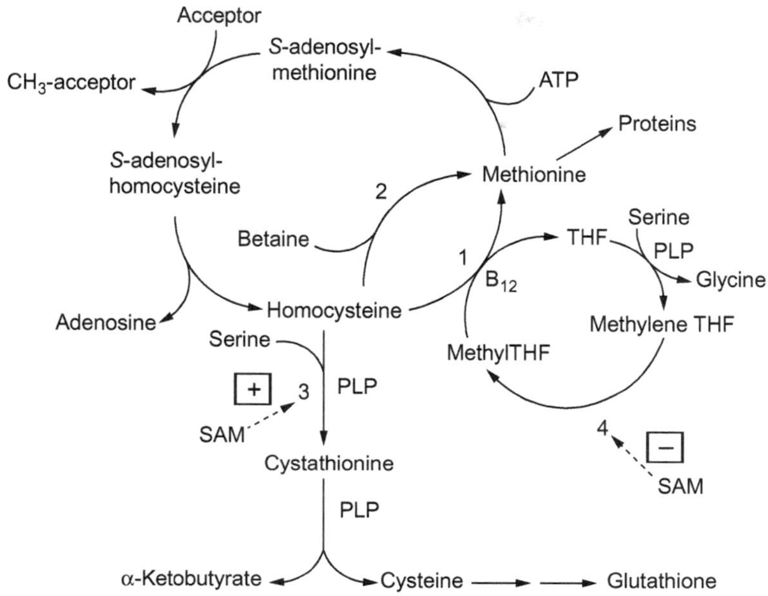

Fig. 3 The methionine cycle and transsulfuration pathway in homocysteine metabolism. Enhanced and inhibitory reactions regulated by SAM are represented by + and − signs. Enzymes involved in the methionine cycle and transsulfuration pathway: 1—methyltetrahydrofolate-homocysteine methyltransferase or methionine synthase; 2—betaine-homocysteine methyltransferase; 3—cystathionine beta-synthase; 4—MTHFR. *From Miller, J. W. (2013). Homocysteine. In B. Caballero, L. Allen, & A. Prentice (Eds.),* Encyclopedia of human nutrition *(pp. 424–430). New York: Elsevier. Permission obtained from Elsevier.*

methyltetrahydrofolate acts as the methyl donor and vitamin B12 is the cofactor. Although this mechanism is the primary homocysteine remethylation pathway in all mammalian cells, homocysteine may also be remethylated in a folate- and vitamin B12-independent reaction using betaine as the methyl donor with catalyzation by betaine-homocysteine methyltransferase. This alternative remethylation reaction however, occurs primarily within the liver and brain, and less likely within the kidney (Imbard et al., 2015; Pérez-Miguelsanz et al., 2017). Transsulfuration of homocysteine occurs in a condensation reaction with serine, during which cystathionine is synthesized. This reaction is catalyzed by cystathionine beta-synthase, and requires vitamin B6 in the form of pyridoxal-50-phosphate (PLP) being the cofactor. During a further PLP-dependent reaction catalyzed by cystathionase, cleaving of synthesized cystathionine forms alpha-ketobutyrate and cysteine. Cysteine metabolism ultimately results in formation of glutathione or inorganic sulfate.

Homocysteine metabolism is subject to allosteric control (Miller, 2013; Selhub, 1999). SAM acts as both an activator of cystathionine beta-synthase and an inhibitor of methylenetetrahydrofolate reductase (MTHFR), outside of serving as the universal methyl donor. MTHFR is a key catalyzer in converting 5,10-methenyltetrahydrofolate into 5-methyltetrahydrofolate, the main circulating form of folate, and therefore plays a major role in determining the extent in which homocysteine is recycled to form methionine or catabolized to form cystathionine. When there is high dietary supply of methionine, which is often the case after a meal rich in protein sources, intracellular SAM levels increase. High concentration of SAM activates cystathionine beta-synthase and inhibits MTHFR, thereby promoting homocysteine catabolism and limiting homocysteine remethylation. These allosteric functions serve to reduce additional recycling of homocysteine when there is a satisfactory supply of methionine. In reverse, when there is low dietary intake of methionine such as in fasting states, intracellular SAM levels reduces. Cystathionine beta-synthase will not be activated and there will be minimal inhibition of MTHFR, thereby promoting greater activity of homocysteine remethylation. This homeostatic mechanism maintains intracellular methionine levels when there is limited dietary supply. Oxidative stress exerts an additional level of control toward homocysteine metabolism (Moroni et al., 2010; Perna, Ingrosso, & De Santo, 2003). Oxidative stress inhibits methionine synthase activity, through oxidative inactivation of the vitamin B12 cofactor or by oxidation of cysteine residues which are required for zinc binding. Oxidative stress diverts homocysteine toward cystathionine instead of methionine synthesis, and thereby increases glutathione synthesis. Glutathione is a product of homocysteine metabolism through the transsulfuration pathway and an important intracellular antioxidant (Vitvitsky, Mosharov, Tritt, Ataullakhanov, & Banerjee, 2003). Other homeostatic alterations to homocysteine metabolism could be influenced by physiological changes relating to menopause, activity of estrogen, insulin, thyroxine, and thyroid stimulating hormone. Dysregulation in homocysteine metabolism could be observed in conditions such as diabetes mellitus and hypothyroidism as well as CKD (Wijekoon, Brosnan, & Brosnan, 2007).

Hydrogen Sulfide (H2S) plays an important role in the process of vitamin B12-dependent methyl group transfer (Pushpakumar, Kundu, & Sen, 2014; Toohey, 2017). H2S is a side product conversion of homocysteine to cysteine by cystathionine beta-synthase and cystathionine gamma-lyase. Plasma H2S levels are lowered with increased plasma homocysteine levels, a result of downregulation of cystathionine beta-synthase and cystathionine gamma-lyase, as is the case in CKD. H2S is an angiogenic agent with antioxidant and vasodilating properties. Similar to nitric oxide and carbon monoxide, it is also an endogenous gaseous mediator involved in vascular smooth muscle relaxation and inhibition of vascular smooth muscle proliferation, as well as lowering blood pressure. Impairment in the normal physiological pathway of H2S metabolism will accelerate development of uremia-associated atherosclerosis, commonly observed among the diabetic nephropathy population (Li, Feng, Zhang, & Wang, 2014; Zhou, Feng, Zhan, & Chen, 2014).

Kidney elimination of homocysteine is minimal in normal physiological conditions (Perna et al., 2010). Only nonprotein bound homocysteine is subject to glomerular filtration, before majority of the nonprotein bound homocysteine is reabsorbed in the tubule and oxidized to carbon dioxide and sulfate in the kidney cells. Homocysteine would be fully transsulfurated with well-functioning kidneys.

2.3 Vitamin B12 deficiency and hyperhomocysteinemia in CKD

Adequacy of vitamin B12, or cobalamin intake is a major factor which determines plasma homocysteine levels (Capelli et al., 2019; Long & Nie, 2016; Yeh et al., 2018). Cobalamin is metabolized into two active forms in target tissues, adenosylcobalamin in the mitochondria and methylcobalamin in the cytosol. Methylcobalamin is a methyl-transferring cofactor to the enzyme methionine synthase, which allows the process of homocysteine remethylation to methionine. Metabolic dysregulation in CKD and ESKD resulting in acidosis, systemic inflammation and disruption to hormone homeostasis would subsequently lead to malnutrition. B-vitamin deficiency, in particular vitamin B6, folate and vitamin B12 deficiency is commonly observed. Vitamin B6, folate and vitamin B12 deficiency may be exacerbated by surrounding comorbidities such as anorexia and gastroparesis. Dialysis-dependent patients may suffer from micronutrient loss including vitamin B12, as a result of long-term dialysis. Vitamin B12 deficiency in the background of CKD and ESKD leads to insufficient renal transsulfuration of homocysteine, contributing to increased plasma homocysteine levels.

In well-functioning kidneys, vitamin B12 binds in a complex with the plasma transport protein transcobalamin in the bloodstream following endocytosis via cubilin from the distal ileum enterocytes, and filtered by the kidney glomerulus before being reabsorbed in the proximal tubule with minimal urinary excretion. Vitamin B12 in the blood is primarily protein-bound. Approximately only 20% of circulating vitamin B12 will be bound to transcobalamin, in a biologically active form that can undergo cellular uptake while the remaining will be bound to haptocorrin (McMahon et al., 2015). Patients with CKD and ESKD usually display increased transcobalamin levels with an impaired tissue uptake of vitamin B12. There will be greater transcobalamin losses in the urine and reduced absorption in the proximal tubule, leading to functional vitamin B12 deficiency. Even with normal levels of total vitamin B12, there may be a paradoxical rise in cellular homocysteine levels (Ermens, Vlasveld, & Lindemans, 2003). This mechanism suggests downstream metabolites such as methylmalonic acid and homocysteine are more representative indicators of functional vitamin B12 status, rather than total vitamin B12 levels.

2.4 MTHFR gene polymorphisms

Individuals with certain MTHFR gene polymorphisms have greater susceptibility toward homocysteine-mediated disorders, namely myocardial infarction, cancer, diabetes mellitus and CKD (Cristalli et al., 2017). Multiple studies evaluated the association between MTHFR gene polymorphisms and clinical outcomes in the CKD and ESKD population (Chang et al., 2020; Jamison et al., 2009). Among the most frequently characterized MTHFR gene polymorphisms are the single nucleotide polymorphism at position 677 (MTHFR 677 C>T) and position 1298 (MTHFR 1298 A>C) (Cristalli et al., 2017). Both MTHFR 677 C>T and MTHFR 1298 A>C have been shown to be associated with increased cardiovascular risk among ESKD patients, whether they were receiving long-term dialysis or managed without dialysis (Poduri et al., 2008; Wrone et al., 2004). Patients with diabetic nephropathy as well as MTHFR 677 C>T and MTHFR 1298 A>C polymorphisms were shown to have greater likelihood of CKD progression, particularly if they are of South Indian or East Asian heritage (Ramanathan, Harichandana, Kannan, Elumalai, & Sfd, 2019). Despite these findings, the suggested associations between MTHFR gene polymorphisms and pathological consequences in CKD remain controversial and requires further validation. Contrary to earlier studies, a recent observational study concluded that MTHFR gene polymorphisms displayed a protective role toward kidney function. The study results found a lower frequency of MTHFR 677 A>C and MTHFR

1298 A > C among ESKD patients when compared to the non-ESKD group (Trovato et al., 2015). Other studies disregarded any significant association between MTHTR gene polymorphisms and long-term CKD outcomes (Chang et al., 2020; Rady et al., 2002).

Direct links between MTHFR gene polymorphisms and vitamin B12 levels have been evaluated. Patients with a MTHFR 677 C>T polymorphism C677 mutation and TT genotype were found to be greatly susceptible to vitamin B12 deficiency (Achour et al., 2016; Ni et al., 2017; Shiran et al., 2015). Coexistence of homozygosity for the C677T mutation or TT genotype and vitamin B12 deficiency significantly increases risks of endothelial dysfunction and cardiovascular risks (Shiran et al., 2015). Although the influence of MTHFR gene polymorphisms on increasing vitamin B12 levels and reducing plasma homocysteine have been recognized, conflicting results remain on whether TT, CC or CT MTHFR genotype achieve the best response (Dierkes et al., 1999; Tremblay et al., 2000). Nevertheless, the appropriateness to associate MTHFR gene polymorphisms with vitamin B12 intake requirements in CKD is debatable. The potential toxic effects of regular vitamin B12 supplementation in CKD patients are acknowledged. High-dose cyanocobalamin metabolism in CKD may lead to greater generation of cyanide with reduced renal clearance (Koyama et al., 1997).

2.5 Homocysteine-mediated and vitamin B12-mediated tissue damage in CKD

Homocysteine, being a thiol amino acid, contribute toward tissue and organ damage by advancing endothelial and vascular damage and atherosclerosis progression in CKD patient groups (Ma, Peng, Liu, Huang, & Luo, 2017; Wu et al., 2017). Hyperhomocysteinemia induces local oxidative stress through generation of reactive oxygen species (ROS), which rapidly undergoes autoxidation with oxygen and metal ion present. Further ROS production may be promoted by hyperhomocysteinemia from increased nicotinamide adenine dinucleotide phosphate oxidase (NAPDH) activity. Endothelial dysfunction may be severed in hyperhomocysteinemia with increased impairment of nitric oxide (NO) metabolism, which enhances vascular damage together with ROS-induced local microinflammation in CKD (Koyama et al., 1997; Li et al., 2019). Significant association between hyperhomocysteinemia and upregulated monocyte chemotactic protein 1 and interleukin-8 production resulting in endothelial monocyte adhesion, has been identified by evaluating cultured endothelial cell samples in basic science studies (Poddar, Sivasubramanian, DiBello, Robinson, & Jacobsen, 2001). The link between hyperhomocysteinemia and production of inflammatory factors is explained by the activated transcription factor nuclear factor kappa light chain enhancer of activated B-cells (NF-kappa B). NF-kappa B activation is a critical mechanism in causing endothelial reticulum stress with increased plasma homocysteine levels, by elevating endothelial inflammatory and apoptosis activity (Zhao et al., 2017). Endothelial cells do not express cystathionine beta-synthase nor betaine-homocysteine methyltransferase, and therefore lacks ability to catalyze the remethylation pathway within the methionine cycle (Zhang & Kaufman, 2008). Therefore, they are susceptible to further endothelial damage with hyperhomocysteinemia and CKD.

Hyperhomocysteinemia induces vascular smooth muscle cell proliferation through promotion of adhesion molecule, chemokine and vascular smooth muscle cell mitogen expression, triggering interactions with platelets, clotting factors and lipids (Jiang et al., 2021; Li et al., 2002). This may stimulate scavenger receptor-mediated macrophage uptake of oxidized low-density lipoprotein (LDL), increasing foam cell formation in atherosclerotic plaques with aggregation and accumulation of cholesterol (Colì et al., 2011; Pecoits-Filho, Lindholm, & Stenvinkel, 2002). Fibronectin is also involved in the process of protein homocysteinylation, and contribute to the results of extracellular matrix remodeling and development of sclerotic processes (Garcia-Fernandez et al., 2020). Progressive reduction of vascular wall elasticity in hyperhomocysteinemia is explained by a vascular remodeling process involving activation of metalloproteinase and intensified collagen synthesis (Böttiger, Hurtig-Wennlöf, Sjöström, Yngve, & Nilsson, 2007).

Effects of kidney tissue damage from hyperhomocysteinemia have been described. Hyperhomocysteinemia can cause sclerosis in glomerular cells and induce AKI through reducing plasma and tissue adenosine levels. Reduction in plasma adenosine levels enhances vascular smooth muscle cell proliferation, advances vascular sclerosis and mesangial expansion within the glomeruli which may eventually exacerbate podocyte dysfunction and kidney fibrosis (Perna et al., 2009; Prathapasinghe, Siow, & Karmin, 2007). Glomerular endothelial structure and function may be damaged over time by endothelial-derived NO and inducible NO synthase, resulting in exponentially increased urinary albumin excretion and acceleration of CKD progression (Jager et al., 2001).

A direct relationship between B-vitamins and adverse cardiovascular outcomes has been shown within the CKD population through the homocysteine-mediated pathogenic pathway and also other mechanisms unrelated to homocysteine metabolism. CKD patients in chronic inflammatory states, such as those in ESKD receiving hemodialysis, will likely have reduced production of transcobalamin II due to impaired intake of vitamin B12 from the circulation to peripheral tissues. As a result, there is increased synthesis of transcobalamin I and III which elevates the plasma vitamin B12 levels (Park et al., 2014; Seetharam & Li, 2000). Despite increased plasma vitamin B12 levels, there is still vitamin B12 deficiency within peripheral tissues due to defects in the uptake mechanism, which may lead to hyperhomocysteinemia and increased cardiovascular tissue damage (Sviri et al., 2012).

3. Vitamin B12 and chronic kidney disease in clinical practice

3.1 Defining and measuring vitamin B12 levels in CKD

Most healthy adults within the general population would have a 3–6 year supply of vitamin B12 in the body (Steiber & Kopple, 2011). This suggests that if the individual does not consume adequate quantities of vitamin B12 over 3 years, for example, then they would not strictly be classified as vitamin B12 deficient. The Recommended Dietary Allowance (RDA) for vitamin B12 is determined by maintenance of hematological status and on assumption that at least 50% of dietary vitamin B12 intake will be absorbed. According to the National Institutes of Health recommendations, the current RDA of vitamin B12 is 2.4 μg/day for the general population (Institute of Medicine, 2000). Food sources naturally rich in vitamin B12 include liver, beef, chicken, eggs, trout and salmon. These foods are protein-rich, and individuals with no vitamin B12 supplementation who consume a low-protein diet for extended periods are likely vitamin B12 deficient. Individuals age >50 are advised to consume food sources fortified with vitamin B12 such as cereal, or take vitamin B12 supplements to meet their RDA requirements, since up to 30% of people over age 50 are unable to absorb natural food sources of vitamin B12. Those with vitamin B12 deficiency due to lack of intrinsic factor require long-term medical treatment (Combs, 1998; Institute of Medicine, 2000).

Although the major pathophysiological mechanisms relating to increased risk of vitamin B12 deficiency in CKD are known, there is not even a paucity of data which quantifies body storage amounts of vitamin B12 within the CKD population, compared to that of the healthy general adult population. Discrepant views regarding routine screening of vitamin B12 levels for CKD patients remain, due to emerging evidence that total vitamin B12 levels may not be reflective of functional vitamin B12 status (Capelli et al., 2019; Ermens et al., 2003). Nevertheless, the updated KDOQI clinical practice guideline for nutrition in CKD recommended that homocysteine and vitamin B12 levels should be monitored as appropriate for each individual patient. The guidelines acknowledge that individuals receiving folic acid supplementation may have masked signs and symptoms of pernicious anemia or neurologic deficits. Therefore, regular monitoring of vitamin B12 levels is indicated within this context (Ikizler et al., 2020).

Whether to employ routine measurements of plasma homocysteine levels for CKD patients or not is currently unestablished and require further study. While it has been recognized that elevated plasma homocysteine levels is useful to determine true functional vitamin B12 deficiency, there is conflicting evidence on the clinical reliability of hyperhomocysteinemia to prognosticate outcomes in the CKD and ESKD population (refer to Sections 2.3 and 3.2) (Homocysteine Studies Collaboration, 2002; Xie et al., 2015).

3.2 Vitamin B12 levels as a prognostic marker of CKD outcomes

There is seldom evidence evaluating the reliability of vitamin B12 levels as a prognostic marker of outcomes in the CKD population. Previously, a retrospective observational study was conducted over a 5-year period for over 20,000 hemodialysis patients to determine the association between folic acid, vitamin B12 levels and all-cause mortality in a large cohort of hemodialysis patients. Interestingly, the authors concluded that having higher plasma vitamin B12 levels (>550 pg/mL) was strongly predictive of higher mortality risks after adjustment for sociodemographic and other laboratory variables (Soohoo et al., 2017). Further study on the prognostic value of vitamin B12 levels on CKD outcomes is required.

On the other hand, homocysteine levels as a prognostic marker in CKD and ESKD was well-investigated across retrospective and prospective studies, albeit with conflicting findings. Hyperhomocysteinemia was shown to be associated with left ventricular hypertrophy, left ventricular dysfunction and vascular disease, as well as CKD progression in a cross-sectional study conducted by Ye et al. for 1042 CKD stage I–V patients (Ye et al., 2016). Various studies demonstrated hyperhomocysteinemia to have good prognostic value for mortality in the CKD population, though Kalantar-Zadeh et al. demonstrated through a prospective observational study of 367 hemodialysis patients that lowered homocysteine levels was a strong predictor of mortality (Amin, El-Sayed, & Leheta, 2016; Ducloux, Bresson-Vautrin, Kribs, Abdelfatah, & Chalopin, 2002; Kalantar-Zadeh, Block, Humphreys, McAllister, & Kopple, 2004; Levi et al., 2014; Mallamaci et al., 2002). There were also multiple sources suggesting that homocysteine levels did not have strong predictive ability of CKD outcomes, such as cardiovascular risk and mortality (Bayés et al., 2003; London et al., 2004; Nair et al., 2005).

3.3 Vitamin B12 supplementation in nondialysis and dialysis dependent CKD

Numerous interventional trials were performed to investigate the effects of B complex vitamin supplementation on CKD progression and outcomes, in particular folic acid and vitamin B12 supplementation (Table 1).

The HOST (Homocysteinemia in Kidney and End-Stage Kidney Disease) randomized controlled trial (RCT) is double-blinded and placebo-controlled, which recruited 2056 patients with advanced CKD or dialysis-dependent ESKD and elevated homocysteine levels. The primary end point of the study was reduction of CKD progression and all-cause mortality while the secondary end point was reduction of cardiovascular mortality, amputation and thrombosis of vascular access. Patients were randomized to receive a combined therapy of vitamin B6, folic acid and vitamin B12 or placebo. Following a median follow-up of 3.2 years, the investigators noted significant reduction of homocysteine levels in the interventional group, but there were no significant differences between the interventional and placebo groups to reduce CKD progression and all-cause mortality. Treatment with high doses of vitamin B12, alongside vitamin B6 and folic acid, failed to delay the time to dialysis initiation in patients with previously nondialysis advanced CKD. The study also failed to reach its secondary end point (Jamison et al., 2007). Negative findings in the HOST RCT were attributed to high prevalence of cardiovascular comorbidities and suboptimal adherence to treatment among study participants. The HOST trial was underpowered to distinguish the CKD and ESKD populations for separate, individual analyses, which is a major limitation of this study. Heinz et al. conducted a multicenter RCT for 650 ESKD patients on chronic hemodialysis. Patients were randomized to receiving postdialysis—a multivitamin regime of vitamin B6 20 mg, folic acid 5 mg and vitamin B12 50 μg or placebo three times a week over 2 years. At the

Table 1 Major clinical trials investigating the effects of B complex vitamin (vitamin B6, folic acid and vitamin B12) supplementation on CKD progression and outcomes.

Study	Design and intervention	Study participants (*n*)	Primary end-point	Follow-up (years)	Study findings
Jamison et al. (2007)	Double-blind RCT (**HOST**) Vitamin B6 100 mg + folic acid 40 mg + vitamin B12 2 mg vs placebo	2056 CKD (eGFR ≤30 mL/min) or hemodialysis-dependent patients *Folic acid fortification*	CKD progression All-cause mortality	3.2	No significant differences between intervention and placebo groups
Heinz et al. (2010)	Double-blind RCT Vitamin B6 20 mg + folic acid 5 mg + vitamin B12 50 μg vs placebo	650 Hemodialysis-dependent patients *Folic acid fortification status not specified*	All-cause mortality Fatal and nonfatal cardiovascular event	2	No significant differences between intervention and placebo groups
House et al. (2010)	Double-blind RCT (**DIVINe**) Vitamin B6 25mg + folic acid 2.5 mg + vitamin B12 1 mg vs placebo	238 Patients with diabetic nephropathy and eGFR >30 mL/min *Folic acid fortification*	CKD progression Cardiovascular and cerebrovascular events	2.6	Greater GFR reduction and more frequent cardiovascular and cerebrovascular events in the intervention group
Xu et al. (2016)	Double-blind RCT (**prespecified renal substudy of CSPPT**) Enalapril 10 mg + folic acid vs enalapril alone	15,104 Patients with eGFR ≥30 mL/min *No folic acid fortification*	CKD progression	4.4	Delay in CKD progression in the Enalapril + Folic Acid group
Li, Bahnson, et al. (2020); Li, Spence, et al. (2020)	Double-blind RCT (**posthoc analysis for prespecified renal substudy of CSPPT**) Enalapril 10 mg + folic acid vs enalapril alone	1374 Patients with eGFR between 30 and 60 mL/min and hypertension *No folic acid fortification*	Modifying effects of plasma vitamin B12 levels on CKD progression	4.4	Delay in CKD progression in the Enalapril + Folic Acid group with higher plasma vitamin B12 levels

end of follow-up, there were no significant differences between the two groups for all-cause mortality, fatal and nonfatal cardiovascular events, although a posthoc analysis of the same cohort identified a significant reduction in unstable angina episodes and revascularization procedures (Heinz et al., 2010).

The Diabetic Intervention with Vitamins to Improve Nephropathy (DIVINe) RCT is a double-blinded study which randomized 238 diabetic kidney disease patients to receive high-dose B vitamin supplementation (vitamin B6 25 mg, folic acid 2.5 mg and vitamin B12 1 mg) or placebo. The mean eGFR at commencement of the study was 64 mL/min for the intervention group and 58 mL/min for the placebo group. Over a follow-up of 2.6 years, interventional treatment with vitamin B6, folic acid and vitamin B12 demonstrated significant associations with eGFR decline and increased risk of cardiovascular and cerebrovascular events (House et al., 2010).

The China Stroke Primary Prevention Trial (CSPPT) remains the largest double-blinded, placebo-controlled RCT evaluating the outcomes of folic acid supplementation. CSPPT recruited 20,702 patients with hypertension without previous folic acid fortification or vitamin B12 supplementation, and who did not have medical history of myocardial infarction or stroke. Recruited patients were randomized to receive enalapril, an angiotensin-converting enzyme inhibitor medication for hypertension with folic acid 0.8 mg in a daily basis or enalapril alone. The original primary analysis showed adjusted risk reduction of first stroke and composite cardiovascular events over a median follow-up of 4.5 years for the enalapril-folic acid compared to enalapril only group (Qin et al., 2016). A prespecified renal substudy of the CSPPT involving 15,104 patients with baseline eGFR >30 mL/min showed a 21% reduction in the adjusted odds of the primary outcome of CKD progression from enalapril-folic acid compared to enalapril alone. Another subgroup analysis of study patients with baseline eGFR between 30 and 60 mL/min indicated a 55% reduction in risk of CKD progression and 33% reduced risk for rapid eGFR decline with enalapril-folic acid compared to enalapril alone. There were no significant clinical benefits of folic acid supplementation alongside enalapril in individuals without CKD. Findings suggest that effects of folic acid supplementation may be modified by eGFR status (Xu et al., 2016).

The relevance of CSPPT in this discussion stems from published results in a recent posthoc cross-sectional analysis. Li et al. explored the modifying effects of plasma vitamin B12 levels on the association between folic acid supplementation and CKD progression for 1374 patients with hypertension and baseline eGFR between 30 and 60 mL/min. As expected, patients with higher plasma vitamin B12 levels had lower plasma homocysteine levels. Enalapril-folic acid treatment had no effect on plasma vitamin B12 levels for each baseline stratum defined from this posthoc study. Plasma vitamin B12 levels however, appeared as a modifying factor for the treatment effects of folic acid. For patients with baseline plasma vitamin B12 levels ≥248 pmol/L, enalapril-folic acid displayed an 83% odds reduction of CKD progression compared to administering enalapril alone. The upper stratum of baseline plasma vitamin B12 levels was associated with 58% reduced risk for a composite outcome of CKD progression and mortality and 55% reduced risk of rapid eGFR decline. For the lower stratums of plasma vitamin B12, enalapril-folic acid treatment had no significant effect compared to enalapril alone on lowering risks of CKD progression, mortality and rapid eGFR decline (Li, Spence, et al., 2020).

Contrasting results from the aforementioned RCTs have been acknowledged (Capelli et al., 2019; Wang, 2020). From the folic acid supplementation perspective, baseline folic acid levels were different between the three major RCTs (HOST 16.5 ng/mL, DIVINe 15 ng/mL and CSPPT CKD substudy 7.7 ng/mL), supporting the hypothesis that beneficial effects of folic acid supplementation in CKD may be more apparent when there are low baseline folic acid levels. Another explanation for these contrasting results are the different dosages administered in the B vitamin supplementation regime (CSPPT used folic acid 0.8 mg/day, HOST and DIVINe used folic acid up to 40 mg/day and vitamin B12 up to 2000 mg/day). There were massive differences in folate blood levels reached by the HOST study (2000 ng/mL) compared with CSPPT (23 ng/mL), for instance. These findings allowed postulation of potential plasma toxicity from accumulation of unmetabolized folic acid. Contrasting results from these RCTs could also be explained by the different levels of CKD severity of study participants between these studies. The HOST population comprised of patients with advanced CKD and a high comorbidity burden, which may reduce study power when determining CKD outcomes from this patient cohort. In CSPPT, the significant benefits of folic acid supplementation to reduce risks of kidney damage could be related to choosing a study population with mild-moderate CKD. In addition, CSPPT was carried out in China, in a population without folic acid grain fortification, while the study population in HOST and DIVINe were based in countries with folic acid fortification programs.

Various systematic reviews and meta-analysis were conducted to summarize findings from the RCTs in this topic area. A meta-analysis from Heinz et al. involving 5 RCTs and 1642 dialysis patients suggested those receiving vitamin B6, folic acid and vitamin B12 compared to placebo may achieve significant cardiovascular risk reduction, despite no reduced mortality risks (Heinz, Kropf, Luley, & Dierkes, 2009). A larger scale meta-analysis involving 14 RCTs conducted by Ji et al. demonstrated reduced stroke events among CKD patients receiving vitamin B6, folic acid and vitamin B12 supplementation, resulting in lowering of homocysteine levels (Ji et al., 2013). Pan et al. conducted a meta-analysis involving 10 RCTs which suggested otherwise, noting that homocysteine-lowering therapy in the form of multivitamin supplementation was not associated with significant reduction in cardiovascular event and stroke risks, as well as all-cause mortality (Pan et al., 2012). Its conclusion was echoed by another systematic review and meta-analysis conducted by Nigwekar et al. of 6 studies and 2452 patients on chronic hemodialysis, all of which were included in the meta-analyses from the aforementioned publications (Nigwekar et al., 2016).

Few studies specifically addressed the effect of vitamin B12 supplementation alone on CKD progression and outcomes. The role of vitamin B12 supplementation in CKD progression and outcomes remains unclear with mixed results in literature, and subject to further investigation. Although advised vitamin B12 intake should be primarily based from RDA recommendations for the general population, updated KDOQI guidelines advocate for greater individualization of vitamin B12 supplementation based on clinical judgment. The benefits of routine vitamin B12 administration in CKD and ESKD are not yet well-established. There is recognition for a need to increase vitamin B12 supplementation in the older CKD population, due to higher prevalence of atrophic gastritis (Ikizler et al., 2020).

Other key issues regarding vitamin B12 supplementation in CKD relate to potassium and phosphate dietary restriction, drug interaction and polypharmacy. Potassium and phosphate intake may interfere with the vitamin B12 absorption and excess levels may contribute to vitamin B12 deficiency. In early CKD, strict potassium and phosphate dietary intake is not indicated. Decline in potassium and phosphate clearance is observed in advanced CKD. Low potassium and phosphate diet is recommended to reduce electrolyte complications secondary to CKD, such as cardiovascular arrhythmias and mineral bone disease. For patients with CKD or ESKD requiring dialysis, a cautious approach to vitamin B12 supplementation with meticulous monitoring of serum electrolyte levels is required to reduce risks of secondary complications (Capelli et al., 2019; Kalantar-Zadeh et al., 2015; Kelly, Rossi, Johnson, & Campbell, 2017). In addition to potassium and phosphate intake, numerous medications may disrupt vitamin B12 absorption. Medications which reduce absorption of vitamin B12 supplements include colchicine, metformin, aminoglycoside-class antibiotics such as gentamicin, antiepileptic medications such as phenytoin and heartburn medications such as ranitidine. Patients with advanced CKD or ESKD may have multiple medical comorbidities which indicate the need for polypharmacy. Awareness of potential drug interactions and consideration of alternative treatment options may ensure greater adequacy of vitamin B12 supplementation (O'Leary & Samman, 2010; Patel, Rana, Suthar, Malhotra, & Patel, 2014).

3.4 Vitamin B12 and kidney transplantation

Multiple factors influence kidney transplantation outcomes, including the immunosuppression received, chronicity of immunosuppression, transplant failure, metabolic dysregulation and inflammatory abnormalities, and malignant complications, in addition to nutritional intake adequacy (La Manna et al., 2010). Conventionally, decline in homocysteine levels following kidney transplantation is observed, though in many cases, hyperhomocysteinemia may persist. Results from previous studies indicate that kidney transplant recipients administered with multivitamins containing high-dose vitamin B12, alongside vitamin B6 and folic acid, can lower plasma homocysteine levels (Beaulieu et al., 1999; Bostom et al., 1999; Friedman, Rosenberg, Selhub, Levey, & Bostom, 2002).

The effect of vitamin B12 supplementation on kidney transplantation outcomes has been seldom discussed and available results are controversial. In the Folic Acid for Vascular Outcome Reduction in Transplantation (FAVORIT) trial, the effect of B-vitamin supplementation including vitamin B6, folic acid and vitamin B12 on cardiovascular risk and all-cause mortality was investigated among 4110 kidney transplant recipients. 2056 study participants were randomized to the high dose supplementation group (vitamin B6 50 mg, folic acid 5.0 mg, vitamin B12 1.0 mg) while 2054 study participants were randomized to the low dose supplementation group (vitamin B6 1.4 mg, no folic acid, vitamin B12 2.0 μg). FAVORIT demonstrated the high dose supplementation group being more effective in lowering plasma homocysteine levels compared to the low dose supplementation group. However, the incidence of cardiovascular disease, onset of dialysis-dependent kidney failure and all-cause mortality did not differ between the high dose and low dose supplementation groups, and FAVORIT was terminated following interim analysis (Bostom et al., 2011). In a longitudinal ancillary study of the FAVORIT trial, the high dose supplementation regime only provided modest cognitive benefit for kidney transplant recipients with baseline hyperhomocysteinemia, compared with the low dose supplementation regime (Scott et al., 2017).

3.5 Vitamin B12 and AKI

Since Sevitt and Hoffbrand's initial study in 1969 showing potential associations between elevated plasma vitamin B12 levels and AKI, there is scarce evidence over the ensuing decades which evaluated the association between vitamin B12 and AKI to a greater extent (Sevitt & Hoffbrand, 1969). There are conflicting findings and the physiological relationship between vitamin B12 and AKI remain unclear.

In a retrospective propensity score matching analysis study, Park et al. aimed to determine the predictive role of vitamin B12 in AKI among 591 patients who underwent elective living donor liver transplantation. The study noted that higher preoperative plasma vitamin B12 levels were observed in patients who developed AKI following liver transplant. Propensity scoring matching analyses concluded that posttransplant high plasma vitamin B12 levels (defined by >900 ng/mL) was significantly associated with risk for AKI. Patients in the high plasma vitamin B12 levels group had a 2.8-fold risk for AKI higher than those in the normal plasma vitamin B12 levels group. In fact, increase in plasma vitamin B12 levels is also associated with the severity of AKI stage. Associations between vitamin B12 levels and inflammatory factors such as C reactive protein, white blood cells and albumin was demonstrated. In the context of this study, preoperative and intraoperative factors may have significantly contributed to inflammation and increased risk for AKI, confounding the effects of elevated vitamin B12 levels (Park et al., 2020).

Given kidney ischemia/reperfusion injury (IRI) being the leading cause of AKI, a mouse study was conducted by Li et al. to investigate the effects of oral high dose vitamin B12 supplementation on post-IRI outcomes. C57BL/6J wild type mice with IRI were either administered with vitamin B12 via drinking water at a dose 50 mg/L or left untreated. Results showed that treated mice displayed near normal kidney function and morphology.

Furthermore, IRI-induced changes in RNA, protein markers of inflammation, fibrosis, apoptosis and DNA damage response were significantly attenuated by at least 50% in the treated IRI mice group compared to the untreated mice group. The presence of vitamin B12 at 0.3 μM in the culture medium of mouse proximal tubular cells subjected to 3 h hypoxia followed by 1 h reperfusion period in vitro showed significant protective effects, of which there is increased cell viability and decreased reactive oxygen species level. The study investigators concluded that high dose of vitamin B12 is protective against perfusion injury both in vivo and in vitro in mice, without noticeable detrimental effects. Therefore, this suggests the potential use of vitamin B12 supplementation as a beneficial treatment measure for IRI-mediated AKI (Li, Bahnson, et al., 2020; Li, Spence, et al., 2020).

4. Vitamin B12 and chronic kidney disease: Evidence gaps and future directions

Several major evidence gaps in our knowledge base of this topic are still to be addressed. A program to inform recommended daily allowance of vitamin B12 for individual patients is to be established, considering various stages of CKD and different comorbidity status. Designing a screening program to measure plasma vitamin B12 levels in the CKD population may help, as is more complete knowledge on the prevalence of vitamin B12 deficiency across various stages of CKD. Further research is needed to clarify the prognostic value of vitamin B12, folate and homocysteine in the context of kidney disease. Well-designed trials are required to investigate whether and to what extent vitamin B12 supplementation improves outcomes in CKD patients and slows down CKD progression. Future trials should limit study inclusion to a certain baseline status, or statistically adjust for baseline status during analysis of results. Dose-response studies for vitamin B12 intake should be considered, especially for the ESKD population receiving maintenance dialysis and CKD patients with polypharmacy taking medications which interfere with the absorption efficacy of vitamin B12. Dietary interventions which target increased micronutrient intake should be evaluated, particularly its effects on relevant CKD outcomes (Ikizler et al., 2020). Going forward, greater abundance of high-level evidence to guide an approach toward vitamin B12 measurement, monitoring and supplementation in the CKD population may contribute to improved clinical outcomes.

References

Achour, O., Elmtaoua, S., Zellama, D., Omezzine, A., Moussa, A., Rejeb, J., et al. (2016). The C677T MTHFR genotypes influence the efficacy of B9 and B12 vitamins supplementation to lowering plasma total homocysteine in hemodialysis. *Journal of Nephrology, 29*, 691–698.

Amin, H. K., El-Sayed, M. I. K., & Leheta, O. F. (2016). Homocysteine as a predictive biomarker in early diagnosis of renal failure susceptibility and prognostic diagnosis for end stages renal disease. *Renal Failure, 38*, 1267–1275.

Bayés, B., Pastor, M. C., Bonal, J., Junca, J., Hernandez, J. M., Riutort, N., et al. (2003). Homocysteine, C-reactive protein, lipid peroxidation and mortality in haemodialysis patients. *Nephrology Dialysis Transplantation, 18*, 106–112.

Beaulieu, A. J., Gohh, R. Y., Han, H., Hakas, D., Jacques, P. F., Selhub, J., et al. (1999). Enhanced reduction of fasting total homocysteine levels with supraphysiological versus standard multivitamin dose folic acid supplementation in renal transplant recipients. *Arteriosclerosis, Thrombosis, and Vascular Biology, 19*, 2918–2921.

Bostom, A. G., Carpenter, M. A., Kusek, J. W., Levey, A. S., Hunsicker, L., Pfeffer, M. A., et al. (2011). Homocysteine-lowering and cardiovascular disease outcomes in kidney transplant recipients: Primary results from the Folic Acid for Vascular Outcome Reduction in Transplantation trial. *Circulation, 123*, 1763–1770.

Bostom, A. G., Gohh, R. Y., Liaugaudas, G., Beaulieu, A. J., Han, H., Jacques, P. F., et al. (1999). Prevalence of mild fasting hyperhomocysteinemia in renal transplant versus coronary artery disease patients after fortification of cereal grain flour with folic acid. *Atherosclerosis, 145*, 221–224.

Böttiger, A. K., Hurtig-Wennlöf, A., Sjöström, M., Yngve, A., & Nilsson, T. K. (2007). Association of total plasma homocysteine with methylenetetrahydrofolate reductase genotypes 677C> T, 1298A> C, and 1793G> A and the corresponding haplotypes in Swedish children and adolescents. *International Journal of Molecular Medicine, 19*, 659–665.

Brown, K. L. (2005). Chemistry and enzymology of vitamin B12. *Chemical Reviews, 105*, 2075–2150.

Capelli, I., Cianciolo, G., Gasperoni, L., Zappulo, F., Tondolo, F., Cappuccilli, M., et al. (2019). Folic acid and vitamin B12 administration in CKD, why not? *Nutrients, 11*, 383.

Carrero, J. J., Stenvinkel, P., Cuppari, L., Ikizler, T. A., Kalantar-Zadeh, K., Kaysen, G., et al. (2013). Etiology of the protein-energy wasting syndrome in chronic kidney disease: A consensus statement from the International Society of Renal Nutrition and Metabolism (ISRNM). *Journal of Renal Nutrition, 23*, 77–90.

Cattaneo, M. (1999). Hyperhomocysteinemia, atherosclerosis and thrombosis. *Thrombosis and Haemostasis, 81*, 165–176.

Cesari, M., Zanchetta, M., Burlina, A., Pedon, L., Maiolino, G., Sticchi, D., et al. (2005). Hyperhomocysteinemia is inversely related with left ventricular ejection fraction and predicts cardiovascular mortality in high-risk coronary artery disease hypertensives. *Arteriosclerosis, Thrombosis, and Vascular Biology, 25*, 115–121.

Chang, H. L., Chen, G. R., Hsiao, P. J., Chiu, C. C., Tai, M. C., Kao, C. C., et al. (2020). Decisive evidence corroborates a null relationship between MTHFR C677T and chronic kidney disease: A case–control study and a meta-analysis. *Medicine, 99*(29), e21045.

Clegg, A., Young, J., Iliffe, S., Rikkert, M. O., & Rockwood, K. (2013). Frailty in elderly people. *The Lancet, 381*, 752–762.

Colì, L., Donati, G., Cappuccilli, M. L., Cianciolo, G., Comai, G., Cuna, V., et al. (2011). Role of the hemodialysis vascular access type in inflammation status and monocyte activation. *The International Journal of Artificial Organs, 34*, 481–488.

Combs, G. (1998). *The vitamins: Fundamental aspects in nutrition and health*. San Diego, CA: Academic Press.

Crews, D. C., Kuczmarski, M. F., Miller, E. R., III, Zonderman, A. B., Evans, M. K., & Powe, N. R. (2015). Dietary habits, poverty, and chronic kidney disease in an urban population. *Journal of Renal Nutrition*, *25*, 103–110.

Cristalli, C. P., Zannini, C., Comai, G., Baraldi, O., Cuna, V., Cappuccilli, M., et al. (2017). Methylenetetrahydrofolate reductase, MTHFR, polymorphisms and predisposition to different multifactorial disorders. *Genes & Genomics*, *39*, 689–699.

De Koning, A. L., Werstuck, G. H., Zhou, J., & Austin, R. C. (2003). Hyperhomocysteinemia and its role in the development of atherosclerosis. *Clinical Biochemistry*, *36*, 431–441.

Dhonukshe-Rutten, R. A. M., De Vries, J. H. M., De Bree, A., Van Der Put, N., Van Staveren, W. A., & De Groot, L. C. P. G. M. (2009). Dietary intake and status of folate and vitamin B12 and their association with homocysteine and cardiovascular disease in European populations. *European Journal of Clinical Nutrition*, *63*, 18–30.

Dierkes, J., Domröse, U., Ambrosch, A., Schneede, J., Guttormsen, A. B., Neumann, K. H., et al. (1999). Supplementation with vitamin B12 decreases homocysteine and methylmalonic acid but also serum folate in patients with end-stage renal disease. *Metabolism*, *48*, 631–635.

Ducloux, D., Bresson-Vautrin, C., Kribs, M., Abdelfatah, A., & Chalopin, J. M. (2002). C-reactive protein and cardiovascular disease in peritoneal dialysis patients. *Kidney International*, *62*, 1417–1422.

Ermens, A. A. M., Vlasveld, L. T., & Lindemans, J. (2003). Significance of elevated cobalamin (vitamin B12) levels in blood. *Clinical Biochemistry*, *36*, 585–590.

Foreman, K. J., Marquez, N., Dolgert, A., Fukutaki, K., Fullman, N., McGaughey, M., et al. (2018). Forecasting life expectancy, years of life lost, and all-cause and cause-specific mortality for 250 causes of death: Reference and alternative scenarios for 2016–40 for 195 countries and territories. *The Lancet*, *392*, 2052–2090.

Fouque, D., Pelletier, S., Mafra, D., & Chauveau, P. (2011). Nutrition and chronic kidney disease. *Kidney International*, *80*, 348–357.

Friedman, A. N., Rosenberg, I. H., Selhub, J., Levey, A. S., & Bostom, A. G. (2002). Hyperhomocysteinemia in renal transplant recipients. *American Journal of Transplantation*, *2*, 308–313.

Garcia-Fernandez, N., Jacobs-Cachá, C., Mora-Gutiérrez, J. M., Vergara, A., Orbe, J., & Soler, M. J. (2020). Matrix metalloproteinases in diabetic kidney disease. *Journal of Clinical Medicine*, *9*, 472.

Green, R., & Miller, J. W. (2014). Vitamin B_{12}. In J. Zempleni, J. W. Suttie, J. F. Gregory III,, & P. J. Stover (Eds.), *Handbook of vitamins* (5th ed.). New York: CRC Press.

Guarnieri, G., Antonione, R., & Biolo, G. (2003). Mechanisms of malnutrition in uremia. *Journal of Renal Nutrition*, *13*, 153–157.

Heinz, J., Kropf, S., Domröse, U., Westphal, S., Borucki, K., Luley, C., et al. (2010). Vitamins and the risk of total mortality and cardiovascular disease in end-stage renal disease results of a randomized controlled trial. *Circulation*, *121*, 1432–1438.

Heinz, J., Kropf, S., Luley, C., & Dierkes, J. (2009). Homocysteine as a risk factor for cardiovascular disease in patients treated by dialysis: A meta-analysis. *American Journal of Kidney Diseases*, *54*, 478–489.

Homocysteine Studies Collaboration. (2002). Homocysteine and risk of ischemic heart disease and stroke. *Journal of the American Medical Association*, *288*, 2015–2022.

House, A. A., Eliasziw, M., Cattran, D. C., Churchill, D. N., Oliver, M. J., Fine, A., et al. (2010). Effect of B-vitamin therapy on progression of diabetic nephropathy: A randomized controlled trial. *Journal of the American Medical Association*, *303*, 1603–1609.

Ikizler, T. A., Burrowes, J. D., Byham-Gray, L. D., Campbell, K. L., Carrero, J. J., Chan, W., et al. (2020). KDOQI clinical practice guideline for nutrition in CKD: 2020 update. *American Journal of Kidney Diseases, 76*, S1–S107.

Imbard, A., Benoist, J. F., Esse, R., Gupta, S., Lebon, S., De Vriese, A. S., et al. (2015). High homocysteine induces betaine depletion. *Bioscience Reports, 35*, e00222.

Inker, L. A., Astor, B. C., Fox, C. H., Isakova, T., Lash, J. P., Peralta, C. A., et al. (2014). KDOQI US commentary on the 2012 KDIGO clinical practice guideline for the evaluation and management of CKD. *American Journal of Kidney Diseases, 63*, 713–735.

Institute of Medicine. (2000). *For thiamine, riboflavin, niacin, vitamin B6, folate, B12, pantothenic acid, biotin/choline*. Washington, DC: National Academy Press.

Jager, K. J., Merkus, M. P., Huisman, R. M., Boeschoten, E. W., Dekker, F. W., Korevaar, J. C., et al. (2001). Nutritional status over time in hemodialysis and peritoneal dialysis. *Journal of the American Society of Nephrology, 12*, 1272–1279.

Jamison, R. L., Hartigan, P., Kaufman, J. S., Goldfarb, D. S., Warren, S. R., Guarino, P. D., et al. (2007). Effect of homocysteine lowering on mortality and vascular disease in advanced chronic kidney disease and end-stage renal disease: A randomized controlled trial. *Journal of the American Medical Association, 298*, 1163–1170.

Jamison, R. L., Shih, M. C., Humphries, D. E., Guarino, P. D., Kaufman, J. S., Goldfarb, D. S., et al. (2009). Effect of the MTHFR C677T and A1298C polymorphisms on survival in patients with advanced CKD and ESRD: A prospective study. *American Journal of Kidney Diseases, 53*, 779–789.

Ji, Y., Tan, S., Xu, Y., Chandra, A., Shi, C., Song, B., et al. (2013). Vitamin B supplementation, homocysteine levels, and the risk of cerebrovascular disease: A meta-analysis. *Neurology, 81*, 1298–1307.

Jiang, Q., Wang, L., Si, X., Tian, J. L., Zhang, Y., Gui, H. L., et al. (2021). Current progress on the mechanisms of hyperhomocysteinemia-induced vascular injury and use of natural polyphenol compounds. *European Journal of Pharmacology, 905*, 174168.

Kalantar-Zadeh, K., Block, G., Humphreys, M. H., McAllister, C. J., & Kopple, J. D. (2004). A low, rather than a high, total plasma homocysteine is an indicator of poor outcome in hemodialysis patients. *Journal of the American Society of Nephrology, 15*, 442–453.

Kalantar-Zadeh, K., Jafar, T. H., Nitsch, D., Neuen, B. L., & Perkovic, V. (2021). Preserving kidney function in people with chronic kidney disease. *The Lancet, 398*, 786–802.

Kalantar-Zadeh, K., Tortorici, A. R., Chen, J. L., Kamgar, M., Lau, W. L., Moradi, H., et al. (2015). Dietary restrictions in dialysis patients: Is there anything left to eat? *Seminars in Dialysis, 28*, 159–168.

Kelly, J. T., Rossi, M., Johnson, D. W., & Campbell, K. L. (2017). Beyond sodium, phosphate and potassium: Potential dietary interventions in kidney disease. *Seminars in Dialysis, 30*, 197–202.

Kim, J. C., Kalantar-Zadeh, K., & Kopple, J. D. (2013). Frailty and protein-energy wasting in elderly patients with end stage kidney disease. *Journal of the American Society of Nephrology, 24*, 337–351.

Kim, H., Kim, G., Jang, W., Kim, S. Y., & Chang, N. (2014). Association between intake of B vitamins and cognitive function in elderly Koreans with cognitive impairment. *Nutrition Journal, 13*, 1–11.

Kovesdy, C. P., & Kalantar-Zadeh, K. (2009). Why is protein–energy wasting associated with mortality in chronic kidney disease? *Seminars in Nephrology, 29*, 3–14.

Koyama, K., Yoshida, A., Takeda, A., Morozumi, K., Fujinami, T., & Tanaka, N. (1997). Abnormal cyanide metabolism in uraemic patients. *Nephrology Dialysis Transplantation, 12*, 1622–1628.

Kumar, A., Palfrey, H. A., Pathak, R., Kadowitz, P. J., Gettys, T. W., & Murthy, S. N. (2017). The metabolism and significance of homocysteine in nutrition and health. *Nutrition & Metabolism, 14*, 1–12.

La Manna, G., Cappuccilli, M. L., Cianciolo, G., Conte, D., Comai, G., Carretta, E., et al. (2010). Cardiovascular disease in kidney transplant recipients: The prognostic value of inflammatory cytokine genotypes. *Transplantation, 89*, 1001–1008.

Levi, A., Cohen, E., Levi, M., Goldberg, E., Garty, M., & Krause, I. (2014). Elevated serum homocysteine is a predictor of accelerated decline in renal function and chronic kidney disease: A historical prospective study. *European Journal of Internal Medicine, 25*, 951–955.

Levin, A., Stevens, P. E., Bilous, R. W., Coresh, J., De Francisco, A. L., De Jong, P. E., et al. (2013). Kidney Disease: Improving Global Outcomes (KDIGO) CKD Work Group. KDIGO 2012 clinical practice guideline for the evaluation and management of chronic kidney disease. *Kidney International Supplements, 3*, 1–150.

Li, F., Bahnson, E. M., Wilder, J., Siletzky, R., Hagaman, J., Nickekeit, V., et al. (2020). Oral high dose vitamin B12 decreases renal superoxide and post-ischemia/reperfusion injury in mice. *Redox Biology, 32*, 101504.

Li, H., Feng, S. J., Zhang, G. Z., & Wang, S. X. (2014). Correlation of lower concentrations of hydrogen sulfide with atherosclerosis in chronic hemodialysis patients with diabetic nephropathy. *Blood Purification, 38*, 188–194.

Li, H., Lewis, A., Brodsky, S., Rieger, R., Iden, C., & Goligorsky, M. S. (2002). Homocysteine induces 3-hydroxy-3-methylglutaryl coenzyme a reductase in vascular endothelial cells: A mechanism for development of atherosclerosis? *Circulation, 105*, 1037–1043.

Li, S., Qiu, B., Lu, H., Lai, Y., Liu, J., Luo, J., et al. (2019). Hyperhomocysteinemia accelerates acute kidney injury to chronic kidney disease progression by downregulating heme oxygenase-1 expression. *Antioxidants & Redox Signaling, 30*, 1635–1650.

Li, Y., Spence, J. D., Wang, X., Huo, Y., Xu, X., & Qin, X. (2020). Effect of vitamin B12 levels on the association between folic acid treatment and CKD progression: A post hoc analysis of a folic acid interventional trial. *American Journal of Kidney Diseases, 75*, 325–332.

London, G. M., Pannier, B., Agharazii, M., Guerin, A. P., Verbeke, F. H., & Marchais, S. J. (2004). Forearm reactive hyperemia and mortality in end-stage renal disease. *Kidney International, 65*, 700–704.

Long, Y., & Nie, J. (2016). Homocysteine in renal injury. *Kidney Diseases, 2*, 80–87.

Ma, Y., Peng, D., Liu, C., Huang, C., & Luo, J. (2017). Serum high concentrations of homocysteine and low levels of folic acid and vitamin B 12 are significantly correlated with the categories of coronary artery diseases. *BMC Cardiovascular Disorders, 17*, 1–7.

Mallamaci, F., Zoccali, C., Tripepi, G., Fermo, I., Benedetto, F. A., Cataliotti, A., et al. (2002). Hyperhomocysteinemia predicts cardiovascular outcomes in hemodialysis patients. *Kidney International, 61*, 609–614.

McMahon, G. M., Hwang, S. J., Tanner, R. M., Jacques, P. F., Selhub, J., Muntner, P., et al. (2015). The association between vitamin B12, albuminuria and reduced kidney function: An observational cohort study. *BMC Nephrology, 16*, 1–8.

Miller, J. W. (2013). Homocysteine. In B. Caballero, L. Allen, & A. Prentice (Eds.), *Encyclopedia of human nutrition* (pp. 424–430). New York: Elsevier.

Moroni, G., Novembrino, C., Quaglini, S., De Giuseppe, R., Gallelli, B., Uva, V., et al. (2010). Oxidative stress and homocysteine metabolism in patients with lupus nephritis. *Lupus, 19*, 65–72.

Nair, A. P., Nemirovsky, D., Kim, M., Geer, E. B., Farkouh, M. E., Winston, J., et al. (2005). Elevated homocysteine levels in patients with end-stage renal disease. *The Mount Sinai Journal of Medicine, New York, 72*, 365–373.

Ni, J., Zhang, L., Zhou, T., Xu, W. J., Xue, J. L., Cao, N., et al. (2017). Association between the MTHFR C677T polymorphism, blood folate and vitamin B12 deficiency, and elevated serum total homocysteine in healthy individuals in Yunnan Province, China. *Journal of the Chinese Medical Association, 80*, 147–153.

Nigwekar, S. U., Kang, A., Zoungas, S., Cass, A., Gallagher, M. P., Kulshrestha, S., et al. (2016). Interventions for lowering plasma homocysteine levels in dialysis patients. *Cochrane Database of Systematic Reviews, 5*, 1–49.

Nixon, A. C., Bampouras, T. M., Pendleton, N., Woywodt, A., Mitra, S., & Dhaygude, A. (2018). Frailty and chronic kidney disease: Current evidence and continuing uncertainties. *Clinical Kidney Journal, 11*, 236–245.

O'Leary, F., & Samman, S. (2010). Vitamin B12 in health and disease. *Nutrients, 2*, 299–316.

Pan, Y., Guo, L. L., Cai, L. L., Zhu, X. J., Shu, J. L., Liu, X. L., et al. (2012). Homocysteine-lowering therapy does not lead to reduction in cardiovascular outcomes in chronic kidney disease patients: A meta-analysis of randomised, controlled trials. *British Journal of Nutrition, 108*, 400–407.

Park, J., Ahmadi, S. F., Streja, E., Molnar, M. Z., Flegal, K. M., Gillen, D., et al. (2014). Obesity paradox in end-stage kidney disease patients. *Progress in Cardiovascular Diseases, 56*, 415–425.

Park, J., Choi, J. H., Choi, H. J., Hong, S. H., Park, C. S., Choi, J. H., et al. (2020). Predictive role of vitamin B12 in acute kidney injury in living donor liver transplantation: A propensity score matching analysis. *BMJ Open, 10*, e038990.

Patel, P. S., Rana, D. A., Suthar, J. V., Malhotra, S. D., & Patel, V. J. (2014). A study of potential adverse drug-drug interactions among prescribed drugs in medicine outpatient department of a tertiary care teaching hospital. *Journal of Basic and Clinical Pharmacy, 5*, 44–48.

Pecoits-Filho, R., Lindholm, B., & Stenvinkel, P. (2002). The malnutrition, inflammation, and atherosclerosis (MIA) syndrome—The heart of the matter. *Nephrology Dialysis Transplantation, 17*, 28–31.

Pérez-Miguelsanz, J., Vallecillo, N., Garrido, F., Reytor, E., Pérez-Sala, D., & Pajares, M. A. (2017). Betaine homocysteine S-methyltransferase emerges as a new player of the nuclear methionine cycle. *Biochimica et Biophysica Acta (BBA). Molecular Cell Research, 1864*, 1165–1182.

Perna, A. F., Ingrosso, D., & De Santo, N. G. (2003). Homocysteine and oxidative stress. *Amino Acids, 25*, 409–417.

Perna, A. F., Ingrosso, D., Violetti, E., Luciano, M. G., Sepe, I., Lanza, D., et al. (2009). Progress in uremic toxin research: Hyperhomocysteinemia in uremia—A red flag in a disrupted circuit. *Seminars in Dialysis, 22*, 351–356.

Perna, A. F., Sepe, I., Lanza, D., Capasso, R., Di Marino, V., De Santo, N. G., et al. (2010). The gasotransmitter hydrogen sulfide in hemodialysis patients. *Journal of Nephrology, 23*, S92–S96.

Poddar, R., Sivasubramanian, N., DiBello, P. M., Robinson, K., & Jacobsen, D. W. (2001). Homocysteine induces expression and secretion of monocyte chemoattractant protein-1 and interleukin-8 in human aortic endothelial cells: Implications for vascular disease. *Circulation, 103*, 2717–2723.

Poduri, A., Mukherjee, D., Sud, K., Kohli, H. S., Sakhuja, V., & Khullar, M. (2008). MTHFR A1298C polymorphism is associated with cardiovascular risk in end stage renal disease in North Indians. *Molecular and Cellular Biochemistry, 308*, 43–50.

Prathapasinghe, G. A., Siow, Y. L., & Karmin, O. (2007). Detrimental role of homocysteine in renal ischemia-reperfusion injury. *American Journal of Physiology. Renal Physiology, 292*, F1354–F1363.

Pushpakumar, S., Kundu, S., & Sen, U. (2014). Endothelial dysfunction: The link between homocysteine and hydrogen sulfide. *Current Medicinal Chemistry, 21*, 3662–3672.

Qin, X., Li, J., Spence, J. D., Zhang, Y., Li, Y., Wang, X., et al. (2016). Folic acid therapy reduces the first stroke risk associated with hypercholesterolemia among hypertensive patients. *Stroke, 47*, 2805–2812.

Rady, P. L., Szucs, S., Grady, J., Hudnall, S. D., Kellner, L. H., Nitowsky, H., et al. (2002). Genetic polymorphisms of methylenetetrahydrofolate reductase (MTHFR) and methionine synthase reductase (MTRR) in ethnic populations in Texas; a report of a novel MTHFR polymorphic site, G1793A. *American Journal of Medical Genetics, 107*, 162–168.

Ramanathan, G., Harichandana, B., Kannan, S., Elumalai, R., & Sfd, P. (2019). Association between end-stage diabetic nephropathy and MTHFR (C677T and A1298C) gene polymorphisms. *Nephrology, 24*, 155–159.

Scott, T. M., Rogers, G., Weiner, D. E., Livingston, K., Selhub, J., Jacques, P. F., et al. (2017). B-Vitamin therapy for kidney transplant recipients lowers homocysteine and improves selective cognitive outcomes in the randomized FAVORIT Ancillary Cognitive Trial. *The Journal of Prevention of Alzheimer's Disease, 4*, 174–182.

Seetharam, B., & Li, N. (2000). Transcobalamin II and its cell surface receptor. In G. Litwack (Ed.), *Vitamins & hormones* (pp. 337–366). New York: Elsevier.

Selhub, J. (1999). Homocysteine metabolism. *Annual Review of Nutrition, 19*, 217–246.

Sevitt, L. H., & Hoffbrand, A. V. (1969). Serum folate and vitamin B12 levels in acute and chronic renal disease. Effect of peritoneal dialysis. *British Medical Journal, 2*, 18–21.

Shen, L., & Ji, H. F. (2015). Associations between homocysteine, folic acid, vitamin B12 and Alzheimer's disease: Insights from meta-analyses. *Journal of Alzheimer's Disease, 46*, 777–790.

Shiran, A., Remer, E., Asmer, I., Karkabi, B., Zittan, E., Cassel, A., et al. (2015). Association of vitamin B12 deficiency with homozygosity of the TT MTHFR C677T genotype, hyperhomocysteinemia, and endothelial cell dysfunction. *The Israel Medical Association Journal, 17*, 288–292.

Soohoo, M., Ahmadi, S. F., Qader, H., Streja, E., Obi, Y., Moradi, H., et al. (2017). Association of serum vitamin B12 and folate with mortality in incident hemodialysis patients. *Nephrology Dialysis Transplantation, 32*, 1024–1032.

Steiber, A. L., & Kopple, J. D. (2011). Vitamin status and needs for people with stages 3-5 chronic kidney disease. *Journal of Renal Nutrition, 21*, 355–368.

Strain, J. J., Dowey, L., Ward, M., Pentieva, K., & McNulty, H. (2004). B-vitamins, homocysteine metabolism and CVD. *Proceedings of the Nutrition Society, 63*, 597–603.

Supakul, S., Chabrun, F., Genebrier, S., N'Guyen, M., Valarche, G., Derieppe, A., et al. (2020). Diagnostic performances of urinary methylmalonic acid/creatinine ratio in vitamin B12 deficiency. *Journal of Clinical Medicine, 9*, 2335.

Sviri, S., Khalaila, R., Daher, S., Bayya, A., Linton, D. M., Stav, I., et al. (2012). Increased vitamin B12 levels are associated with mortality in critically ill medical patients. *Clinical Nutrition, 31*, 53–59.

Toohey, J. I. (2017). Possible involvement of hydrosulfide in B12-dependent methyl group transfer. *Molecules, 22*, 582.

Tremblay, R., Bonnardeaux, A., Geadah, D., Busque, L., Lebrun, M., Ouimet, D., et al. (2000). Hyperhomocysteinemia in hemodialysis patients: Effects of 12-month supplementation with hydrosoluble vitamins. *Kidney International, 58*, 851–858.

Trovato, F. M., Catalano, D., Ragusa, A., Martines, G. F., Pirri, C., Buccheri, M. A., et al. (2015). Relationship of MTHFR gene polymorphisms with renal and cardiac disease. *World Journal of Nephrology, 4*, 127–137.

United States Renal Data System. (2020). *USRDS 2020 annual data report*. Bethesda, MD: National Institute of Diabetes and Digestive and Kidney Diseases.

Verhoef, P., Stampfer, M. J., Buring, J. F., Gaziano, J. M., Allen, R. H., Stabler, S. P., et al. (1996). Homocysteine metabolism and risk of myocardial infarction: Relation with vitamins B6, B12, and folate. *American Journal of Epidemiology, 143*, 845–859.

Vitvitsky, V., Mosharov, E., Tritt, M., Ataullakhanov, F., & Banerjee, R. (2003). Redox regulation of homocysteine-dependent glutathione synthesis. *Redox Report, 8*, 57–63.

Walston, J., Hadley, E. C., Ferrucci, L., Guralnik, J. M., Newman, A. B., Studenski, S. A., et al. (2006). Research agenda for frailty in older adults: Toward a better understanding of physiology and etiology: Summary from the American Geriatrics Society/National Institute on Aging Research Conference on Frailty in Older Adults. *Journal of the American Geriatrics Society, 54*, 991–1001.

Wang, A. Y. M. (2020). Does vitamin B12 delay CKD progression? *American Journal of Kidney Diseases, 75*, 317–319.

Wijekoon, E. P., Brosnan, M. E., & Brosnan, J. T. (2007). Homocysteine metabolism in diabetes. *Biochemical Society Transactions, 35*, 1175–1179.

Wrone, E. M., Hornberger, J. M., Zehnder, J. L., McCann, L. M., Coplon, N. S., & Fortmann, S. P. (2004). Randomized trial of folic acid for prevention of cardiovascular events in end-stage renal disease. *Journal of the American Society of Nephrology, 15*, 420–426.

Wu, G. H., Kong, F. Z., Dong, X. F., Wu, D. F., Guo, Q. Z., Shen, A. R., et al. (2017). Association between hyperhomocysteinemia and stroke with atherosclerosis and small artery occlusion depends on homocysteine metabolism-related vitamin levels in Chinese patients with normal renal function. *Metabolic Brain Disease, 32*, 859–865.

Xie, D., Yuan, Y., Guo, J., Yang, S., Xu, X., Wang, Q., et al. (2015). Hyperhomocysteinemia predicts renal function decline: A prospective study in hypertensive adults. *Scientific Reports, 5*, 1–10.

Xu, X., Qin, X., Li, Y., Sun, D., Wang, J., Liang, M., et al. (2016). Efficacy of folic acid therapy on the progression of chronic kidney disease: The renal substudy of the China Stroke Primary Prevention Trial. *JAMA Internal Medicine, 176*, 1443–1450.

Ye, Z., Zhang, Q., Li, Y., Wang, C., Zhang, J., Ma, X., et al. (2016). High prevalence of hyperhomocysteinemia and its association with target organ damage in Chinese patients with chronic kidney disease. *Nutrients, 8*, 645.

Yeh, E. L., Huang, Y. C., Tsai, S. F., Yu, T. M., Wu, M. J., & Chen, C. H. (2018). Relationship between plasma levels of homocysteine and the related B vitamins in patients with hemodialysis adequacy or inadequacy. *Nutrition, 53*, 103–108.

Zhang, K., & Kaufman, R. J. (2008). From endoplasmic-reticulum stress to the inflammatory response. *Nature, 454*, 455–462.

Zhao, J., Chen, H., Liu, N., Chen, J., Gu, Y., Chen, J., et al. (2017). Role of hyperhomocysteinemia and hyperuricemia in pathogenesis of atherosclerosis. *Journal of Stroke and Cerebrovascular Diseases, 26*, 2695–2699.

Zhou, X., Feng, Y., Zhan, Z., & Chen, J. (2014). Hydrogen sulfide alleviates diabetic nephropathy in a streptozotocin-induced diabetic rat model. *Journal of Biological Chemistry, 289*, 28827–28834.

CHAPTER THIRTEEN

Inherited defects of cobalamin metabolism

David Watkins* **and David S. Rosenblatt**
Department of Human Genetics, McGill University, The Research Institute of the McGill University Health Centre, Montreal, QC, Canada
*Corresponding author: e-mail address: david.watkins@mcgill.ca

Contents

Abstract

Cobalamin (vitamin B_{12}) is required for activity of the enzymes methylmalonyl-CoA mutase and methionine synthase in human cells. Inborn errors affecting cobalamin uptake or metabolism are characterized by accumulation of the substrates for these enzymes, methylmalonic acid and homocysteine, in blood and urine. Inborn errors affecting synthesis of the adenosylcobalamin coenzyme required by methylmalonyl-CoA mutase (*cblA* and *cblB*) result in isolated methylmalonic aciduria; inborn errors affecting synthesis of the methylcobalamin coenzyme required by methionine synthase (*cblE* and *cblG*) result in isolated homocystinuria. Combined methylmalonic aciduria and homocystinuria is seen in patients with impaired intestinal cobalamin absorption (intrinsic factor deficiency, Imerslund-Gräsbeck syndrome) and with defects affecting synthesis of both cobalamin coenzymes (*cblC*, *cblD*, *cblF* and *cblJ*). A series of disorders caused by pathogenic variant mutations affecting gene regulators (transcription factors) of the *MMACHC* gene have recently been described (HCFC1 [*cblX* disorder] and deficiencies of THAP11, and ZNF143 [the *cblK* disorder]).

Vitamins and Hormones, Volume 119
ISSN 0083-6729
https://doi.org/10.1016/bs.vh.2022.01.010

Abbreviations

AdoCbl adenosylcobalamin
Cbl cobalamin
CNCbl cyanocobalamin
MeCbl methylcobalamin
OHCbl hydroxocobalamin

1. Introduction

Cobalamin (vitamin B_{12}) is required for the activity of two enzymes in human cells. The mitochondrial enzyme methylmalonyl-CoA mutase, which catalyzes the isomerization of methylmalonyl-CoA, formed during catabolism of branched-chain amino acids (isoleucine, valine and threonine) and odd-chain fatty acids, to succinyl-CoA, requires adenosylcobalamin (AdoCbl) for activity. A second cobalamin derivative, methylcobalamin (MeCbl), is generated during the catalytic cycle of the cytoplasmic enzyme methionine synthase, which remethylates homocysteine to form methionine (Fig. 1). Deficiency of these cobalamin coenzymes results in accumulation of the substrates of the two reactions, methylmalonic acid and homocysteine, in the blood and urine. Methylmalonic aciduria, homocystinuria, or combined methylmalonic aciduria and homocystinuria, is the hallmark of genetic disorders affecting cobalamin uptake and metabolism.

Cobalamin derivatives exist with the central cobalt atom at different redox states (cob(I)alamin, with a fully reduced Co^{+}, cob(II)alamin (Co^{++}) and cob(III)alamin, with a fully oxidized Co^{+++}) as well as with different upper axial ligands (AdoCbl, MeCbl, hydroxocobalamin (OHCbl)). Cellular cobalamin metabolism involves uptake of transcobalamin-bound cobalamin (holotranscobalamin) by carrier-mediated endocytosis, dissociation of cobalamin from transcobalamin in lysosomes, transport across the lysosomal membrane into the cytoplasm, removal of the upper axial ligand and conversion to cob(II)alamin, partition between the cytoplasm and mitochondria and addition of the appropriate upper axial ligand to form AdoCbl or MeCbl. Because the intracellular concentration of cobalamin is extremely low, and because cobalamins are labile and reactive, it is believed that intracellular cobalamin exists bound to proteins, either chaperones or enzymes, at all times.

Inborn errors have been described that affect all known steps in cobalamin handling in the body, including intestinal absorption, transport in the blood, uptake by peripheral cells and cellular metabolism. All of these genetic

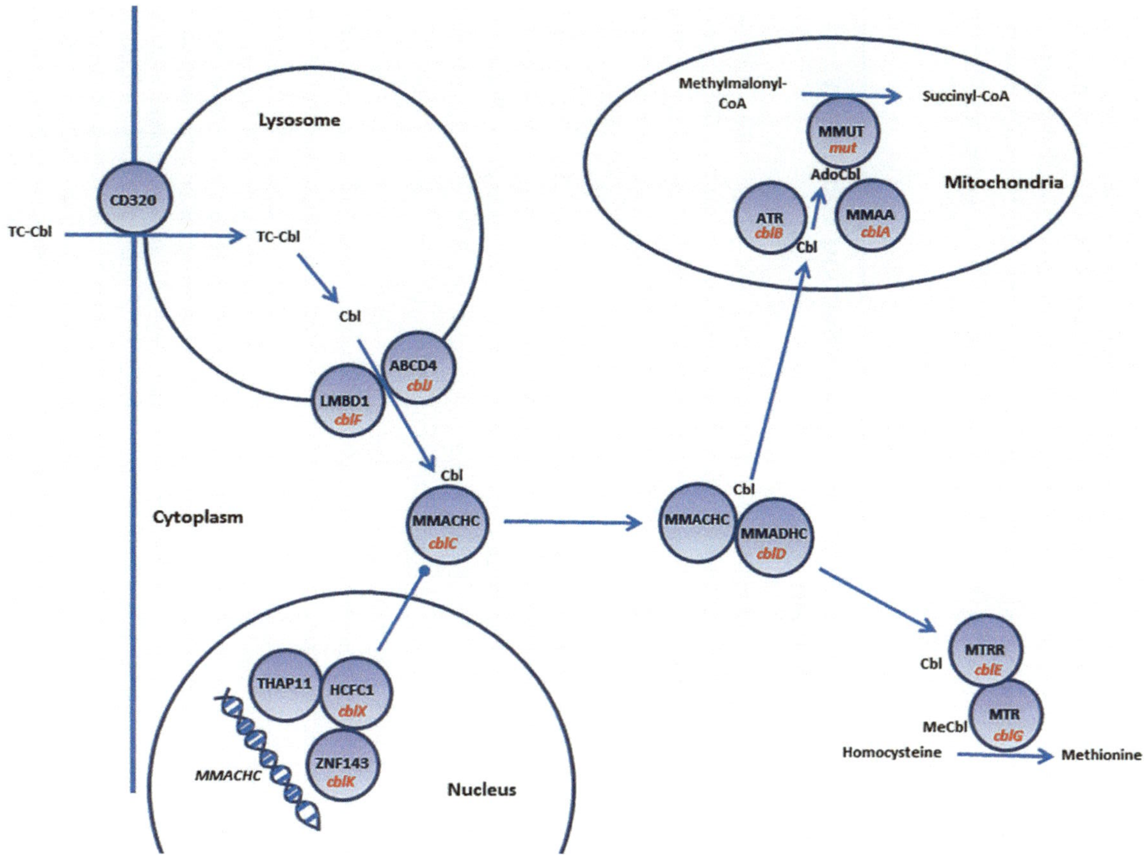

Fig. 1 See figure legend on next page.

conditions are rare, although there is considerable variation in their frequencies. The *cblC* disorder is by far the most common, having been reported in over 1200 patients worldwide; its frequency has been estimated at between 1:200,000 and 1:100,000 in North America, while it appears considerably more frequent in some regions of China (Zhou, Li, Wang, Wang, & Gu, 2019). On the other hand, THAP11 deficiency and ZNF143 deficiency have each been reported in a single patient. All of the recognized disorders are inherited as autosomal recessive traits, with the exception of the *cblX* disorder, which is X-linked recessive. The disorders and affected genes are listed in Table 1.

2. Diagnosis of inborn errors affecting cobalamin metabolism

Inborn errors affecting cobalamin are suspected in patients with elevated levels of methylmalonic acid or homocysteine, or both, in the absence of dietary cobalamin deficiency. Disorders that affect intestinal absorption result in decreased blood cobalamin levels, while most other conditions are associated with serum levels within the reference range. Historically, the inborn errors of cellular cobalamin metabolism were diagnosed by studies of patient fibroblasts, including measurement of methylmalonyl-CoA mutase and methionine synthase function as well as the ability of cells to convert exogenous radiolabeled cyanocobalamin to AdoCbl and MeCbl. Specific diagnosis was achieved by complementation analysis. Patient fibroblasts are mixed with cells from patients with established diagnoses and fused to generate heterokaryons containing nuclei from both cell lines; the metabolic defect is corrected in heterokaryons when the mutations in the two

Fig. 1 Cellular cobalamin metabolism: Uptake of holotranscobalamin (TC-Cbl at the cell surface is followed by release of free cobalamin (Cbl) in lysosomes and its transport into the cytoplasm where it becomes associated with the cytoplasmic enzyme methionine synthase or the mitochondrial enzyme methylmalonyl-CoA mutase. Expression of the MMACHC protein is modulated by the interaction of the transcription regulators HCFC1, THAP11 and ZNF143 with the *MMACHC* promotor. The names of the proteins/enzymes are shown in black; the names of the diseases associated with the protein are shown in red. ABCD4 = ABCD4 protein; ATR = cobalamin adenosyltransferase (product of the *MMAB* gene); CD320 = transcobalamin receptor (product of the *CD320* gene); HCFC1 = HCFC1 transcription co-regulator; LMBD1 = LMBD1 protein; MMAA = MMAA protein; MMACHC = MMACHC protein; MMADHC = MMADHC protein; MMUT = methylmalonyl-CoA mutase (product of the *MMUT* gene); MTR = methionine synthase (product of the *MTR* gene); MTRR = methionine synthase reductase (product of the *MTRR* gene); THAP11 = THAP11 transcription regulator; ZNF143 = ZNF143 transcription regulator.

Table 1 Inborn errors of cobalamin uptake and metabolism

Disorder	MIM number	Gene	Location	Gene number	Alternative names
Intrinsic factor deficiency	261000	*CBLIF*	11q12.1	609342	
Imerslund-Gräsbeck syndrome 1	26110	*CUBN*	10p13	602997	Hereditary megaloblastic anemia 1
Imerslund-Gräsbeck syndrome 2	218882	*AMN*	14q32.32	605799	Hereditary megaloblastic anemia 1
Transcobalamin deficiency	275350	*TCN2*	22q12.2	613441	
Haptocorrin deficiency	193090	*TCN1*	11q12.1	189905	
Transcobalamin receptor deficiency		*CD320*	19p13.2	606,475	
cblA	251100	*MMAA*	4q31.21	607481	Cobalamin A
cblB	251110	*MMAB*	12q24.11	607568	Cobalamin B
cblC	277400	*MMACHC*	1p34.1	609831	Cobalamin C
"Epi-*cblC*"	277400	*PRDX1*	1p34.1	176763	
cblD	277410	*MMADHC*	2q23.2	611935	Cobalamin D
cblE	236270	*MTRR*	5p15.31	602568	Cobalamin E
cblF	277380	*LMBRD1*	6q13	612625	Cobalamin F
cblG	250940	*MTR*	1q43	156570	Cobalamin G
cblJ	614857	*ABCD4*	14q24.3	603214	Cobalamin J
cblX	309541	*HCFC1*	Xq28	300019	Cobalamin X
THAP11 deficiency		*THAP11*	16q22.1	609119	
ZNF143 deficiency		*ZNF143*	11p15.4	603433	*cblK*
Methylmalonyl-CoA deficiency	251000	*MMUT*	6p12.3	609058	

cell lines affect separate loci, but not when a single locus is affected. Using this procedure, a series of complementation classes designated *cblA*-*cblG* and *cblJ* were defined in the absence of knowledge of the genes and proteins affected in each disorder (Coelho et al., 2012; Gravel, Mahoney, Ruddle, & Rosenberg, 1975; Watkins & Rosenblatt, 1986, 1988; Willard, Mellman, & Rosenberg, 1978). With the identification of the genes underlying all of these disorders, targeted gene sequencing has become the first line approach to diagnosis, and next generation gene panels are available that analyze all of the genes known to play a role in cobalamin metabolism. Nevertheless, classical somatic cell genetics remains valuable in cases where variants in candidate genes cannot be identified, or in cases where one or more variants of uncertain significance are detected.

3. Inborn errors affecting intestinal cobalamin absorption

Uptake of ingested cobalamin involves dissociation of cobalamin from food protein in the acid environment of the stomach and binding of the released vitamin to intrinsic factor, a binding protein secreted by gastric parietal cells. The intrinsic factor-cobalamin complex subsequently binds to the receptor cubam on the surface of epithelial cells in the distal ileum, initiating endocytosis. Cubam is composed of two proteins: cubilin, a large multi-ligand binding protein, and amnionless, a smaller transmembrane protein that anchors cubilin to the epithelial membrane. Ultimately, ingested cobalamin appears in the portal circulation bound to another cobalamin binding protein, transcobalamin.

Patients have been identified with genetic defects affecting synthesis of both intrinsic factor and cubam. Both result in impaired intestinal absorption and decreased serum cobalamin levels. Intrinsic factor deficiency is associated with mutations in the *CBLIF* gene (Tanner et al., 2005); Imerslund-Gräsbeck syndrome can be caused by mutations the genes encoding either component of cubam, *CUBN* or *AMN* (Aminoff et al., 1999; Tanner et al., 2003). Nearly 100 patients have been reported with intrinsic factor deficiency, and at least 300 patients with Imerslund-Gräsbeck syndrome. Patients with both conditions usually come to medical attention between 1 and 5 years of age, when liver stores of cobalamin accumulated *in utero* are exhausted, although later onset has been reported. Typically there is megaloblastic anemia, failure to thrive and increased susceptibility to infection. In cases where diagnosis is delayed or effective therapy is not instituted, neurological problems can occur (Gräsbeck & Tanner, 2011; Tanner, Sturm, Baack, Liyanarachchi, & de la Chapelle, 2012). There are increased serum levels of methylmalonic acid and total homocysteine, although levels are usually lower than those observed in patients with inborn errors affecting cellular cobalamin metabolism. Patients with Imerslund-Gräsbeck syndrome frequently present with proteinuria that does not progress over time and is not accompanied by other signs of impaired renal function. This appears to be due to the role of cubam in the reabsorption of various filtered proteins including albumin in the proximal renal tubule.

Therapy of these disorders involves parenteral administration of cobalamin, bypassing the intestinal absorption system. This leads to improvement in metabolic, hematologic and neurologic findings, but has no effect on proteinuria in Imerslund-Gräsbeck syndrome patients. Neurological symptoms may not completely respond if therapy is not started early enough.

Approximately 20 different pathogenic variants in *CBLIF* have been reported in patients with intrinsic factor deficiency. Patients with Imerslund-Gräsbeck syndrome were initially reported from Scandinavia, where Norwegian patients were found to have predominantly variants in *AMN* while Finnish patients had *CUBN* variants. Subsequently, most identified patients have been of Mediterranean or Middle Eastern origin, with mutations in both genes present. Overall, over 30 *AMN* variants and a similar number of *CUBN* variants have been reported. Cubilin mutations associated with Imerslund-Gräsbeck syndrome affect the first 1600 amino acids of the protein (which has a total length of 3597 amino acids); this includes the binding site that recognizes the intrinsic factor cobalamin complex. More distal mutations have been associated with albuminuria in the presence of normal cobalamin absorption (Ovunc et al., 2011).

4. Defects affecting cobalamin transport

Cobalamin circulates in blood bound to two different proteins, haptocorrin (originally named transcobalamin I), and transcobalamin (originally transcobalamin II). Cobalamin bound to transcobalamin (approximately 10%–20% of circulating cobalamin) is available for uptake by most cells in the body, which express the CD320 receptor on their cell surface. The majority of circulating cobalamin is bound to haptocorrin and is unavailable for uptake by cells outside of the liver. Binding of holotranscobalamin to CD320 initiates endocytosis. The complex dissociates in lysosomes with release of free cobalamin that is subsequently transported across the lysosomal membrane into the cytoplasm.

More than 50 patients with congenital deficiency in transcobalamin have been reported. These patients usually come to medical attention in the first months of life with severe megaloblastic anemia, failure to thrive, weakness and diarrhea (Trakadis et al., 2014). In several cases, the presence of immature white cell precursors in a hypocellular bone marrow has led to an erroneous diagnosis of leukemia (Schiff et al., 2010). Hypogammaglobulinemia and immune dysfunction have been reported in several patients. Neurological sequelae of cobalamin deficiency may develop if diagnosis is delayed. There is usually elevation of methylmalonic acid and total homocysteine in blood. Circulating cobalamin levels are typically within the reference range, since the majority of circulating cobalamin at any time is bound to haptocorrin, which is unaffected in the transcobalamin deficiency. Total unsaturated cobalamin binding capacity is decreased, but must be measured prior to institution of cobalamin therapy.

Treatment involves administration of pharmacologic doses of cobalamin parenterally or orally, with the aim of maintaining high circulating levels of the vitamin. This will usually result in correction of biochemical and hematologic problems; neurological findings may become irreversible if treatment is delayed.

Patients with transcobalamin deficiency in many cases have no detectable circulating transcobalamin, but cases of proteins capable of binding cobalamin but unable to support its internalization have been described (Haurani, Hall, & Rubin, 1979). More than 30 different variants have been detected in the *TCN2* gene in patients with this disorder (Trakadis et al., 2014).

A small number of individuals with severely decreased levels of circulating cobalamin as the result of variants in the *TCN1* gene encoding haptocorrin have been reported (Carmel, 1983; Carmel, Parker, & Kelman, 2009). It is not clear if there is any clinical effect of this disorder, since the level of transcobalamin-bound cobalamin available for cellular uptake is not affected. This condition is likely underreported since the reduced circulating cobalamin level is usually detected by accident in a patient undergoing testing for an unrelated problem. Some affected individuals have had neurological findings that did not respond to therapy with cobalamin and were likely unrelated to the haptocorrin deficiency.

5. Cellular cobalamin uptake

Uptake of holotranscobalamin at the cell membrane is mediated by the transcobalamin receptor, encoded by the *CD320* gene. Binding of holotranscobalamin initiates endocytosis; the endocytosed transcobalamin enters the lysosomal compartment where the protein is degraded to release cobalamin.

A number of individuals homozygous for a 3-base-pair deletion in the *CD320* gene, resulting in deletion of a glutamate residue in the extracellular domain of the transcobalamin receptor, have been identified following abnormal newborn screens for C3 acylcarnitines (Quadros et al., 2010). Fibroblasts from these individuals have decreased ability to take up holo-transcobalamin, although they are able to convert the cobalamin to AdoCbl and MeCbl once it is taken up. There may be mildly elevated circulating or urine levels of methylmalonic acid and serum total homocysteine, although levels are not as high as those seen in patients with other inborn errors of cobalamin metabolism. It is not clear that patients homozygous for the deletion are clinically affected. Bilateral retinal artery occlusions were identified in one such homozygote (Karth, Singh, Kim, & Costakos, 2012), but it is not clear that these were the result of the deletion. Long-term follow-up of two affected individuals demonstrated normal development at 6 and 7 years of age (Hannah-Shmouni, Cruz, Schulze, & Mercimek-Andrews, 2018). The variant has been annotated as a polymorphism in an Irish population, where it occurred at a frequency of approximately 2% (Pangilinan et al., 2010).

6. Lysosomal membrane cobalamin transport

Free cobalamin exits the lysosome to the cytoplasm in a process that requires the presence of two lysosomal membrane proteins, ABCD4 and LMBD1. ABCD4, an ATP binding cassette half-transporter, appears to be the actual cobalamin transporter (Kitai et al., 2021). LMBD1 may be required to direct ABCD4 to the lysosomal membrane (Kowaguchi, Okamoto, Morita, & Imanaka, 2016); both proteins must be present for either to reach the lysosomal membrane.

Transport of cobalamin across the lysosomal membrane is impaired in patients with the *cblF* and the *cblJ* disorders. Studies of cultured patient fibroblasts demonstrated accumulation unmetabolized cyanocobalamin in the lysosomal compartment following incubation with exogenous radiolabeled cyanocobalamin in both diseases. Synthesis of the coenzyme derivatives AdoCbl and MeCbl was decreased, and function of both methylmalonyl-CoA mutase and methionine synthase reduced (Rosenblatt et al., 1986; Coelho et al., 2012), Radioautography in *cblF* fibroblasts demonstrated trapping of cobalamin in the lysosome, consistent with inability to transport cobalamin across the lysosomal membrane (Vassiliadis, Rosenblatt, Cooper, & Bergeron, 1991). The *cblF* disorder was shown to be caused by variants in *LMBRD1,* which encodes LMBD1 (Rutsch et al., 2009), while *cblJ* is caused by *ABCD4* variants (Coelho et al., 2012).

Patients typically present in the first months of life. Clinical presentation has been variable; findings seen in more than one patient have included megaloblastic anemia, small for gestational age, feeding difficulties, failure to thrive, developmental delay, stomatitis, facial dysmorphology and congenital heart defects. Three patients with later onset and a milder presentation, with skin hyperpigmentation and gray hair in early childhood as major complaints, were homozygous for the same *ABCD4* variant (Kim et al., 2012; Takeichi et al., 2015).

Treatment has usually involved parenteral cyanocobalamin or hydroxocobalamin. Long-term follow-up of two *cblF* patients indicated that, with treatment, they had developed normally with no signs of delay (Alfadhel, Lillquist, Davis, Junker, & Stockler-Iposiroglu, 2011; Oladipo et al., 2011). On the other hand, a patient who was not diagnosed until 6 years of age had severe developmental delay that did not respond to cobalamin therapy (Constantinou et al., 2016).

Several patients have had reduced serum cobalamin levels, which may reflect a role of lysosomal uptake in intestinal absorption following internalization of the intrinsic factor-cobalamin complex by epithelial cells; a defect in absorption was demonstrated by abnormal Schilling tests in *cblF* patients (Laframboise, Cooper, & Rosenblatt, 1992; Alfadhel et al., 2011). This test was not readily available at the time the *cblJ* disorder was identified.

7. The *cblC* disorder

Following transport of cobalamin across the lysosomal membrane, it becomes associated with the cytoplasmic protein MMACHC. This protein is thought to function as a chaperone, binding cobalamin in the partially reduced cob(II)alamin form. It also catalyzes the removal of the upper axial ligand when one is present to generate protein-bound cob(II)alamin: The cyanide group of cyanocobalamin is removed by reductive decyanation, while removal of the upper ligand of alkylcobalamins, such as AdoCbl and MeCbl, occurs by a glutathione-dependent reaction (Kim, Gherasim, & Banerjee, 2008; Kim, Hannibal, Gherasim, Jacobsen, & Banerjee, 2009).

The *cblC* disorder is by far the most common disorder of cobalamin metabolism, with over 1200 patients reported. Patients have elevated serum and urine levels of both methylmalonic acid and homocysteine. Studies of patient fibroblasts showed decreased synthesis of AdoCbl and MeCbl and decreased function of both methylmalonyl-CoA mutase and methionine synthase. In contrast to *cblF* and *cblJ* fibroblasts, intracellular cobalamin levels are decreased, apparently reflecting inability of cells to retain cobalamin when it is not enzyme-bound. The disorder is caused by mutations in the *MMACHC* gene (Lerner-Ellis et al., 2006).

Clinical findings are variable, with both early-onset (within the first year of life) and later-onset forms reported (Carrillo-Carrasco, Chandler, & Venditti, 2012; Carrillo-Carrasco & Venditti, 2012; Fischer et al., 2014). The early-onset form is characterized by megaloblastic anemia, failure to thrive, feeding difficulties, lethargy, and developmental delay or regression. Neurological findings are usually present, including cerebral or cerebellar atrophy, white matter disease, microcephaly and seizures. Ophthalmologic findings may include a bull's eye maculopathy and a progressive salt-and-pepper retinopathy, nystagmus, strabismus, and in a few instances optic nerve atrophy (Brooks et al., 2016). Thrombotic microangiopathy, possibly the result of elevated homocysteine levels, may result in an atypical hemolytic uremic syndrome or pulmonary arterial hypertension or both (Beck, van Spronsen, Diepstra, Berger, & Kömhoff, 2016). Congenital heart defects, in some cases mild, have been detected in several patients (Profitlich, Kirmse, Wasserstein, Diaz, & Srivastava, 2009). Late-onset cases are often characterized by neurological problems, including developmental regression, gait and posture abnormalities, seizures, dementia or psychosis, but cases without neurological symptoms have also been reported.

Treatment typically includes parenteral hydroxocobalamin (cyanocobalamin appears to be less effective) and oral betaine; the latter is substrate for betaine-homocysteine methyltransferase, an enzyme present in liver and kidney that can convert homocysteine to methionine in the absence of cobalamin. Extremely high dosages of hydroxocobalamin (in excess of 20 mg/day in some cases) may be required to reduce methylmalonic acid and homocysteine levels to their reference ranges (Carrillo-Carrasco, Sloan, Valle, Hamosh, & Venditti, 2009; Scalais et al., 2019). Even in the presence of metabolic correction, there may be residual neurologic damage. Ophthalmologic symptoms have usually been resistant to therapy. Best results have been obtained with prenatal diagnosis of the disorder and prenatal therapy of the mother with hydroxocobalamin followed by therapy of the infant from birth (Huemer et al., 2005; Zhang et al., 2008; Trefz et al., 2016).

Over 100 *MMACHC* pathogenic variants have been identified in *cblC* patients. In cohorts of patients of European origin, the most common variant is c.271dupA, which accounts for as many as half of identified disease associated alleles in this population; this variant is rare in Chinese patients, among whom c.609G > A (p.W203X) is the most frequent allele. Both these alleles are associated with severe early-onset *cblC* disease when present in homozygosity. Several other cases of specific mutations associated with different populations have been observed (Morel, Lerner-Ellis, & Rosenblatt, 2006).

Epimutation in *MMACHC* secondary to variants in the *PRDX1* gene has been described in a small number of patients with *cblC* disease. *PRDX1* (which encodes peroxiredoxin 1, a protein with no role in cobalamin metabolism) is located immediately 3′ to *MMACHC*, in the opposite orientation. The variants that have been implicated in *cblC* affect the splice acceptor site for the final exon of *PRDX1*, resulting in a message lacking a polyadenylation signal that spans the entire *MMACHC* gene and causes methylation of a CpG island in its 5′ region, silencing *MMACHC* expression (Guéant et al., 2018). This secondary epimutation is heritable and behaves as a typical MMACHC pathogenic allele.

8. Regulation of MMACHC expression

Investigation of patients with complementation-verified diagnosis of *cblC* disease in whom no *MMACHC* variants could be detected, has resulted in identification of patients with pathogenic variants in genes encoding proteins involved in regulation of *MMACHC* expression. These patients represent the first known examples of an inborn error of metabolism caused by aberrant regulation of a gene encoding a metabolic enzyme instead of mutations in metabolic genes themselves. Sixteen male patients were identified with hemizygous variants of the *HCFC1* gene on the X chromosome (Yu et al., 2013; Gérard et al., 2015; Scalais et al., 2017). This disorder has been called "cblX" although it does not represent a novel complementation class, since fibroblasts from all of the patients behaved as *cblC* on complementation analysis.

The host factor 1 protein encoded by *HCFC1* is an expression co-regulator; it does not bind directly to DNA but acts by interacting with a number of proteins, including THAP11 and ZNF143, that bind DNA. While HCFC1 can in this way affect the expression of thousands of genes, RNA sequencing indicates that expression of only a small subset of these, including *MMACHC*, is affected in *cblX* patient fibroblasts. HCFC1 pathogenic variants associated with *cblX* affect the Kelch domain of HCFC1, which is believed to mediate protein–protein interactions (Yu et al., 2013). Overall, *cblX* patients appear to have a milder biochemical phenotype than the classical *cblC* disorder but have a more severe neurological phenotype, including intractable epilepsy, choreoathetosis, microcephaly and severe intellectual impairment. Patients with *HCFC1* mutations outside of the Kelch domain have been reported; they have had neurological findings in the absence of any defect in cobalamin metabolism (Koufaris, Alexandrou, Tanteles, Anastasiadou, & Sismani, 2016; Wongkittichote, Wagner, & Srinawi, 2021).

A single patient has been described with a homozygous variant in the *THAP11* gene, which encodes one of the DNA-binding proteins known to interact with HCFC1 (Quintana et al., 2017). As with the *cblX* patients, RNA sequencing demonstrated decreased *MMACHC* expression in patient fibroblasts. THAP11 engineered to express the patient's mutation was demonstrated to have decreased binding to the *MMACHC* promoter in HEK-293 cells (Dehaene, Praz, Lhôte, Lopes, & Herr, 2020). The patient had elevated methylmalonic acid levels in the presence of apparently normal homocysteine levels (although at the time the patient was ascertained modern assays for serum total homocysteine had not been developed). Complementation analysis gave a diagnosis of *cblC*. He had encephalopathy, tetraplegia, seizures that did not respond to therapy, and profound intellectual disability; he died at the age of 10 years, decades before THAP11 deficiency was diagnosed using preserved cultured fibroblasts.

A patient has been reported with two pathogenic variants in the *ZNF143* gene, which encodes another DNA-binding protein known to interact with HCFC1 (Pupavac et al., 2016). The patient had bilateral cleft palate, microcephaly, a ventricular septal defect, intractable epilepsy, progressive encephalopathy, combined immune deficiency, cortical blindness and sensorineural deafness; he died at 2 years of age from hypercapnic respiratory failure. Unlike patients with *HCFC1* or *THAP11* variants, fibroblasts from this patient complemented cells from patients in all known classes of inborn error of cobalamin metabolism, including *cblC*. This patient therefore represents a novel complementation class, which can be designated *cblK*. There was decreased *MMACHC* expression on RNA sequencing, and elevated levels of methylmalonic acid and serum total homocysteine. However, fibroblast studies showed accumulation of transcobalamin-bound cobalamin, suggesting a defect in cobalamin metabolism affecting an earlier step prior to dissociation of transcobalamin and cobalamin in the lysosomes, in addition to decreased MMACHC function.

9. Partitioning cobalamin between the cytoplasm and mitochondria

The *cblD* complementation class is heterogeneous in its clinical presentation: although the first patients described were similar to *cblC* patients with combined methylmalonic aciduria and homocystinuria (Goodman, Moe, Hammond, Mudd, & Uhlendorf, 1970), some subsequent patients have presented with isolated methylmalonic aciduria or isolated homocystinuria (Suormala et al., 2004). This suggests that the protein affected in this disorder is in some way involved in partitioning of cobalamin between the mitochondrial and cytoplasmic compartments of the cell. The *MMADHC* gene encodes a protein of unknown function that has been shown to form a complex with MMACHC following binding of cob(II)alamin to the latter protein (Yamada, Gherasim, Banerjee, & Koutmos, 2015; Froese et al., 2015). Although most studies failed to show binding of cobalamin to MMADHC (Deme et al., 2012; Gherasim, Hannibal, Rajagopalan, Jacobsen, & Banerjee, 2013; Gherasim, Lofgren, & Banerjee, 2013), a recent study suggested binding occurred between the cobalt of cob(II)alamin and a cysteine residue of MMADHC when care was taken to prevent oxidation of exposed sulfhydryl groups (Li et al., 2020). This study found formation of a complex between cob(II)alamin, MMACHC and MMADHC with simultaneous binding of cobalamin to both proteins. The MMACHC-MMADHC-cobalamin complex can bind to methionine synthase, suggesting a role in directing cobalamin to the latter enzyme (Fofou-Caillierez et al., 2013). How MMADHC can direct cobalamin to the mitochondria is not understood.

Twenty-seven *cblD* families have been identified, including 10 with the classic presentation, 9 with isolated methylmalonic aciduria, and 8 with isolated homocystinuria. The identification of the underlying gene, *MMADHC*, and mapping of pathogenic variants to its sequence, has suggested a possible explanation for the phenotypic variation (Coelho et al., 2008). Mutations associated with isolated methylmalonic aciduria create a stop signal near the 5′-terminus of the gene, allowing re-initiation of transcription at one of two alternative in-frame start sites and generating a protein lacking a mitochondrial targeting signal; this protein can apparently support transfer of cobalamin to methionine synthase but not transport of cobalamin into the mitochondria. Mutations associated with isolated homocystinuria affect the 3′-domain of MMADHC, creating a protein that can support translocation of cobalamin to the mitochondria but does not permit transfer of cobalamin to methionine synthase. Nonsense mutations distal to the alternative start sites produce no functional protein and are associated with inability to transfer cobalamin to methionine synthase or the mitochondria (Coelho et al., 2008).

10. Methylmalonyl-CoA isomerization in the mitochondria

AdoCbl-dependent isomerization of methylmalonyl-CoA to form succinyl-CoA in the mitochondria requires activity of three proteins: methylmalonyl-CoA mutase itself, encoded by the *MMUT* gene; cobalamin adenosyltransferase, encoded by the *MMAB* gene, which catalyzes transfer of an adenosyl group from ATP to cob(I)alamin, to form AdoCbl, and transfer of AdoCbl to mutase; and MMAA, encoded by the *MMAA* gene (Gherasim, Hannibal, et al., 2013; Gherasim, Lofgren, & Banerjee, 2013). MMAA is a chaperone protein with GTPase activity that ensures that cobalamin transferred from cobalamin adenosyltransferase to methylmalonyl-CoA mutase is in the AdoCbl form, as well as protecting mutase-bound AdoCbl from degradation and releasing damaged AdoCbl from mutase so it can be replaced. Mutations affecting any of these proteins can result in decreased methylmalonyl-CoA isomerization and accumulation of methylmalonic acid in blood and urine.

The complementation classes associated with mutations in these genes—*mut* caused by variants in *MMUT*, *cblA* caused by variants in *MMAA*, and *cblB* caused by variants in *MMAB* – are clinically similar. The mut class has been subdivided into the mut^0 class, with complete absence of mutase, and the mut^- class with expression of mutant enzyme. Patients most often present in the first months of life, often within the first hours or days, with metabolic acidosis, although cases with onset later in childhood have been reported (Sloan, Carrillo, Adams, & Venditti, 2018; Forny et al., 2021). They have repeated episodes of metabolic acidosis, often accompanied by hyperammonemia, that can be life threatening. Long-term consequences can include growth and developmental retardation, or mild microcephaly. There may be metabolic stroke, affecting predominantly the globus pallidus (Baker et al., 2015); this appears to represent sensitivity to methylmalonic acid or its metabolites in specific neural cell populations. Patients frequently develop progressive renal disease that can develop into end stage renal failure (Morath, Hörster, & Sauer, 2013). Liver disease has been described in approximately a third of patients (Imbard et al., 2018), and several patients have developed hepatic neoplasms (Forny et al., 2019). The severity of the clinical picture in patients with isolated methylmalonic aciduria is related to the underlying metabolic defect: patients with *mut*^{0} tend to be the most severely affected, with *cblB* patients also seriously affected; patients with the *cblA* or the *mut*$^{-}$ disorder are the least severely affected (Matsui, Mahoney, & Rosenberg, 1983; Hörster et al., 2007; Cosson et al., 2009). An exception to this generalization is metabolic stroke, which seems to occur at a similar frequency in all classes (Baker et al., 2015).

Treatment has generally depended on a protein-restricted diet to decrease the levels of branched chain amino acids, which are the primary metabolic source of methylmalonyl-CoA. Parenteral cobalamin supplementation may be used in patients (primarily *cblA*) who show decreased levels of methylmalonic acid in response to the vitamin. Carnitine supplementation can replenish the intracellular CoA pool and help in elimination of toxic derivatives of methylmalonyl-CoA, and antibiotics may be prescribed to limit production of propionic acid by gut bacteria. The most promising therapy is been liver transplantation. This has been shown to prevent or decrease the frequency of episodes of metabolic decompensation (Niemi et al., 2015; Brassier et al., 2020), although methylmalonic acid levels remain somewhat elevated since its metabolism in tissues outside the liver continues to be affected, and neurologic and renal deterioration may still occur.

The *mut* disorder is the most common of these disorders, with over 1000 patients and approximately 400 mutations in *MMUT* reported. Approximately 200 *cblA* patients (more than 80 different *MMAA* mutations) and nearly 100 *cblB* patients (more than 35 *MMAB* mutations) have also been reported.

11. Homocysteine remethylation

Methionine synthase catalyzes the transfer of a methyl group from 5-methyltetrahydrofolate to homocysteine to generate methionine. This reaction proceeds by way of a MeCbl intermediate: a methyl group from 5-methyltetrahydrofolate is first transferred to an enzyme-bound fully-reduced cob(I)alamin molecule to generate MeCbl; the methyl group is then transferred to homocysteine, with regeneration of cob(I)alamin. Methionine synthase function therefore is characterized by shuttling of the enzyme-bound cobalamin coenzyme between cob(I)alamin and MeCbl. Oxidation of the enzyme-bound cobalamin to the cob(II)alamin form occurs occasionally. A second protein, methionine synthase reductase, acts in conjunction with methionine synthase to catalyze the reductive methylation of cob(II)alamin to MeCbl, using S-adenosylmethionine as a methyl group donor.

Decreased methionine synthase function resulting in accumulation of its homocysteine substrate in blood and urine occurs in the *cblE* and *cblG* disorders (Watkins & Rosenblatt, 1988). Clinical findings are similar in both disorders. Patients usually present in the first year of life, although later onset cases have been reported. Patients usually have megaloblastic anemia; neutropenia and thrombocytopenia may also occur. There is an array of neurological findings including developmental delay, cerebral atrophy, microcephaly, hypotonia, ataxia, seizures and EEG anomalies (Watkins & Rosenblatt, 1989; Huemer et al., 2015). Some patients have presented with atypical hemolytic uremic syndrome. Treatment involves parenteral cobalamin together with betaine to reduce homocysteine levels. Biochemical and hematologic findings are usually corrected, but neurologic problems may not completely respond.

Studies of patient fibroblasts demonstrated decreased synthesis of MeCbl from exogenous cyanocobalamin and decreased methionine synthase function in the presence of adequate AdoCbl synthesis and methylmalonyl-CoA mutase function (Watkins & Rosenblatt, 1988). Complementation analysis demonstrated the presence of two complementation classes that were designated *cblE* and *cblG*. The specific activity of methionine synthase was within the reference range in extracts of *cblE* fibroblasts when the assay was performed under standard conditions, including exogenous reducing agents. However, decreased specific activity was present when the assay was performed in the presence of suboptimal concentrations of reducing agent, implicating a defect in a reducing activity associated with the enzyme. Specific activity of *cblG* fibroblast extracts was decreased under all assay conditions.

The *cblG* disorder is caused by mutations in the *MTR* gene, which encodes methionine synthase itself (Gulati et al., 1996; Leclerc et al., 1996), while the *cblE* disorder is caused by mutations in *MTRR*, encoding methionine synthase reductase (Leclerc et al., 1998). Approximately 50 patients have been reported with the *cblG* disorder, with 44 pathogenic variants in *MTR* described. Approximately 35 patients have been reported with the *cblE* disorder, with 31 pathogenic variants in *MTRR* described.

References

Alfadhel, M., Lillquist, Y. P., Davis, C., Junker, A. K., & Stockler-Iposiroglu, S. (2011). Eighteen-year follow-up of a patient with cobalamin F disease (cblF): Report and review. *American Journal of Medical Genetics. Part A*, *155A*, 2571–2577.

Aminoff, M., Carter, J. E., Chadwick, R. B., Johnson, C., Gräsbeck, R., Abdelaal, M. A., et al. (1999). Mutations in CUBN, encoding the intrinsic factor-vitamin B_{12} receptor, cubilin, cause hereditary megaloblastic anaemia 1. *Nature Genetics*, *21*, 309–313.

Baker, E. H., Sloan, J. L., Hauser, N. S., Gropman, A. L., Adams, D. R., Toro, C., et al. (2015). MRI characteristics of globus pallidus infarcts in isolated methylmalonic acidemia. *American Journal of Neuroradiology, 36*, 194–201.

Beck, B. B., van Spronsen, F. J., Diepstra, A., Berger, R. M. F., & Kömhoff, M. (2016). Renal thrombotic microangiopathy in patients with cblC defect: Review of an under-recognized entity. *Pediatric Nephrology, 32*, 733–741.

Brassier, A., Krug, P., Lacaille, F., Pontoizeau, C., Krid, S., Sissaoui, S., et al. (2020). Long-term outcome of methylmalonic aciduria after kidney, liver, or combined liver-kidney transplantation: The French experience. *Journal of Inherited Metabolic Disease, 43*, 234–243.

Brooks, B. P., Thompson, A. H., Sloan, J. L., Manoli, I., Carrillo-Carrasco, N., Zein, W. M., et al. (2016). Ophthalmic manifestations and long-term visual outcome in patients with cobalamin C deficiency. *Ophthalmology, 123*, 571–582.

Carmel, R. (1983). R-binder deficiency. A clinically benign cause of cobalamin pseudodeficiency. *JAMA, 250*, 1886–1890.

Carmel, R., Parker, J., & Kelman, Z. (2009). Genomic mutations associated with mild and severe deficiencies of transcobalamin I (haptocorrin) that cause mildly and severely low serum cobalamin levels. *British Journal of Haematology, 147*, 386–391.

Carrillo-Carrasco, N., Chandler, R. J., & Venditti, C. P. (2012). Combined methylmalonic acidemia and homocystinuria, cblC type. I. Clinical presentation, diagnosis and management. *Journal of Inherited Metabolic Disease, 35*, 91–102.

Carrillo-Carrasco, N., Sloan, J., Valle, D., Hamosh, A., & Venditti, C. P. (2009). Hydroxocobalamin dose escalation improves metabolic control in cblC. *Journal of Inherited Metabolic Disease, 32*, 728–731.

Carrillo-Carrasco, N., & Venditti, C. P. (2012). Combined methylmalonic acidemia and homocystinuria, cblC type. II. Complications, pathophysiology, and outcomes. *Journal of Inherited Metabolic Disease, 35*, 103–114.

Coelho, D., Kim, J. C., Miousse, I. R., Fung, S., Du Moulin, M., Buers, I., et al. (2012). Mutations in *ABCD4* cause a new inborn error of vitamin B_{12} metabolism. *Nature Genetics, 44*, 1152–1155.

Coelho, D., Suormala, T., Stucki, M., Lerner-Ellis, J. P., Rosenblatt, D. S., Newbold, R. F., et al. (2008). Gene identification for the cblD defect of vitamin B_{12} metabolism. *The New England Journal of Medicine, 358*, 1454–1464.

Constantinou, P., D'Allesandro, M., Lochead, P., Samant, S., Bisset, W. M., Hauptfleisch, C., et al. (2016). A new, atypical case of cobalamin F disorder diagnosed by whole exome sequencing. *Molecular Syndromology, 6*, 254–258.

Cosson, M. A., Benoist, J. F., Touati, G., Déchaux, M., Royer, N., Grandin, L., et al. (2009). Long-term outcome in methylmalonic aciduria: A series of 30 French patients. *Molecular Genetics and Metabolism, 97*, 172–178.

Dehaene, H., Praz, V., Lhôte, P., Lopes, M., & Herr, W. (2020). $THAP11_{F80L}$ cobalamin disorder-associated mutation reveals normal and pathogenic THAP11 functions in gene expression and cell proliferation. *PLoS One, 15*, e0224646.

Deme, J. C., Miousse, I. R., Plesa, M., Kim, J. C., Hancock, M. A., Mah, W., et al. (2012). Structural features of recombinant MMADHC isoforms and their interactions with MMACHC, proteins of mammalian vitamin B_{12} metabolism. *Molecular Genetics and Metabolism, 107*, 352–362.

Fischer, S., Huemer, M., Baumgartner, M., Deodato, F., Ballhausen, D., Boneh, A., et al. (2014). Clinical presentation and outcome in a series of 88 patients with the cblC defect. *Journal of Inherited Metabolic Disease, 37*, 831–840.

Fofou-Caillierez, M. B., Mrabet, N. T., Chéry, C., Dreumont, N., Flayac, J., Pupavac, M., et al. (2013). Interaction between methionine synthase isoforms and MMACHC: Characterization in cblG-variant, cblG and cblC inherited causes of megaloblastic anemia. *Human Molecular Genetics, 22*, 4591–4601.

Forny, P., Hochuli, M., Rahman, Y., Deheragoda, M., Weber, A., Baruteau, J., et al. (2019). Liver neoplasms in methylmalonic aciduria: An emerging complication. *Journal of Inherited Metabolic Disease, 42*, 793–802.

Forny, P., Hörster, F., Ballhausen, D., Chakrapani, A., Chapman, K. A., Dionisi-Vici, C., et al. (2021). Guidelines for the diagnosis and management of methylmalonic acidaemia and propionic acidaemia: First revision. *Journal of Inherited Metabolic Disease, 44*, 566–592.

Froese, D. S., Kopec, J., Fitzpatrick, F., Schuller, M., McCorvie, T. J., Chalk, R., et al. (2015). Structural insights into the MMACHC-MMADHC protein complex involved in vitamin B_{12} trafficking. *The Journal of Biological Chemistry, 290*, 29167–29177.

Gérard, M., Morin, G., Bourillon, A., Colson, C., Mathieu, S., Rabier, D., et al. (2015). Multiple congenital anomalies in two boys with mutations in *HCFC1* and cobalamin disorder. *European Journal of Medical Genetics, 58*, 148–153.

Gherasim, C., Hannibal, L., Rajagopalan, D., Jacobsen, D. W., & Banerjee, R. (2013). The C-terminal domain of CblD interacts with CblC and influences intracellular cobalamin partitioning. *Biochimie, 95*, 1023–1032.

Gherasim, C., Lofgren, M., & Banerjee, R. (2013). Navigating the B_{12} road: Assimilation, delivery and disorders of cobalamin. *The Journal of Biological Chemistry, 288*, 13186–13193.

Goodman, S. I., Moe, P. G., Hammond, K. B., Mudd, S. H., & Uhlendorf, B. W. (1970). Homocystinuria with methylmalonic aciduria: Two cases in a sibship. *Biochemical Medicine, 4*, 500–515.

Gräsbeck, R., & Tanner, S. M. (2011). Juvenile selective vitamin B_{12} malabsorption: 50 Years after its description—10 Years of genetic testing. *Pediatric Research, 70*, 222–228.

Gravel, R. A., Mahoney, M. J., Ruddle, F., & Rosenberg, L. E. (1975). Genetic complementation in heterokaryons of human fibroblasts defective in cobalamin metabolism. *Proceedings of the National Academy of Sciences of the United States of America, 72*, 3181–3185.

Guéant, J. L., Chéry, C., Oussalah, A., Nadaf, J., Coelho, D., Josse, T., et al. (2018). A *PRDX*1 mutant allele causes a MMACHC secondary epimutation in cblC patients. *Nature Republic, 9*, 67.

Gulati, S., Baker, P., Li, Y. N., Fowler, B., Kruger, W., Brody, L. C., et al. (1996). Defects in human methionine synthase in cblG patients. *Human Molecular Genetics, 5*, 1859–1865.

Hannah-Shmouni, F., Cruz, V., Schulze, A., & Mercimek-Andrews, S. (2018). Transcobalamin receptor defect: Identification of two new cases through positive newborn screening for propionic/methylmalonic aciduria and long-term outcome. *American Journal of Medical Genetics. Part A, 176*, 1411–1415.

Haurani, F. I., Hall, C. A., & Rubin, R. (1979). Megaloblastic anemia as a result of an abnormal transcobalamin II (Cardeza). *The Journal of Clinical Investigation, 64*, 1253–1259.

Hörster, F., Baumgartner, M. R., Viardot, C., Suormala, T., Burgard, P., Fowler, B., et al. (2007). Long-term outcome in methylmalonic acidurias is influenced by the underlying defect (mut^0, mut^-, cblA, cblB). *Pediatric Research, 62*, 225–230.

Huemer, M., Bürer, C., Jesina, P., Kozich, V., Landolt, M. A., Suormala, T., et al. (2015). Clinical onset and course, response to treatment and outcome in 24 patients with the cblE or cblG remethylation defect complemented by genetic and in vitro enzyme study data. *Journal of Inherited Metabolic Disease, 38*, 957–967.

Huemer, M., Simma, B., Fowler, B., Suormala, T., Bodamer, O. A., & Sass, J. O. (2005). Prenatal and postnatal treatment in cobalamin C defect. *The Journal of Pediatrics, 147*, 469–472.

Imbard, A., Segarra, N. G., Tardieu, M., Broué, P., Bouchereau, J., Pichard, S., et al. (2018). Long-term liver disease in methylmalonic and propionic acidemias. *Molecular Genetics and Metabolism, 123*, 433–440.

Karth, P., Singh, R., Kim, J., & Costakos, D. (2012). Bilateral central retinal artery occlusions in an infant with hyperhomocysteinemia. *Journal of AAPOS, 16*, 398–400.

Kim, J., Gherasim, C., & Banerjee, R. (2008). Decyanation of vitamin B_{12} by a trafficking chaperone. *Proceedings of the National Academy of Sciences of the United States of America, 105*, 14551–14554.

Kim, J., Hannibal, L., Gherasim, C., Jacobsen, D. W., & Banerjee, R. (2009). A human B_{12} trafficking protein uses glutathione transferase activity for processing alkylcobalamins. *The Journal of Biological Chemistry, 284*, 33418–33424.

Kim, J. C., Lee, N. C., Hwu, P. W. L., Chien, Y. H., Fahiminiya, S., Majewski, J., et al. (2012). Late onset of symptoms in an atypical patient with the *cblJ* inborn error of vitamin B_{12} metabolism: Diagnosis and novel mutation revealed by exome sequencing. *Molecular Genetics and Metabolism, 107*, 664–668.

Kitai, K., Kawaguchi, K., Tomohiro, T., Morito, M., So, T., & Imanaka, T. (2021). The lysosomal protein ABCD4 can transport vitamin B12 across lysosomal membranes *in vitro*. *The Journal of Biological Chemistry*. https://doi.org/10.1016/j.jbc.2021.100654.

Koufaris, C., Alexandrou, A., Tanteles, G. A., Anastasiadou, V., & Sismani, C. (2016). A novel *HCFC1* variant in male siblings with intellectual disability and microcephaly in the absence of cobalamin disorder. *Biomedical Reports, 4*, 215–218.

Kowaguchi, K., Okamoto, T., Morita, M., & Imanaka, T. (2016). Translocation of the ABC transporter ABCD4 from the endoplasmic reticulum to lysosomes requires the escort protein LMBD1. *Scientific Reports, 6*, 30183.

Laframboise, R., Cooper, B. A., & Rosenblatt, D. S. (1992). Malabsorption of vitamin B_{12} from the intestine in a child with *cblF* disease: Evidence for lysosomal-mediated absorption. *Blood, 80*, 291–292.

Leclerc, D., Campeau, E., Goyette, P., Adjalla, C. E., Christensen, B., Ross, M., et al. (1996). Human methionine synthase: cDNA cloning and identification of mutations in patients of the *cblG* complementation group of folate/cobalamin disorders. *Human Molecular Genetics, 5*, 1867–1874.

Leclerc, D., Wilson, A., Dumas, R., Gafuik, C., Song, D., Watkins, D., et al. (1998). Cloning and mapping of a cDNA for methionine synthase reductase, a flavoprotein defective in patients with homocystinuria. *Proceedings of the National Academy of Sciences of the United States of America, 95*, 3059–3064.

Lerner-Ellis, J. P., Tirone, J. C., Pawelek, P. D., Doré, C., Atkinson, J. L., Watkins, D., et al. (2006). Identification of the gene responsible for methylmalonic aciduria and homocystinuria, cblC type. *Nature Genetics, 38*, 93–100.

Li, Z., Mascarenhas, R., Twahir, U. T., Kallon, A., Deb, A., Yaw, M., et al. (2020). An interprotein Co-S coordination complex in the B12-trafficking pathway. *Journal of the American Chemical Society, 142*, 16334–16340.

Matsui, S. M., Mahoney, M. J., & Rosenberg, L. E. (1983). The natural history of the inherited methylmalonic acidemias. *The New England Journal of Medicine, 308*, 857–861.

Morath, M. A., Hörster, F., & Sauer, S. W. (2013). Renal dysfunction in methylmalonic acidurias: Review for the pediatric nephrologist. *Pediatric Nephrology, 28*, 227–235.

Morel, C. F., Lerner-Ellis, J. P., & Rosenblatt, D. S. (2006). Combined methylmalonic aciduria and homocystinuria (*cblC*): Phenotype-genotype correlations and ethnic-specific observations. *Molecular Genetics and Metabolism, 88*, 315–321.

Niemi, A. K., Kim, I. K., Krueger, C. E., Cowan, T. M., Baugh, N., Farrell, R., et al. (2015). Treatment of methylmalonic acidemia by liver or combined liver-kidney transplantation. *The Journal of Pediatrics, 166*, 1455–1461.

Oladipo, O., Rosenblatt, D. S., Watkins, D., Miousse, I. R., Sprietsma, L., Dietzen, D. J., et al. (2011). Cobalamin F disease detected by newborn screening and follow-up on a 14-year-old patient. *Pediatrics, 128*, e1636–e1640.

Ovunc, B., Otto, E. A., Vega-Warner, V., Saisawat, P., Ashraf, S., Ramaswami, G., et al. (2011). Exome sequencing reveals cubilin mutation as a single-gene cause of proteinuria. *Journal of the American Society of Nephrology, 22*, 1815–1820.

Pangilinan, F., Mitchell, A., VanderMeer, J., Molloy, A. M., Troendle, J., Conley, M., et al. (2010). Transcobalamin II receptor polymorphisms are associated with increased risk for neural tube defects. *Journal of Medical Genetics*, *47*, 677–685.

Profitlich, L. E., Kirmse, B., Wasserstein, M. P., Diaz, G. A., & Srivastava, S. (2009). High prevalence of structural heart disease in children with cblC-type methylmalonic aciduria and homocystinuria. *Molecular Genetics and Metabolism*, *98*, 344–348.

Pupavac, M., Watkins, D., Petrella, F., Fahiminiya, S., Janer, A., Cheung, W., et al. (2016). Inborn error of cobalamin metabolism associated with the intracellular accumulation of transcobalamin-bound cobalamin and mutations in *ZNF143*, which codes for a transcriptional activator. *Human Mutation*, *37*, 976–982.

Quadros, E. V., Lai, S. C., Nakayama, Y., Sequeira, J. M., Hannibal, L., Wang, S., et al. (2010). Positive newborn screen for methylmalonic aciduria identifies the first mutation in TCblR/*CD320*, the gene for cellular uptake of transcobalamin-bound vitamin B_{12}. *Human Mutation*, *31*, 924–929.

Quintana, A. M., Yu, H. C., Brebner, A., Pupavac, M., Geiger, E. A., Watson, A., et al. (2017). Mutations in *THAP11* cause an inborn error of cobalamin metabolism and developmental abnormalities. *Human Molecular Genetics*, *26*, 2838–2849.

Rosenblatt, D. S., Laframboise, R., Pichette, J., Langevin, P., Cooper, B. A., & Costa, T. (1986). New disorder of vitamin B_{12} metabolism (cobalamin F) presenting as methylmalonic aciduria. *Pediatrics*, *78*, 51–54.

Rutsch, F., Gailus, S., Miousse, I. R., Suormala, T., Sagné, C., Reza Toliat, M., et al. (2009). Identification of a putative lysosomal cobalamin exporter mutated in the cblF inborn error of vitamin B_{12} metabolism. *Nature Genetics*, *41*, 234–239.

Scalais, E., Osterheld, E., Geron, C., Pierron, C., Chafai, R., Schlesser, V., et al. (2019). Parenteral hydroxocobalamin dose intensification in five patients with different types of early onset intracellular cobalamin defects: Clinical and biochemical responses. *JIMD Reports*, *49*, 70–79.

Scalais, E., Osterheld, E., Weitzel, C., De Meirleir, L., Mataigne, F., Martens, G., et al. (2017). X-linked cobalamin disorder (HCFC1) mimicking non-ketotic hyperglycinemia with increased both cerebrospinal fluid glycine and methylmalonic acid. *Pediatric Neurology*, *71*, 65–69.

Schiff, M., Ogier de Baulny, H., Bard, G., Barlogis, V., Hamel, C., Moat, S. J., et al. (2010). Should transcobalamin deficiency be treated aggressively? *Journal of Inherited Metabolic Disease*, *33*, 223–229.

Sloan, J. L., Carrillo, N., Adams, D., & Venditti, C. P. (2018). Disorders of intracellular cobalamin metabolism. In M. P. Adam, H. H. Ardinger, & R. A. Pagon (Eds.), *GeneReviews*. Seattle: University of Washington.

Suormala, T., Baumgartner, M. R., Coelho, D., Zavadakova, P., Kozich, V., Koch, H. G., et al. (2004). The cblD defect causes either isolated or combined deficiency of methylcobalamin and adenosylcobalamin synthesis. *The Journal of Biological Chemistry*, *279*, 42742–42749.

Takeichi, T., Hsu, C. K., Yang, H. S., Chen, H. Y., Wong, T. W., Tsai, W. L., et al. (2015). Progressive hyperpigmentation in a Taiwanese child due to an inborn error of vitamin B_{12} metabolism (cblJ). *The British Journal of Dermatology*, *172*, 1111–1115.

Tanner, S. M., Aminoff, M., Wright, F. A., Liyanarachchi, S., Kuronen, M., Saarinen, A., et al. (2003). Amnionless, essential for mouse gastrulation, is mutated in recessive hereditary megaloblastic anemia. *Nature Genetics*, *33*, 426–429.

Tanner, S. M., Li, Z., Perko, J. D., Öner, C., Çetin, M., Altay, Ç., et al. (2005). Hereditary juvenile cobalamin deficiency caused by mutations in the intrinsic factor gene. *Proceedings of the National Academy of Sciences of the United States of America*, *102*, 4130–4133.

Tanner, S. M., Sturm, A. C., Baack, E. C., Liyanarachchi, S., & de la Chapelle, A. (2012). Inherited cobalamin malabsorption. Mutations in three genes reveal functional and ethnic patterns. *Orphanet Journal of Rare Diseases*, 7, 56.

Trakadis, Y. J., Alfares, A., Bodamer, O. A., Boyukavci, M., Christodoulou, J., Connor, P., et al. (2014). Update on transcobalamin deficiency: Clinical presentation, treatment and outcome. *Journal of Inherited Metabolic Disease*, *37*, 461–473.

Trefz, F., Scheible, D., Frauendienst-Egger, G., Huemer, M., Suormala, T., Fowler, B., et al. (2016). Successful intrauterine treatment of a patient with cobalamin C defect. *Molecular Genetics and Metabolism Reports*, *6*, 55–59.

Vassiliadis, A., Rosenblatt, D. S., Cooper, B. A., & Bergeron, J. J. M. (1991). Lysosomal cobalamin accumulation in fibroblasts from a patient with an inborn error of cobalamin metabolism (cblF complementation group): Visualization by electron microscope radioautography. *Experimental Cell Research*, *195*, 295–302.

Watkins, D., & Rosenblatt, D. S. (1986). Failure of lysosomal release of vitamin B_{12}: A new complementation group causing methylmalonic aciduria. *American Journal of Human Genetics*, *39*, 404–408.

Watkins, D., & Rosenblatt, D. S. (1988). Genetic heterogeneity among patients with methylcobalamin deficiency. Definition of two complementation groups, cblE and cblG. *The Journal of Clinical Investigation*, *81*, 1690–1694.

Watkins, D., & Rosenblatt, D. S. (1989). Functional methionine synthase deficiency (cblE and cblG): Clinical and biochemical heterogeneity. *American Journal of Medical Genetics*, *34*, 427–434.

Willard, H. F., Mellman, I. S., & Rosenberg, L. E. (1978). Genetic complementation among inherited deficiencies of methylmalonyl-CoA mutase activity: Evidence for a new class of human cobalamin mutant. *American Journal of Human Genetics*, *30*, 1–13.

Wongkittichote, P., Wagner, D. J., & Srinawi, M. S. (2021). Novel exon-skipping variant disrupting the basic domain of *HCFC1* causes intellectual disability without metabolic abnormalities in both male and female patients. *Journal of Human Genetics*, *66*, 717–724.

Yamada, K., Gherasim, C., Banerjee, R., & Koutmos, M. (2015). Structure of human CblD reveals molecular mimicry and identifies a new subfamily of nitro-FMN reductases. *The Journal of Biological Chemistry*, *290*, 29155–29166.

Yu, H. C., Sloan, J. L., Scharer, G., Brebner, A., Quintana, A., Achilly, N. P., et al. (2013). An X-linked cobalamin disorder caused by mutations in transcriptional coregulator *HCFC1*. *American Journal of Human Genetics*, *93*, 506–514.

Zhang, Y., Yang, Y., Hasegawa, Y., Yamaguchi, S., Shi, C., Song, J., et al. (2008). Prenatal diagnosis of methylmalonic aciduria by analysis of organic acids and total homocysteine in amniotic fluid. *Chinese Medical Journal*, *121*, 216–219.

Zhou, W., Li, H., Wang, C., Wang, X., & Gu, M. (2019). Newborn screening for methylmalonic acidemia in a Chinese population: Molecular genetic confirmation and genotype phenotype correlations. *Frontiers in Genetics*, *9*, 726.

CHAPTER FOURTEEN

Behavioral profile of vitamin B_{12} deficiency: A reflection of impaired brain development, neuronal stress and altered neuroplasticity

Grégory Pourié[a,*], Jean-Louis Guéant[a,b], and Edward V. Quadros[c]
[a]Université de Lorraine, Inserm, UMRS 1256, NGERE—Nutrition, Genetics, and Environmental Risk Exposure, Nancy, France
[b]CHRU-Nancy, National Center of Inborn Errors of Metabolism, Nancy, France
[c]Department of Medicine, SUNY Downstate Medical Center, Brooklyn, NY, United States
*Corresponding author: e-mail address: gregory.pourie@univ-lorraine.fr

Contents

Abstract

Our understanding of brain biology and function is one of the least characterized and therefore, there are no effective treatments for most of neurological disorders. The influence of vitamins, and particularly vitamin B_{12}, in neurodegenerative disease is demonstrated but largely unresolved. Behaviors are often quantified to attest brain dysfunction

Vitamins and Hormones, Volume 119
ISSN 0083-6729
https://doi.org/10.1016/bs.vh.2022.02.002

alone or in parallel with neuro-imaging to identify regions involved. Nevertheless, attention should be paid to extending observations made in animal models to humans, since, first, behavioral tests have to be adjusted in each model to address the initial question and second, because brain analysis should not be conducted for a whole organ but rather to specific sub-structures to better define function. Indeed, cognitive functions such as psychiatric disorders and learning and memory are often cited as the most impacted by a vitamin B_{12} deficiency. In addition, differential dysfunctions and mechanisms could be defined according sub-populations and ages. Vitamin B_{12} enters the cell bound to Transcobalamin, through the Transcobalamin Receptor and serves in two cell compartments, the lipid metabolism in the mitochondrion and the one-carbon metabolism involved in methylation reactions. Dysfunctions in these mechanisms can lead to two majors outcomes; axons demyelinisation and upregulation of cellular stress involving mislocalization of RNA binding proteins such as the ELAVL1/HuR or the dysregulation of pro- or anti-oxidant NUDT15, TXNRD1, VPO1 and ROC genes. Finally, it appears that apart from developmental problems that have to be identified and treated as early as possible, other therapeutic approaches for behavioral dysfunctions should investigate cellular methylation, oxidative and endoplasmic reticulum stress and mitochondrial function.

1. Introduction and background

Our knowledge regarding neurology and psychology advanced in the early 1950s when medicine and sciences converged to investigate and understand mental retardation and related disorders. In the early 20th century, a majority of cases, presented with aberrant behaviors and low IQ scores and were sequestered in hospices or were rejected from society, especially because no specific treatment was available. What was evident in these disorders was mainly behavioral aberrations and external traits such as aggression, mood imbalance and social isolation. About this time, pharmaceuticals were being developed through psychotropic agents for both treatment and research, with the hope that it would lead to a better understanding of brain function. In parallel to this increasing general knowledge of functional neurology, evidence gathered pointed to environmental influences on brain development and functional outcomes. Vitamin-B_{12} status was especially proposed as a determinant of brain health (Batata, Spray, Bolton, Higgins, & Wollner, 1967; Holmes, 1956). A direct relationship was proposed between a B_{12} deficiency and neonatal failures during pregnancy; highlighting the importance of the one-carbon metabolism for fetal development (Woodward & Newberne, 1967). Finally, in the 1970s, nutritional approaches confirmed that neuropathies were associated with low levels of B_{12} and that timely treatment could reverse the neurological manifestations with dramatic effects (Agamanolis et al., 1978).

2. Neuropathology and behavioral abnormalities in human disorders

Two major causes lead to intracellular B_{12} dysregulation; (i) nutritional deficiency due to inappropriate intake and/or malabsorption (ii) inherited disorders of intracellular metabolism. Both of these are characterized by neurologic manifestations that vary according to age of onset and are sometimes but not always related to functional abnormalities of the brain. The behavioral disabilities often represent a major cue of this neurological symptomatology (see Table 1).

Child development under a low B_{12} status is correlated in all cohorts, case-reports and cross-data analysis studied with impairments of various behaviors, and one of the major consequence observed is a lower academic level scored compared to non-B_{12} deficient persons (Bjørke-Monsen & Ueland, 2011; Strand et al., 2013; Venkatramanan et al., 2016). As a direct link, the cognitive functions are always cited as the most impacted. Infact, a high correlation was reported between plasma B_{12} and the mental developmental index score, but not with psychomotor scores, in children before 2 years of age. Moreover, an inverse correlation was also established using homocysteine and methylmalonic acid markers, reinforcing a strong influence of the B_{12} status on neurofunctional outcomes in the perinatal period (Strand et al., 2013). But among this wide category of functions named "cognition," a higher function attributed to mammals, different sub-functions need to be differentiated, such as attention, learning, memory and psychologic status like depression, especially in growing infants until adulthood. In addition to the metabolic parameters, early environmental influences have been proposed to influence the mental construction, and under negative conditions in a deficiency state of B_{12} with, depression-like behaviors emerging in adulthood (Black, 2008).

Inherited disorders of intracellular B_{12} metabolism are good illustrations of the influence of B_{12} on behavior in children. The manifestations of inherited disorders of intracellular B_{12} metabolism produce hypotonia, cognitive abnormalities with developmental delay, seizures and microcephaly, in the first year of life. In later life, the behavioral disorders, including feeding

Table 1 Causes and behavioral characteristics in human vitamin B_{12} deficiencies.

Age category	Behavioral deficit	Related markers	Mechanism/gene	Reference
Infants <1 year	Hypotonia, delayed acquisition of posture (head control and sitting/turning)	B_{12} gestational/nutritional intake	Lower myelination and cortical atrophy	Bjørke-Monsen and Ueland (2011)
Infants <2 years	Mental developmental score (Bayley Scales of Infant Development 2nd) = cognition	Twofold increase of plasma Cbl correlates with 1.3-fold increase of mental dev. score Twofold increase of homocysteine/MMA correlates with twofold decrease in the mental dev. score	(speculative) Reduced methylation Lower myelination	Strand et al. (2013)
Infants deficiency lead to mid-aged consequences	Psychic health Depression	B_{12} and folate deficiency in early ages	Myelination Synaptic connectivity Inflammation	Black (2008)
Infants deficiency lead to mid-aged (children, adolescent, young adult) consequences	Attention Perception Memory Academic performance Intelligence quotient	B_{12} deficiency in early ages + related biomarkers (MMA, tHcy, holotranscobalamin)		Venkatramanan, Armata, Strupp, and Finkelstein (2016)
Infants and Children	Hypotonia Ataxia	Behavioral defects depends of MTHFR residual activity	MTHFR mutation and related mutations in the	Huemer et al. (2016, 2017)

	Cognitive impairments Mental retardation Epilepsy	Markers: plasma tHcy and methionine, urine Hcy	homocysteine remethylation cycle	
Elderly	Cognitive impairments	Low plasma B_{12} and holoTC + high plasma tHcy and MMA	Accelerated brain atrophy	Vogiatzoglou et al. (2013)
Elderly	Mild cognitive impairment slowed	Supplementation with $B_9/B_6/B_{12}$	Brain atrophy	Smith et al. (2010)
Elderly	Dementia (Alzheimer-like)	Plasma tHcy	Brain atrophy White matter damages	Smith and Refsum (2016)
Elderly	Cognitive impairment Depression	Plasmatic low folate, B_{12} and high Hcy	Brain injury	Heok and Ho (2008)
50–70 years adults	Cognitive assessment (MMSE and MoCA scores)	Plasmatic B_6, B_9 and B_{12}	Low global DNA methylation Foal hypermethylation of anti-oxidant genes NUDT15 and TXNRD1	An et al. (2019)

disorders, are evidenced in more than 10% and psychiatric manifestations in about one-third of cases with cblC mutation. This disorder is the most frequent cause of inherited disorders of intracellular metabolism of vitamin-B_{12} (Huemer et al., 2019). It is produced by a mutation in MMACHC, a gene involved in the delivery of the B_{12} coenzymes to the cytoplasmic methionine synthase and mitochondrial methylmalonylCoA mutase. Other less frequent disorders include mutations in MTR, MTRR and MMUT which encode methionine synthase, methionine synthase reductase and mitochondrial methylmalonylCoA mutase, respectively.

The best example of the key influence of B_{12} on behavior in adults is pernicious anemia, which often presents with depression and memory impairment. Historical presentation of pernicious anemia with dramatic and prolonged B_{12} deficiency included dementia as a major neurological outcome (Green, Griffiths, & Almond, 2017). Low blood level of vitamin B_{12} is associated with a more rapid rate of brain atrophy and mild cognitive decline in elderly subjects, whereas folate level purported to be unrelated (Vogiatzoglou et al., 2013). The accelerated rate of brain atrophy can be slowed by homocysteine-lowering treatment with B_{12}, B_9 and B_6 vitamins (Smith et al., 2010). Trials in high-risk subjects showed a slowing of cognitive decline and of atrophy in critical brain regions, including hippocampus and is consistent with modification of the Alzheimer's disease process by B_{12} and folate treatment (Smith & Refsum, 2016).

Vitamin B_{12} deficiency is frequently associated with a folate deficiency, and further, with altered one carbon metabolism (Agrawal et al., 2015; Kim et al., 2013). Subsequently, the apparition of psychiatric disorders in adults is linked to low concentration of S-adenosine-methyonine (SAM) in both the blood and the cerebrospinal fluid, a universal methyl donor for transmethylation reactions in the cell. This also led to the successful treatment of depression using SAM (Black, 2008). In aging humans, a B_{12} deficiency is also linked to psychiatric disorders. In some populations, it was proposed to monitor this metabolic risk factor for an early diagnosis, in order to avoid dramatic psychiatric outcomes such as suicide in elderly patients (Heok & Ho, 2008). But, in light of the knowledge of both human neurology and the research in animal models, the so-called psychiatric disorders have to be defined because they primarily belong to the limbic system and the frontal and pre-frontal cortex. These brain circuits are particularly well developed in mammals and much more in primates with psychiatric functions spanning motivation, pain, depression and cognition.

3. Behavioral abnormalities in animal models

3.1 Defining behaviors affected by a B_{12} deficiency

Much of our knowledge on the consequences B_{12} deficiency is derived from human studies. It appears that most of the major brain functions affected by a low vitamin B_{12} status is represented by psychiatric disturbances and/or cognitive deficits. These two functions are strongly interconnected and affected in numerous neurologic disorders. From a neurobiologist's perspective, cognition and psychosis could be considered as separate entities. Thus, it appears important to clearly define what was investigated, especially under the influence of environmental factors such as nutrition and vitamin supplements. The questions are not often focused on vitamin B_{12} deficiency exclusively and studies generally consider B_{12} with folate and the consecutive hyperhomocysteinemia, a consequence of a down regulation of the one-carbon metabolism. Nevertheless, using an animal model, and considering a global metabolic dysregulation or a specific reduction of a single vitamin level, it is important to define the quantified behaviors. The brain functions could be investigated in living animal models using behavioral tools, but only if the chosen method fits a specific brain function. Otherwise, only a global functional status, or simple correlations could be evoked in the best case. Thus, in order to obtain a mechanistic answer to propose specific diagnosis or therapeutic approaches for humans, the design of each behavioral test should be defined to investigate a specific brain circuit as much as possible, in a so-called structure–function approach. Considering animal models and in an ideal scientific approach, the measurements of specific behavioral tests should provide information about specific brain sub-regions such as the hippocampus which is strongly implicated in situation analysis, learning and memory formation, items of cognitive functions, or the frontal cortex implicated in motivation, anxiety and depressive-like behaviors (Blaise et al., 2007; Gupta et al., 2016; Hassan et al., 2019).

Moreover, in such behavioral investigations, attention should be paid to the strict correspondence between scientific questions asked or objectives and the methodology of the chosen tests. In the majority of cases, a behavioral test could identify significant traits or items that are known to be characteristic of defined behaviors. Sometimes the methodology of a test could also activate functions other than the target one. An example is provided by the unexpected emotional activation during a learning and memory test, that can modulate the result obtained (Fig. 1; Conrad, 2010; Galliot, Levaillant,

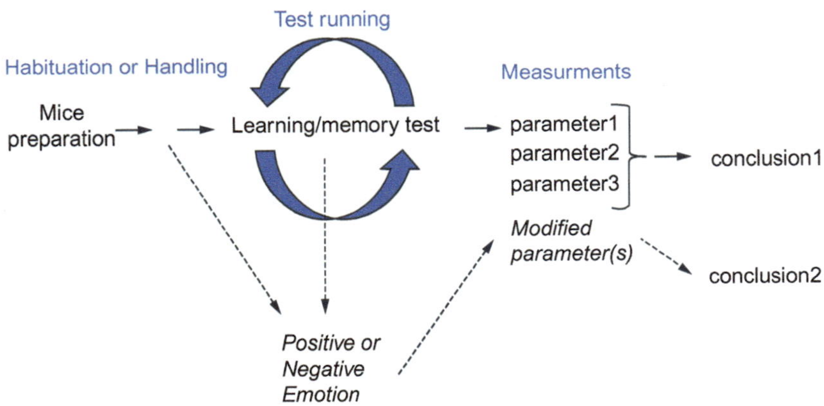

Fig. 1 schematic representation of putative modification of the conclusion after a behavioral experiment. Controlled or not controlled environmental factors could provide modified recorded parameters leading to various conclusions. Indeed, behavioral experiments have to be clearly defined previously (see Table 1) and the conditions surrounding the experiment should be evaluated or controlled as much as possible in order to avoid wrong conclusion compared to the initial scientific question.

Beard, Millot, & Pourié, 2010). Also, the behavioral tests has to be correctly set up to fit the objective, otherwise the results and conclusion would be difficult to interpret. For example, the learning and memory exploration could activate various diverse circuits, such as hippocampus-dependant or -independant ones. Thus, a classical visuo-spatial learning (i.e., like the Morris Water Maze) would activate the hippocampal circuitry, while an associative learning (i.e., a stimulus or an action is learned by the mouse tested) would not (Table 2, modified from Beck & Pourié, 2013).

3.2 The CD320 mouse: A model of brain B_{12} deficiency

When evaluating animal models of human cobalamin deficiency, the majority of the studies dealing with brain functional consequences describe a global dysfunction of the one carbon metabolism instead of a specific B_{12} deficiency. A systemic decrease of cobalamin highlights at least two questions. First, nowadays, it is quite easy to lower blood cobalamin levels using manufactured specific diets for rodents to mimic the defects of the stomach absorption using surgical approaches (Aires Neto et al., 2005; Mutti et al., 2011; Scalabrino et al., 1990). This model produces dramatic effects with weight loss, multiple nutrient deficiency inflammation, global DNA hypomethylation and mortality. It is not therefore adapted to the study of specific effects of B_{12} deficiency in brain (Brunaud et al., 2003). It appears

Table 2 Classification of the major behavioral tests used to investigate brain or peripheral circuits dysfunctions.

Tests	Recommended for[a]	Early description/late application and discussion
Hippocampus dependant		
Morris Water Maze (MWM)	Visuo-spatial memory	Morris (1984)
T-mazes	Spatial memory	Gentry, Brown, and Lee (1984), Raffaele et al. (1987)
Radial maze	Working memory	Olton (1987), Mei et al. (2020)
Passive avoidance test	Associative memory	van der Poel (1967), Nasehi, Sharifi, and Zarrindast (2012)
Hippocampus independant		
Object recognition	Discrimination memory	Rothblat and Kromer (1991)
Taste-aversion conditioning	Associative memory	Krane, Sinnamon, and Thomas (1976), Huang, Yu, He, and Ou (2020)
Other tests		
Elevated plus maze (EPM)	Anxiety status	Pellow, Chopin, File, and Briley (1985), Knight et al. (2021)
Open field	Locomotion activity, arousal, cues for stress status[b]	Walsh and Cummins (1979), Knight et al. (2021)
Rota-Rod	Locomotion, muscular abilities	Fanini, Palumbo, Giorgi and Pantaleoni (1990), Bertollo et al. (2006)
Social test (two or more individuals)	Social interactions, aggressively, sexual interactions, anxiety	File and Hyde (1978), Ceschim, Sumarán, Borges, Girardi, and Suchecki (2021)
Forced Swim Test (Porsolt)	Depressive-like behavior	Kitada, Miyauchi, Satoh, and Satoh (1981), Nomura, Shimizu, Kinjo, Kametani, and Nakazawa (1982), Unal and Canbeyli (2019), Kraeuter, Guest, and Sarnyai (2019)

[a]Some variations in the protocol could lead to assess a different neuronal function than the mentioned.
[b]The open field test allows measuring some cues that only suggest an evaluation of the stress and anxiety status (needs confirmation with a more appropriate test such as EPM).
A description of the initial test is proposed and modern discussions may lead to differential conclusions when animal models are used in the light of human pathologies, especially cognition.

difficult to obtain a balanced standardization of the level of deficiency between animals in the same experimental group, and to assure a significant deficiency in the nervous tissue, since the metabolic cross-talk with the liver could modulate the brain intake of cobalamin. Second, if a multi-nutrients systemic model (i.e., a nutritional model concerning cobalamin, folate and other related B vitamins) is chosen to target the one carbon metabolism, the scientific conclusions will be difficult to classify as due to cobalamin deficiency.

Indeed, a more specific model presenting a brain cobalamin deficiency appears more appropriate to investigate the neurofunctional outcomes in terms of behavioral deficits. The transcobalamin receptor/*Cd320* KO mouse model (TCblR/*Cd320*) was developed by Quadros and collaborators in 2013. These homozygous *Cd320* −/− mice, obtained by natural heterozygote crossing do not present a systemic cobalamin deficiency in early life. They develop, are carried to term and grow normally, but rapidly loose B_{12} in nervous system post birth (Lai et al., 2013). The reduction of B_{12} in the brain and the spinal cord is dramatic with concentrations reaching undetectable levels. These mice provided a whole new perspective for functional investigation of B_{12} deficiency, since they did not present typical neurodevelopment problems, such as spina bifida, reported in classical nutritional models (Blaise et al., 2007; Copp, Stanier, & Greene, 2013; Geoffroy et al., 2019). Indeed, the putative behavioral results could be directly linked to a real depletion of brain B_{12} levels while mice reach the adult life, and not to an aberrant neurodevelopment.

Exploring the behaviors of these TCblR/*Cd320* KO mice, it appeared that different cognitive and/or psychiatric traits were particularly affected in terms of elevation of anxiety and learning disabilities for memory formation. Gait and locomotion were also significantly affected. This last function is not directly relevant of cognitive abilities and is rather classified in vegetative functions, but appears necessary to express all behaviors. Moreover, exploring the brain sub-regions, the specific measurements of TCblR/*Cd320* mRNA levels were close to zero. Thus, considering a "structure–function" approach, the cognitive and psychiatric defects could be related to the hippocampus and cortex B_{12} down regulation, and the reduction of locomotion, to the cerebellum and spinal cord, as reported (Arora, Sequeira, Hernández, Alarcon, & Quadros, 2017; Arora, Sequeira, & Quadros, 2017).

Surprisingly, if a longer time is allowed to KO mice for memory formation—what is commonly named consolidation—in a place avoidance test using foot-shocks, the functional deficit appeared reduced in terms of global memory capacities.

Indeed, two major conclusions can be drawn from behavioral investigations in the TCblR/*Cd320* KO mice. First, this model of CNS specific B_{12} deficiency, independent of developmental aberrations, provides parallel results compared to human symptoms. Thus behavioral performances are reduced rather than nonexistent. This is particularly true for learning and memory functions of mice that could be compared to the human disorder. This observation even fits meta-analysis or population studies dealing with a lower IQ or academic performance obtained in B_{12} deficient humans (Golding et al., 2021; Youness et al., 2020). Thus, we can conclude that if a brain is depleted of B_{12} after the developmental phase (i.e., when circuits are fully formed and drive functions), neurons do not systematically die or undergo dramatic structural changes, but are rather disturbed in their activities such as the formation of action potential and second messenger pathways (see Section 4). Second, the testing of KO mice in a difficult learning and memory procedure and their ability to ultimately learn the task (i.e., much more difficult than an Y-maze sometimes used for learning abilities—but nevertheless informative) could be due to the use of an aversive stimulus such as foot-shocks that activates a stress response. It appears that a mild to moderate elevation of stress could activate hypothalamus hormonal axis and up-regulate the arousal/attention response leading to better learning and memory performance (Conrad, 2010). It appears that, the principal effect of a diminution of B_{12} intake in a mature brain is functional rather than structural; and that such an affected function could be partially compensated by the influence of parallel or collateral neuro-circuits.

3.3 Other genetically engineered models of impaired cellular B_{12} metabolism

Besides the *Cd320* mouse, other genetically engineered mice have been produced to model human pathologies related to B_{12} deficiency, but few of them have been fully studied for behavioral disorders (Peng, Dreumont, Coelho, Guéant, & Arnold, 2016).

Methionine synthase encoded by MTR is the cytoplasmic target enzyme of B_{12}, that remethylates homocysteine into methionine, the immediate precursor of SAM (Guéant, Guéant-Rodriguez, Kosgei, & Coelho, 2021). The systemic KO mice are not viable (see for review Peng et al., 2016). Nevertheless, a mouse model with conditional inactivation of *Mtr* in the brain was proposed. These mice have dramatic behavioral alterations in a water maze test, showing a slight visuo-spatial learning in successive sessions, but significantly poorer than wild-type mice (Ghemrawi et al., 2019).

This behavioral cognitive profile of MTR-KO mice in the brain mimics the behavioral traits observed in a lot of patients.

Methionine synthase reductase is encoded by MTRR and is necessary to maintain methionine synthase in a functional state by regenerating the reduced form of the B_{12} cofactor. The *Mtrr* gene-trap knockout mice have short-term memory impairment and spend more time exploring the familiar object when compared with control animals. $Mtrr^{gt/gt}$ mice made fewer alternations, i.e., they did not remember which arms of the maze they had previously entered, in the Y-maze test. This observation also reflects an alteration of memory capacities (Jadavji et al., 2014).

Oleosin(O) is a plant protein localized to lipid droplets and endoplasmic reticulum of plant cells and transcobalamin (TC) is the B_{12} binding blood protein that delivers B_{12} to cells. A cellular model deficient in vitamin B_{12} has been produced by expressing anchoring transcobalamin-oleosin chimera in transfected cells (Battaglia-Hsu et al., 2009; Pons et al., 2009). Two chimera have been produced, TCO, in which TC is fused to O via its N-terminal end and OTC, where O is fused to TC via its N-terminal end. The TCO, but not the OTC fused protein binds and sequesters B_{12} in the ER. Neurotensin polyplex system was used to target the transfection with either TCO, OTC or empty plasmids in the substantia nigra of adult rats (Orozco-Barrios et al., 2009). The expression of TCO produced apoptosis with increased cleaved Caspase-3 in cells immunoreactive to tyrosine hydroxylase of the substantia nigra. These cellular effects were not observed with the control OTC or empty plasmids. Rats transfected with TCO plasmid presented with Parkinson-like phenotype, with a significantly higher number of Methamphetamine-Induced ipsilateral turns, compared with those transfected with the other plasmids. No difference among the different groups of transfected rats was reported in the open field test. These data show that intracellular B_{12} status is critical in dopaminergic cells, with apoptosis produced by intracellular B_{12} deficiency (Orozco-Barrios et al., 2009). It is noticeable that few studies have evaluated the Parkinson phenotype of animal models, despite this evidence.

3.4 Nutritional models of B_{12} deficiency

A nutritional rodent model of a strict B_{12} deficiency appears difficult to realize since the amount of B_{12} necessary for normal functions is relatively low and that organs such as the liver can regulate the physiological amounts of B_{12} needed. In addition B_{12} is provided with other vitamins by coprophagy.

Guéant and collaborators have developed a fetal programming model of pups born to mothers subjected to a diet deficient in B_{12} and folate before mating and during pregnancy and lactation (Guéant, Namour, Guéant-Rodriguez, & Daval, 2013). This model produces increased homocysteine and low serum B_{12} and folate and behavior alterations throughout life even after recovery on a normal diet after weaning. The underlying mechanisms are related to altered brain development, cell stress and impaired neuroplasticity in particular in hippocampus and cerebellum (Blaise et al., 2007; Pourié et al., 2015).

A few studies indicate that a reduction of dietary B_{12} only leads to circulating amounts under 200 pg/mL; and not a drastic deficiency. Indeed, during pregnancy, the physiological needs increase and a severe B_{12} deficiency could be considered despite a B_{12} concentration over several 100 pg/mL. These pregnant mice presented a disturbance in their maternal care, elevated anxiety and depressive-like behavior. A correlation with the oxidative stress and epigenetic modification in the cortex was presented, but it is noted that the severity of the modification of the psychological behaviors under a dietary Cbl deficiency appears proportional to the level of deficiency (Ghosh et al., 2020; Rathod, Khaire, Kale, & Joshi, 2016).

Studies using various multiple B-vitamins nutritional deficiencies have reported dramatic disturbance of behaviors, but several categories of effects or sub-populations appeared related to age. In a perinatal context, the neuronal developmental consequences of multi-B-vitamins deficiency are correlated to transient deficits in motor behavior (i.e., vegetative reflex, locomotion, balance control), but persistent deficits in psychological and cognitive behaviors (i.e., learning, memory retrieval, anxiety, depressive-like behaviors, social interactions) (Blaise et al., 2007; Hassan et al., 2019; Kemse, Kale, Chavan-Gautam, & Joshi, 2018; Pourié et al., 2015). If a multi-B-vitamins deficiency is effective during adulthood or aging, the structural consequences in neuronal circuits appear weak or relative to other tissues such as vascular. However, the principal functional effects observed related to psychological and cognitive alterations rather than basic locomotor activities (Troen et al., 2008). Thus it seems that neocortical functions, which are the most complex, are very sensitive to dysregulation of the one-carbon metabolism leading rather to a functional cellular stress or a global cellular health with trophic factors, than a structural change with apoptosis of the neuronal circuits (Mutti, Lildballe, Kristensen, Birn, & Nexo, 2013; Roy et al., 2014; Sable, Kale, & Joshi, 2013).

4. Regions of the brain involved and the physiologic basis of behavioral alterations

As mentioned above, various brain functional outcomes could be affected by a B_{12} deficiency. Nevertheless two groups of behaviors are often listed in both human and animal models. These loss or reductions of functions could affect the quality of life since they appear in simple muscular movements (vegetative functions) until high analytic cerebral capabilities (cognitive and psychological). In order to focus on the biological origin of these disabilities, a anatomical approach must be conducted. In the CNS, several brain sub-areas and the spinal cord are particularly affected by B_{12} deficiency, but rather slightly than dramatically, mostly in the circuits functioning than in the number of dyeing neurons or cell integrity. Due to this concept, the markers of axonal projections and synaptic contacts (white matter) must be considered to be important than the strict neuroanatomy (gray matter), represented by cell bodies. Indeed, the nature of neuroprojections and synapses have to be considered and investigated such as cholinergic, glutamatergic, monoaminergic, GABAergic and others, inside brain sub-regions.

4.1 Does a vitamin B_{12} deficiency influence neurotransmitters or other synaptic actors?

Since a dysregulation in the neurotransmitter metabolism has been identified as a major cause in many neurologic diseases leading to cognitive and psychiatric disorders, hypothesis for such pathogenic mechanisms concerning monoamines were proposed under a B_{12} deficiency (Folstein et al., 2007). Nevertheless, experiments conducted with the CNS specific B_{12} deficient mice, TCblR/*Cd320* KO, Arora et al. (2019) clearly showed that the decrease of B_{12} intake in nervous tissues led to no major effects concerning the metabolism of serotonin and dopamine. The only significant result obtained was an increase in the catabolism of the serotonin neurotransmitter restricted to the cerebellum, and not in other sub-regions such as the cortex and hippocampus. Thus, this result in the serotonin turn-over in the cerebellum fits with previous functional results obtained with the same mice indicating significant behavioral problems in motor and coordination, partly dependent on the control of the cerebellum circuits (Arora, Sequeira, Hernández, et al., 2017). But, apart from such punctual motor problems, no evidence was presented for a direct link between the neuronal or glial cobalamin availability, neurotransmitter metabolism and behavioral outcomes, especially for cognitive or psychiatric disorders in these mice. Thus, the behavioral disabilities observed under a decreased B_{12} status, both in animal models and Humans, are likely to be researched in the accessory environment of synaptic function, rather than in the neurotransmitter metabolism or the level of neuro-receptor expression. These so-called accessory actors are proteins involved in the dynamics of neurotransmitter vesicles, such as synapsins, synaptotagmins, or stabilization proteins such as Post-Synaptic-Density-95 (PSD-95) and proteins of the post-receptor related pathways. These proteins could be down-regulated through indirect epigenomic mechanisms under a low B_{12} status with protein-arginine-methyltransferase (PRMT) and/or PPAR-gamma co-activator-1-alpha (PGC1-alpha) pathways (Dreumont et al., 2021; Pourié et al., 2015).

4.2 Defective myelin synthesis and the decrease of nerve function

In the case of a specific B_{12} deficiency, several scientific cues obtained from different studies converge though a pool of mechanistic dysregulations in neurons, and probably also in glial cells, rather than huge structural manifestations such as neurodegenerative changes (except during the late aging process). When B_{12} enters the cell though the transcobalamin receptor (*Cd320*), it serves for at least two major groups of reactions; the cytoplasmic one carbon metabolism linked to epigenetic/epigenomic pathways, and the mitochondrial metabolism linked to lipid-related molecules. Both of these reactions are of great importance for neuronal functions. In this way, a reduction in the cobalamin intake in the nervous circuits leads to lower (i) the first neuronal function namely the formation and conduction of the action potential and (ii) the secondary or accessory events in neurons or glial cells that reinforce the global status of nerve circuits (cell reparation, myelination, etc.). Thus, it can be pointed out that a cobalamin deficiency occurring during the post-developmental period of life contributes to negative insidious cellular mechanisms by which the consequences are seen in the behaviors (Fig. 2).

Arora et al. (2019) showed the importance of a potential secondary mechanism that modulates what we consider as "the first function of a neuronal cell," the action potential. The myelination of axons is known to favor a powerful and rapid action potential, and a succession of

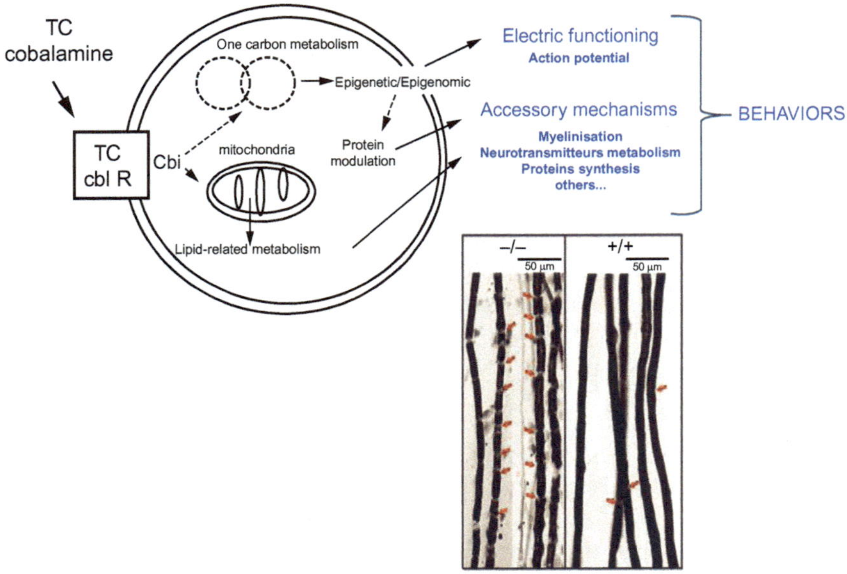

Fig. 2 Summary of intra-cellular cobalamin pathways mainly leading to mechanisms involved in the neuronal functioning and behaviors. The photograph illustrates one of the cellular outcomes (demyelinisation) under a brain vitamin B_{12} deficiency in the mouse model from E. Quadros team. Sudan black staining of myelin sheaths on nerve fibers showed smaller segments of various lengths with increased nodes of Ranvier (arrows) in the cd3202/2 mice (−/− is the brain transcobalamin receptor KO mouse and +/+ is the WT). *From Arora, K., Sequeira, J.M., Alarcon, J.M., Wasek, B., Arning, E., Bottiglieri, T., Quadros, E.V. (2019). Neuropathology of vitamin B12 deficiency in the Cd320−/− mouse.* The FASEB Journal, 33, *2563–2573.*

action potentials within a nerve circuits that leads to a codification that defines a behavior. A link is established between a specific brain B_{12} deficiency and a lack of myelination that could reveal a reduction of axonal repair. Indeed, the subsequent action potentials were found reduced in intensity. B_{12} and other B-vitamins are recommended in the treatment of myelin disorders such as multiple sclerosis in combination with other therapeutic agents (Nemazannikova, Mikkelsen, Stojanovska, Blatch, & Apostolopoulos, 2018).

In these conditions, it appears that a cascade of linked mechanisms beginning with the reduction of cellular B_{12} uptake can lead to the modification of neuronal outcomes represented by behaviors. As a perspective, it is also known that neurons exposed to degradation conditions and reduced functioning, could trigger a death process, necrosis or apoptosis (McCully, 2016).

4.3 Cellular and molecular mechanism of neuronal dysfunction leading to behavioral abnormalities

Consistent with the clinical outcomes of patients and behavioral studies of animal models, the behavioral abnormalities produced by impaired B_{12} metabolism and/or deficiency are related to cellular and molecular mechanisms that vary according to age. They include altered proliferation/differentiation and apoptosis of neurons, and altered neuroplasticity. These mechanisms have been delineated in a neuronal cell model made deficient in B_{12} through the stable expression of the TCO chimeric protein in the endoplasmic reticulum (Battaglia-Hsu et al., 2009; Pons et al., 2009). In this model cellular B_{12} deficiency produces a slower proliferation and a speedier differentiation of cells through interacting signaling pathways that are related to increased expression of PP2A, proNGF, TACE and protein phosphatase 2A. It promotes cellular differentiation by improving initial outgrowth of neurites and, at the molecular level, by augmenting the levels of proNGF and p75(NTR). Vitamin B_{12} has a strong modulating influence on ER stress in N1E115 dopaminergic neuron-like cells. The impaired cellular availability in B_{12} induces irreversible ER stress by greater acetylation of HSF1 through decreased SIRT1 expression, whereas adding B_{12} produces protective effects in cells subjected to ER stress stimulation (Ghemrawi et al., 2019). The transcriptomic changes of genes are related to the subcellular mislocalization of several RNA binding proteins, including the ELAVL1/HuR protein implicated in neuronal stress. These data have been confirmed in fibroblasts from patients with inborn errors of B_{12} metabolism and in *Cd320* knockout mice (Battaglia-Hsu et al., 2018). The decreased interaction of ELAVL1/HuR with the CRM1/exportin protein of the nuclear pore complex and its subsequent mislocalization results from decreased methylation by protein methyl transferase CARM1 and dephosphorylation by increased protein phosphatase PP2A. The mislocalization of ELAVL1/HuR triggers the decreased expression of SIRT1 deacetylase and genes involved in brain development, neuroplasticity, myelin formation, and brain aging. In particular dramatic changes were observed for genes involved in the serotoninergic, GABAergic, dopaminergic and cholinergic synaptic and postsynaptic pathways (Battaglia-Hsu et al., 2018). The SIRT1 activating compound SRT1720 inhibited ER stress and rescued RBP and mRNA mislocalization and IRF3 splicing. Treatment with this SIRT1 agonist prevented all these hallmarks in patient fibroblasts and also improved the deficient hippocampo-dependent learning ability of brain-*Mtr* conditional knockout mice. Sirt1, protein methyl transferases and SAM play an important role in rat brain manifestations of fetal programming produced by B_{12} and folate deficiency during pregnancy and lactation. In this model of fetal programming, the imbalanced methylation and SIRT-dependent acetylation of PGC1α leads to PGC1α inactivation and dissociation from nuclear receptors such as ERRα, which is involved in the expression of synapsins (Pourié et al., 2015). Similarly, the decreased methylation status in the brain of *Cd320−/−* mouse, leads to altered synaptic activity with decreased NMDA glutamatergic receptor/PSD95 interaction, and lower expression of synapsin 1 through decreased methylation of GR and PGC1α, and decreased dimerization and interaction of GR with PGC1α. The importance of this GR stress-related pathway in *Cd320−/−* mouse behavior is highlighted by the rescue of the molecular changes and improvement in learning memory by Intraperitoneal injection of hydrocortisone (Dreumont et al., 2021).

Considering the organism level, behaviors are controlled by a balance between synaptic activation (EPSP for excitatory post synaptic potential) and inhibition (IPSP for inhibitory post synaptic potential). Thus, all behaviors can be obtained in a range of intensity with neither aberrant excessive overlapping nor absence of function, both characteristic to pathogenic situations. These potentials are directly dependent on the molecular actors of the synapse; the neurotransmitters, the receptors and their associated molecules. Considering the modification of metabolism of several amino acids by B_{12} (and other related vitamins) deficiency, it has been proposed that the imbalance in the amount and dynamics of such amino acids could modulate glutamate synapses (Santhosh-Kumar, Hassell, Deutsch, & Kolhouse, 1994). Both homocysteic and cysteic acids have the potential of being excitatory ligands, while taurine, a decarboxylated product of cysteic acid, could act as inhibitory transmitter. These modifications under a B_{12} deficient status were proposed as direct mechanisms that promote an imbalance of the principal cognitive *N*-methyl-D-aspartate glutamatergic receptor (NMDAR) involved in psychiatric and neurologic disorders.

During the past few decades, a lot of molecular investigations assessed that a cellular cobalamin deficiency can affect a great variety of protein interactions leading to reduced cellular function or integrity. Nevertheless, concerning neurons, no clear results have linked the level of expression of synaptic receptors as a direct consequence of a low B_{12} status. However, several important receptors for cognitive functions, such as NMDAR, GABAR and DopaminergicR, have been listed as targets for epigenetic modifications in neurologic and psychiatric disorders including numerous pathologies in which behaviors are modified (Snyder & Gao, 2020; Sweatt, 2016).

Beside the level of receptors expression, other proteins and pathways related to the neuronal function could be modified and influence the resulting behaviors. Using the TCblR/*Cd320* KO mice, it has been demonstrated that parallel pathways could target the same actors of the synaptic dynamics and then modulate the neuronal outcome in both reduction and activation of function. Synaptic proteins such as Synapsins, acting as chaperones of the neurotransmitter vesicles that exocytose, could be down- or up-regulated by hormonal receptors pathways such as the Glucocorticoid receptor (Dreumont et al., 2021). In such a mechanism, it appears that a cross-talk between an epigenomic down-regulation of Synapsins in the presence of a B_{12} deficiency, and the restoration of the same Synapsins by the GR pathway may play a role. Moreover, these results suggest that the B_{12} deficient mice did not behaviorally present an absence of learning function, but rather a reduction of learning performances. Therefore, if a longer time and stimulation of corticoid release (i.e., mild stress stimulus) are provided to the mice, learning capacities are restored despite the B_{12} deficiency in the CNS. Thus, several neuro-molecular pathways have to be considered under a B_{12} deficiency, and not only those known to be directly activated for the action potentials leading to measured behavioral changes.

5. Genes and pathways involved

Investigating the direct implication of specific genes in behavioral disabilities, appears to provide two groups of results. Indeed, using human cohort meta-analysis, a link is established between the dysregulation of genes involved in B_{12} metabolism along with pro- or anti-oxidant genes in an aging background. On the other hand, studies dealing with cellular and animal models highlight effects on genes related to neurotrophic factors under a developmental or post-developmental time frame.

In humans, correlations between biomarkers and functional or behavioral abilities highlighted several genes to be dysregulated. Nevertheless, the causal context is often related to the methylation status, one-carbon metabolism and the subsequent hyperhomocysteinemia. One of the main mechanisms linking B_{12} deficiency and cognitive impairments, revealed by reduced mental scores, is the elevation of the oxidative stress. An et al. (2019) reported links between a global DNA hypomethylation with a concomitant focal hypermethylation of some anti-oxidant genes such as

NUDT15 and TXNRD1 in MCI/AD in patients. Also, a folate supplementation under a B_{12} deficiency could reduce the oxidative stress though DNA methylation of the VPO1 and ROC genes (Cui et al., 2018). Thus, in aging patients, these mechanistic links between the methylation status of specific genes and the subsequent functional neuronal outcomes can strongly impact the everyday behaviors, since numerous cognitive functions such as attention, auditory and verbal learning, working memory are depleted.

Beside aging, another period appears critical for B_{12} deficiency, since the perinatal background could affect long term functions even without neurodevelopmental damages such as neural tube closure defects (NTD). In fact, some insidious mechanisms that take place in young ages could be linked to behavioral abnormalities in adult life, in a so-called fetal programming effect. Investigating in vitro and animal studies, it has been shown that the depletion of the methylation status though a cobalamin and/or a folate deficiencies induced a reduction of the expression of specific neurotrophic factors in the cortex and the hippocampus. This is particularly true for both the ligand BDNF and the Ntrk-2 gene encoding for the BDNF-receptor TRKB, and partially for the neurotrophic growth factor NGF (Battaglia-Hsu et al., 2009; Sable et al., 2013; Trautmann, Bock, Urbach, Hübner, & Engmann, 2020). Related studies pointed out that BDNF and the PI3K/Akt pathway are able to suppress apoptotic mechanisms of neurons alternated by a low B_{12} status (Li et al., 2019). These studies provide concordant information because on the one hand, it is known for a long time that the neuronal life and death balance is weak and under the control of healthy cellular and extracellular trophic conditions. On the other hand, the specific cortical and hippocampus brain structures that may be particularly sensitive to low B_{12} condition, fit very well with cognitive impairments. The frontal and pre-frontal cortex are implicated in motivation, decision making, consciousness, memory retention, and the hippocampus is one of the major brain sub-region of the limbic system which controls almost all the neurologic mechanisms (i.e., inputs and outputs) leading to cognition, such as emotion, events comparisons, previous experiences and learning.

Finally, the inborn errors of cobalamin metabolism, named CblC (MMACHC gene mutation) and CblG (Methionine synthase gene mutation) of the major cases in humans, present a combined deficits of the above evoked behavioral abnormalities. Typically, neurodevelopmental abnormalities seen in affected newborns rapidly lead to vegetative and motor impairments in children. It also appears that other behavioral milestones that must normally be acquired in growing children and teenagers such as cognitive abilities and mental representations, are not corrected by an early treatment (Fischer et al., 2014; Huemer et al., 2016). Thus it appears that besides the dysregulation of specific genes and disruption of functional and behavioral outcomes, temporal windows of time are important for both (i) the necessity for temporal gene expression, especially during the perinatal period, and (ii) the correspondence between sensitive brain sub-regions, intracellular mechanisms (genes of trophic factors and oxidative stress pathways) to develop normal behaviors extending from the young to advanced age.

6. Conclusions and future directions

Since it is known for decades that B_{12} deficiency is a risk factor for neurodevelopmental abnormalities and behavioral dysfunctions, the correction of circulating B_{12} using intramuscular injection was tried as a treatment, especially in growing infants. Because of its direct link, folate supplementation was also tried in order to stimulate the one-carbon metabolism (Stanger et al., 2009). Nevertheless several studies in both animal models and humans reported a partial or no correction of neurofunctional disabilities despite a successful restoration of biochemical parameters (Arora et al., 2019; Blaise et al., 2007; Fischer et al., 2014; Huemer et al., 2019). This suggests that the functional refinement of the nervous system may not be amenable simply to correction of methylation dysregulation but that other factors, such as the cellular stress mechanisms may be involved (Arora, Sequeira & Quadros, 2017; Battaglia-Hsu et al., 2018). Indeed, it has been shown that the mislocalization of some RNA binding proteins (RBP) could be reversed by the pharmacological activation of the deacetylase SIRT1 in fibroblasts of Cbl deficient patients and that leads to an improvement in learning disabilities in a mouse model (Ghemrawi et al., 2019). Thus, if the CNS structure is not dramatically affected in B_{12} deficiency, as reported in neurodevelopmental disorders like spina-bifida related pathologies, where the circuits leading to behaviors are functionally conversed, then, interventional protocols may be tried to correct functional deficits. In contrary to what was established by early scientific and medical knowledge in the 19th and early 20th centuries, the brain is able to react to negative conditions; this was named neuroplasticity. This concept could be defined as the ability of the brain to adapt to changes in normal function and recover from pathologic events. This concept of neurology was described in the 1950s in the last century with the "*Hebb postulate*" (Brown & Donald, 2020) and nowadays split into two mechanisms. (i) structural changes such as neuronal regeneration or collateral sprouting and (ii) functional reorganization without structural changes, such as modification/adaptation of synaptic power (Puderbaugh & Emmady, 2021). The question here is relative to the possibility for nervous circuits to compensate some deleterious effects of B_{12} deficiency using different forms of cell rescue (i.e., structural changes), since functional plasticity was already presented with epigenomic mechanism (Dreumont et al., 2021).

Scalabrino in 2009, indicated that glial cells are the major cell-type affected in case of a brain B_{12} deficiency. Moreover, recent studies confirm the strong link between B_{12} and the repair of neural circuits through glial functions such as remyelination and neuronal events such as microtubule stabilization or mechanisms of apoptosis related to the endoplasmic reticulum stress (Wu et al., 2019). Considering such brain resistance to putative damages or brain repair, synergistic mechanisms between neurons and glial cells could lead to functional restoration. This was specifically shown for peripheral nerves with the demonstration of walking behaviors (Arora et al., 2019; Sun et al., 2012). In light of regeneration studies, it appears that B_{12} acts to ameliorate the micro-environmental conditions to repair damaged neuro-muscular tissues. Thus, B_{12} is responsible of the up-regulation/expression of trophic factors, growth associated proteins leading to structural repair using microtubules for nerve/neuronal repair (Liao, Wang, Huang, & Tseng, 2013; Sun et al., 2012). Such reactive or compensatory mechanisms were demonstrated for both sensitive and motor functions, two neuronal functions indispensible for correct behavior expressions (Sloan et al., 2020; Xu et al., 2020). The question posed now is to know if these positive effects on damaged peripheral nerve tissues could also take place in the central nervous system in which the neuronal network is much more tight, complex with multiple connections. An additional challenge would be to enhance the B_{12} intake in the all brain cell-types, and targeting molecular actors of the cellular stress demonstrated in many studies.

Acknowledgments

Authors acknowledge the support of the i-SITE Lorraine University of Excellence (LUE); the French National Agency for Research (ANR); Region of Lorraine (France); "Etude des mécanismes moléculaires à l'origine des pathologies associées à un dysfonctionnement du métabolisme des monocarbones" that is co-funded by the European Union within the framework "FEDER-ESF Lorraine et Massif des Vosges 2014–20."

References

Agamanolis, D. P., Victor, M., Harris, J. W., Hines, J. D., Chester, E. M., & Kark, J. A. (1978). An ultrastructural study of subacute combined degeneration of the spinal cord in vitamin B12-deficient rhesus monkeys. *Journal of Neuropathology and Experimental Neurology*, *37*, 273–299.

Agrawal, A., Ilango, K., Singh, P. K., Karmakar, D., Singh, G. P., Kumari, R., et al. (2015). Age dependent levels of plasma homocysteine and cognitive performance. *Behavioural Brain Research*, *283*, 139–144.

Aires Neto, T., Cavalcante, J. F., Brandão-Neto, J., Almeida, M., de Rezende, A. A., Egito, E. S., et al. (2005). Total gastrectomy with substitution of stomach by jejunal pouch with and without duodenal passage. Study in rats. *Acta Cirurgica Brasilia*, *20*, 167–172.

An, Y., Feng, L., Zhang, X., Wang, Y., Wang, Y., Tao, L., et al. (2019). Dietary intakes and biomarker patterns of folate, vitamin B(6), and vitamin B(12) can be associated with cognitive impairment by hypermethylation of redox-related genes NUDT15 and TXNRD1. *Clinical Epigenetics*, *11*, 139.

Arora, K., Sequeira, J. M., Alarcon, J. M., Wasek, B., Arning, E., Bottiglieri, T., et al. (2019). Neuropathology of vitamin B_{12} deficiency in the $Cd320^{-/-}$ mouse. *The FASEB Journal*, *33*, 2563–2573.

Arora, K., Sequeira, J. M., Hernández, A. I., Alarcon, J. M., & Quadros, E. V. (2017). Behavioral alterations are associated with vitamin B12 deficiency in the transcobalamin receptor/CD320 KO mouse. *PLoS One*, *12*, e0177156.

Arora, K., Sequeira, J. M., & Quadros, E. V. (2017). Maternofetal transport of vitamin B_{12}: Role of TCblR/*CD320* and megalin. *The FASEB Journal*, *31*, 3098–3106.

Batata, M., Spray, G. H., Bolton, F. G., Higgins, G., & Wollner, L. (1967). Blood and bone marrow changes in elderly patients, with special reference to folic acid, vitamin B12, iron, and ascorbic acid. *British Medical Journal*, *2*, 667–669.

Battaglia-Hsu, S. F., Akchiche, N., Noel, N., Alberto, J. M., Jeannesson, E., Orozco-Barrios, C. E., et al. (2009). Vitamin B12 deficiency reduces proliferation and promotes differentiation of neuroblastoma cells and up-regulates PP2A, proNGF, and TACE. *Proceedings of the National Academy of Sciences of the United States of America*, *106*, 21930–21935.

Battaglia-Hsu, S. F., Ghemrawi, R., Coelho, D., Dreumont, N., Mosca, P., Hergalant, S., et al. (2018). Inherited disorders of cobalamin metabolism disrupt nucleocytoplasmic transport of mRNA through impaired methylation/phosphorylation of ELAVL1/HuR. *Nucleic Acids Research*, *46*, 7844–7857.

Beck, B., & Pourié, G. (2013). Ghrelin, neuropeptide Y, and other feeding-regulatory peptides active in the hippocampus: Role in learning and memory. *Nutrition Review*, *71*, 541–561.

Bertollo, C. M., Oliveira, A. C., Rocha, L. T., Costa, K. A., Nascimento, E. B., Jr., & Coelho, M. M. (2006). Characterization of the antinociceptive and anti-inflammatory activities of riboflavin in different experimental models. *European Journal of Pharmacology*, *547*, 184–191.

Bjørke-Monsen, A. L., & Ueland, P. M. (2011). Cobalamin status in children. *Journal of Inherited Metabolism Disorders*, *34*, 111–119.

Black, M. M. (2008). Effects of vitamin B12 and folate deficiency on brain development in children. *Food and Nutrition Bulletin*, *29*, S126–S131.

Blaise, S. A., Nédélec, E., Schroeder, H., Alberto, J. M., Bossenmeyer-Pourié, C., Guéant, J. L., et al. (2007). Gestational vitamin B deficiency leads to homocysteine-associated brain apoptosis and alters neurobehavioral development in rats. *American Journal of Pathology*, *170*, 667–679.

Brown, R. E., & Donald, O. (2020). Hebb and the organization of behavior: 17 years in the writing. *Molecular Brain, 13*, 55.
Brunaud, L., Alberto, J. M., Ayav, A., Gérard, P., Namour, F., Antunes, L., et al. (2003). Vitamin B12 is a strong determinant of low methionine synthase activity and DNA hypomethylation in gastrectomized rats. *Digestion, 68*, 133–140.
Ceschim, V. C., Sumarán, P., Borges, A. A., Girardi, C. E. N., & Suchecki, D. (2021). Maternal deprivation during early infancy in rats increases oxytocin immunoreactivity in females and corticosterone reactivity to a social test in both sexes without changing emotional behaviour. *Hormones and Behaviour, 129*, 104928.
Conrad, C. D. (2010). A critical review of chronic stress effects on spatial learning and memory. *Progress in Neuropsychopharmacology and Biological Psychiatry, 34*, 742–755.
Copp, A. J., Stanier, P., & Greene, N. D. (2013). Neural tube defects: Recent advances, unsolved questions, and controversies. *Lancet Neurology, 12*, 799–810.
Cui, S., Lv, X., Li, W., Li, Z., Liu, H., Gao, Y., et al. (2018). Folic acid modulates VPO1 DNA methylation levels and alleviates oxidative stress-induced apoptosis in vivo and in vitro. *Redox Biology, 19*, 81–91.
Dreumont, N., Mimoun, K., Pourié, C., Quadros, E. V., Alberto, J. M., Umoret, R., et al. (2021). Glucocorticoid receptor activation restores learning memory by modulating hippocampal plasticity in a mouse model of brain vitamin B_{12} deficiency. *Molecular Neurobiology, 58*, 1024–1035.
Fanini, D., Palumbo, G., Giorgi, R., & Pantaleoni, G. (1990). Behavioral effects of PCBs in mice. *Behavioral Pharmacology, 1*, 505–510.
File, S. E., & Hyde, J. R. (1978). Can social interaction be used to measure anxiety? *British Journal of Pharmacology, 62*, 19–24.
Fischer, S., Huemer, M., Baumgartner, M., Deodato, F., Ballhausen, D., Boneh, A., et al. (2014). Clinical presentation and outcome in a series of 88 patients with the cblC defect. *Journal of Inherited Metabolism Disorders, 37*, 831–840.
Folstein, M., Liu, T., Peter, I., Buell, J., Arsenault, L., Scott, T., et al. (2007). The homocysteine hypothesis of depression. *American Journal of Psychiatry, 164*, 861–867.
Galliot, E., Levaillant, M., Beard, E., Millot, J. L., & Pourié, G. (2010). Enhancement of spatial learning by predator odor in mice: Involvement of amygdala and hippocampus. *Neurobiology of Learning and Memory, 93*, 196–202.
Gentry, G., Brown, W. L., & Lee, H. (1984). Spatial location in the learning of a multiple-T maze. *Journal of Comparative Physiology and Psychology, 41*, 312–318.
Geoffroy, A., Saber-Cherif, L., Pourié, G., Helle, D., Umoret, R., Guéant, J. L., et al. (2019). Developmental impairments in a rat model of methyl donor deficiency: Effects of a late maternal supplementation with folic acid. *International Journal of Molecular Sciences, 20*, 973.
Ghemrawi, R., Arnold, C., Battaglia-Hsu, S. F., Pourié, G., Trinh, I., Bassila, C., et al. (2019). SIRT1 activation rescues the mislocalization of RNA-binding proteins and cognitive defects induced by inherited cobalamin disorders. *Metabolism, 101*, 153992.
Ghosh, S., Sinha, J. K., Khandelwal, N., Chakravarty, S., Kumar, A., & Raghunath, M. (2020). Increased stress and altered expression of histone modifying enzymes in brain are associated with aberrant behaviour in vitamin B12 deficient female mice. *Nutritional Neuroscience, 23*, 714–723.
Golding, J., Gregory, S., Clark, R., Iles-Caven, Y., Ellis, G., Taylor, C. M., et al. (2021). Maternal prenatal vitamin B12 intake is associated with speech development and mathematical abilities in childhood. *Nutrition Research, 86*, 68–78.
Green, B., Griffiths, E., & Almond, S. (2017). Neuropsychiatric symptoms following metal-on-metal implant failure with cobalt and chromium toxicity. *BMC Psychiatry, 17*, 33.
Guéant, J. L., Guéant-Rodriguez, R. M., Kosgei, V. J., & Coelho, D. (2021). Causes and consequences of impaired methionine synthase activity in acquired and inherited disorders of vitamin B_{12} metabolism. *Critical Reviews in Biochemistry and Molecular Biology*, 1–23.

Guéant, J. L., Namour, F., Guéant-Rodriguez, R. M., & Daval, J. L. (2013). Folate and fetal programming: A play in epigenomics? *Trends in Endocrinology and Metabolism, 24*, 279–289.

Gupta, L., Gupta, R. K., Gupta, P. K., Malhotra, H. S., Saha, I., & Garg, R. K. (2016). Assessment of brain cognitive functions in patients with vitamin B12 deficiency using resting state functional MRI: A longitudinal study. *Magnetic Resonance Imaging, 34*, 191–196.

Hassan, Z., Coelho, D., Kokten, T., Alberto, J. M., Umoret, R., Daval, J. L., et al. (2019). Brain susceptibility to methyl donor deficiency: From fetal programming to aging outcome in rats. *International Journal of Molecular Sciences, 20*, 5692.

Heok, K. E., & Ho, R. (2008). The many faces of geriatric depression. *Current Opinion in Psychiatry, 21*, 540–545.

Holmes, J. M. (1956). Cerebral manifestations of vitamin-B12 deficiency. *British Medical Journal, 2*, 1394–1398.

Huang, A. C. W., Yu, Y. H., He, A. B. H., & Ou, C. Y. (2020). Interactions between prelimbic cortex and basolateral amygdala contribute to morphine-induced conditioned taste aversion in conditioning and extinction. *Neurobiology of Learning and Memory, 172*, 107248.

Huemer, M., Diodato, D., Martinelli, D., Olivieri, G., Blom, H., Gleich, F., et al. (2019). Phenotype, treatment practice and outcome in the cobalamin-dependent remethylation disorders and MTHFR deficiency: Data from the E-HOD registry. *Journal of Inherited Metabolism Disorders, 42*, 333–352.

Huemer, M., Diodato, D., Schwahn, B., Schiff, M., Bandeira, A., Benoist, J. F., et al. (2017). Guidelines for diagnosis and management of the cobalamin-related remethylation disorders cblC, cblD, cblE, cblF, cblG, cblJ and MTHFR deficiency. *Journal of Inherited Metabolism Diseases, 40*, 21–48.

Huemer, M., Mulder-Bleile, R., Burda, P., Froese, D. S., Suormala, T., Zeev, B. B., et al. (2016). Clinical pattern, mutations and in vitro residual activity in 33 patients with severe 5, 10 methylenetetrahydrofolate reductase (MTHFR) deficiency. *Journal of Inherited Metabolism Disorders, 39*, 115–124.

Jadavji, N. M., Bahous, R. H., Deng, L., Malysheva, O., Grand'maison, M., Bedell, B. J., et al. (2014). Mouse model for deficiency of methionine synthase reductase exhibits short-term memory impairment and disturbances in brain choline metabolism. *Biochemical Journal, 461*, 205–212.

Kemse, N., Kale, A., Chavan-Gautam, P., & Joshi, S. (2018). Increased intake of vitamin B_{12}, folate, and omega-3 fatty acids to improve cognitive performance in offspring born to rats with induced hypertension during pregnancy. *Food and Function, 9*, 3872–3883.

Kim, G., Kim, H., Kim, K. N., Son, J. I., Kim, S. Y., Tamura, T., et al. (2013). Relationship of cognitive function with B vitamin status, homocysteine, and tissue factor pathway inhibitor in cognitively impaired elderly: A cross-sectional survey. *Journal of Alzheimer's Disease, 33*, 853–862.

Kitada, Y., Miyauchi, T., Satoh, A., & Satoh, S. (1981). Effects of antidepressants in the rat forced swimming test. *European Journal of Pharmacology, 72*, 145–152.

Knight, P., Chellian, R., Wilson, R., Behnood-Rod, A., Panunzio, S., & Bruijnzeel, A. W. (2021). Sex differences in the elevated plus-maze test and large open field test in adult Wistar rats. *Pharmacology Biochemistry and Behavior, 204*, 173168.

Kraeuter, A. K., Guest, P. C., & Sarnyai, Z. (2019). The forced swim test for depression-like behavior in rodents. *Methods in Molecular Biology, 1916*, 75–80.

Krane, R. V., Sinnamon, H. M., & Thomas, G. J. (1976). Conditioned taste aversions and neophobia in rats with hippocampal lesions. *Journal of Comparative Physiology and Psychology, 90*, 680–693.

Lai, S. C., Nakayama, Y., Sequeira, J. M., Wlodarczyk, B. J., Cabrera, R. M., Finnell, R. H., et al. (2013). The transcobalamin receptor knockout mouse: A model for vitamin B12 deficiency in the central nervous system. *The FASEB Journal, 27*, 2468–2475.

Li, E. Y., Zhao, P. J., Jian, J., Yin, B. Q., Sun, Z. Y., Xu, C. X., et al. (2019). Vitamin B1 and B12 mitigates neuron apoptosis in cerebral palsy by augmenting BDNF expression through MALAT1/miR-1 axis. *Cell Cycle*, *18*, 2849–2859.

Liao, W. C., Wang, Y. J., Huang, M. C., & Tseng, G. F. (2013). Methylcobalamin facilitates collateral sprouting of donor axons and innervation of recipient muscle in end-to-side neurorrhaphy in rats. *PLoS One*, *8*, e76302.

McCully, K. S. (2016). Homocysteine, infections, polyamines, oxidative metabolism, and the pathogenesis of dementia and atherosclerosis. *Journal of Alzheimer's Disease*, *54*, 1283–1290.

Mei, J., Kohler, J., Winter, Y., Spies, C., Endres, M., Banneke, S., et al. (2020). Automated radial 8-arm maze: A voluntary and stress-free behavior test to assess spatial learning and memory in mice. *Behavioral Brain Research*, *381*, 112352.

Morris, R. (1984). Developments of a water-maze procedure for studying spatial learning in the rat. *Journal of Neuroscience Methods*, *11*, 47–60.

Mutti, E., Lildballe, D. L., Kristensen, L., Birn, H., & Nexo, E. (2013). Vitamin B_{12} dependent changes in mouse spinal cord expression of vitamin B_{12} related proteins and the epidermal growth factor system. *Brain Research*, *1503*, 1–6.

Mutti, E., Magnaghi, V., Veber, D., Faroni, A., Pece, S., Di Fiore, P. P., et al. (2011). Cobalamin deficiency-induced changes of epidermal growth factor (EGF)-receptor expression and EGF levels in rat spinal cord. *Brain Research*, *1376*, 23–30.

Nasehi, M., Sharifi, S., & Zarrindast, M. R. (2012). Involvement of the cholinergic system of CA1 on harmane-induced amnesia in the step-down passive avoidance test. *Journal of Psychopharmacology*, *26*, 1151–1161.

Nemazannikova, N., Mikkelsen, K., Stojanovska, L., Blatch, G. L., & Apostolopoulos, V. (2018). Is there a link between vitamin B and multiple sclerosis? *Medicinal Chemistry*, *14*, 170–180.

Nomura, S., Shimizu, J., Kinjo, M., Kametani, H., & Nakazawa, T. (1982). A new behavioral test for antidepressant drugs. *European Journal of Pharmacology*, *83*, 171–175.

Olton, D. S. (1987). The radial arm maze as a tool in behavioral pharmacology. *Physiology and Behaviors*, *40*, 793–797.

Orozco-Barrios, C. E., Battaglia-Hsu, S. F., Arango-Rodriguez, M. L., Ayala-Davila, J., Chery, C., Alberto, J. M., et al. (2009). Vitamin B12-impaired metabolism produces apoptosis and Parkinson phenotype in rats expressing the transcobalamin-oleosin chimera in substantia nigra. *PLoS One*, *4*, e8268.

Pellow, S., Chopin, P., File, S. E., & Briley, M. (1985). Validation of open:closed arm entries in an elevated plus-maze as a measure of anxiety in the rat. *Journal of Neuroscience Methods*, *14*, 149–167.

Peng, L., Dreumont, N., Coelho, D., Guéant, J. L., & Arnold, C. (2016). Genetic animal models to decipher the pathogenic effects of vitamin B12 and folate deficiency. *Biochimie*, *126*, 43–51.

Pons, L., Battaglia-Hsu, S. F., Orozco-Barrios, C. E., Ortiou, S., Chery, C., Alberto, J. M., et al. (2009). Anchoring secreted proteins in endoplasmic reticulum by plant oleosin: The example of vitamin B12 cellular sequestration by transcobalamin. *PLoS One*, *4*, e6325.

Pourié, G., Martin, N., Bossenmeyer-Pourié, C., Akchiche, N., Guéant-Rodriguez, R. M., Geoffroy, A., et al. (2015). Folate- and vitamin B12-deficient diet during gestation and lactation alters cerebellar synapsin expression via impaired influence of estrogen nuclear receptor α. *The FASEB Journal*, *29*, 3713–3725.

Puderbaugh, M., & Emmady, P. D. (2021). *Neuroplasticity*. Treasure Island (FL): StatPearls Publishing.

Raffaele, K., Hughey, D., Wenk, G., Olton, D., Modrow, H., & McDonough, J. (1987). Long-term behavioral changes in rats following organophosphonate exposure. *Pharmacology Biochemistry and Behavior*, *27*, 407–412.

Rathod, R. S., Khaire, A. A., Kale, A. A., & Joshi, S. R. (2016). Effect of vitamin B12 and omega-3 fatty acid supplementation on brain neurotrophins and cognition in rats: A multigeneration study. *Biochimie, 128-129*, 201–208.
Rothblat, L. A., & Kromer, L. F. (1991). Object recognition memory in the rat: The role of the hippocampus. *Behavioral Brain Research, 42*, 25–32.
Roy, S., Sable, P., Khaire, A., Randhir, K., Kale, A., & Joshi, S. (2014). Effect of maternal micronutrients (folic acid and vitamin B12) and omega 3 fatty acids on indices of brain oxidative stress in the offspring. *Brain Development, 36*, 219–227.
Sable, P. S., Kale, A. A., & Joshi, S. R. (2013). Prenatal omega 3 fatty acid supplementation to a micronutrient imbalanced diet protects brain neurotrophins in both the cortex and hippocampus in the adult rat offspring. *Metabolism, 62*, 1607–1622.
Santhosh-Kumar, C. R., Hassell, K. L., Deutsch, J. C., & Kolhouse, J. F. (1994). Are neuropsychiatric manifestations of folate, cobalamin and pyridoxine deficiency mediated through imbalances in excitatory sulfur amino acids? *Medical Hypotheses, 43*, 239–244.
Scalabrino, G., Monzio-Compagnoni, B., Ferioli, M. E., Lorenzini, E. C., Chiodini, E., & Candiani, R. (1990). Subacute combined degeneration and induction of ornithine decarboxylase in spinal cords of totally gastrectomized rats. *Laboratory Investigations, 62*, 297–304.
Sloan, J. L., Achilly, N. P., Arnold, M. L., Catlett, J. L., Blake, T., Bishop, K., et al. (2020). The vitamin B12 processing enzyme, mmachc, is essential for zebrafish survival, growth and retinal morphology. *Human Molecular Genetics, 29*, 2109–2123.
Smith, A. D., & Refsum, H. (2016). Homocysteine, B vitamins, and cognitive impairment. *Annual Review of Nutrition, 36*, 211–239.
Smith, A. D., Smith, S. M., de Jager, C. A., Whitbread, P., Johnston, C., Agacinski, G., et al. (2010). Homocysteine-lowering by B vitamins slows the rate of accelerated brain atrophy in mild cognitive impairment: A randomized controlled trial. *PLoS One, 5*, e12244.
Snyder, M. A., & Gao, W. J. (2020). NMDA receptor hypofunction for schizophrenia revisited: Perspectives from epigenetic mechanisms. *Schizophrenia Research, 217*, 60–70.
Stanger, O., Fowler, B., Piertzik, K., Huemer, M., Haschke-Becher, E., Semmler, A., et al. (2009). Homocysteine, folate and vitamin B12 in neuropsychiatric diseases: Review and treatment recommendations. *Expert Review of Neurotherapeutics, 9*, 1393–1412.
Strand, T. A., Taneja, S., Ueland, P. M., Refsum, H., Bahl, R., Schneede, J., et al. (2013). Cobalamin and folate status predicts mental development scores in north Indian children 12-18 mo of age. *American Journal of Clinical Nutrition, 97*, 310–317.
Sun, H., Yang, T., Li, Q., Zhu, Z., Wang, L., Bai, G., et al. (2012). Dexamethasone and vitamin B(12) synergistically promote peripheral nerve regeneration in rats by upregulating the expression of brain-derived neurotrophic factor. *Archives of Medical Science, 8*, 924–930.
Sweatt, J. D. (2016). Dynamic DNA methylation controls glutamate receptor trafficking and synaptic scaling. *Journal of Neurochemistry, 137*, 312–330.
Trautmann, C., Bock, A., Urbach, A., Hübner, C. A., & Engmann, O. (2020). Acute vitamin B12 supplementation evokes antidepressant response and alters Ntrk-2. *Neuropharmacology, 171*, 108112.
Troen, A. M., Chao, W. H., Crivello, N. A., D'Anci, K. E., Shukitt-Hale, B., Smith, D. E., et al. (2008). Cognitive impairment in folate-deficient rats corresponds to depleted brain phosphatidylcholine and is prevented by dietary methionine without lowering plasma homocysteine. *Journal of Nutrition, 138*, 2502–2509.
Unal, G., & Canbeyli, R. (2019). Psychomotor retardation in depression: A critical measure of the forced swim test. *Behavioral Brain Research, 372*, 112047.
van der Poel, A. M. (1967). Ethological study of the behaviour of the albino rat in a passive-avoidance test. *Acta Physiologica et Pharmacologia Neerlandica, 14*, 503–505.

Venkatramanan, S., Armata, I. E., Strupp, B. J., & Finkelstein, J. L. (2016). Vitamin B-12 and cognition in children. *Advances in Nutrition*, *7*, 879–888.

Vogiatzoglou, A., Smith, A. D., Nurk, E., Drevon, C. A., Ueland, P. M., Vollset, S. E., et al. (2013). Cognitive function in an elderly population: Interaction between vitamin B12 status, depression, and apolipoprotein E ε4: The Hordaland homocysteine study. *Psychosomatic Medicine*, *75*, 20–29.

Walsh, R. N., & Cummins, R. A. (1979). The open-field test: A critical review. *Psychological Bulletin*, *83*, 482–504.

Woodward, J. C., & Newberne, P. M. (1967). The pathogenesis of hydrocephalus in newborn rats deficient in vitamin B12. *Journal of Embryology and Experimental Morphology*, *17*, 177–187.

Wu, F., Xu, K., Liu, L., Zhang, K., Xia, L., Zhang, M., et al. (2019). Vitamin B_{12} enhances nerve repair and improves functional recovery after traumatic brain injury by inhibiting ER stress-induced neuron injury. *Frontiers in Pharmacology*, *10*, 406.

Xu, G., Zhou, C. S., Tang, W. Z., Xu, J., Xu, G., Cheng, C., et al. (2020). Local administration of methylcobalamin for subacute ophthalmic herpetic neuralgia: A randomized, phase III clinical trial. *Pain Practice*, *20*, 838–849.

Youness, E. R., Shady, M. M. A., Abd Elaziz, A., Galal, E., El-Sonbaty, M., El-Sonbaty, M. M., et al. (2020). Association of folic acid, vitamin B12, and intelligence scores in epileptic children. *Applied Neuropsychology: The Child*, 1–5.

CHAPTER FIFTEEN

Vitamin B12 deficiency

Ralph Green[a,*] and Joshua W. Miller[b]
[a]Department of Pathology and Laboratory Medicine, University of California Davis Medical Center, Sacramento, CA, United States
[b]Department of Nutritional Sciences, Rutgers University, New Brunswick, NJ, United States
*Corresponding author: e-mail address: rgreen@ucdavis.edu

Contents

Abstract

Of the water-soluble vitamins, vitamin B12 (B12) has the lowest daily requirement. It also has several unique properties including a complex pathway for its absorption and assimilation requiring intact gastric and terminal small intestinal function, an enterohepatic pathway, and several dedicated binding proteins and chaperons. The many causes of B12 deficiency include malabsorption and defects in cellular delivery and uptake, as well as limited dietary intake. B12 is required as a cofactor for only two reactions in humans, the cytosolic methionine synthase reaction and the mitochondrial methymalonyl CoA mutase reaction. Disruption of either of these reactions gives rise to B12 deficiency. Although more common with advancing age, because of the higher prevalence of malabsorptive disorders in the elderly, B12 deficiency is widely distributed across all age groups particularly where food insecurity occurs. The consequences and severity of B12 deficiency are variable depending on the degree of deficiency and its duration. Major organ systems affected include the blood, bone marrow and nervous system. Megaloblastic anemia results from a defect in thymidine and therefore DNA synthesis in rapidly dividing cells. Nervous system involvement is varied, some of which

Vitamins and Hormones, Volume 119
ISSN 0083-6729
https://doi.org/10.1016/bs.vh.2022.02.003

results from defective myelin synthesis and repair. Cognitive impairment and psychosis may also occur. Diagnosis of B12 deficiency rests on clinical suspicion followed by laboratory testing, which consists of a panel of tests, that together provide clinically reliable predictive indices. B12 metabolism and deficiency is closely intertwined with folate, another B-vitamin. This chapter explores the various aspects of a unique and fascinating micronutrient.

Abbreviations

^{14}C	carbon-14
AdoCbl	adenosylcobalamin
B12	vitamin B12 and cobalamin
CNCbl	cyanocobalamin
HoloTC	holotranscobalamin
IF	intrinsic factor
LRRK2	leucine-rich repeat kinase 2
MCV	mean corpuscular volume
MeCbl	methycobalamin
MRI	magnetic resonance imaging
N_2O	nitrous oxide
NHANES	National Health and Examination Survey
NTDs	neural tube defects
OHCbl	hydroxocobalamin
PA	pernicious anemia
RDW	red cell distribution width
SALSA	Sacramento Area Latino Study on Aging
SAM	S-adenosylmethionine

1. Overview

As is the case for all the vitamins, the discovery of vitamin B12 (B12) arose from the study of individuals suffering from a particular constellation of symptoms that could be alleviated by the administration of a food substance, followed by a diligent search for and subsequent purification and chemical characterization of a particular molecule. In the case of B12, the condition was pernicious anemia (PA), first described by Thomas Addison in 1843, but doubtless the cause of much disease and death stretching back to the very beginnings of humankind. Then, in the just over 20 years that elapsed from 1926, when Minot, Murphy and Whipple reported their groundbreaking discovery that patients with PA were cured by feeding them raw liver, and following an intense and laborious quest for the active substance in liver, Dorothy Hodgkin using X-ray crystallography characterized crystals of the substance that came to be known as "vitamin B12," and this otherwise fatal disease became curable (Green & Miller, 2014).

For conservation of the tight body economy of B12 with a daily requirement of 2–3 μg (Institute of Medicine, 2000) and an estimated total body content of 1–5 mg (Doets et al., 2013), the assimilation of B12 requires a complex set of mechanisms involving several protein chaperones. Clearly, the evolution of such pathways, speaks to the precious nature and relative rarity of this nutrient. The disruption in cellular metabolism and consequent morbidity resulting from B12 deficiency is remarkable given that B12 is known to function as a cofactor for only two enzymatic reactions in human and other eukaryotic animal cells. The serious and sometimes life-threatening complications of B12 deficiency coupled with the widespread and growing prevalence of the problem around the world and the many and varied manifestations of insufficiency of this micronutrient has led to renewed interest in B12 deficiency. Cogent factors have driven this focus at the bookends of life. The recognition that nutritional inadequacy of B12 occurs among pregnant mothers, infants and children, particularly in vegan and vegetarian communities, has uncovered several neurodevelopmental and metabolic consequences thereof (Allen, 2009; Yajnik et al., 2006). At the other end of the age scale, the association of B12 deficiency with the risk of neurocognitive decline among the elderly has been another reason for re-examining its causes, consequences, and correction (Christine et al., 2020; McCaddon, 2013; Smith & Refsum, 2009).

2. Basics of vitamin B12 biochemistry and physiology

2.1 Nomenclature and structure

The preferred, biochemical term for vitamin B12 is "cobalamin," and the basic structure of the generic cobalamin molecule consists of a planar backbone comprised of a corrin ring with a central cobalt atom (Fig. 1). The parent corrin ring contains four reduced daughter pyrrole rings that bind the central cobalt atom whose two remaining coordination positions are occupied by a 5,6-dimethylbenzimidazolyl group below the ring, and interchangeable ligands, which include -OH (hydroxocobalamin or OHCbl) and -CN (cyanocobalamin or CNCbl). (Technically, CNCbl is the form that is considered to be "vitamin B12." However, "vitamin B12" or "B12" is often used to refer generically to the vitamin in all its forms

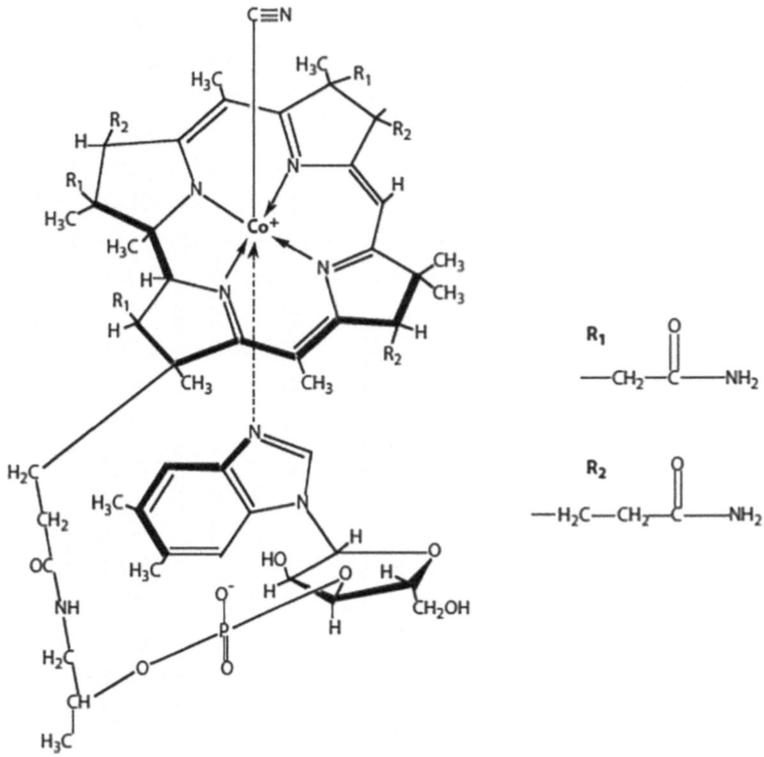

Fig. 1 Structure of vitamin B12. The family of molecules with vitamin B12 activity share common structures consisting of a planar backbone comprised of a corrin ring with a central cobalt atom coordinated with four reduced pyrrole rings. The two remaining coordination positions of the cobalt atom are occupied by a 5,6-dimethylbenzimidazolyl group below the ring, and interchangeable ligands above the ring. The different forms of the vitamin are distinguished by these interchangeable ligands, which include cyano (depicted here), hydroxo, methyl, and 5′-deoxyadenosyl. *Reproduced with permission Green, R., & Miller, J. W. (2014). Vitamin B12. In: Zempleni J, Suttie JW, Gregory JF, Stover PJ, eds. Handbook of vitamins, 5th Edition. Boca Raton, FL: CRC Press, pp. 447–489.*

(Green & Miller, 2014), as is done in this chapter.) Other cobalamins arise from alkyl substitutions in the upper axial position that are synthesized from OHCbl, two of which serve as coenzymes. In one, adenosylcobalamin (AdoCbl), a 5′-deoxyadenosyl group replaces OH as the cobalt ligand above the ring (Lenhert & Hodgkin, 1961). In the second, methylcobalamin (MeCbl), the upper ligand is a methyl group. MeCbl is the major form of cobalamin in human blood plasma (Lindstrand, 1964).

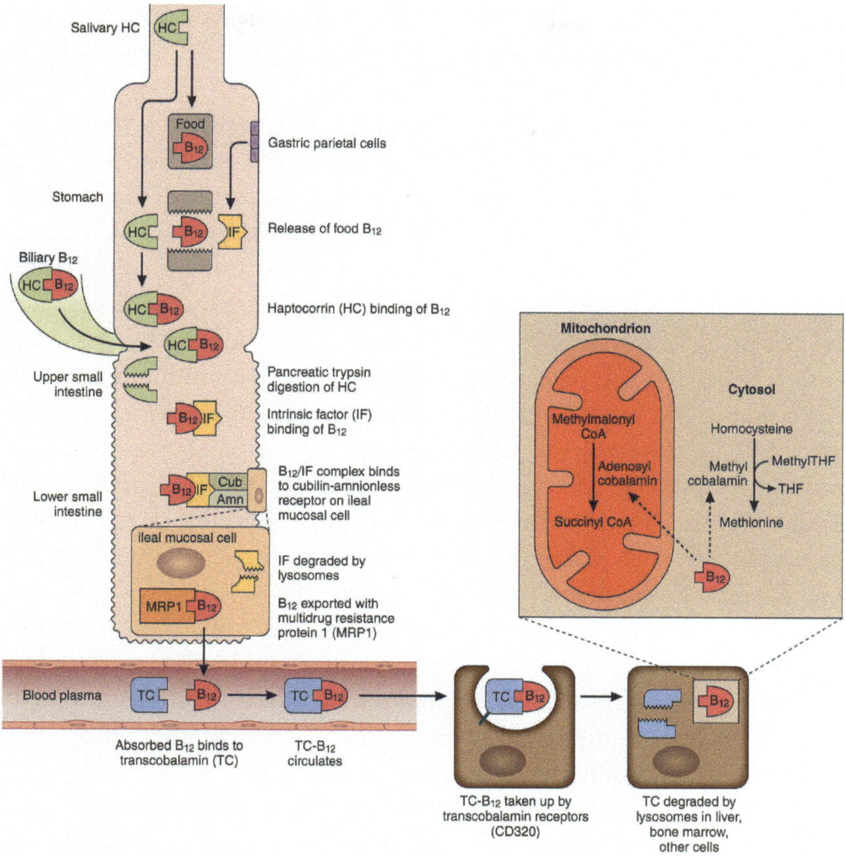

Fig. 2 **Vitamin B12 absorption, transport, and cellular uptake and coenzyme functions.** ***Reproduced with permission Green, R. (2017). Vitamin B12 deficiency from the perspective of a practicing hematologist. Blood, 129(19), 2603–2611.***

The processes of B12 absorption, transport, and assimilation are fairly complex and involve several B12 transport proteins, membrane receptors, and intracellular enzymes ultimately leading to conversion of dietary B12 into AdoCbl and MeCbl. In order to understand the pathophysiology of its deficiency, it is first important to review the normal physiology of B12 absorption, transport, and cellular uptake and metabolism (Fig. 2).

2.2 Absorption

There are two processes by which orally consumed B12 is absorbed. The first is passive diffusion of non-protein bound B12, which can occur at all absorptive surfaces along the length of the gastrointestinal tract. Passive diffusion is inefficient, with no more than 1–2% of a given B12 bolus absorbed in this manner (Chanarin, 1969).

The second process is a highly complex active transport system. Upon oral consumption, B12 in foods (specifically animal source foods such as meats, fish, eggs, and dairy products, as well as fortified foods) travels through the acidic milieu of the stomach along with salivary R binder, a B12 transport protein that belongs to a group of proteins called haptocorrins. The B12 in the foods is released from proteins by the direct digestive action of stomach acid (hydrochloric acid produced by the gastric parietal cells) and the proteolytic enzyme, pepsin (a product of the acid hydrolysis of the proenzyme, pepsinogen, produced by the gastric chief cells). Also produced by the parietal cells is a second B12 transport protein, intrinsic factor (IF). After its release from foods, the B12 preferentially binds to salivary R binder under the low pH conditions in the stomach (approximately 1–2 pH units), and then enters the small intestine as an R-binder-B12 complex along with the unbound IF.

In the small intestine, the higher pH (around 5–6 pH units) causes the release of the B12 from the R binder and subsequent binding to IF to form an IF-B12 complex. The IF-B12 complex then travels to the terminal ileum where it encounters the IF-B12 receptor, which is composed of two proteins, cubilin and amnionless (Nielsen, Rasmussen, Andersen, Nexø, & Moestrup, 2012). The IF-B12 complex is then absorbed across the apical membrane of the ileal cells via calcium-dependent receptor-mediated endocytosis. Within lysosomes, the IF is degraded and the B12 is released into the cytoplasm. Some of the B12 is retained by the ileal cells and converted to the active coenzyme forms of B12. The majority of the B12 is transported across the basolateral membrane of the ileal cells into the portal circulation where it is bound to a third B12 transport protein, transcobalamin. There is some debate over whether the B12 binds transcobalamin within the ileal cell before entry into the portal circulation, or if the B12 is first transported unbound by multidrug resistance protein 1 before binding to unbound transcobalamin in the blood (Beedholm-Ebsen, van de Wetering, Hardlei, et al., 2010).

An important aspect of B12 absorption is its enterohepatic circulation, a process which serves to conserve the vitamin. A portion of hepatic B12 (~0.5 to 5 μg/day) is secreted into bile and released into the duodenum (el Kholty et al., 1991; Green, Jacobsen, Van Tonder, Kew, & Metz, 1982). This B12 is available to be bound to IF coming from the stomach, and thus a large proportion of the biliary B12 is retained by IF-mediated absorption in the ileum.

2.3 Transport

Transcobalamin (originally known as transcobalamin II) is a β-globulin protein produced by ileal enterocytes, endothelial cells, macrophages, and hepatocytes. Newly absorbed B12 bound to transcobalamin (holo-transcobalamin or holoTC) travels through the portal circulation to the liver and to the rest of the bodily organs where it is taken up by receptor-mediated endocytosis via the CD320 receptor protein. Because the CD320 receptor is expressed in essentially all tissues, holoTC is sometimes referred to as "active-B12." The half-life of circulating holoTC is approximately 90–120 min (Finkler & Hall, 1967; Hom, 1967), reflecting the efficiency of its uptake into tissues.

B12 also travels through the circulation bound to a second B12 transport protein, haptocorrin (originally thought to be two proteins known as transcobalamin I and transcobalamin III). Haptocorrin is a glycoprotein produced by neutrophil-specific granules, with different forms found in various bodily fluids, including serum, saliva (the R-binder described above), gastric juice, bile, and breast milk. These forms differ based on the extent and nature of the carbohydrate moieties attached to the molecule. The half-life of circulating B12 bound to haptocorrin (holohaptocorrin) is ~10 days (Finkler & Hall, 1967; Hom, 1967). Moreover, unlike holoTC, holohaptocorrin is only taken up by non-specific asialoglycoprotein receptors in the liver. Thus, the B12 bound to haptocorrin in circulation reflects a form of B12 that is not directly available to all tissues of the body except the liver.

When assessing B12 status, the standard clinical measurement is serum total B12, i.e., the total of the B12 bound to both transcobalamin and haptocorrin. Because of its efficient uptake into tissues and its short serum half-life, holoTC typically constitutes about 20–30% of serum total B12 (Heyssel, Bozian, Darby, & Bell, 1966; Refsum, Johnston, Guttormsen, & Nexo, 2006), with the remainder present in the form of holohaptocorrin. In recent years, measurement of holoTC has become more common with some, but not all studies finding that it is a more sensitive and specific indicator of B12 status than serum total B12 (Bor, Nexø, & Hvas, 2004; Nexo, Hvas, Bleie, et al., 2002; Obeid & Herrmann, 2007; von Castel-Roberts, Morkbak, Nexo, et al., 2007).

2.4 Cellular metabolism

Within the cell and after release from transcobalamin within lysosomes, B12 undergoes a series of enzyme-catalyzed processing steps to form AdoCbl and MeCbl. AdoCbl serves as coenzyme for the mitochondrial enzyme, methylmalonyl CoA mutase, which catalyzes the isomerization of methylmalonyl CoA to succinyl CoA, an intermediate step in the oxidation of odd-chain fatty acids and the catabolism of ketogenic amino acids, including valine, isoleucine, and methionine. During the isomerization reaction, a hydrogen on the methyl carbon of the substrate exchanges places with the –COSCoA group. AdoCbl serves as an intermediate hydrogen carrier, accepting the hydrogen from the substrate in the initial phase of the reaction and returning it to the product after migration of the –COSCoA group.

MeCbl serves as coenzyme for the cytoplasmic enzyme, N5-methyltetrahydrofolate-homocysteine methyltransferase (also known as methionine synthase), which catalyzes the conversion of the sulfur amino acid homocysteine to methionine. S-adenosylmethionine (SAM) and a second enzyme, methionine synthase reductase, are required for the activity of this methyltransferase (Taylor & Weissbach, 1967). Methionine synthase reductase converts oxidized cobalt from the +3 to the +1 state (Co^{3+} to Co^{1+}), which then avidly accepts a methyl group from SAM. This reaction serves to recycle methionine, an essential amino acid, for incorporation into proteins and for the synthesis of SAM, which is not only required for N5-methyltetrahydrofolate-homocysteine methyltransferase, but also for methylation reactions involving many substrates, including DNA, RNA, histones, proteins, membrane phospholipids, neurotransmitters, guanidinoacetate (creatine precursor), and environmental toxins such as arsenic. This pathway also serves as the key mechanism for tetrahydrofolate regeneration through methyl group removal from N5-methyltetrahydrofolate. Tetrahydrofolate is a key substrate for folate polyglutamation, which allows for the retention of folates within cells, and for the synthesis of one-carbon adducts of folate that are required for the synthesis of DNA and RNA (Shane, 2010).

In addition to its critical coenzymatic functions, which require relatively low concentrations in the picomolar to nanomolar concentration ranges, B12 may also function in non-canonical metabolic roles. Examples include the ability of OHCbl to avidly bind cyanide, a property used for treatment of cyanide toxicity. Tobacco and certain foods (tubers, fruits, beans, and nuts) contain cyanide in the form of thiocyanate. Although conclusive evidence is lacking, vitamin B12 is believed to play a role in neutralizing cyanide taken in via ingestion of these substances with possible deleterious effects on B12 metabolic pathways. More recently, reports have appeared that B12 can inhibit overexpression of leucine-rich repeat kinase 2 (LRRK2), which has been incriminated in Parkinson disease (Schaffner et al., 2019). It has been postulated that the frequently low B12 serum concentrations encountered in Parkinson disease (Christine et al., 2018) may reflect consumption and increased demand for the vitamin (Green & Christine, 2019).

3. Pathophysiology and pathobiochemistry of vitamin B12 deficiency

3.1 Causes of deficiency

Reflecting the limited dietary availability of B12, and the complexity of its absorptive and assimilative pathways, the causes of B12 deficiency are many and varied as shown in Table 1.

Vitamin B12 deficiency is usually caused by impaired absorption of the micronutrient and the severity and rate of onset of the deficiency depends on the progression of the underlying disease process as well as the degree of disruption of the absorptive pathway. For example, total gastrectomy would lead to immediate cessation of the physiologic pathway for B12 absorption and without replacement of the vitamin would cause rapid onset of deficiency. In general, so long as the intrinsic factor-dependent physiological pathway remains intact, the onset of deficiency will be relatively slow. Causes of malabsorption of vitamin B12 can generally be considered to be either gastric or ileal. Descriptions of the varied causes of vitamin B12 deficiency are provided here:

3.1.1 Pernicious anemia

The disease PA is paradigmatic among the gastric causes of B12 deficiency and was the index disorder that led to the discovery of the vitamin. A more detailed description of this fascinating disorder is provided in the chapter

Table 1 Causes of vitamin B12 deficiency.

A. Impaired Vitamin B12 Absorption
   1. Gastric Causes
      a. Pernicious anemia
      b. Gastrectomy/Gastric reduction surgery
      c. Hyperchlorhydria (Zollinger-Ellison syndrome)
   2. Intestinal Causes
      a. Small intestine disease (ileal resection, ileitis, ileal lymphoma, tropical sprue)
      b. Blind loop syndrome
      c. Fish tapeworm infestation (*Diphyllobothrium latum*)
      d. Acquired immunodeficiency syndrome
   3. Pancreatic insufficiency
   4. Inherited disorders of B12 absorption (Imerslund-Gräsbeck syndrome)
   5. Drugs (proton pump inhibitors, H2-receptor antagonists, metformin)
B. Metabolic and Chemical Causes
   1. Inherited disorders of B12 metabolism (*mut*, *cblA* through *cblH*)
C. Low Vitamin B12 Intake
   1. Veganism/Vegetarianism
   2. Low animal source food intake due to socioeconomic factors

"Vitamin B12 absorption and malabsorption" by Guéant et al. in this volume. PA is an autoimmune gastritis, with onset generally in middle age but occurring at any age, which leads to the destruction of gastric parietal cells and consequent failure of intrinsic factor production, associated with gastric achlorhydria. Previously considered to be a disease confined to Caucasians, it is now known that PA occurs in all ethnic groups (Green et al., 2017). Susceptibility to PA is associated with human leukocyte antigen types A2, A3, B7, and B12 (Ungar, Mathews, Tait, & Cowling, 1981), and with blood group A (Hoskins, Loux, Britten, & Zamcheck, 1965), and also often occurs in association with other autoimmune diseases such as autoimmune thyroiditis, rheumatoid arthritis, type 1 diabetes, and vitiligo. There is a connection of PA with infection of the gastric mucosa with *Helicobacter pylori*, a microorganism known to be associated with gastritis and peptic ulceration. However, the exact role of *H. pylori* in the pathogenesis of PA remains obscure. Previous studies have shown a low incidence of *H. pylori* infection in PA patients (Karnes, Samloff, Siurala, et al., 1991). An interesting hypothesis has been put forward that in susceptible individuals, chronic infection with *H. pylori* may trigger an autoimmune process directed against the host H^+/K^+-ATPase protein present on gastric parietal cells ultimately leading to PA (Green, 2006; Hershko, Ronson, Souroujon, et al., 2006).

3.1.2 Gastrectomy syndromes

Gastric resections, as well Roux-en-Y gastric bypass surgery for morbid obesity, lead to multiple micronutrient deficiencies. Iron deficiency anemia is the most common anemia to occur after gastric surgery. However, B12 deficiency can also occur due to the lack of intrinsic factor coupled with hypochlorhydria. Reduced amounts of gastric hydrochloric acid output interfere with release of B12 from food B12, a condition known as "food B12 malabsorption," which can also develop in simple atrophic gastritis without the autoimmune component and that is common among the elderly. B12 deficiency following gastric resection usually develops 5–6 years following surgery with the gradual depletion of the body B12 stores (Maclean & Sundberg, 1956). Following partial gastrectomy, B12 deficiency can develop due to a combination of factors including both loss of functioning gastric mucosa with atrophy of mucosa in the gastric remnant (Lees & Grandjean, 1958) and bacterial overgrowth, classically known as "Blind Loop Syndrome" (Chen, Krishnamurthy, Mohamed, & Green, 2013), which is due to intestinal colonization of a *cul de sac* of small bowel with bacteria which thrive on B12.

3.1.3 Hyperchlorhydria

In contrast to conditions associated with reduced gastric acid, in the Zollinger-Ellison Syndrome, a gastrin-secreting tumor most commonly occurring in the pancreas, stimulates the gastric mucosa to secrete immense amounts of hydrochloric acid. The acidification of the duodenal contents prevents the binding of B12 to intrinsic factor (Shimoda & Rubin, 1968).

3.1.4 Diseases of the small intestine

The lower third of the ileum is the site for B12 absorption, so intestinal disorders like ileal resection (Kennedy, Callender, Truelove, & Warner, 1982), ileitis of any cause (inflammatory bowel disease, radiation) (Anderson, Walton, & Chanarin, 1981), lymphoma arising in the terminal ileum, drugs such as colchicine and para-amino salicylate (Waxman, Corcino, & Herbert, 1970), or tropical sprue (Sheehy, Perez-Santiago, & Rubini, 1961) can lead to B12 deficiency.

3.1.5 Blind loop syndrome

As discussed above under gastric reduction surgery, following partial gastrectomy, B12 deficiency can develop due to intestinal colonization with bacteria, which thrive on vitamin B12. This cause of B12 deficiency is amenable to B12 replacement in conjunction with antibiotic therapy (Murphy, Sourial, Burman, et al., 1986).

3.1.6 Fish tapeworm

A somewhat exotic cause of B12 deficiency is infestation with the fish tapeworm *Diphyllobothrium latum*. The prevalence of this condition is highest in the Baltic region and parts of Scandinavia, Canada, and Alaska where raw or undercooked fish is consumed. B12 deficiency results from competition between the worm and the host for ingested cobalamin (Nyberg, 1960). The clinical picture of *D. latum* infestation ranges from being asymptomatic to florid megaloblastic anemia with neurologic changes. The infestation is diagnosed by finding tapeworm ova in the feces and treatment consists of the use of vermicides in addition to B12 replacement.

3.1.7 Acquired immunodeficiency syndrome

A substantial number of patients with AIDS have low serum B12 levels with associated evidence of B12 malabsorption (Harriman, Smith, Horne, et al., 1989). Individuals testing seropositive for HIV infection may also have low serum B12 and evidence of B12 malabsorption (Harriman et al., 1989). The cause of the malabsorption may be intestinal or gastric or a combination of both (Herzlich, Schiano, Moussa, et al., 1992; Remacha & Cadafalch, 1999).

3.1.8 Pancreatic insufficiency

Vitamin B12 malabsorption has been demonstrated in 50–70% of patients with exocrine pancreatic insufficiency (Guéant, Champigneulle, Gaucher, & Nicolas, 1990). B12 malabsorption in pancreatic insufficiency is caused by a deficiency in pancreatic proteases, which result in a partial failure to release B12 from the haptocorrin-B12 complex, required for transfer of the precious B12 cargo to intrinsic factor in the duodenum (Toskes, Deren, & Conrad, 1973). Pancreatic insufficiency rarely causes clinically significant B12 deficiency (Henderson, Simpson, Warwick, & Shearman, 1972).

3.1.9 Inherited disorders of vitamin B12 absorption

Imerslund-Gräsbeck syndrome (IGS), also known as selective vitamin B12 malabsorption or megaloblastic anemia type 1, is a rare genetic pediatric disorder affecting the terminal ileum (Gräsbeck & Tanner, 2011). Found mainly in Scandinavia, where it was first described and also in parts of the Middle East, it generally becomes clinically evident during the first 6 years of life and usually results in megaloblastic anemia with or without neurological deficits of variable severity as well as proteinuria. The disorder results from defects in one of two genes (*cub* or *amn*) that together code for the cubilin protein comprising the receptor for the IF-B12 complex (Kozyraki et al., 1998).

3.1.10 Congenital IF deficiency

Congenital IF deficiency is another inherited cause of vitamin B12 malabsorption and deficiency. This condition closely resembles autoimmune PA in that both conditions are characterized by a lack of IF production. However, in congenital IF deficiency, gastric acid production is unaffected. There are various forms of congenital IF deficiency in which the gene is deleted or mutated resulting in a functionally inert protein (Carmel, Green, Rosenblatt, & Watkins, 2003; Rosenblatt & Fenton, 2001; Whitehead, 2006).

3.1.11 Drugs

Gastric acid-lowering drugs, including proton pump inhibitors and H_2-receptor antagonists, used for the treatment of dyspepsia, peptic ulcer disease, gastroesophageal reflux disease, Barrett's esophagus, and Zollinger-Ellison syndrome, are known to lower B12 status (as indicated by serum total B12 concentrations). This occurs because the suppression of gastric acid reduces the capacity to digest and release vitamin B12 from foods, thus reducing its bioavailability. The drug, metformin, used to treat diabetes, pre-diabetes, and polycystic ovary syndrome, also affects vitamin B12 absorption likely by inhibiting the calcium-dependent binding of the IF-B12 complex to the cubilin-amnionless receptor in the ileum. Whether or not the effects of these drugs on serum total B12 has been clearly demonstrated, it is unclear whether the effects on B12 status are significant enough to cause biochemical or clinical deficiency unless the individual already has low B12 status due to other causes such as low intake (Miller, 2018).

3.1.12 Low dietary intake of vitamin B12

Dietary deficiency of vitamin B12 was previously considered uncommon and restricted largely to vegans or total vegetarians who also do not consume dairy products and eggs (Gilois, Wierzbicki, Hirani, et al., 1992). Low serum

B12 levels occur in 50–60% of such persons. The onset of vitamin B12 deficiency in vegans is slower than in conditions associated with B12 malabsorption; it may take 10–20 years for an individual consuming a vegan diet to manifest features of vitamin B12 deficiency (Ford, 1980; Green, Jacobsen, et al., 1982). This is because the enterohepatic pathway for biliary B12 absorption remains intact, which conserves body B12 stores (Green, Jacobsen, et al., 1982; Green, Jacobsen, van Tonder, et al., 1981). Breastfed infants of vegan mothers also may develop B12 deficiency (Michaud, Lemieux, Ogier, & Lambert, 1992). B12 deficiency in vegans can present with mild megaloblastic anemia, glossitis, and neurologic disturbances. In addition to vegans, however, there is mounting evidence of B12 inadequacy in children and young adults in developing countries that cannot be explained on the basis of B12 malabsorption, but rather is attributable to inadequate dietary intake of animal source foods due to socioeconomic factors (Allen, 2009).

3.1.13 Metabolic and chemical causes of vitamin B12 deficiency

In addition to the inherited causes of vitamin B12 deficiency caused by disruptions of the normal pathway of B12 absorption described above, there are several genetic causes of impaired B12 assimilation and conversion to the active forms of cobalamin (MeCbl and AdoCbl) that can produce serious and life-threatening disturbances in the activation and function of B12, which are depicted in Fig. 3. According to the underlying defect, these disorders, collectively known as cobalamin defects are known and distinguished by sequential letter designations represented as *mut* and *cblA* through *cblH*. These have been described in a number of excellent reviews (Carmel et al., 2003; Rosenblatt & Fenton, 2001; Whitehead, 2006).

Exposure to the gas nitrous oxide (N_2O), used as an anesthetic and recreationally, can cause varying degrees of vitamin B12 deficiency. N_2O impairs the methionine synthase reaction by oxidizing cob(*I*)alamin (a catalytic intermediate in the methyltransferase reaction) to cob(*II*)alamin. This reaction depletes methylcobalamin and produces a vitamin B12 deficiency-like state. The severity of the deficiency will depend on the duration of exposure, as well as the prior amount of whole body B12 stores. This topic has been extensively reviewed recently (Oussalah et al., 2019; Paulus, Wijnhoven, Maessen, Blankensteijn, & van der Heyden, 2021).

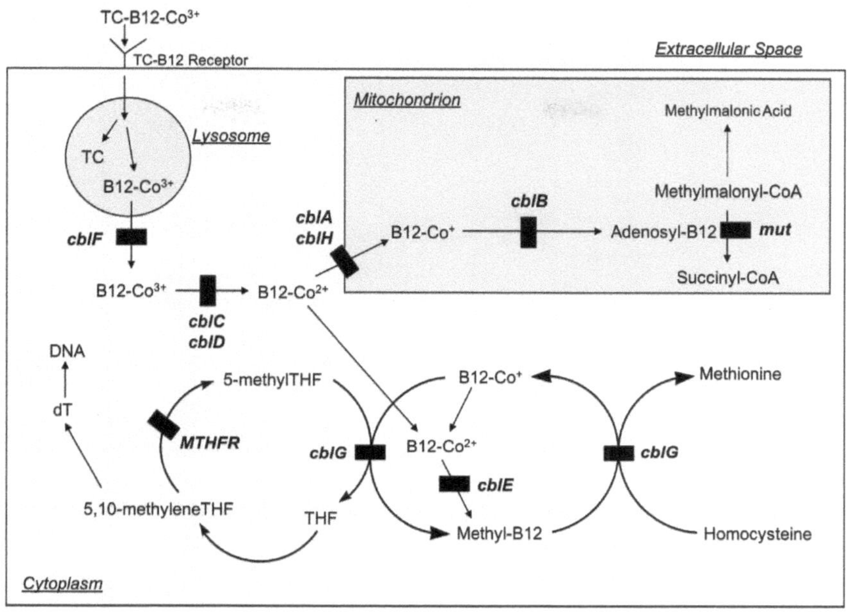

Fig. 3 Intracellular metabolism of vitamin B12 and inherited defects. Vitamin B12 bound to transcobalamin is taken up into cells by CD320 (TC-B12 receptor) via receptor-mediated endocytosis. The B12 then is dissociated from transcobalamin in lysosomes and metabolized in the cytoplasm and mitochondria to the active cofactor forms of the vitamin, methyl-B12 and adenosyl-B12. Numerous genetic defects of B12 metabolism have been identified that inhibit various steps in these metabolic processes, depicted by the black bars. These defects have been designated as *mut* and *cblA* through *cblH. Reprinted with permission Green, R., & Miller, J. W. (2014). Vitamin B12. In: Zempleni J, Suttie JW, Gregory JF, Stover PJ, eds. Handbook of vitamins, 5th Edition. Boca Raton, FL: CRC Press, pp. 447–489.*

3.2 Clinical vitamin B12 deficiency

The clinical manifestations of vitamin B12 deficiency are many and varied both in terms of type, severity, and frequency. As such, they may be considered as either typical or atypical (Green, 1994). The clinical features of vitamin B12 deficiency reflect the consequences of the underlying biochemical disturbances and most can be explained on that basis. Thus, the broad categories of those disturbances largely arise through interdiction of the methionine synthase reaction or, to a lesser extent, the methylmalonyl CoA mutase reaction. Disruption of the methionine synthase pathway has two major consequences. The first results from a shortage of reduced tetrahydrofolate through trapping of folate in the form of

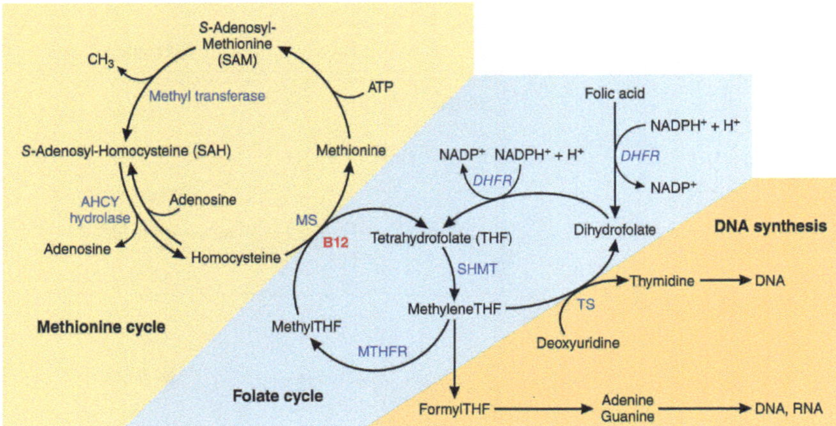

Fig. 4 Role of vitamin B12 in the folate and methionine cycles. Vitamin B12, in the form of methyl B12, serves as a cofactor in the conversion of homocysteine to methionine and 5-methyltetrahydrofolate (MethylTHF) to tetrahydrofolate (THF). *Reproduced with permission Green, R. (2017). Vitamin B12 deficiency from the perspective of a practicing hematologist. Blood, 129(19), 2603–2611.*

methyltetrahydrofolate (the "methylfolate trap") (Herbert & Zalusky, 1962). In the normal intertwining folate and methionine cycles (Fig. 4), tetrahydrofolate acquires a one-carbon fragment from serine to form methylenetetrahydrofolate, which is required for conversion of deoxyuridine to thymidine. Inadequate thymidine synthesis results in defective DNA synthesis affecting rapidly dividing cells in the bone marrow and elsewhere. Consequently, a typical major clinical manifestation of B12 deficiency is anemia. The other important product of the methionine cycle is SAM, the universal methyl donor, responsible for providing substrate for large numbers of methyltransferase enzymes, many of which occur in the nervous system for myelin synthesis and for neurotransmitter metabolism. It is generally believed that failure of methylation reactions is responsible for the neurological complications seen in vitamin B12 deficiency (Dinn et al., 1980).

The anemia seen in vitamin B12 deficiency is generally macrocytic characterized by the presence of fewer but larger red cells, which are the products of the underlying defect in DNA synthesis in the bone marrow, termed megaloblastic change. The anemia develops slowly, providing the opportunity for cardiopulmonary compensatory changes (Fernandes-Costa, Green, & Torrance, 1978). There are therefore few symptoms until the blood hemoglobin concentration is severely depressed. Symptoms, when they appear, are those of anemia: weakness, palpitation, fatigue, pallor, light-headedness, and shortness of breath. The combination of severe pallor and mild jaundice caused by concomitant hemolysis often produces a lemon-yellow skin tint. Although leukocyte and platelet counts may also be low, they rarely cause clinical problems.

Laboratory features of vitamin B12 deficiency-associated anemia include both blood and bone marrow manifestations. Red blood cells vary markedly in size and shape, often large and oval and in severe cases nucleated red cells, normally only present in the marrow, may be present in the blood. The more severe the anemia, the more pronounced the morphologic changes in the red cells. The anemia is usually macrocytic (mean corpuscular volume [MCV] is 100–150 fL or more), although coexisting iron deficiency, thalassemia trait (Green, Kuhl, Jacobson, et al., 1982), or inflammation can prevent macrocytosis (Spivak, 1982). Because of the progressive nature of gradual replacement of normocytic red cells with the macrocytic progeny of a megaloblastic marrow, the earliest observable change in red cell indices is an increase in the red cell distribution width (RDW), reflecting an increase in the size distribution of the cells (Green & Dwyre, 2015).

With respect to white cells, neutrophil nuclei often have more than the usual 3–5 lobes (Lindenbaum & Nath, 1980). Neutrophils may contain six or more lobes, a morphology rarely seen in normal neutrophils, but also not pathognomonic cells of megaloblastic hematopoiesis. Platelets are often reduced in number and slightly smaller than normal with a wider variation in size (increased platelet distribution width) (Bessman, Williams, & Gilmer, 1982).

The bone marrow shows increased cellularity with megaloblastic changes, most evident in the erythroid series with erythroblasts containing nuclei that possess less mature, open nuclear chromatin than their normal counterparts. There is a preponderance of earlier erythroblasts over more mature forms creating the impression of a maturation arrest (Green & Datta Mitra, 2017). In severe cases, an unusually large number of mitotic figures are present. Megaloblastic features in the granulocytic series are also usually present with giant forms containing large horseshoe-shaped nuclei.

With respect to the neurologic abnormalities of vitamin B12 deficiency, it is not clear whether they are attributable to disordered metabolism of myelin lipids caused by impaired methylmalonyl CoA mutase reaction (Frenkel, 1973), or whether they result from deranged methyl group metabolism through interdiction of the methionine synthase reaction. The neurologic syndrome of vitamin B12 deficiency is particularly dangerous because it can develop insidiously and in isolation (Beck, 1991), with no associated megaloblastic anemia to suggest a lack of vitamin B12 (Lindenbaum, Healton, Savage, et al., 1988; Victor & Lear, 1956), and because the neurological complications cannot always be reversed by treatment once far advanced. The syndrome usually begins with paresthesias in the feet and fingers as a result of early peripheral neuropathy and disturbances in other sensory modalities, most notably vibratory sense and proprioception. The earliest signs, which precede other neurologic findings by months, are loss of position sense in the toes and loss of vibration sense in the lower limbs (Di Lazzaro, Restuccia, Fogli, et al., 1992; Herbert, 1985). Left untreated, the neurologic disorder progresses to spastic ataxia resulting from demyelination of the dorsal and lateral columns of the spinal cord, so-called combined system disease (Fraser, 1960).

The peripheral nerves, the spinal cord, and the brain are affected by vitamin B12 deficiency. Somnolence, perversion of taste and smell, and vision loss with occasional optic atrophy are accompanied by slow waves on the electroencephalogram. Varying degrees of dementia mimicking Alzheimer disease may develop (Fraser, 1960). Functionally, nerve conduction velocity and both visual and auditory evoked potentials are reduced. There is evidence linking low vitamin B12 status with brain volume loss and cerebral white matter lesions (de Lau, Smith, Refsum, et al., 2009; Vogiatzoglou, Refsum, Johnston, et al., 2008). Psychological derangements, including psychotic depression and paranoid schizophrenia, can also occur (Shulman, 1967). Frank psychosis in vitamin B12 deficiency has been termed *megaloblastic madness* (Smith, 1960).

The neurologic lesions of cobalamin deficiency can be detected by magnetic resonance imaging (MRI). Demyelination appears as T2-weighted hyperintensity of the white matter (Scherer, 2003; Stojsavljević, Lević, Drulović, & Dragutinović, 1997). MRI is particularly useful for confirming the diagnosis of a neurologic disorder resulting from cobalamin deficiency. MRI also has been used to follow the progress of neurologic abnormalities during treatment of cobalamin-deficient patients (Stojsavljević et al., 1997).

In addition to the neurological manifestations, vitamin B12 deficiency also is a risk factor for neural tube defects (NTDs). The success of folic acid fortification of the food supply in reducing the incidence of NTDs in the United States, Canada, and other countries is well documented (Honein, Paulozzi, Mathews, et al., 2001; Ray, 2004). Because of the interrelated biochemical functions of folate and B12 in the conversion of homocysteine to methionine, it was predicted that B12 status might also be an important factor in neural tube development. The strongest support for a role of B12 in neural tube development comes from case–control studies; mean total B12 level is consistently lower in the amniotic fluid from NTD pregnancies than normal pregnancies (Ray & Blom, 2003). In contrast, the majority of studies find no difference in mean serum B12 between cases and controls (Ray & Blom, 2003). These observations suggest that women susceptible to an NTD-affected birth have a higher requirement for B12 to support normal neural tube development, thus implicating genetic differences in delivery of B12 to the fetus, B12 metabolism, or cofactor function that may contribute to NTDs. In support of this contention are data indicating that a common polymorphic variant in methionine synthase reductase (A66G) influences the risk of NTDs, particularly when maternal B12 status is low (Wilson et al., 1999). Therefore, with folic acid fortification, it can be hypothesized that some of the residual NTD incidence is related to a combination of genetic factors and low B12 nutrition. The effect of polymorphisms in this situation may arise either as a result of maternal or fetal SNPs, or the coexistence of both.

3.3 Laboratory changes and diagnosis

Several plasma biomarkers are available for identifying vitamin B12 deficiency that depend on either directly measuring the concentrations of serum total vitamin B12 or its components in the circulation or the levels of metabolites that accumulate in the blood as a result of the vitamin B12 deficient state. In addition in some diseases, notably PA, the presence of antibodies directed against intrinsic factor or gastric parietal cells can provide evidence of the underlying cause of the deficiency. Descriptions of the biomarkers of vitamin B12 status and absorptive capacity are provided here:

3.3.1 Serum or plasma vitamin B12

Serum or plasma vitamin B12 is low in most but not all patients with vitamin B12 deficiency (Green, 1995). Misleadingly normal vitamin B12 levels may be present in deficiency caused by exposure to nitrous oxide (Oussalah et al., 2019), deficiency of the plasma B12 binding protein transcobalamin, and other inborn errors of vitamin B12 metabolism (Whitehead, 2006). Levels also may be normal in B12-deficient patients with high levels of the other B12 binding protein, haptocorrin, which can occur in myeloproliferative diseases like chronic myelogenous leukemia (Green, 1994). Conversely, plasma vitamin B12 levels may be low in the presence of normal tissue B12 status, in vegetarians, in individuals taking large amounts of ascorbic acid supplements (Herbert, Jacob, Wong, et al., 1978), in some pregnant women (~25%), in the presence of haptocorrin deficiency (Carmel, 1983, 2003), and in about one-third of patients with megaloblastic anemia resulting from folate deficiency (Green, 1995). Notably, plasma folate may be high in vitamin B12 deficiency because of the methylfolate trap, which inhibits polyglutamation of intracellular folates and consequently allows methylfolate to exit cells and accumulate in the blood (Herbert, 1962).

3.3.2 Serum holotranscobalamin

Only 10–30% of the total vitamin B12 in serum is bound to the transcobalamin binding protein. This fraction (holoTC) is the functionally important fraction of circulating B12, and is a better reflection of the vitamin B12 absorptive status than total vitamin B12 (Bor et al., 2004; Lindgren, Kilander, Bagge, & Nexø, 1999; Nexo et al., 2002; Obeid & Herrmann, 2007; von Castel-Roberts et al., 2007). The remaining major fraction of serum B12 is bound to haptocorrin (holohaptocorrin), which appears to be functionally inert and therefore less relevant for the consideration of vitamin status. Consequently, and with the development of assays to measure the TC-bound fraction of the plasma B12, increasing evidence has accumulated to support the utility of measuring holoTC (Green & Miller, 2014).

3.3.3 Urinary methylmalonic acid

Methylmalonic aciduria is a generally reliable indicator of B12 deficiency, except when caused by an inborn metabolic error (Kahn, Williams, Barness, et al., 1965). Normal individuals excrete only traces of methylmalonic acid (0–3.4 mg/day). In vitamin B12 deficiency, urinary methylmalonic acid usually is elevated (Norman & Morrison, 1993). Vitamin B12 therapy restores excretion to normal in a few days. Another possible advantage of measurement of urine rather than plasma methylmalonic acid is that in impaired renal function, when plasma methylmalonic acid may give misleadingly elevated levels, measurement of the metabolite in urine when correlated with creatinine obviates this problem (Norman, Martelo, & Denton, 1982).

3.3.4 Serum or plasma methylmalonic acid and homocysteine

Elevated serum or plasma methylmalonic acid and homocysteine levels are indicators of *tissue* vitamin B12 deficiency. Levels of these metabolites are high in more than 90% of vitamin B12-deficient patients and rise before serum or plasma vitamin B12 falls to subnormal levels (Green, 1995; Lindenbaum, Savage, Stabler, & Allen, 1990). In the absence of any congenital metabolic disorder, elevated serum or plasma methylmalonic acid and/or elevated homocysteine are both indicators of vitamin B12 deficiency. Of the two metabolites, methylmalonic acid measurement is both more sensitive and more specific, and elevated methylmalonic acid will persist for several days after vitamin B12 treatment is instituted. Homocysteine levels are less specific since they rise in folate and pyridoxine deficiencies, as well as in hypothyroidism (Green, 1995). Methylmalonic acid elevation occurs only in vitamin B12 deficiency (Green, 1995). In renal diseases, however, both homocysteine and methylmalonic acid levels are frequently elevated. Furthermore, intestinal bacteria synthesize propionate, a precursor of methylmalonic acid, and in conditions of bacterial overgrowth, microbially-derived methylmalonic acid may contribute to elevations in serum or plasma methylmalonic acid (Green, 1996; Lindenbaum et al., 1990). Although measurement of these metabolites may be used for population screening for evidence of vitamin B12 deficiency, the finding of an isolated elevation of plasma methylmalonic acid cannot be taken as a priori evidence of clinically attributable vitamin B12 deficiency, absent any demonstration of a therapeutic response to the administration of vitamin B12 (Green, 1996; Solomon, 2005). A common polymorphism in the gene for 3-hydroxyisobutyryl-CoA hydrolase (*HIBCH*) has been identified which results in higher methylmalonic acid concentrations unrelated to B12 status (Molloy, Pangilinan, Mills, et al., 2016). It is not clear whether this could be clinically important with regard to defining cutpoints for vitamin B12 deficiency in individuals with this single nucleotide polymorphism.

More recently, reports have indicated that a combined indicator using measurements of several tests (vitamin B12, holoTC, methylmalonic acid and homocysteine) provides improved clinical predictive value for the confirmation of presumed vitamin B12 deficiency (Fedosov, 2013; Fedosov, Brito, Miller, et al., 2015).

3.3.5 Assays of vitamin B12 absorption

The previous "gold standard" test for assessment of vitamin B12 absorption was the Schilling test (Schilling, 1953), which assessed absorption by measuring urine radioactivity after orally administering a physiologic dose of radiolabeled ^{57}Co-cyanocobalamin followed 2 h later by injection of a large "flushing" dose of unlabeled vitamin B12. The Schilling test has become obsolete as a consequence of reduced availability of the test components.

Replacements for the Schilling test are needed and some attempts have been made to do so. One approach uses measurement of the change in holoTC following oral administration of non-radiolabeled cobalamin (Bor, Cetin, Aytaç, et al., 2005; Green & Miller, 2014). A different approach involves the use of accelerator mass spectrometry and microbially produced cyanocobalamin labeled with carbon-14 (^{14}C) (Carkeet et al., 2006; Garrod, Buchholz, et al., 2019; Garrod, Rossow, et al., 2019). In this approach, ^{14}C is measured in blood at the time of peak appearance 6–8 h following the dose using accelerator mass spectrometry. Both methods show promise, but have not been approved or validated for routine clinical use. More recently, in a modification of the ^{14}C-label approach, a method using carbon-13 has been described (Devi et al., 2020; Miller & Green, 2020).

Since the Schilling Test for vitamin B12 absorption became obsolete several years ago, no replacement test for assessing absorptive capacity has been validated clinically. Consequently, the only available test for diagnosing PA is the test for the presence of circulating anti-IF antibodies. While the test has a very high specificity with few false positive results, it unfortunately has poor sensitivity, being positive in only 60% of patients with PA (Green et al., 2017). Further discussion of the usefulness of this test is discussed in the chapter "Vitamin B12 absorption and malabsorption" by Guéant et al. in this volume.

3.4 Low B12/high folate interaction

In recent years, there has been renewed interest and debate regarding what is termed the "low B12/high folate interaction." The origins of this putative interaction come from clinical observations from the late 1940s when clinicians began treating patients with PA with high doses of folic acid (often 10s of milligrams per day) (Meyer, 1947). Such high doses of folic acid were found to alleviate the megaloblastic anemia and other hematological abnormalities observed in these patients. As described above, the severe B12 deficiency caused by malabsorption of the B12 in PA leads to the trapping of folate in the form of methyltetrahydrofolate (the "methylfolate trap") (Herbert & Zalusky, 1962), thus sequestering folate away from its role in thymidine synthesis and the support of hematopoiesis. Importantly, this secondary deficiency of folate can be metabolically overcome by providing high oral doses of folic acid, the metabolism of which bypasses the methylfolate trap and thus supplies sufficient folate to restore hematopoiesis and alleviate the megaloblastic anemia and other hematological abnormalities. However, in the early 1950s it was observed that the efficacy of long-term high-dose folic acid in alleviating the hematological abnormalities was often temporary or insufficient (Conley & Krevans, 1951). Moreover, there was suggestion that the neurological manifestations of PA would progress, or even be exacerbated, with high-dose folic acid treatment (Conley & Krevans, 1951). By the early 1970s, it had been worked out that the proper treatment for PA was parenteral supplementation with B12, and the practice of treating such patients with high-dose folic acid was discontinued.

Renewed interest in the low B12/high folate interaction came shortly before 2010 when it was observed in the 1999–2002 cohort of the National Health and Nutrition Examination Survey (NHANES) that older adults (mean age 70 year) with evidence of B12 deficiency (serum B12 $<$ 148 pmol/L or plasma methylmalonic acid $\geq$210 nmol/L) and elevated folate status (serum folate $\geq$59 nmol/L) were at higher risk of having anemia and cognitive impairment than those who had B12 deficiency and non-elevated folate status (serum folate $<$59 nmol/L) (Morris, Jacques, Rosenberg, & Selhub, 2007). In a follow-up study of the same NHANES cohort, it was observed that both serum methylmalonic acid and plasma homocysteine were elevated in B12 deficiency, but that the elevations were enhanced with increasing serum folate concentrations (Selhub, Morris, & Jacques, 2007). A similar finding of accentuated elevation of methylmalonic acid and homocysteine in subjects with low B12 status (plasma B12 $<$148 pmol/L) and elevated folate (plasma folate $\geq$45.3 nmol/L) was observed in the Sacramento Area Latino Study on Aging (SALSA) (Miller, Garrod, Allen, Haan, & Green, 2009). Moreover, in the SALSA study serum holoTC concentrations were low in the subjects with low B12 status, but the low concentrations of holoTC were accentuated in those with elevated folate ($\sim$50% lower than in low B12 subjects with non-elevated folate).

The clinical observations from the late 1940s and early 1950s and the more recent epidemiological cohort studies cited above (Miller et al., 2009; Morris et al., 2007; Selhub et al., 2007) and others (Bailey et al., 2020; Castillo-Lancellotti et al., 2015; Moore et al., 2014) suggest that high folate status, perhaps specifically as a result of high folic acid intake, exacerbates vitamin B12 deficiency. This is a particularly important consideration in the era of mandatory food folic acid fortification that was first instituted in the United States and Canada in the mid-to-late 1990s and now is prevalent in over 80 countries around the world (Miller, 2011). Coupled with common use of folic acid supplements, many individuals are exposed to relatively high intakes of folic acid, and thus may be at risk of exacerbating B12 deficiency. This is a hotly debated issue, however, mainly because a mechanism for this interaction has not previously been adequately described or tested. Recently, in this regard, a hypothesis has been put forward that high-dose folic acid in B12-deficient individuals causes depletion of serum holoTC by mobilizing the holoTC into the bone marrow to support folic acid-induced hematopoiesis or by blocking reuptake of holoTC from the renal tubules and increasing its urinary excretion (Selhub, Miller, Troen, Mason, & Jacques, 2021). The resulting reduction in serum holoTC reduces the availability of B12 to other tissues leading to accentuation of the elevations of B12-deficiency-induced methylmalonic acid and homocysteine in blood and promotion or exacerbation of neurological sequelae. The predictions of this hypothesis and demonstration of its features remain to be tested.

4. Prevention and treatment of B12 deficiency

In any disease caused by insufficiency of an essential nutrient two considerations apply, namely, how to prevent deficiency and how to treat such deficiency. Generally, the proactive approach of preventing vitamin B12 deficiency would only apply to situations in which the cause of the deficiency is dietary inadequacy of intake. However, there is an additional preventive category to consider in the case of vitamin B12 and that is when surgical intervention is carried out for treatment of one of several diseases in which removal of parts on the bowel occur that are critical for normal vitamin B12 absorption, such as the stomach or terminal ileum.

The conventional treatment of vitamin B12 deficiency consists of parenteral vitamin B12 in the form of either cyanocobalamin or hydroxocobalamin to replace daily losses and restore vitamin B12 stores, which normally contain 2–5 mg of vitamin B12. Side-effects are highly unusual, and there is no defined upper limit to vitamin B12 intake (Institute of Medicine, 2000). A recent nested case–control study found an association of high levels of serum vitamin B12 with lung cancer risk (Fanidi, Carreras-Torres, Larose, et al., 2019). Another Dutch study suggested that higher levels of plasma vitamin B12 were associated with increased risk of all-cause mortality (Flores-Guerrero et al., 2020). This association received wide media publicity, with resulting deleterious effect on the compliance of patients who require regular supplemental vitamin B12 for maintenance of health. A subsequent study reporting on 24,262 subjects in the NHANES cohort has somewhat mitigated the alarm caused by the Dutch study. This study concluded that there was a small but significant increase in cardiovascular mortality in the groups with either low or high serum vitamin B12. However, high intake of vitamin B12 in the form of supplements was not associated with any adverse effect on mortality and therefore can be regarded as safe (Wolffenbuttel, Heiner-Fokkema, Green, & Gans, 2020). Doses of vitamin B12 exceeding 100 mcg saturate the vitamin B12 binding proteins (transcobalamin and haptocorrin), and the excess is lost in the urine.

In a patient with demonstrated vitamin B12 deficiency, a typical treatment schedule consists of 1000 μg vitamin B12 intramuscularly per day for 2 weeks, then weekly until the hematocrit is normal or symptoms improve, and then monthly for life. For neurologic manifestations, 1000 μg every 2 weeks for 6 months is recommended. Higher doses are given for certain inherited disorders (e.g., transcobalamin deficiency). Transfusion occasionally is required in an anemic patient when the hematocrit is $<15\%$ or if the patient is debilitated, infected, or in heart failure. In such instances, packed red blood cells should be given slowly to avoid circulatory overload and pulmonary edema. Infections can impair the response to vitamin B12 and must be treated vigorously.

Key to definitively confirming B12 deficiency is achieved by monitoring the response to treatment or a therapeutic trial. Following parenteral administration of vitamin B12 to deficient patients, elevated plasma bilirubin, iron, and lactate dehydrogenase levels, all features of megaloblastic anemia, fall rapidly (Coleman, Donohue, Finch, et al., 1956). Within 12 h, the marrow begins to change from megaloblastic to normoblastic, a process that is complete within 2–3 days. Consequently, morphologic diagnosis may be difficult after treatment is initiated (Green & Datta Mitra, 2017). An increase in the reticulocyte count begins on days 3–5 and peaks on days 4–10 (Hillman, Adamson, & Burka, 1968). The new red cells come from newly formed normoblasts, and not from the old megaloblasts, most of which die before leaving the marrow (Koury, Horne, Brown, et al., 1997). Blood hemoglobin concentration becomes normal within 1–2 months. If normal values are not achieved by 2 months, another cause of anemia should be sought.

Other changes include: (a) prompt and dramatic improvement in the sense of well-being; (b) normalization of leukocyte and platelet counts, although neutrophil hypersegmentation may persist for 10–14 days; and (c) rise in serum vitamin B12 and folate and return to normal of raised plasma homocysteine and methylmalonate concentrations.

In patients with megaloblastic anemia, since folate deficiency produces a megaloblastic anemia that is indistinguishable from that produced by vitamin B12 deficiency, to avoid the risk of masking an underlying vitamin B12 deficiency by inducing hematological recovery in response to folate, doses in excess of 1 mg folic acid daily should be avoided until an underlying vitamin B12 deficiency has been ruled out (Institute of Medicine, 2000).

Vitamin B12 supplementation is warranted prophylactically or therapeutically in certain special circumstances. Following total gastrectomy, vitamin B12 should always be given. Vitamin B12 administration is not necessary after partial gastrectomy, but patients need to be monitored for megaloblastic anemia, bearing in mind that this anemia can be masked by deficiency of other micronutrients such as iron (Green, 1996; Sumner, Chin, Abrahm, et al., 1996). Also, in gastric reduction surgery for morbid obesity, copper deficiency can cause a syndrome closely resembling the bone marrow and neurological changes of vitamin B12 deficiency (Chen et al., 2013). The anemia of the blind loop syndrome can be treated by parenteral vitamin B12 therapy. It also responds after approximately 1 week to oral broad-spectrum antibiotics (cephalexin monohydrate [Keflex] 250 mg QID plus metronidazole 250 mg TID for 10 days) (Paulk & Farrar, 1964), which usually restores vitamin B12 absorption. Successful surgical correction of a surgically induced anatomic lesion may also ameliorate the condition. Treatment of B12 deficiency due to infestation with the fish tapeworm (*D. latum*) consists of a single oral dose of vermicide in the form of 50 mg/kg of niclosamide or a dose of 5–10 mg/kg of praziquantel, along with parenteral B12 supplementation to build up body stores.

While parenteral B12 supplementation is the standard therapy for PA patients and other B12 malabsorption disorders, interest has been rekindled (Kuzminski, Del Giacco, Allen, et al., 1998) regarding the possibility of treating vitamin B12 deficiency with oral vitamin B12 as had been proposed previously (Crosby, 1980). Oral vitamin B12 can be used not only for treatment of dietary vitamin B12 deficiency that occurs in vegans and in patients with very severe general malnutrition, but also for patients with food vitamin B12 malabsorption (typically caused by achlorhydria resulting from atrophic gastritis) (Andrès, Kurtz, Perrin, et al., 2001) and for patients with

PA, provided the patients are followed carefully (Lederle, 1998). In patients lacking IF, only approximately 1–2% of an oral dose of the vitamin crosses the intestinal epithelium by mass action. Therefore, 1000–2000 μg/day of oral vitamin B12 supplies most PA patients with their daily vitamin B12 requirement without the need for injections and their accompanying pain and expense. Vitamin B12 should be given by mouth to patients with dietary vitamin B12 deficiency and patients (e.g., hemophiliacs, the frail elderly) who cannot tolerate intramuscular injections.

There have been reports refractoriness to treatment with vitamin B12. Though anecdotal, some numbers of patients diagnosed with vitamin B12 deficiency complain of persistent, mainly neurological, symptoms unless the frequency of replacement of injected vitamin B12 is more frequent than the standard protocol (i.e., more frequent than monthly injections). This apparent refractoriness has no known scientific basis but requires further investigation.

5. Conclusion and future directions

The current state of knowledge regarding vitamin B12 deficiency, though extensive, is not comprehensive and is far from complete. As is so often the case, much information has been gleaned from study of inborn errors affecting vitamin B12 metabolism, representing what has been called "experiments of Nature." Some important questions still remain unanswered which range from the fundamental to the arcane. For example, the reason why some patients with vitamin B12 deficiency manifest predominantly or exclusively hematological complications whereas others show no hematological changes and have only neurological abnormalities remains obscure. Nether genetic nor environmental causes for this heterogeneity have thus far been identified.

The gradual expansion of knowledge on vitamin B12 and its deficiency has resulted from technological advances and the applications of ever more sophisticated methods to characterize subtle differences between individuals and to tease out subtle nuances in metabolic pathways. The use of these techniques coupled with the ingenuity of investigators challenged by the unanswered questions surrounding the pathobiology of vitamin B12 deficiency has led to the gradual expansion of knowledge in this field.

In addition to the well-established role of vitamin B12 in human health and disease and the more traditional views concerning the pathophysiology of the known disorders caused by its deficiency, there are several more controversial issues concerning how lack of vitamin B12 may give rise to the tissue changes observed as a result of deficiency of this micronutrient. Subtle genetic and epigenetic factors coupled with nutrient-nutrient interactions will, with the help of artificial intelligence and machine learning, doubtlessly clarify some of the remaining enigmas concerning vitamin B12 deficiency. In <100 years, the state of knowledge has advanced from a fatal disease to an eminently curable and preventable one. On a global scale, much remains to be done to prevent vitamin B12 deficiency, prevalent among vegans and vegetarians which will alleviate disease burden, primarily neurological, caused by a nutritional deficiency. To this end, consideration of vitamin B12 fortification of the food supply is currently under consideration (Allen et al., 2018; Green, 2009).

References

Allen, L. H. (2009). How common is vitamin B-12 deficiency? *The American Journal of Clinical Nutrition*, *89*(2), 693S–696S.

Allen, L. H., Miller, J. W., de Groot, L., Rosenberg, I. H., Smith, A. D., Refsum, H., et al. (2018). Biomarkers of nutrition for development (BOND): Vitamin B-12 review. *The Journal of Nutrition*, *148*(suppl_4), 1995S–2027S.

Anderson, C., Walton, K., & Chanarin, I. (1981). Megaloblastic anaemia after pelvic radiotherapy for carcinoma of the cervix. *Journal of Clinical Pathology*, *34*(2), 151–152.

Andrès, E., Kurtz, J., Perrin, A., et al. (2001). Oral cobalamin therapy for the treatment of patients with food-cobalamin malabsorption. *The American Journal of Medicine*, *111*(2), 126–129.

Bailey, R. L., Jun, S., Murphy, L., Green, R., Gahche, J. J., Dwyer, J. T., et al. (2020). High folic acid or folate combined with low vitamin B-12 status: Potential but inconsistent association with cognitive function in a nationally representative cross-sectional sample of US older adults participating in the NHANES. *The American Journal of Clinical Nutrition*, *112*(6), 1547–1557.

Beck, W. (1991). Neuropsychiatric consequences of cobalamin deficiency. *Advances in Internal Medicine*, *36*, 33–56.

Beedholm-Ebsen, R., van de Wetering, K., Hardlei, T., et al. (2010). Identification of multidrug resistance protein 1 (MRP1/ABCC1) as a molecular gate for cellular export of cobalamin. *Blood*, *115*(8), 1632–1639.

Bessman, J., Williams, L., & Gilmer, P. J. (1982). Platelet size in health and hematologic disease. *American Journal of Clinical Pathology*, *78*(2), 150–153.

Bor, M., Cetin, M., Aytaç, S., et al. (2005). Nonradioactive vitamin B12 absorption test evaluated in controls and in patients with inherited malabsorption of vitamin B12. *Clinical Chemistry*, *51*(11), 2151–2155.

Bor, M. V., Nexø, E., & Hvas, A.-M. (2004). Holo-transcobalamin concentration and transcobalamin saturation reflect recent vitamin B12 absorption better than does serum vitamin B12. *Clinical Chemistry*, *50*(6), 1043–1049.

Carkeet, C., Dueker, S. R., Lango, J., Buchholz, B. A., Miller, J. W., Green, R., et al. (2006). Human vitamin B12 absorption measurement by accelerator mass spectrometry using specifically labeled 14C-cobalamin. *Proceedings of the National Academy of Sciences*, *103*, 5694–5699.

Carmel, R. (1983). R-binder deficiency. A clinically benign cause of cobalamin pseudodeficiency. *JAMA*, *250*(14), 1886–1890.
Carmel, R. (2003). Mild transcobalamin I (haptocorrin) deficiency and low serum cobalamin concentrations. *Clinical Chemistry*, *49*(8), 1367–1374.
Carmel, R., Green, R., Rosenblatt, D., & Watkins, D. (2003). Update on cobalamin, folate, and homocysteine. *Hematology. American Society of Hematology. Education Program*, 62–81.
Castillo-Lancellotti, C., Margozzini, P., Valdivia, G., Padilla, O., Uauy, R., Rozowski, J., et al. (2015). Serum folate, vitamin B12 and cognitive impairment in Chilean older adults. *Public Health Nutrition*, *18*, 2600–2608.
Chanarin, I. (1969). *The megaloblastic Anaemias*. Blackwell Scientific.
Chen, M., Krishnamurthy, A., Mohamed, A. R., & Green, R. (2013). Hematological disorders following gastric bypass surgery: Emerging concepts of the interplay between nutritional deficiency and inflammation. *BioMed Research International*, *205467*.
Christine, C. W., Auinger, P., Joslin, A., Yelpaala, Y., Green, R., & Parkinson Study Group-DATATOP Investigators. (2018). Vitamin B12 and homocysteine levels predict different outcomes in early Parkinson's disease. *Movement Disorders*, *33*(5), 762–770.
Christine, C. W., Auinger, P., Saleh, N., Tian, M., Bottiglieri, T., Arning, E., et al. (2020). Relationship of cerebrospinal fluid vitamin B12 status markers with Parkinson's disease progression. *Movement Disorders*, *35*(8), 1466–1471.
Coleman, D., Donohue, D., Finch, C., et al. (1956). Erythrokinetics in pernicious anemia. *Blood*, *11*(9), 807–820.
Conley, C. L., & Krevans, J. R. (1951). Development of neurological manifestations of pernicious anemia during multivitamin therapy. *The New England Journal of Medicine*, *245*, 529–531.
Crosby, W. (1980). Improvisation revisited. Oral cyanocobalamin without intrinsic factor for pernicious anemia. *Archives of Internal Medicine*, *140*(12), 1582.
de Lau, L., Smith, A., Refsum, H., et al. (2009). Plasma vitamin B12 status and cerebral white-matter lesions. *Journal of Neurology, Neurosurgery, and Psychiatry*, *80*(2), 149–157.
Devi, S., Pasanna, R. M., Shamshuddin, Z., Bhat, K., Sivadas, A., Mandal, A. K., et al. (2020). Measuring vitamin B-12 bioavailability with [13C]-cyanocobalamin in humans. *The American Journal of Clinical Nutrition*, *112*(6), 1504–1515.
Di Lazzaro, V., Restuccia, D., Fogli, D., et al. (1992). Central sensory and motor conduction in vitamin B12 deficiency. *Electroencephalography and Clinical Neurophysiology*, *84*(5), 433–439.
Dinn, J. J., Weir, D. G., McCann, S., Reed, B., Wilson, P., & Scott, J. M. (1980). Methyl group deficiency in nerve tissue: A hypothesis to explain the lesion of subacute combined degeneration. *Irish Journal of Medical Science*, *149*(1), 1–4.
Doets, E. L., In't Veld, P. H., Szczecińska, A., Dhonukshe-Rutten, R. A., Cavelaars, A. E., van't Veer, P., et al. (2013). Systematic review on daily vitamin B12 losses and bioavailability for deriving recommendations on vitamin B12 intake with the factorial approach. *Annals of Nutrition & Metabolism*, *62*(4), 311–322.
el Kholty, S., Gueant, J. L., Bressler, L., Djalali, M., Boissel, P., Gerard, P., et al. (1991). Portal and biliary phases of enterohepatic circulation of corrinoids in humans. *Gastroenterology*, *101*(5), 1399–1408.
Fanidi, A., Carreras-Torres, R., Larose, T. L., et al. (2019). Is high vitamin B12 status a cause of lung cancer? *International Journal of Cancer*, *145*(6), 1499–1503.
Fedosov, S. N. (2013). Biochemical markers of vitamin B12 deficiency combined in one diagnostic parameter: The age-dependence and association with cognitive function and blood hemoglobin. *Clinica Chimica Acta*, *422*, 47–53.
Fedosov, S. N., Brito, A., Miller, J. W., et al. (2015). Combined indicator of vitamin B12 status: Modification for missing biomarkers and folate status and recommendations for revised cut-points. *Clinical Chemistry and Laboratory Medicine*, *53*(8), 1215–1225.

Fernandes-Costa, F., Green, R., & Torrance, J. (1978). Increased erythrocytic diphosphoglycerate in megaloblastic anaemia. A compensatory mechanism? *South African Medical Journal*, *53*(18), 709–712.

Finkler, A. E., & Hall, C. A. (1967). Nature of the relationship between vitamin B12 binding and cell uptake. *Archives of Biochemistry and Biophysics*, *120*(1), 79–85.

Flores-Guerrero, J. L., Minovic, I., Groothof, D., Gruppen, E. G., Riphagen, I. J., Kootstra-Ros, J., et al. (2020). Association of plasma concentration of vitamin B12 with all-cause mortality in the general population in the Netherlands. *JAMA Network Open*, *3*(1), e1919274.

Ford, M. (1980). Megaloblastic anaemia in a vegetarian. *The British Journal of Clinical Practice*, *34*(7), 222.

Fraser, T. (1960). Cerebral manifestations of Addisonian pernicious anaemia. *Lancet*, *2*(7148), 458–459.

Frenkel, E. (1973). Abnormal fatty acid metabolism in peripheral nerves of patients with pernicious anemia. *The Journal of Clinical Investigation*, *52*(5), 1237–1245.

Garrod, M. G., Buchholz, B. A., Miller, J. W., Haack, K. W., Green, R., & Allen, L. H. (2019). Vitamin B12 added as a fortificant to flour retains high bioavailability when baked in bread. *Nuclear Instruments and Methods in Physics Research B*, *438*, 136–140.

Garrod, M. G., Rossow, H. A., Calvert, C. C., Miller, J. W., Green, R., Buchholz, B. A., et al. (2019). 14C-cobalamin absorption from endogenously labeled chicken eggs assessed in humans using accelerator mass spectrometry. *Nutrients*, *11*(9), E2148.

Gilois, C., Wierzbicki, A., Hirani, N., et al. (1992). The hematological and electrophysiological effects of cobalamin. Deficiency secondary to vegetarian diets. *Annals of the New York Academy of Sciences*, *669*, 345–348.

Gräsbeck, R., & Tanner, S. M. (2011). Juvenile selective vitamin B_{12} malabsorption: 50 years after its description-10 years of genetic testing. *Pediatric Research*, *70*(3), 222–228.

Green, R. (1994). Typical and atypical manifestations of pernicious anemia. In M. Besser, R. Bhatt, V. H. T. James, & H. Keen (Eds.), *Vol. 1. Thomas Addison and his diseases: 200 years on* (pp. 377–390). Bristol: Journal of Endocrinology Ltd.

Green, R. (1995). Metabolite assays in cobalamin and folate deficiency. *Baillière's Clinical Haematology*, *8*(3), 533–566.

Green, R. (1996). Screening for vitamin B12 deficiency: Caveat emptor. *Annals of Internal Medicine*, *124*(5), 509–511.

Green, R. (2006). Protean *H pylori*: Perhaps "pernicious" too? *Blood*, *107*(4), 1247.

Green, R. (2009). Is it time for vitamin B-12 fortification? What are the questions? *The American Journal of Clinical Nutrition*, *89*(2), 712S–716S.

Green, R., Allen, L. H., Bjørke Monsen, A.-L., Brito, A., Gueant, J.-L., Miller, J. W., et al. (2017). Vitamin B12 deficiency. *Nature Reviews. Disease Primers*, *3*, 17054.

Green, R., & Christine, C. W. (2019). Linking vitamin B12 and a trembling disorder. *Cell Research*, *29*(5), 343–344.

Green, R., & Datta Mitra, A. (2017). Megaloblastic anemias: Nutritional and other causes. *The Medical Clinics of North America*, *101*(2), 297–317.

Green, R., & Dwyre, D. M. (2015). Evaluation of macrocytic anemias. *Seminars in Hematology*, *52*(4), 279–286.

Green, R., Jacobsen, D. W., Van Tonder, S. V., Kew, M. C., & Metz, J. (1982). Absorption of biliary cobalamin in baboons following total gastrectomy. *The Journal of Laboratory and Clinical Medicine*, *100*(5), 771–777.

Green, R., Jacobsen, D., van Tonder, S., et al. (1981). Enterohepatic circulation of cobalamin in the nonhuman primate. *Gastroenterology*, *81*(4), 773–776.

Green, R., Kuhl, W., Jacobson, R., et al. (1982). Masking of macrocytosis by alpha-thalassemia in blacks with pernicious anemia. *The New England Journal of Medicine*, *307*(21), 1322–1325.

Green, R., & Miller, J. W. (2014). Vitamin B12. In J. Zempleni, J. W. Suttie, J. F. Gregory, & P. J. Stover (Eds.), *Handbook of vitamins* (5th Edition, pp. 447–489). Boca Raton, FL: CRC Press.

Guéant, J., Champigneulle, B., Gaucher, P., & Nicolas, J. (1990). Malabsorption of vitamin B12 in pancreatic insufficiency of the adult and of the child. *Pancreas, 5*(5), 559–567.

Harriman, G., Smith, P., Horne, M., et al. (1989). Vitamin B12 malabsorption in patients with acquired immunodeficiency syndrome. *Archives of Internal Medicine, 149*(9), 2039–2041.

Henderson, J., Simpson, J., Warwick, R., & Shearman, D. (1972). Does malabsorption of vitamin B 12 occur in chronic pancreatitis? *Lancet, 2*(7771), 241–243.

Herbert, V. (1962). Experimental nutritional folate deficiency in man. *Transactions of the Association of American Physicians, 75*, 307–320.

Herbert, V. (1985). Megaloblastic anemias. *Laboratory Investigation, 52*(1), 3–19.

Herbert, V., Jacob, E., Wong, K. T., et al. (1978). Low serum vitamin B12 levels in patients receiving ascorbic acid in megadoses: Studies concerning the effect of ascorbate on radioisotope vitamin B12 assay. *The American Journal of Clinical Nutrition, 31*(2), 253–258.

Herbert, V., & Zalusky, R. (1962). Interrelations of vitamin B12 and folic acid metabolism: Folic acid clearance studies. *The Journal of Clinical Investigation, 41*, 1263–1276.

Hershko, C., Ronson, A., Souroujon, M., et al. (2006). Variable hematologic presentation of autoimmune gastritis: Age-related progression from iron deficiency to cobalamin depletion. *Blood, 107*(4), 1673–1679.

Herzlich, B., Schiano, T., Moussa, Z., et al. (1992). Decreased intrinsic factor secretion in AIDS: Relation to parietal cell acid secretory capacity and vitamin B12 malabsorption. *The American Journal of Gastroenterology, 87*(12), 1781–1788.

Heyssel, R. M., Bozian, R. C., Darby, W. J., & Bell, M. C. (1966). Vitamin B12 turnover in man. The assimilation of vitamin B12 from natural foodstuff by man and estimates of minimal daily dietary requirements. *The American Journal of Clinical Nutrition, 18*(3), 176–184.

Hillman, R., Adamson, J., & Burka, E. (1968). Characteristics of vitamin B12 correction of the abnormal erythropoiesis of pernicious anemia. *Blood, 31*(4), 419–432.

Hom, B. L. (1967). Plasma turnover of 57cobalt-vitamin B12 bound to transcobalamin I and II. *Scandinavian Journal of Haematology, 4*(5), 321–332.

Honein, M., Paulozzi, L., Mathews, T., et al. (2001). Impact of folic acid fortification of the US food supply on the occurrence of neural tube defects. *JAMA, 285*(23), 2981–2986.

Hoskins, L., Loux, H., Britten, A., & Zamcheck, N. (1965). Distribution of ABO blood groups in patients with pernicious anemia, gastric carcinoma and gastric carcinoma associated with pernicious anemia. *The New England Journal of Medicine, 273*(12), 633–637.

Institute of Medicine. (2000). Vitamin B12. In *Dietary reference intakes for thiamin, riboflavin, niacin, vitamin B6, folate, vitamin B12, pantothenic acid, biotin, and choline* (pp. 306–348). Washington, DC: The National Academies Press.

Kahn, S. B., Williams, W. J., Barness, L. A., et al. (1965). Methylmalonic acid excretion: A sensitive indicator of vitamin B12 deficiency. *Translational Research, 66*(1), 75–83.

Karnes, W. J., Samloff, I., Siurala, M., et al. (1991). Positive serum antibody and negative tissue staining for *helicobacter pylori* in subjects with atrophic body gastritis. *Gastroenterology, 101*(1), 167–174.

Kennedy, H., Callender, S., Truelove, S., & Warner, G. (1982). Haematological aspects of life with an ileostomy. *British Journal of Haematology, 52*(3), 445–454.

Koury, M., Horne, D., Brown, Z., et al. (1997). Apoptosis of late-stage erythroblasts in megaloblastic anemia: Association with DNA damage and macrocyte production. *Blood, 89*(12), 4617–4623.

Kozyraki, R., Kristiansen, M., Silahtaroglu, A., Hansen, C., Jacobsen, C., Tommerup, N., et al. (1998). The human intrinsic factor-vitamin B12 receptor, cubilin: Molecular characterization and chromosomal mapping of the gene to 10p within the autosomal recessive megaloblastic anemia (MGA1) region. *Blood*, *91*(10), 3593–3600.

Kuzminski, A., Del Giacco, E., Allen, R., et al. (1998). Effective treatment of cobalamin deficiency with oral cobalamin. *Blood*, *92*(4), 1191–1198.

Lederle, F. (1998). Oral cobalamin for pernicious anemia: Back from the verge of extinction. *Journal of the American Geriatrics Society*, *46*(9), 1125–1127.

Lees, F., & Grandjean, L. (1958). The gastric and jejunal mucosae in healthy patients with partial gastrectomy. *A.M.A. Archives of Internal Medicine*, *101*(5), 943–951.

Lenhert, P., & Hodgkin, D. (1961). Structure of the 5,6-dimethyl-benzimidazolylcobamide coenzyme. *Nature*, *192*, 937–938.

Lindenbaum, J., Healton, E., Savage, D., et al. (1988). Neuropsychiatric disorders caused by cobalamin deficiency in the absence of anemia or macrocytosis. *The New England Journal of Medicine*, *318*(26), 1720–1728.

Lindenbaum, J., & Nath, B. J. (1980). Megaloblastic anaemia and neutrophil hypersegmentation. *British Journal of Haematology*, *44*(3), 511–513.

Lindenbaum, J., Savage, D., Stabler, S., & Allen, R. (1990). Diagnosis of cobalamin deficiency: II. Relative sensitivities of serum cobalamin, methylmalonic acid, and total homocysteine concentrations. *American Journal of Hematology*, *34*(2), 99–107.

Lindgren, A., Kilander, A., Bagge, E., & Nexø, E. (1999). Holotranscobalamin—A sensitive marker of cobalamin malabsorption. *European Journal of Clinical Investigation*, *29*(4), 321–329.

Lindstrand, K. (1964). Isolation of methylcobalamin from natural source material. *Nature*, *204*, 188–189.

Maclean, L., & Sundberg, R. (1956). Incidence of megaloblastic anemia after total gastrectomy. *The New England Journal of Medicine*, *254*(19), 885–893.

McCaddon, A. (2013). Vitamin B12 in neurology and ageing; clinical and genetic aspects. *Biochimie*, *95*(5), 1066–1076.

Meyer, L. M. (1947). Folic acid in the treatment of pernicious anemia. *Blood*, *2*, 50–62.

Michaud, J., Lemieux, B., Ogier, H., & Lambert, M. (1992). Nutritional vitamin B12 deficiency: Two cases detected by routine newborn urinary screening. *European Journal of Pediatrics*, *151*(3), 218–220.

Miller, J. W. (2011). Folic acid fortification. In W. Herrmann, & R. Obeid (Eds.), *Vitamins in the prevention of human diseases* (pp. 273–293). Berlin: De Gruyter.

Miller, J. W. (2018). Proton pump inhibitors, H_2-receptor antagonists, metformin, and vitamin B-12 deficiency: Clinical implications. *Advances in Nutrition*, *9*, 511S–518S.

Miller, J. W., Garrod, M. G., Allen, L. H., Haan, M. N., & Green, R. (2009). Metabolic evidence of vitamin B-12 deficiency, including high homocysteine and methylmalonic acid and low holotranscobalamin, is more pronounced in older adults with elevated plasma folate. *The American Journal of Clinical Nutrition*, *90*(6), 1586–1592.

Miller, J. W., & Green, R. (2020). Assessing vitamin B-12 absorption and bioavailability: Read the label. *The American Journal of Clinical Nutrition*, *112*(6), 1420–1421.

Molloy, A. M., Pangilinan, F., Mills, J. L., et al. (2016). A common polymorphism in HIBCH influences methylmalonic acid concentrations in blood independently of cobalamin. *American Journal of Human Genetics*, *98*(5), 869–882.

Moore, E. M., Ames, D., Mander, A. G., Carne, R. P., Brodaty, H., Woodward, M. C., et al. (2014). Among vitamin B12 deficient older people, high folate levels are associated with worse cognitive function: Combined data from three cohorts. *Journal of Alzheimer's Disease*, *39*(3), 661–668.

Morris, M. S., Jacques, P. F., Rosenberg, I. H., & Selhub, J. (2007). Folate and vitamin B-12 status in relation to anemia, macrocytosis, and cognitive impairment in older Americans in the age of folic acid fortification. *American Journal of Clinical Nutrition*, *85*(1), 193–200.

Murphy, M., Sourial, N., Burman, J., et al. (1986). Megaloblastic anaemia due to vitamin B12 deficiency caused by small intestinal bacterial overgrowth: Possible role of vitamin B12 analogues. *British Journal of Haematology, 62*(1), 7–12.

Nexo, E., Hvas, A., Bleie, Ø., et al. (2002). Holo-transcobalamin is an early marker of changes in cobalamin homeostasis. A randomized placebo-controlled study. *Clinical Chemistry, 48*(10), 1768–1771.

Nielsen, M. J., Rasmussen, M. R., Andersen, C. B., Nexø, E., & Moestrup, S. K. (2012). Vitamin B12 transport from food to the body's cells- -a sophisticated, multistep pathway. *Nature Reviews. Gastroenterology & Hepatology, 9*(6), 345–354.

Norman, E., Martelo, O., & Denton, M. (1982). Cobalamin (vitamin B12) deficiency detection by urinary methylmalonic acid quantitation. *Blood, 59*(6), 1128–1131.

Norman, E., & Morrison, J. (1993). Screening elderly populations for cobalamin (vitamin B12) deficiency using the urinary methylmalonic acid assay by gas chromatography mass spectrometry. *The American Journal of Medicine, 94*(6), 589–594.

Nyberg, W. (1960). The influence of *Diphyllobothrium latum* on the vitamin B12-intrinsic factor complex. I. in vivo studies with Schilling test technique. *Acta Medica Scandinavica, 167*, 185–187.

Obeid, R., & Herrmann, W. (2007). Holotranscobalamin in laboratory diagnosis of cobalamin deficiency compared to total cobalamin and methylmalonic acid. *Clinical Chemistry and Laboratory Medicine, 45*(12), 1746–1750.

Oussalah, A., Julien, M., Levy, J., Hajjar, O., Franczak, C., Stephan, C., et al. (2019). Global burden related to nitrous oxide exposure in medical and recreational settings: A systematic review and individual patient data meta-analysis. *Journal of Clinical Medicine, 8*(4), 551.

Paulk, E. J., & Farrar, W. J. (1964). Diverticulosis of the small intestine and megaloblastic anemia: Intestinal microflora and absorption before and after tetracycline administration. *The American Journal of Medicine, 37*, 473–480.

Paulus, M. C., Wijnhoven, A. M., Maessen, G. C., Blankensteijn, S. R., & van der Heyden, M. A. G. (2021). Does vitamin B12 deficiency explain psychiatric symptoms in recreational nitrous oxide users? A narrative review. *Clinical Toxicology (Philadelphia), 59*(11), 947–955.

Ray, J. G. (2004). Folic acid food fortification in Canada. *Nutrition Reviews, 62*(6 Pt 2), S35–S39.

Ray, J. G., & Blom, H. J. (2003). Vitamin B12 insufficiency and the risk of fetal neural tube defects. *QJM, 96*(4), 289–295.

Refsum, H., Johnston, C., Guttormsen, A. B., & Nexo, E. (2006). Holotranscobalamin and total transcobalamin in human plasma: Determination, determinants, and reference values in healthy adults. *Clinical Chemistry, 52*(1), 129–137.

Remacha, A., & Cadafalch, J. (1999). Cobalamin deficiency in patients infected with the human immunodeficiency virus. *Seminars in Hematology, 36*(1), 75–87.

Rosenblatt, D. S., & Fenton, W. A. (2001). Inherited disorders of folate and cobalamin transport and metabolism. In C. R. Scriver, A. L. Beaudet, W. S. Sly, et al. (Eds.), *The metabolic and molecular bases of inherited metabolic disease* (8th ed., pp. 3897–3933). McGraw-Hill.

Schaffner, A., Li, X., Gomez-Llorente, Y., Leandrou, E., Memou, A., Clemente, N., et al. (2019). Vitamin B12 modulates Parkinson's disease LRRK2 kinase activity through allosteric regulation and confers neuroprotection. *Cell Research, 29*(4), 313–329.

Scherer, K. (2003). Images in clinical medicine. Neurologic manifestations of vitamin B12 deficiency. *The New England Journal of Medicine, 348*(22), 2208.

Schilling, R. F. (1953). Intrinsic factor studies. II. The effect of gastric juice on the urinary excretion of radioactivity after the oral administration of radioactive vitamin B12. *The Journal of Laboratory and Clinical Medicine, 42*(6), 860–866.

Selhub, J., Miller, J. W., Troen, A. M., Mason, J. B., & Jacques, P. F. (2021). Perspective: The high-folate-low-vitamin B-12 interaction is a novel cause of vitamin B-12 depletion with a specific etiology-a hypothesis. *Advances in Nutrition* (Oct 11:nmab106 (published online ahead of print).

Selhub, J., Morris, M. S., & Jacques, P. F. (2007). In vitamin B12 deficiency, higher serum folate is associated with increased total homocysteine and methylmalonic acid concentrations. *Proceedings of the National Academy of Sciences of the United States of America*, *104*(50), 19995–20000.

Shane, B. (2010). Folate chemistry and metabolism. In L. B. Bailey (Ed.), *Folate in health and disease* (2nd edition, pp. 1–24). Boca Raton, FL: CRC Press.

Sheehy, T., Perez-Santiago, E., & Rubini, M. (1961). Tropical sprue and vitamin B12. *The New England Journal of Medicine*, *265*, 1232–1236.

Shimoda, S., & Rubin, C. (1968). The Zollinger-Ellison syndrome with steatorrhea. I. Anticholinergic treatment followed by total gastrectomy and colonic interposition. *Gastroenterology*, *55*(6), 695–704.

Shulman, R. (1967). Psychiatric aspects of pernicious anaemia: A prospective controlled investigation. *British Medical Journal*, *3*(5560), 266–270.

Smith, A. D. M. (1960). Megaloblastic madness. *British Medical Journal*, *2*(5216), 1840–1845.

Smith, A. D., & Refsum, H. (2009). Vitamin B-12 and cognition in the elderly. *The American Journal of Clinical Nutrition*, *89*(2), 707S–711S.

Solomon, L. R. (2005). Cobalamin-responsive disorders in the ambulatory care setting: Unreliability of cobalamin, methylmalonic acid, and homocysteine testing. *Blood*, *105*(3), 978–985 (author reply 1137).

Spivak, J. L. (1982). Masked megaloblastic anemia. *Archives of Internal Medicine*, *142*(12), 2111–2114.

Stojsavljević, N., Lević, Z., Drulović, J., & Dragutinović, G. (1997). A 44-month clinical-brain MRI follow-up in a patient with B12 deficiency. *Neurology*, *49*(3), 878–881.

Sumner, A., Chin, M., Abrahm, J., et al. (1996). Elevated methylmalonic acid and total homocysteine levels show high prevalence of vitamin B12 deficiency after gastric surgery. *Annals of Internal Medicine*, *124*(5), 469–476.

Taylor, R., & Weissbach, H. (1967). Enzymic synthesis of methionine: Formation of a radioactive cobamide enzyme with N5-methyl-14C-tetrahydrofolate. *Archives of Biochemistry and Biophysics*, *119*(1), 572–579.

Toskes, P., Deren, J., & Conrad, M. (1973). Trypsin-like nature of the pancreatic factor that corrects vitamin B12 malabsorption associated with pancreatic dysfunction. *The Journal of Clinical Investigation*, *52*(7), 1660–1664.

Ungar, B., Mathews, J., Tait, B., & Cowling, D. (1981). HLA-DR patterns in pernicious anaemia. *British Medical Journal (Clinical Research Ed.)*, *282*(6266), 768–770.

Victor, M., & Lear, A. (1956). Subacute combined degeneration of the spinal cord; current concepts of the disease process; value of serum vitamin B12; determinations in clarifying some of the common clinical problems. *The American Journal of Medicine*, *20*(6), 896–911.

Vogiatzoglou, A., Refsum, H., Johnston, C., et al. (2008). Vitamin B12 status and rate of brain volume loss in community-dwelling elderly. *Neurology*, *71*(11), 826–832.

von Castel-Roberts, K., Morkbak, A., Nexo, E., et al. (2007). Holo-transcobalamin is an indicator of vitamin B-12 absorption in healthy adults with adequate vitamin B-12 status. *The American Journal of Clinical Nutrition*, *85*(4), 1057–1061.

Waxman, S., Corcino, J., & Herbert, V. (1970). Drugs, toxins and dietary amino acids affecting vitamin B12 or folic acid absorption or utilization. *The American Journal of Medicine*, *48*(5), 599–608.

Whitehead, V. (2006). Acquired and inherited disorders of cobalamin and folate in children. *British Journal of Haematology*, *134*(2), 125–136.

Wilson, A., Platt, R., Wu, Q., Leclerc, D., Christensen, B., Yang, H., et al. (1999). A common variant in methionine synthase reductase combined with low cobalamin (vitamin B12) increases risk for spina bifida. *Molecular Genetics and Metabolism, 67*(4), 317–323.

Wolffenbuttel, B. H. R., Heiner-Fokkema, M. R., Green, R., & Gans, R. O. B. (2020). Relationship between serum B12 concentrations and mortality: Experience in NHANES. *BMC Medicine, 18*(1), 307.

Yajnik, C. S., Deshpande, S. S., Lubree, H. G., Naik, S. S., Bhat, D. S., Uradey, B. S., et al. (2006). Vitamin B12 deficiency and hyperhomocysteinemia in rural and urban Indians. *The Journal of the Association of Physicians of India, 54*, 775–782.

Further reading

Green, R. (2017). Vitamin B12 deficiency from the perspective of a practicing hematologist. *Blood, 129*(19), 2603–2611.

CHAPTER SIXTEEN

Metabolism-mediated thrombotic microangiopathy and B12

Hadi Goubran[a,*], Gaafar Ragab[b], and Waleed Sabry[a,*]
[a]College of Medicine, University of Saskatchewan & Saskatoon Cancer Centre, SK, Canada
[b]Internal Medicine Department, Rheumatology and Clinical Immunology Unit, Faculty of Medicine, Cairo University, Egypt
[*]Corresponding authors: e-mail address: hadi.goubranmessiha@saskcancer.ca; waleed.sabry@saskcancer.ca

Contents

Abstract

Thrombotic microangiopathies (TMAs) are a group of life-threatening conditions requiring urgent management and characterized by a clinical triad of microangiopathic hemolytic anemia, thrombocytopenia, and ischemic tissue injury. Severe vitamin B12 (Cobalamin—Cbl) deficiency or defective cobalamin metabolism, particularly defects in intracellular B12 metabolism, may lead to a TMA-like picture. The latter has been termed metabolism-mediated TMA (MM-TMA). This confusing picture is mediated partly by ineffective erythropoiesis with significant red cell fragmentation resulting in a hemolytic pattern, coupled with reduced platelet production and endothelial injury with organ damage resulting from accumulated toxic byproducts of B12 dysmetabolism. However, unlike in classic thrombotic thrombocytopenic purpura, where therapeutic plasma exchange has to be initiated promptly, cases of MM-TMA can be treated, if diagnosed properly, with adequate B12 replacement.

Vitamins and Hormones, Volume 119
ISSN 0083-6729
https://doi.org/10.1016/bs.vh.2022.01.002

Abbreviations

AdoHcy	adenosylhomocysteine
aHUS	atypical hemolytic uremic syndrome
Cbl	cobalamin
IM	intramuscular
LDH	lactate dehydrogenase
MMA	methyl malonic acid
MM-TMA	metabolism-mediated thrombotic microangiopathy
TMA	thrombotic microangiopathy
TPE	therapeutic plasma exchange
TTP	thrombotic thrombocytopenic purpura

1. Introduction

Thrombotic microangiopathies (TMAs) are a group of disorders characterized by a clinical triad of microangiopathic hemolytic anemia, thrombocytopenia and ischemic tissue injury (Tsai, 2013). The main pathological hallmarks are endothelial damage and platelet activation resulting in arteriolar and capillary thrombosis (Laszik & Silva, 2007; George & Nester, 2014). It is a lesion with multiple etiologies. The presentation depends on the cause (Lusco, Fogo, Najafian, & Alpers, 2016). Though rare, thrombotic microangiopathies are life-threatening conditions that require urgent management (Arnold, Patriquin, & Nazy, 2017). The etiologic spectrum of TMA ranges from abnormalities in the ADAMTS13, with the accumulation of ultra-large von Willebrand factor multimers in the life-threatening thrombotic thrombocytopenic purpura (TTP), to a long list of complementopathies including atypical hemolytic uremic syndrome (aHUS). It also includes mechanical etiologies (Sabry, Elemary, Burnouf, Seghatchian, & Goubran, 2020). Severe vitamin B12 (Cobalamin—Cbl) deficiency or defective cobalamin metabolism, particularly defects in intracellular B12 metabolism, may lead to a TMA-like picture. The latter has been termed metabolism-mediated TMA (MM-TMA) (Geraghty et al., 1992). However, given the life-threatening nature of TMAs, empiric treatment is often initiated pending the results of complicated and advanced laboratory testing. In contrast, cobalamin-mediated TMA could be treated by simple replacement of vitamin B12. Therefore, it is crucial to promptly identify these cases to avoid exposure to expensive, potentially risky procedures.

2. TTP-TMA

Currently, the diagnosis of TTP rests on a constellation of clinical and laboratory findings. It includes a Coomb's negative hemolytic anemia with schistocytes observed on the peripheral blood smear, thrombocytopenia reflecting the consumptive process and end-organ damage with renal impairment or neurological findings, Together with fever in a proportion of patients. Diagnosis of TTP is challenging because of its diverse clinical manifestations, overlap in clinical presentation with other thrombotic microangiopathies, and limited availability of ADAMTS13 testing (Chiasakul & Cuker, 2018). The plasmic score, resting on seven variables including, a platelet count of less than 30 × 10e9/L, a creatinine of less than 2 mg/dl, an INR of less than 1.5, MCV of less than 90 fl, evidence of hemolysis (Reticulocytosis or raised bilirubin, low haptoglobin) with no previous bone marrow or solid organ transplantation and no evidence of active malignancy, was correlating to an acceptable extent with ADAMTS13 values. PLASMIC scores of 0–4, 5, and 6–7 correspond to low, intermediate, and high TTP probabilities, respectively (Jamme & Rondeau, 2017). A recent systematic review pointed out that a PLASMIC score threshold of ≥5 is associated with high sensitivity and negative predictive value and maybe a helpful screening tool for identifying patients who are unlikely to have TTP and do not require therapeutic plasma exchange (TPE). A PLASMIC score <6 appears to have insufficient sensitivity to rule out TTP and the need for TPE (Paydary, Banwell, Tong, Chen, & Cuker, 2020). In 13 eligible studies, collectively enrolling 970 patients, the sensitivity and specificity of a PLASMIC score ≥ 5 was 0.99 and 0.57, respectively.

In contrast, a PLASMIC score ≥ 6 was associated with a sensitivity and specificity of 0.85 and 0.89, respectively (Paydary et al., 2020). However, the International Society on Thrombosis and Hemostasis pointed out a lack of high-quality evidence to support solid recommendations for the initial diagnosis and management of a suspected TTP. The panel emphasized the importance of obtaining ADAMTS13 testing in a proper clinical context. Future research should focus on monitoring and acting on ADAMTS13 levels during remission (Zheng et al., 2020).

aHUS, related to complement disorders, presents similarly, often with less profound thrombocytopenia and more renal involvement.

3. B12 dysmetabolism and TMA-like syndrome

This confusing picture is mediated partly by ineffective erythropoiesis with significant red cell fragmentation resulting in a hemolytic pattern, coupled with reduced platelet production and endothelial injury with organ damage resulting from accumulated toxic byproducts of B12 dysmetabolism (Sabry et al., 2020). Tran & Tran, 2018 reviewed the literature and identified 41 cases. Cobalamin deficiency was noted in 35/40 cases (87.5%), but fold increases in the mean methylmalonic acid and homocysteine levels were 30 and 6, respectively. The etiology was pernicious anemia in 28/41 (68%) cases. Anemia occurred in all cases and was severe with hemoglobin levels of 55 g/L (47–66). Hypersegmented neutrophils were noted in 23/37 cases (62%), schistocytes in 29/38 cases (76%), and median LDH levels 3981 U/L (2004–5467 U/L). The reticulocytes were <3.0% in the majority of patients. Thrombocytopenia occurred in 33/41 (80.5%) with a median platelet count of 91×10^9/L (42–112). Platelet counts $<31 \times 10^9$/L, reticulocyte count of >3%, or indirect (unconjugated) bilirubin levels >25.7 mol/L occurred among 5/39 (13%), 11/31 (35%), and 14/19 (74%) of patients, respectively. Plasma infusion or exchange was initiated in 14/41 (34%) with associated complications in 2 cases (Tran & Tran, 2018). Another case report and literature review were published in 2019, alluding to patients with severe vitamin B12 deficiency presenting with features mimicking TTP such as microangiopathic hemolytic anemia, thrombocytopenia and schistocytosis. An early and accurate diagnosis of pseudo-TMA has a critical clinical impact concerning administering the correct treatment with vitamin B12 replacement and avoiding or shortening the duration of unnecessary therapy with plasmapheresis (Fahmawi et al., 2019). Another case was also reported in 2020 (Ito, 2020). Other small series and isolated case reports have also been published since then (Yamanishi, Koba, Jo, Kotera, & Imashuku, 2018; Rao, Colon Hidalgo, Doria Medina Sanchez, Navarrete, & Berg, 2020; Harada, Komori, Morinaga, & Shimizu, 2018; Noël et al., 2013; Osman, Alwasaidi, Al-Hebshi, Almutairi, & Eltabbakh, 2021; Podder, Cervates, & Dey, 2015) including a case of Imerslund-Gräsbeck Syndrome (Gurlek Gokcebay, Akpinar Tekgunduz, & Cavdarli, 2020).

Interestingly, when applying the PLASMIC score on most of these case series and case reports, all patients had an MCV of >90 fl, a platelet count of >30x10e9/L and hence had a score of 5 or less. Furthermore, if the evidence of hemolysis was rephrased to "*Reticulocytosis AND (not OR) raised bilirubin or low haptoglobin,"* the score will be less than 4, ruling out TTP. Of note is that in all case reports and case series, LDH is usually markedly elevated above 5–10 folds the upper limits of normal, to an extent disproportionate to the degree of hemolysis. Thus, around 38.8% of cases with pseudo-TMA are misdiagnosed as thrombotic thrombocytopenic purpura and treated with plasma product therapy (Fahmawi et al., 2019).

In patients presenting with possible TMAs, macrocytosis, moderate rather than severe thrombocytopenia, a disproportionally elevated LDH, and lack of reticulocytosis should alert the treating physician to consider the diagnosis of MM-TMA.

4. The triad of TMA in B12 deficiency and dysmetabolism

As in all TMAs, a triad of non-immune hemolytic anemia with schistocytes, thrombocytopenia, and endothelial dysfunction with end-organ damage is observed in MM-TMA.

4.1 Non-immune hemolytic anemia with schistocytosis

In nearly all the reported cases, anemia was both universal and severe, with hemoglobin levels of 55 g/L (47–66) (Tran & Tran, 2018).

The nature of the anemia is related to both an underproduction from the bone marrow resulting from the lack of essential hematinics and ineffective erythropoiesis, as well as potential peripheral mechanical hemolysis due to the large cell size.

B12- or folate-deficient hemopoietic cells display abnormalities in their morphology under both the light and electron microscope, their cell kinetics, and their capacity to synthesize protein. These abnormalities are maximal in the last dividing cell class and non-dividing cells, presumably because B12 and folate uptake is confined mainly to the most immature erythroid and granulocyte precursors. In patients with moderate or severe anemia due to B12 or folate deficiency, erythropoiesis is markedly ineffective; and intramedullary cell death occurs mainly in the early and late polychromatic megaloblasts. The damaged erythroblasts appear to display neoantigens or usually hidden antigens at their cell surface, and these react with naturally occurring antibodies (Wickramasinghe, 1995). Therefore, the noted hemolytic markers reflect the ineffective, intramedullary hematopoiesis associated with undetectable haptoglobin and raised bilirubin with a disproportionally elevated LDH

(Sabry et al., 2020). Lysis of immature nucleated red cells in the bone marrow contributes to higher LDH levels than peripheral hemolysis (Tran & Tran, 2018).

Most of the reported cases of MM-TMA described blood work showing evidence of a positive non-immune hemolysis panel. Hemolysis is mechanical mainly due to the large red cells and is associated with significant elevation of the lactate dehydrogenase (LDH) >2500 IU/L (Wickramasinghe, 1995). Peripheral smear invariably shows significant macrocytosis, anisocytosis, poikilocytosis, schistocytes and hypersegmented neutrophils (Fahmawi et al., 2019). However, identifying true schistocytes in the presence of substantial macrocytosis and marked poikilocytosis can be challenging (Hall, Mason, Choi, & Holguin, 2017; Koubaissi & Degheili, 2019).

True schistocytes can also be observed due to fragmentation of the large erythrocytes (Koubaissi & Degheili, 2019; Pieralli et al., 2021).

Characteristically, unlike in other hemolytic anemias, MM-TMA hemolysis/pseudo-hemolysis is not associated with reticulocytosis (Gurlek Gokcebay et al., 2020; Harada et al., 2018; Noël et al., 2013; Osman et al., 2021; Podder et al., 2015; Rao et al., 2020; Yamanishi et al., 2018). Reticulocytopenia with a count of <3% is very common (Tran & Tran, 2018). Reticulocytopenia in cobalamin deficiency reflects ineffective erythropoiesis and intramedullary destruction.

4.2 Thrombocytopenia

Unlike in true TMA and microangiopathic hemolytic anemias where thrombocytopenia is due to consumption, in B12 dysmetabolism, it results from bone marrow underproduction.

In a series of 41 case reports, Tran and Tran (2018) documented thrombocytopenia to occur in 33/41 (80.5%) with a median platelet count of 91×10^9/L (42–112) (Tran & Tran, 2018). Once more, in the majority of the case reports, severe thrombocytopenia below 30×10^9/L was uncommon, with most of the reported platelet count in the range of moderate thrombocytopenia (Bailey, Maestas, Betancourt, Mikhael, & Babiker, 2019; Koubaissi & Degheili, 2019; Pieralli et al., 2021; Podder et al., 2015; Tran & Tran, 2018).

It is interesting to note that isolated thrombocytopenia in the absence of anemia is not a common feature of B12 deficiency. For example, in a study where 57 patients underwent assessment for isolated thrombocytopenia and had a B12 assay, only three (5.3%) had low B12 levels (179, 148, 143 ng/L; normal 180–914 ng/L) (DeLoughery et al., 2019). On the other hand, severe thrombocytopenia has been described in infantile severe B12 deficiency (Rajasekaran, Sheriff, Moore, McCay, & Winstanley, 2020).

4.3 Endothelial dysfunction, end-organ damage and defective cobalamin metabolism

The absorption, transport, storage, and intracellular processing of Cbl are regulated and affected by at least 21 genes (Froese & Gravel, 2010). Thus, deficiency of Cbl and/or its metabolites can result from genetic mutations or other conditions interfering with its intake, absorption or storage. Nevertheless, under normal circumstances, the human liver stores of Cbl are sufficient to meet the body's need for several years, even with reduced intake or impaired absorption.

Cbl dysmetabolism results in hyperhomocysteinemia, decreased plasma methionine level and raised methylmalonic acid (MMA) with methylmalonic aciduria. In addition, to their power to generate reactive oxygen species, these byproducts lead to endothelial dysfunction, platelet activation and increased tissue factor expression with subsequent activation of the coagulation cascade (Coppola et al., 2000), with the first link between hyperhomocysteinemia and thrombosis reported in the 1960s (McCully, 1969).

Homocysteine (Hcy) is a sulfur-containing non-proteinogenic amino acid and is considered a risk factor for atherosclerosis and cardiovascular disease, but the molecular basis of these associations remains elusive.

Three meta-analyses demonstrated a modest association of increased venous thrombosis (VT) risk per 0.7-mg/L (5 μmol/L), higher homocysteine concentration or hyperhomocysteinemia (den Heijer, Lewington, & Clarke, 2005; den Heijer, Rosendaal, Blom, Gerrits, & Bos, 1998; Ray, 1998). However, Ospina-Romero et al. (2018) demonstrated that after extensive adjustments for confounding factors, hyperhomocysteinemia was not associated with an increased risk of venous thrombosis among 1689 cases and 1726 controls from the Netherlands (Ospina-Romero et al., 2018). Furthermore, randomized trials of homocysteine-lowering therapy and VT risk have been negative (den Heijer et al., 2007; Ray, Kearon, Yi, et al., 2007).

The role played by Hcy in endothelial damage and vascular toxicity, however, is non-negligible. It interferes with nitric oxide (NO) production, a gaseous master regulator of endothelial homeostasis. Moreover, Hcy deregulates the signaling pathways of hydrogen sulphide and mediates the loss of critical endothelial antioxidant systems. It also increases the intracellular concentration of reactive oxygen species (ROS), yielding oxidative stress. ROS disturb lipoprotein metabolism, contributing to the growth of atherosclerotic vascular lesions. Moreover, excess Hcy may be indirectly incorporated into proteins, a process referred to as protein N-homocysteinylation, inducing vascular damage. Lastly, cellular hypomethylation caused by the build-up of *S*-adenosylhomocysteine (AdoHcy) also contributes to the molecular basis of Hcy-induced vascular toxicity (Esse, Barroso, Tavares de Almeida, & Castro, 2019).

MMA damages the brain and kidney in experimental animal models with changes in p53 and caspase 3 levels (Andrade et al., 2014), contributing significantly to the B12 deficiency-associated neuropathies and bone marrow suppression. Isolated thrombocytopenia, however, is rarely caused by B12 deficiency (DeLoughery et al., 2019).

Cobalamin C disease is a hereditary defect of Cbl metabolism and a rare cause of TMA with more pronounced renal impairment mimicking atypical hemolytic uremic syndrome (aHUS). It is an autosomal recessive disorder typically seen in infants (Geraghty et al., 1992). It results from either a homozygous or a compound heterozygous mutation of methylmalonic aciduria and homocystinuria type C protein (MMACHC) gene, located on chromosome 1 P 34 (Lerner-Ellis et al., 2006). This results in a significant deficiency in methylcobalamin (Coppola et al., 2000). The pathophysiologic link between the mutation and TMA is not fully established. The generation of ROS and endothelial dysfunction caused by hyperhomocysteinemia have been implicated (Coppola et al., 2000; Sharma, Greenberg, Prasad, & Prasad, 2007). Cbl C manifesting as failure to thrive, hypotonia, seizures and a developmental delay occurs mainly in neonates or infants during the 1 st year of life (Fischer et al., 2014; Huemer et al., 2014). Later onset disease was described in some children older than 4 years. Adult-onset disease is rare and may present with neuropsychiatric manifestations (Guigonis et al., 2005; Van Hove, Van Damme-Lombaerts, & Grünewald, 2002). Pulmonary artery hypertension (PAH) and renal TMA have been described in early-onset Cbl C deficiency, whereas neuropsychiatric symptoms prevail in adult-onset forms (Fischer et al., 2014; Huemer et al., 2014). The condition is well known to pediatricians who routinely screen for hyperhomocysteinemia and methylmalonic aciduria in all children suspected of having aHUS. However, this screening is not part of the diagnostic algorithm in adults (Guigonis et al., 2005; Loirat & Frémeaux-Bacchi, 2011; Loirat, Saland, & Bitzan, 2012; Van Hove et al., 2002; Zuber, Fakhouri, Roumenina, Loirat, & Frémeaux-Bacchi, 2012). Lemoine et al. suggested

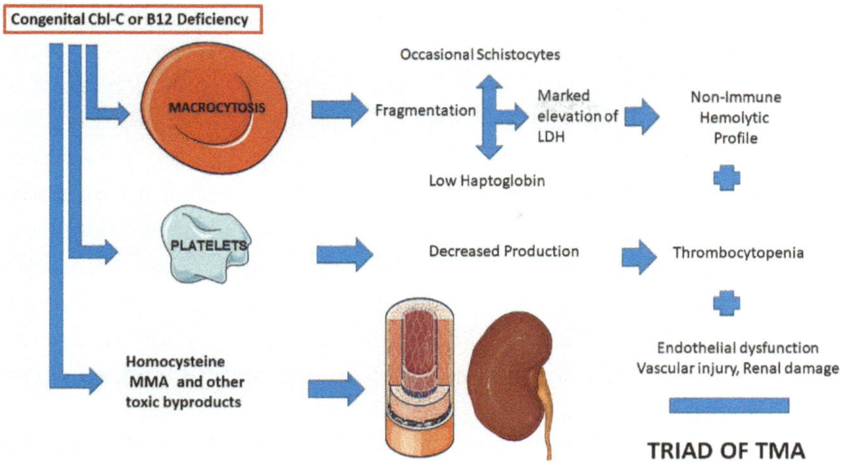

Fig. 1 Cobalamin C or B12 deficiency results in macrocytosis and ineffective erythropoiesis with a picture that mimics non-immune hemolysis with secondary thrombocytopenia due to defective bone marrow platelet production. This, coupled with the endothelial damage affecting organs like the kidneys resulting from accumulated toxic byproducts, leads to a picture that simulates a TMA.

screening all patients with renal TMA for a cobalamin metabolism disorder, regardless of age and even in the absence of neurological symptoms, to rapidly initiate the appropriate treatment (Grangé et al., 2015; Lemoine, Grange, & Guerrot, 2019).

Fig. 1 illustrates how the non-immune hemolytic profile, thrombocytopenia and endothelial dysfunction result in a picture that mimics TMA.

5. Diagnosis

The diagnosis of MM-TMA rests on the findings of a non-immune hemolytic profile with occasional schistocytes seen on the peripheral blood smear and thrombocytopenia with or without evidence of end-organ damage with renal impairment or neurological deficits.

It is imperative to rule out TTP as TPE will have to be initiated upfront. The application of the plasmic score in cases of MM-TMA will likely report a score of 4–5 with a low- intermediate probability for TTP.

The clinical context may help identify patients with MM-TMA as patients with polyglandular autoimmune disorders are more prone to pernicious anemia with severe B12 deficiency. In addition, chronic neurological findings compatible with a cobalamin-mediated disease rather than an abrupt focal or neuropsychiatric change will favor an MM over TTP.

Cutaneous manifestations in the form of vitiligo (Brescoll & Daveluy, 2015) or hyperpigmentation should alert the treating physician to search for an underlying B12 etiology (Bhanja, Sil, Chakraborty, Panigrahi, & Punithakumar, 2020; García-Villa & Cabañuz, 2018; Rao, 2018). Cobalamin deficiency-related hyperpigmentation can be generalized, but characteristically involved areas include oral mucosa and extremities, with accentuation over the interphalangeal joints. This was noted in a few cases and often improved with B12 supplementation (Gunaseelan, Narasingarajan, & Mahadevan, 2016; Malla & Seetharam, 2016; Mishra, Harbada, & Sharma, 2015). The pathophysiology is uncertain, but increased melanocytes are pointed out in the dermis and epidermis (Mori, Ando, & Kukita, 2001).

The current recommendation is to subject patients with suspected TMA/TTP to a panel of tests that include a blood count, peripheral smear for schistocytes, LDH, Haptoglobin, bilirubin, reticulocytotic count and Coomb's test as well as markers of cardiac injury, renal functions, B12 assay, PT, aPTT, D-Dimer and an assay of ADAMTS13 antigen and antibody. A careful look at the peripheral blood smear is always needed to differentiate poikilocytosis from schistocytosis. In addition, the presence of macrocytosis with a disproportionately elevated LDH in the presence of moderate rather than severe thrombocytopenia should prompt for a quick B12 assay which will confirm the diagnosis.

Brain imaging may also be needed in the presence of central nervous system affection.

However, the exclusion of aHUS might be slightly more challenging when renal impairment is more pronounced, particularly in infants and children with Cobalamin C disease. Luckily, the level of awareness is higher among pediatricians, and B12 assay is often part of the initial workup of infants and children with TMA (Chen, Zhuang, Yang, Wang, & Yang, 2017; Guigonis et al., 2005).

A borderline serum level of b12 should trigger confirmatory assays for homocysteine and MMA. However, on occasions, an active B12 level may also be needed to confirm the diagnosis.

6. Treatment

Treatment of MM-TMA rests on a prompt and rapid replacement of B12, usually by parentral supplementation of cobalamin at a dose of 1000 μg by IM injection given initially daily then monthly after that. In most reported cases, supplementation led to improved hemolytic parameters, restoring the platelet count with prompt reticulocytosis (Chen et al., 2017; Fahmawi et al., 2019; Harada et al., 2018; Lemoine et al., 2019; Rao et al., 2020; Yamanishi et al., 2018).

7. Conclusions

The diagnosis is of TMA is often challenging due to its high morbidity and mortality. With the most widely used PLASMIC score, diagnostic algorithms and scores have been developed by many scientific bodies to help clinicians initiate treatment promptly. Unfortunately, TPE is often triggered before confirmatory tests become available on an empirical basis.

Severe B12 deficiency and its various metabolic syndromes do rarely mimic such disorders, presenting as MM-TMA. However, although rare, more and more cases are currently described in the literature, and many clinicians and clinical centres are looking back at their records identifying patients with this disorder.

Early identification of MM-TMA is essential as it could avoid exposing patients to unnecessary rather aggressive procedures like TPE pending the availability of the TTP/aHUS confirmatory test results.

Adding a B12 assay for the patient with suspected TMA, particularly for those with marked poikilocytosis, macrocytosis, and elevated LDH, should therefore be a routine practice.

References

Andrade, V. M., Dal Pont, H. S., Leffa, D. D., Damiani, A. P., Scaini, G., Hainzenreder, G., et al. (2014). Methylmalonic acid administration induces DNA damage in rat brain and kidney. *Molecular and Cellular Biochemistry*, *391*(1–2), 137–145. https://doi.org/10.1007/s11010-014-1996-4.

Arnold, D., Patriquin, C., & Nazy, I. (2017). Thrombotic microangiopathies: A general approach to diagnosis and management. *CMAJ : Canadian Medical Association Journal*, *189*(4), E153–E159. https://doi.org/10.1503/cmaj.160142.

Bailey, M., Maestas, T., Betancourt, R., Mikhael, D., & Babiker, H. M. (2019). A Rare Cause of Thrombotic Thrombocytopenia Purpura- (TTP-) Like Syndrome, Vitamin B12 Deficiency: Interpretation of Significant Pathological Findings. *Case Reports in Hematology*, *2019*, 1529306. https://doi.org/10.1155/2019/1529306.

Bhanja, D. B., Sil, A., Chakraborty, S., Panigrahi, A., & Punithakumar, E. J. (2020). Acral hyperpigmentation due to vitamin B12 deficiency. *The Journal of the Royal College of Physicians of Edinburgh*, *50*(4), 416–417. https://doi.org/10.4997/JRCPE.2020.415.

Brescoll, J., & Daveluy, S. (2015). A review of vitamin B12 in dermatology. *American Journal of Clinical Dermatology*, *16*(1), 27–33. https://doi.org/10.1007/s40257-014-0107-3.

Chen, M., Zhuang, J., Yang, J., Wang, D., & Yang, Q. (2017). Atypical hemolytic uremic syndrome induced by CblC subtype of methylmalonic academia: A case report and literature review. *Medicine (Baltimore)*, *96*(43), e8284. https://doi.org/10.1097/MD.0000000000008284.

Chiasakul, T., & Cuker, A. (2018). Clinical and laboratory diagnosis of TTP: An integrated approach. *Hematology. American Society of Hematology. Education Program, 2018*(1), 530–538. https://doi.org/10.1182/asheducation-2018.1.530.

Coppola, A., Davi, G., De Stefano, V., Mancini, F. P., Cerbone, A. M., & Di Minno, G. (2000). Homocysteine, coagulation, platelet function, and thrombosis. *Seminars in Thrombosis and Hemostasis, 26*, 243–254. https://doi.org/10.1055/s-2000-8469.

DeLoughery, E. P., Ravindran, A., Ashrani, A. A., Begna, K. H., Hook, C. C., Marshall, A. L., et al. (2019). Patterns and utility of vitamin B12 and folate testing in patients with isolated thrombocytopenia. *Annals of Hematology, 98*(8), 1993–1994. https://doi.org/10.1007/s00277-019-03666-2.

den Heijer, M., Rosendaal, F. R., Blom, H. J., Gerrits, W. B., & Bos, G. M. (1998). Hyperhomocysteinemia and venous thrombosis: A meta-analysis. *Thrombosis and Haemostasis, 80*(6), 874–877.

den Heijer, M., Lewington, S., & Clarke, R. (2005). Homocysteine, MTHFR and risk of venous thrombosis: A meta-analysis of published epidemiological studies. *Journal of Thrombosis and Haemostasis, 3*(2), 292–299. https://doi.org/10.1111/j.1538-7836.2005.01141.x.

den Heijer, M., Willems, H. P., Blom, H. J., Gerrits, W. B., Cattaneo, M., Eichinger, S., et al. (2007). Homocysteine-lowering by B vitamins and the secondary prevention of deep vein thrombosis and pulmonary embolism: A randomized, placebo-controlled, double-blind trial. *Blood, 109*(1), 139–144. https://doi.org/10.1182/blood-2006-04-014654.

Esse, R., Barroso, M., Tavares de Almeida, I., & Castro, R. (2019). Contribution of Homocysteine Metabolism Disruption to Endothelial Dysfunction: State-of-the-Art. *International Journal of Molecular Sciences, 20*(4), 867. https://doi.org/10.3390/ijms20040867.

Fahmawi, Y., Campos, Y., Khushman, M., Alkharabsheh, O., Manne, A., Zubair, H., et al. (2019). Vitamin B12 deficiency presenting as pseudo-thrombotic microangiopathy: A case report and literature review. *Clinical Pharmacology, 11*, 127–131. https://doi.org/10.2147/CPAA.S207258.

Fischer, S., Huemer, M., Baumgartner, M., Deodato, F., Ballhausen, D., Boneh, A., et al. (2014). Clinical presentation and outcome in a series of 88 patients with the cblC defect. *Journal of Inherited Metabolic Disease, 37*, 831–840. https://doi.org/10.1007/s10545-014-9687-6.

Froese, D. S., & Gravel, R. A. (2010). Genetic disorders of vitamin B(12) metabolism: Eight complementation groups—Eight genes. *Expert Reviews in Molecular Medicine, 2010:12: e37*. https://doi.org/10.1017/S1462399410001651.

García-Villa, A., & Cabañuz, C. (2018). Cutaneous hyperpigmentation in vitamine B12 deficiency. *Revista Clínica Española, 218*(2), 108–109. https://doi.org/10.1016/j.rce.2017.10.002.

George, J. N., & Nester, C. M. (2014). Syndromes of thrombotic microangiopathy. *The New England Journal of Medicine, 371*, 654–666. https://doi.org/10.1056/NEJMra1312353.

Geraghty, M. T., Perlman, E. F., Martin, L. S., Hayflick, S. J., Casella, J. F., Rosenblatt, D. S., et al. (1992). Cobalamin C defect associated with hemolytic-uremic syndrome. *The Journal of Pediatrics, 120*, 934–937. https://doi.org/10.1016/s0022-3476(05)81967-5.

Grangé, S., Bekri, S., Artaud-Macari, E., Francois, A., Girault, C., Poitou, A. L., et al. (2015). Adult-onset renal thrombotic microangiopathy and pulmonary arterial hypertension in cobalamin C deficiency. *Lancet, 386*, 1011–1012. https://doi.org/10.1016/S0140-6736(15)00076-8.

Guigonis, V., Frémeaux-Bacchi, V., Giraudier, S., Favier, R., Borderie, D., Massy, Z., et al. (2005). Late-onset thrombocytic microangiopathy caused by cblC disease: Association with a factor H mutation. *American Journal of Kidney Diseases, 45*, 588–595. https://doi.org/10.1053/j.ajkd.2004.12.004.

Gunaseelan, P., Narasingarajan, S., & Mahadevan, S. (2016). Melanoderma: uncommon presentation of a common condition. *BML Case Reports*. https://doi.org/10.1136/bcr-2016-217766. bcr2016217766.

Gurlek Gokcebay, D., Akpinar Tekgunduz, S., & Cavdarli, B. (2020). Imerslund-Gräsbeck syndrome presenting with microangiopathic hemolytic anemia in a child. *European Journal of Medical Genetics*, *63*(6), 103880. https://doi.org/10.1016/j.ejmg.2020.103880.

Hall, J. A., Mason, J., Choi, J., & Holguin, M. (2017). B12 deficiency leading to marked poikilocytosis versus true schistocytosis, a pernicious problem. *Transfusion and Apheresis Science*, *56*(4), 576–577. https://doi.org/10.1016/j.transci.2017.06.003.

Harada, Y., Komori, I., Morinaga, K., & Shimizu, T. (2018). Microangiopathic haemolytic anaemia with thrombocytopenia induced by vitamin B12 deficiency long term after gastrectomy. *BML Case Reports*, 2018:bcr2018225915. https://doi.org/10.1136/bcr-2018-225915.

Huemer, M., Scholl-Burgi, S., Hadaya, K., Kern, I., Beer, R., Seppi, K., et al. (2014). Three new cases of late-onset cblC defect and review of the literature illustrating when to consider inborn errors of metabolism beyond infancy. *Orphanet Journal of Rare Diseases*, *161*. https://doi.org/10.1186/s13023-014-0161-1.

Ito, H. (2020). Vitamin B12 deficiency and metabolism-mediated thrombotic microangiopathy (MM-TMA). *Transfusion and Apheresis Science*, *59*(5), 102872. https://doi.org/10.1016/j.transci.2020.102872.

Jamme, M., & Rondeau, E. (2017). The PLASMIC score for thrombotic thrombocytopenic purpura. *Lancet Haematology*, *4*(4), e148–e149. https://doi.org/10.1016/S2352-3026(17)30024-8.

Koubaissi, S. A., & Degheili, J. A. (2019). Severe cobalamin deficiency disguised as schistocytes: A case report. *American Journal of Case Reports*, *17*(20), 1691. https://doi.org/10.12659/AJCR.918807.

Laszik, Z. G., & Silva, F. G. (2007). Hemolytic uremic syndrome, thrombotic thrombocytopenic purpura, and other thrombotic microangiopathies and coagulopathies. In J. C. Jennett, J. L. Olson, M. M. Schwartz, & F. G. Silva (Eds.), *Heptinstall's pathology of the kidney* (6th ed., pp. 699–762). Philadelphia: Lippincott Williams & Wilkins.

Lemoine, M., Grange, S., & Guerrot, D. (2019). Kidney disease in cobalamine C deficiency. *Néphrologie & Thérapeutique*, 201–214. https://doi.org/10.1016/j.nephro.2019.03.011.

Lerner-Ellis, J. P., Tirone, J. C., Pawelek, P. D., Doré, C., Atkinson, J. L., Watkins, D., et al. (2006). Identification of the gene responsible for methylmalonic aciduria and homocystinuria, cblC type. *Nature Genetics*, *38*(1), 93–100. https://doi.org/10.1038/ng1683.

Loirat, C., & Frémeaux-Bacchi, V. (2011). Atypical hemolytic uremic syndrome. *Orphanet Journal of Rare Diseases*, *6*, 60. https://doi.org/10.1186/1750-1172-6-60.

Loirat, C., Saland, J., & Bitzan, M. (2012). Management of hemolytic uremic syndrome. *Presse Médicale*, *41*(3, pt 2), e115–e135. https://doi.org/10.1016/j.lpm.2011.11.013.

Lusco, M., Fogo, A., Najafian, B., & Alpers, C. (2016). Atlas of Renal Pathology II: *Thrombotic Microangiopathy*. *American Journal of Kidney Diseases*, *68*(6), E33–E34. https://doi.org/10.1053/j.ajkd.2016.10.006.

Malla, M., & Seetharam, M. (2016). To treat or not to treat: A rare case of pseudo-thrombotic thrombocytopenic purpura in a Jehovah's witness. *Transfusion*, *56*(1), 160–163. https://doi.org/10.1111/trf.13285.

McCully, K. S. (1969). Vascular pathology of homocysteinemia: Implications for the pathogenesis of arteriosclerosis. *The American Journal of Pathology*, *56*(1), 111–128.

Mishra, V. A., Harbada, R., & Sharma, A. (2015). Vitamin B12 and vitamin d deficiencies: An unusual cause of fever, severe hemolytic anemia and thrombocytopenia. *Journal of Family Medicine and Primary Care*, *4*(1), 145. https://doi.org/10.4103/2249-4863.152276.

Mori, K., Ando, I., & Kukita, A. (2001). Generalized hyperpigmentation of the skin due to vitamin B12 deficiency. *The Journal of Dermatology*, *28*(5), 282–285. https://doi.org/10.1111/j.1346-8138.2001.tb00134.x.

Noël, N., Maigné, G., Tertian, G., Anguel, N., Monnet, X., Michot, J. M., et al. (2013). Hemolysis and schistocytosis in the emergency department: Consider pseudothrombotic microangiopathy related to vitamin B12 deficiency. *QJM*, *106*(11), 1017–1022. https://doi.org/10.1093/qjmed/hct142.

Osman, H., Alwasaidi, T. A., Al-Hebshi, A., Almutairi, N., & Eltabbakh, H. (2021). Vitamin B12 deficiency presenting with microangiopathic hemolytic anemia. *Cureus.*, *13*(1), e12600. https://doi.org/10.7759/cureus.12600.

Ospina-Romero, M., Cannegieter, S. C., den Heijer, M., Doggen, C. J. M., Rosendaal, F. R., & Lijfering, W. M. (2018). Hyperhomocysteinemia and risk of first venous thrombosis: The influence of (unmeasured) confounding factors. *American Journal of Epidemiology*, *187*(7), 1392–1400. https://doi.org/10.1093/aje/kwy004.

Paydary, K., Banwell, E., Tong, J., Chen, Y., & Cuker, A. (2020). Diagnostic accuracy of the PLASMIC score in patients with suspected thrombotic thrombocytopenic purpura: A systematic review and meta-analysis. *Transfusion*, *60*(9), 2047–2057. https://doi.org/10.1111/trf.15954.

Pieralli, F., Milia, A., Fruttuoso, S., Bandini, G., Mercatelli, P., Nozzoli, C., et al. (2021). The doctor who stared at schistocytes: An intriguing case of suspected thrombotic microangiopathic anemia. *Internal and Emergency Medicine*, *16*(2), 437–441. https://doi.org/10.1007/s11739-019-02219-9.

Podder, S., Cervates, J., & Dey, B. R. (2015). Association of acquired thrombotic thrombocytopaenic purpura in a patient with pernicious anaemia. *BML Case Reports*, *2015*. https://doi.org/10.1136/bcr-2015-211989. bcr2015211989.

Rajasekaran, V., Sheriff, J., Moore, H., McCay, H., & Winstanley, M. (2020). Infantile B12 deficiency with severe thrombocytopenia-an under-recognised public health problem? *The New Zealand Medical Journal*, *133*(1516), 93–96.

Rao, V. R. (2018). Vitamin B12 deficiency presenting with hyperpigmentation and pancytopenia. *Journal of Family Medicine and Primary Care*, 7(3), 642–644. https://doi.org/10.4103/jfmpc.jfmpc_347_16.

Rao, S., Colon Hidalgo, D., Doria Medina Sanchez, J. A., Navarrete, D., & Berg, S. (2020). Et Tu, B12? Cobalamin deficiency masquerading as pseudo-thrombotic microangiopathy. *Cureus.*, *12*(7), e9097. https://doi.org/10.7759/cureus.9097.

Ray, J. G. (1998). Meta-analysis of hyperhomocysteinemia as a risk factor for venous thromboembolic disease. *Archives of Internal Medicine*, *158*(19), 2101–2106. https://doi.org/10.1001/archinte.158.19.2101.

Ray, J. G., Kearon, C., Yi, Q., et al. (2007). Homocysteine-lowering therapy and risk for venous thromboembolism: A randomized trial. *Annals of Internal Medicine*, *146*(11), 761–767. https://doi.org/10.7326/0003-4819-146-11-200706050-00157.

Sabry, W., Elemary, M., Burnouf, T., Seghatchian, J., & Goubran, H. (2020). Vitamin B12 deficiency and metabolism-mediated thrombotic microangiopathy (MM-TMA). *Transfusion and Apheresis Science*, *59*(1), 102717. https://doi.org/10.1016/j.transci.2019.102717.

Sharma, A. P., Greenberg, C. R., Prasad, A. N., & Prasad, C. (2007). Hemolytic uremic syndrome (HUS) secondary to cobalamin C (cblC) disorder. *Pediatric Nephrology*, *22*(12), 2097–2103. https://doi.org/10.1007/s00467-007-0604-1.

Tran, P. N., & Tran, M. H. (2018). Cobalamin deficiency presenting with thrombotic microangiopathy (TMA) features a systematic review. *Transfusion and Apheresis Science*, *57*(1), 102–106. https://doi.org/10.1016/j.transci.2018.01.003.

Tsai, H. M. (2013). Untying the knot of thrombotic thrombocytopenic purpura and atypical hemolytic uremic syndrome. *The American Journal of Medicine*, *126*, 200–209. https://doi.org/10.1016/j.amjmed.2012.09.006.

Van Hove, J. L., Van Damme-Lombaerts, R., & Grünewald, S. (2002). Cobalamin disorder Cbl-C presenting with late-onset thrombotic microangiopathy. *American Journal of Medical Genetics*, *111*(2), 195–201. https://doi.org/10.1002/ajmg.10499.

Wickramasinghe, S. N. (1995). Morphology, biology and biochemistry of cobalamin- and folate-deficient bone marrow cells. *Baillière's Clinical Haematology*, *8*(3), 441–459. https://doi.org/10.1016/s0950-3536(05)80215-x.

Yamanishi, M., Koba, S., Jo, T., Kotera, T., & Imashuku, S. (2018). Thrombotic microangiopathy-like hemolysis in Vitamin B12 deficiency-related macrocytic Anemia. *Clinical Laboratory*, *64*(4), 639–643. https://doi.org/10.7754/Clin.Lab.2017.171138.

Zheng, X. L., Vesely, S. K., Cataland, S. R., Coppo, P., Geldziler, B., Iorio, A., et al. (2020). ISTH guidelines for the diagnosis of thrombotic thrombocytopenic purpura. *Journal of Thrombosis and Haemostasis*, *18*(10), 2486–2495. https://doi.org/10.1111/jth.15006.

Zuber, J., Fakhouri, F., Roumenina, L. T., Loirat, C., & Frémeaux-Bacchi, V. (2012). Use of eculizumab for atypical haemolytic uraemic syndrome and C3 glomerulopathies. *Nature Reviews. Nephrology*, *8*(11), 643–657. https://doi.org/10.1038/nrneph.2012.214.

CHAPTER SEVENTEEN

Neuropsychiatric manifestations in vitamin B12 deficiency

Prashant Sahu[a], Harish Thippeswamy[b], and Santosh K. Chaturvedi[c,*]
[a]Prospect Park Hospital, Berkshire Healthcare NHS Foundation Trust, Reading, Berkshire, United Kingdom
[b]National Institute of Mental Health & Neurosciences, Bangalore, India
[c]Leicestershire Partnership NHS Trust, Leicester, United Kingdom
*Corresponding author: e-mail address: skchatur@yahoo.com

Contents

Abstract

Vitamin B12 deficiency can have distressing neuropsychiatric symptoms. It can have an etiological role in clinical presentations like depression, anxiety, psychosis, dementia, and delirium, requiring screening of at-risk populations. Few mechanisms that underlie the neuropsychiatric manifestations of B12 deficiency include alteration in one-carbon metabolism, genetic vulnerability, and alteration in folate metabolism. Maintaining a high serum B12 level in elderly can be protective against Alzheimer's disease (AD). In an established AD, its deficiency is associated with higher cognitive decline and risk for delirium. The other mental changes associated with B12 deficiency include apathy, agitation, impaired concentration, insomnia, persecutory delusions, auditory and visual hallucinations, and disorganized thought-process. Besides serum vitamin B12, plasma methylmalonic acid (MMA) and homocysteine helps in diagnosis. The chapter focuses on early recognition and effective treatment of these neuropsychiatric manifestations of vitamin B12 deficiency.

Vitamins and Hormones, Volume 119
ISSN 0083-6729
https://doi.org/10.1016/bs.vh.2022.01.001

1. Vitamin B12 and neuropsychiatry

Vitamin B12 is linked to a wide range of symptoms. Though the classical presentation of B12 deficiency is hematological, it is important to note that neuropsychiatric manifestations could be the initial symptom of a developing B12 deficiency syndrome or sometimes the sole manifestation of B12 deficiency. Over the last century, several manifestations related to neuropsychiatry have been reported concerning B12 deficiency. It is interesting to note that patients admitted to mental institutions responded well to specific nutritional therapy that included B vitamins (Jolliffe, 1950; Spies, Cogswell, & Vilter, 1944). Psychiatric manifestation with comorbid subclinical deficiency of B12 among elderly, those with a lack of family history of psychiatric illness, and vegans respond well following B12 supplement (Young, Pipingas, White, Gauci, & Scholey, 2019).

2. Neurobiology of vitamin B12 related neuropsychiatric manifestations

Vitamin B12 is an essential cofactor for enzymatic reactions involved in one-carbon (1-C) metabolism. Vitamin B12 deficiency impairs the enzymatic functions of *L*-methyl-malonyl–coenzyme A mutase and methionine synthase leading to elevated plasma levels of accumulation of methylmalonic acid (MMA) and homocysteine, respectively. Hence, the elevated MMA and homocysteine concentration in the peripheral blood could be used as an early and sensitive marker of Vitamin B12 deficiency (Snow, 1999). B12 deficiency can lead to an impairment in folate-mediated one-carbon metabolism on account of methionine synthase dysfunction. Cellular functioning appears to be affected both directly and indirectly due to B12 deficiency. The accumulation of homocysteine, lack of availability of S adenosylmethionine for critical cellular functions including gene expression, and accumulation of 5-methyltetrahydrofolate leading to impairments in the synthesis of purines and thymidylate that are necessary for DNA synthesis are some of the important mechanisms that underlie the clinical manifestations arising due to B12 deficiency (Sugden, 2006).

3. Neuropsychiatric manifestations of vitamin B12 deficiency

As early as the early 20th century, studies have reported psychiatric symptoms in patients with pernicious anemia that is associated with B12 deficiency (Langdon, 1905). The wide range of vitamin B12 related neuropsychiatric symptoms include cognitive deficits, changes in mood, psychotic symptoms, and acute confusional states. An early recognition of the vitamin B12 deficiency followed by adequate supplementation of B12 often is clinically rewarding as patients are likely to recover from the neuropsychiatric symptoms (Lindenbaum et al., 1995.)

3.1 Cognitive impairments including dementia and delirium

The risk of developing Alzheimer's disease (AD) increases among the cognitively intact elderly who are deficient in vitamin B12 levels (Hsu, Huang, Lo, Wang, & Tu, 2016). The incidence of low vitamin B12 levels among AD is reported to be 29% to 47% highlighting the importance of screening and correction of B12 deficiency (Cole & Prchal, 1984; Karnaze & Carmel, 1987). B12 deficiency-related neurodegenerative changes in the brain are evidenced by impaired myelin integrity and related brain volume reduction (Hsu et al., 2016). B12 deficiency directly and indirectly by causing hyperhomocysteinemia leads to reduced concentration of S-adenosylmethionine. By altering the activity of DNA methylases (DNMTs), demethylation of DNA is induced leading to over-expression of genes associated with Alzheimer's disease such as APP, BACE1, and PSEN1 (Marques & Outeiro, 2013). Additionally, there seems to be an association between B12 deficiency with elevated plasma homocysteine levels and the onset of dementia (Health Quality Ontario, 2013). Hyperhomocysteinemia contributes to cerebral microangiopathic changes (Faßbender et al., 1999). It is still uncertain whether B12 deficiency would result in a unique profile of neuropsychological deficits among patients with mild cognitive impairment (Silva et al., 2013) or AD (Larner, Janssen, Cipolotti, & Rossor, 1999) but patients with low vitamin B12 levels appear to experience greater levels of cognitive impairment (Whyte, 2002). There is a direct association between low vitamin B12 concentration and AD, irrespective of plasma homocysteine and folate levels (Chen et al., 2015).

A diet rich in B12 or its supplementation may be protective against AD in a healthy elderly (Mielech, Puścion-Jakubik, Markiewicz-żukowska, & Socha, 2020). There have been case reports of supplementation of B12 improving cognitive functions in mild AD with a history of symptoms less than 2years (Lopes Rocha, Flavia de Morao, & Barbosa, 1995). Vitamin B12 may exhibit immunomodulatory effects on cytokines leading to an improvement in psychotic symptoms in AD (Vakilian et al., 2017). Once a B12 deficiency state is identified, oral vitamin B12 can be as effective as parenteral vitamin B12 (Health Quality Ontario, 2013). However, the RCTs and meta-analysis could not establish the effectiveness of vitamin B12 supplementation in preventing the progression of a mild cognitive impairment (MCI) to dementia (Mccleery et al., 2018) or for improving the cognitive symptoms in an MCI (Markun et al., 2021).

Delirium is characterized by an acute confusional state with fluctuating levels of consciousness, attention deficits, behavioral changes. Vitamin B12 deficiency has been associated with acute cognitive changes with altered mental sensorium (Samson, Swisher, Christian, & Engel, 1952). B12 deficiency can predispose to delirium especially in the presence of risk factors like infection, anemia, older age, and dementia (Buchman, Mendelsson, Lerner, & Kotler, 1999; Lerner & Kanevsky, 2002). Patients with mild-to-moderate dementia with low vitamin b12 levels are at a significantly reduced risk for delirium following B12 supplementation (Kwok, Lee, Lam, & Woo, 2008).

3.2 Depression and anxiety

Poor diet may have a role in the manifestation of depression mediated by serum levels of folate and vitamin B12 (Khosravi, Sotoudeh, Amini, et al., 2020). B12 deficiency can often present with depression (Shulman, 1967), and patients with depression in turn manifest vitamin B12 deficiency (Hutto, 1997). Vitamin B12 deficiency leads to impaired 1-carbon metabolism and hyperhomocysteinemia predisposing to depression (Tiemeier et al., 2002). The elevated levels of homocysteine could be posing a greater risk for depression and anxiety, especially among susceptible individuals (Saraswathy, Ansari, Kaur, Joshi, & Chandel, 2019). Apathy and somnolence could be a manifestation of anemia associated with vitamin B12 deficiency (Shulman, 1967) and therefore a clinician needs to focus on cognitive symptoms to diagnose depression in a patient with low levels of B12. Mental changes due to B12 deficiency can manifest with agitation, irritability, forgetfulness, amnesia, impaired concentration and attention, and insomnia (Holmes, 1956). Supplementation of B vitamins in healthy and at-risk populations appears to have a positive effect on mood and stress but the evidence is not robust and there is a need for further research (Young et al., 2019). In contrast, in a diagnosed depression with B12 deficiency, supplementing B12 along with antidepressant treatments appears to be better than treating with antidepressants alone (Syed, Wasay, & Awan, 2013). Overall, clinical outcomes appear to be better among patients with major depression who have higher baseline serum B12 levels (Hintikka, Tolmunen, Tanskanen, & Viinamäki, 2003).

3.3 Psychosis

B12 deficiency can manifest as psychosis alone or can present as depression with psychosis (Jayaram et al., 2013). B12 associated psychosis could manifest as suspiciousness, persecutory or religious delusions, auditory and visual hallucinations, tangential or incoherent speech, and disorganization in the thought process (Hutto, 1997). B12 deficiency can sometimes present with a fluctuating clinical picture characterized by affective changes, anxiety and restlessness, alteration in sensorium which could be collectively termed as atypical psychosis (Phillips & Kahaner, 1988). Therapeutically, folate plus vitamin B12 supplementation could improve negative symptoms of schizophrenia but response rates vary across individuals (Roffman et al., 2013).

3.4 Somatoform and dissociative disorder

B12 deficient status can present as non-specific sensorimotor symptoms of numbness and deep aching/burning pain in the legs or feet as confirmed by nerve conduction studies (Leishear et al., 2012). B12 deficient states can also manifest as conversion disorder (Ozdemir, Gur, Cokluk, Isik, & Tapan, 2020). In a study from South India, it was found that 94 out of 125 patients presenting with vague ill-defined sensory symptoms were having deficiency of neurotropic vitamins like Vitamin B12 (Keshavan, Isaac, & Kapur, 1980). This deficiency can cause somatosensory symptoms which can be misinterpreted as a cause of a medical disorder, for which no organic cause is found on routine investigations, and are labeled as medically unexplained symptoms or somatoform symptoms.

3.5 Alcohol abuse

Individuals who are dependent on alcohol are particularly vulnerable to nutritional deficiencies. B12 deficiency could be due to a nutritionally deficient diet, poor intestinal absorption, excessive urinary flushing, and altered B12 metabolism could explain the vitamin deficiency in alcoholism (Hoyumpa, 1986). While interpreting B12 levels in patients with alcohol dependence, it is important to be aware that higher serum vitamin B12 levels correlate with liver damage (Himmerich, 2001). The paradoxical higher serum levels could be due limited ability of the damaged liver to absorb and metabolize cobalamin from the serum (Green, 1983; Kanazawa & Herbert, 1985) or by leakage of vitamin B12 from the liver into the circulation due to defective binding and storage (Baker, Leevy, DeAngelis, Frank, & Baker, 1998). Hence, alcoholism predisposes to Vitamin B12 deficiency which may be masked by high normal or elevated serum B12 levels (Himmerich, 2001). Hence, in alcohol dependence, clinicians should correlate serum B12 levels with elevated liver enzymes and also consider plasma MMA and homocysteine to be more sensitive for B12 deficiency.

3.6 Other rare manifestations

The rarer manifestations of B12 deficiency include mania and obsessive-compulsive disorders (OCD) through the association between these manifestations and B12 deficiency is not clear (Upadhyaya & Sharma, 2012). Patients with OCD can have comorbid B12 deficiency (Hermesh, Weizman, Shahar, & Munitz, 1988). There is a serotonergic dysfunction in OCD and B12 is a cofactor for serotonin synthesis (Tables 1 and 2).

Table 1 Neuropsychiatric manifestations in vitamin B12 deficiency.

Memory loss/confabulation/cognitive slowness
Depression/mood instability/irritability/rarely maniacal aggression
Delusions
Hallucinations-visual and auditory
Apathy
Delirium
Dementia

Table 2 Vitamin B12 deficiency presenting as a Psychiatric manifestation.

Author, years	Type of study	Age/sex	Presentations	Other features	Serum. B12 level	Management
Kimoto et al. (2020)	Case report	55/M	Auditory hallucinations, persecutory delusions with delirium.	Dysstasia, Incontinence Renal dysfunction and anemia.	190 pg/mL	Intravenous methylcobalamin 1000 μg alternate days for a week followed by oral administration
Rannelli, Watterson, Pandya, and Leung (2014)	Case report	55/M	Irritability and emotional lability, difficulty with self-care, grooming, and personal hygiene, paranoid, delusional, circumstantial grandiose thoughts and impaired insight.	History of repeated falls, postural dizziness, progressive fatigue, generalized weakness, and significant weight loss	89 pg/mL	Daily subcutaneous injections of vitamin B12 at 1000 μg for 1 week, followed by weekly injections for the next 4 weeks, and subsequently monthly injections continued indefinitely
Jayaram et al. (2013)	Case series	–	Psychosis Depression Bipolar disorder	No physical comorbidity; vegetarians	B12 level <225 pg/mL—78.9%	Intramuscular vitamin B12 injections
Kibirige, Wekesa, Kaddu-Mukasa, and Waiswa (2013)	Case report	44/F	Uncoordinated speech, aggressiveness, reduced sleep, wandering tendency, poor concentration, and visual hallucinations mainly worsening at night	Generalized weakness, intense paraesthesia of the lower limbs, palpitations, and exertional dyspnoea	62 pg/mL	Intramuscular B12 1000 μg daily for 1 week followed by 1000 μg monthly with oral iron and folate therapy for a week.
Tufan, Bilici, Usta, and Erdoğan (2012)	Case report	16/F	Mixed affective state	Glossitis, Ataxia, Cogwheel rigidity, Cerebellar signs Romberg's sign	166 pg/mL	Risperidone 0.5 mg/day and intramuscular Vitamin B12 500 μg/day

Continued

Table 2 Vitamin B12 deficiency presenting as a Psychiatric manifestation.—cont'd

Author, years	Type of study	Age/sex	Presentations	Other features	Serum. B12 level	Management
Sahoo, Avasthi, and Singh (2011)	Case report	27/M	Forgetfulness, social withdrawal, paucity of speech, decreased interest in pleasurable activities, and apathy	Reduced motor strength and the reflexes were symmetrical diminished in the lower extremities with a positive Babinski sign.	50 pg/mL	Intramuscular vitamin B12 injections
Durand, Mary, Brazo, and Dollfus (2003)	Case report	64/F	Severe depression with delusion and Capgras' syndrome, delusion with lability of mood and hypomania successively	Comorbid Hypertension Diabetes	208 pg/mL	Intramuscular Hydroxocobalamin 1000 mg/day for 10 days
Goebels and Soyka (2000)	Case report	64/M	Cognitive impairment, social withdrawal, lack of appetite with significant weight loss; reduced drive, spontaneous movement, and facial expression	Diminished tendon reflexes and sense of posture and vibration of the legs, emphysematous lungs, CT scan and MRI of the brain showed a generalized cerebral atrophy	307 pg/mL	Substitution of vitamin B12 with doxepin and lorazepam
Goebels and Soyka (2000)	Case report	77/M	Delusion of impoverishment, disoriented with depressed mood	Reduced visual acuity, reduced ankle jerks, diminished sense of posture and vibration. CT scan of the brain showed generalized atrophy	203 pg/mL	Vitamin B12 supplement with clomipramine and lorazepam
Catalano, Catalano, Rosenberg, Embi, and Embi (1998)	Case report	59/M	Catatonia	Presented to emergency	1.07 pg/mL	B12 supplementation

4. Treatment of neuropsychiatric manifestation due to vitamin B12 deficiency

Serum B12 estimation is a commonly used laboratory investigation to assess B12 deficiency. Clinicians need to be aware that patients with clinical symptoms of cobalamin deficiency may have serum levels of B12 within the reference range (Devalia, Hamilton, & Molloy, 2014). The World Health Organization (WHO) Expert Committee on biological standardization suggests an International standard for serum vitamin B 12 at 480 pg/mL (Thorpe et al., 2007). It is suggested that psychiatric symptoms can present before neurological or hematological consequences and therefore the threshold should be 660 pg/L (Berry, Sagar, & Tripathi, 2003). However, only serum vitamin B12 may not adequately reflect cellular B12 levels (Lysne et al., 2016). Serum total homocysteine(tHcy) could be a useful screening tool and a value >9 μM indicates the need for further assessment. (Schedvin, Jones, Hultdin, & Nilsson, 2005). Plasma holo-TC (cut off <73 pmol/L) and MMA (cut off >152 nmol/L) are additional biomarkers to assess functional B12 deficiency (Campos et al., 2020; Devalia et al., 2014; Herrmann & Obeid, 2008). Oral vitamin B12 can be as effective as parenteral vitamin B12 (Chatterjee, Yapundich, Palmer, Marson, & Mitchell, 1996). However, the intramuscular (IM) route is preferable for the rapid and adequate correction of B12 deficiency (Lachner, Steinle, & Regenold, 2012). The recommendation for IM dose of crystalline cyanocobalamin ranges from 125 μg/day to 2000 μg/day (Andrès et al., 2007; Vidal-Alaball et al., 2005). A dose of 1000 μg/day is commonly practiced, initially, 1000 μg/day IM for 1 week, then weekly for 1 month, and then monthly thereafter (Andrès et al., 2007). The British National Formulary recommends 1000 μg hydroxy cobalamin intramuscularly (IM) three times a week for 2 weeks to patients without neurological involvement, and 1000 μg i.m. on alternate days to patients with neurological symptoms until there is no further improvement (Charkin, 2014). The NICE guideline advises oral cyanocobalamin 50–150 micrograms daily or hydroxocobalamin 1 mg injection twice yearly; whereas in vegans treatment may be needed life-long (National Institute for Health and Care Excellence, 2019). In patients with pernicious anemia or gastric bypass surgery, intrinsic factor deficiency causes poor absorption of B12. However, a higher oral dose at greater than 350 μg/day can be tried which can completely saturate intestinal B12 receptors. If vitamin B12 insufficiency persists, a parenteral route (IM) of vitamin B12 dose at 1000 μg of vitamin B12 monthly or 1000–3000 μg every

6–12 months is imperative (Ziegler, Sirveaux, Brunaud, Reibel, & Quilliot, 2009). It is to be noted that patients with severe anemia can develop hypokalemia during B12 replacement (Goebels & Soyka, 2000). Plasma vitamin B12 and homocysteine levels should be repeated after 4-to-6 weeks to assess for response followed by serum B12 levels every 6-to-12 months (Lachner et al., 2012).

5. Conclusion

The deficiency of Vitamin B12 can masquerade as a psychiatric presentation. It is essential to identify the underlying cause as psychiatric manifestation could be the initial and only manifestation of B12 deficiency. The elderly, vegetarian diet, chronic mental illness, and alcohol abuse are common predisposing factors for B12 deficiency. In an established psychiatric disorder, it is advisable to maintain the serum B12 levels at a higher normal range. Homocysteine and MMA levels would help interpret the B12 status in the body. Complete restoration of B12 may require long-term treatment. Study of vitamin B12 levels and its relationship with psychiatric symptoms and disorders opens up avenues to revisit the etiopathogenesis of various psychiatric disorders and their management.

References

Andrès, E., Federici, L., Affenberger, S., Vidal-Alaball, J., Loukili, N. H., Zimmer, J., et al. (2007). B12 deficiency: A look beyond pernicious anemia. *Journal of Family Practice*, *56*(7), 537–542.

Campos, A. J., Risch, L., Nydegger, U., Wiesner, J., Van Dyck, M. V., Renz, H., et al. (2020). Diagnostic accuracy of holotranscobalamin, vitamin b12, methylmalonic acid, and homocysteine in detecting B12 deficiency in a large, mixed patient population. *Disease Markers*, *2020*. Article ID 7468506, 11 pages https://doi.org/10.1155/2020/7468506.

Baker, H., Leevy, C. B., DeAngelis, B., Frank, O., & Baker, E. R. (1998). Cobalamin (vitamin B12) and holotranscobalamin changes in plasma and liver tissue in alcoholics with liver disease. *Journal of the American College of Nutrition*, *17*, 235–238.

Berry, N., Sagar, R., & Tripathi, B. M. (2003). Catatonia and other psychiatric symptoms with vitamin B12 deficiency. *Acta Psychiatrica Scandinavica*, *108*(2). https://doi.org/10.1034/j.1600-0447.2003.00089.x.

Buchman, N., Mendelsson, E., Lerner, V., & Kotler, M. (1999). Delirium associated with vitamin B12 deficiency after pneumonia. *Clinical Neuropharmacology*, *22*(6).

Catalano, G., Catalano, M. C., Rosenberg, E. I., Embi, P. J., & Embi, C. S. (1998). Catatonia: Another neuropsychiatric presentation of vitamin B12 deficiency? *Psychosomatics*, *39*(5). https://doi.org/10.1016/S0033-3182(98)71307-6.

Charkin, S. (2014). BNF recommends hydroxocobalamin rather than cyanocobalamin for vitamin B12 deficiency. *BMJ (Online)*, *349*. https://doi.org/10.1136/bmj.g5389.

Chatterjee, A., Yapundich, R., Palmer, C. A., Marson, D. C., & Mitchell, G. W. (1996). Leukoencephalopathy associated with cobalamin deficiency. *Neurology*, *46*(3). https://doi.org/10.1212/WNL.46.3.832.

Chen, H., Liu, S., Ji, L., Wu, T., Ma, F., Ji, Y., et al. (2015). Associations between Alzheimer's disease and blood homocysteine, vitamin B12, and folate: A case-control study. *Current Alzheimer Research*, *12*(1), 88–94. https://doi.org/10.2174/1567205012666141218144035. PMID: 25523421.

Cole, M. G., & Prchal, J. F. (1984). Low serum vitamin B12 in alzheimer-type dementia. *Age and Ageing*, *13*(2). https://doi.org/10.1093/ageing/13.2.101.

Devalia, V., Hamilton, M. S., & Molloy, A. M. (2014). Guidelines for the diagnosis and treatment of cobalamin and folate disorders. *British Journal of Haematology*, *166*(4). https://doi.org/10.1111/bjh.12959.

Durand, C., Mary, S., Brazo, P., & Dollfus, S. (2003). Psychiatric manifestations of vitamin B12 deficiency: A case report. *L'Encephale*, *29*(6).

Faßbender, K., Mielke, O., Bertsch, T., Nafe, B., Fröschen, S., & Hennerici, M. (1999). Homocysteine in cerebral macroangiography and microangiopathy. *The Lancet*, *353*-(9164). https://doi.org/10.1016/S0140-6736(99)00309-8.

Goebels, N., & Soyka, M. (2000). Dementia associated with vitamin B12 deficiency: Presentation of two cases and review of the literature. *Journal of Neuropsychiatry and Clinical Neurosciences*, *12*(3). https://doi.org/10.1176/jnp.12.3.389.

Green, P. H. (1983). Alcohol, nutrition and malabsorption. *Clinics in Gastroenterology*, *12*(2), 563–574.

Health Quality Ontario. (2013). Vitamin B12 and cognitive function: An evidence-based analysis. *Ontario Health Technology Assessment Series*, *13*(23), 1–45.

Hermesh, H., Weizman, A., Shahar, A., & Munitz, H. (1988). Vitamin B12 and folic acid serum levels in obsessive compulsive disorder. *Acta Psychiatrica Scandinavica*, *78*-(1). https://doi.org/10.1111/j.1600-0447.1988.tb06294.x.

Herrmann, W., & Obeid, R. (2008). Causes and early diagnosis of vitamin B12 deficiency. *Deutsches Ärzteblatt International*, *105*(40), 680–685. https://doi.org/10.3238/arztebl.2008.0680.

Himmerich, H. (2001). Vitamin B12 and hepatic enzyme serum levels correlate in male alcohol-dependent patients. *Alcohol and Alcoholism*, *36*(1). https://doi.org/10.1093/alcalc/36.1.26.

Hintikka, J., Tolmunen, T., Tanskanen, A., & Viinamäki, H. (2003). High vitamin B12 level and good treatment outcome may be associated in major depressive disorder. *BMC Psychiatry*, *3*(1). https://doi.org/10.1186/1471-244X-3-17.

Holmes, J. M. D. (1956). Cerebral manifestations of vitamin-B12 deficiency. *British Medical Journal*, *2*(5006). https://doi.org/10.1136/bmj.2.5006.1394.

Hoyumpa, A. M. (1986). Mechanisms of vitamin deficiencies in alcoholism. *Alcoholism: Clinical and Experimental Research*, *10*(6). https://doi.org/10.1111/j.1530-0277.1986.tb05147.x.

Hsu, Y. H., Huang, C. F., Lo, C. P., Wang, T. L., & Tu, M. C. (2016). Vitamin B12 deficiency: Characterization of psychometrics and MRI morphometrics. *Nutritional Neuroscience*, *19*(2). https://doi.org/10.1179/1476830515Y.0000000045.

Hutto, B. R. (1997). Folate and cobalamin in psychiatric illness. *Comprehensive Psychiatry*, *38*-(6). https://doi.org/10.1016/S0010-440X(97)90925-1.

Jayaram, N., Rao, M. G., Narasimha, A., Raveendranathan, D., Varambally, S., Venkatasubramanian, G., et al. (2013). Vitamin B12 levels and psychiatric symptomatology: A case series. *Journal of Neuropsychiatry and Clinical Neurosciences*, *25*(2). https://doi.org/10.1176/appi.neuropsych.12060144.

Jolliffe, N. T. F. F. C. (1950). *Clinical nutrition*. Hoeber.

Kanazawa, S., & Herbert, V. (1985). Total corrinoid, cobalamin (vitamin B12), and cobalamin analogue levels may be normal in serum despite cobalamin in liver depletion in patients with alcoholism. *Laboratory Investigation*, *53*, 108–110.

Karnaze, D. S., & Carmel, R. (1987). Low serum cobalamin levels in primary degenerative dementia: Do some patients harbor atypical cobalamin deficiency states? *Archives of Internal Medicine, 147*(3). https://doi.org/10.1001/archinte.1987.00370030033007.

Keshavan, M. S., Isaac, M., & Kapur, R. L. (1980). I11-defined somatic symptoms in a South Indian rural clinic. Some preliminary clinical observations. *Tropical and Geographical Medicine, 32*(2), 163–168.

Khosravi, M., Sotoudeh, G., Amini, M., et al. (2020). The relationship between dietary patterns and depression mediated by serum levels of folate and vitamin B12. *BMC Psychiatry, 20*, 63. https://doi.org/10.1186/s12888-020-2455-2.

Kibirige, D., Wekesa, C., Kaddu-Mukasa, M., & Waiswa, M. (2013). Vitamin B12 deficiency presenting as an acute confusional state: A case report and review of literature. *African Health Sciences, 13*(3). https://doi.org/10.4314/ahs.v13i3.47.

Kimoto, K., Shibasaki, B., Tamura, N., Takahashi, Y., Maehara, M., Watanabe, N., et al. (2020). A case of vitamin B12 deficiency with various psychiatric symptoms and cognitive impairment. *The Primary Care Companion for CNS Disorders, 22*(1). https://doi.org/10.4088/PCC.19l02490.

Kwok, T., Lee, J., Lam, L., & Woo, J. (2008). Vitamin B12 supplementation did not improve cognition but reduced delirium in demented patients with vitamin B12 deficiency. *Archives of Gerontology and Geriatrics, 46*(3). https://doi.org/10.1016/j.archger.2007.05.001.

Rannelli, L., Watterson, R., Pandya, R., & Leung, A. A. (2014). Vitamin B12 deficiency with combined hematological and neuropsychiatric derangements: A case report. *Journal of Medical Case Reports, 8*(1).

Lachner, C., Steinle, N. I., & Regenold, W. T. (2012). The neuropsychiatry of vitamin B 12 deficiency in elderly patients. *The Journal of Neuropsychiatry and Clinical Neurosciences, 24*(1). https://doi.org/10.1176/appi.neuropsych.11020052.

Langdon, F. W. (1905). Nervous and mental manifestations of pre-pernicious anemia. *Journal of the American Medical Association, XLV*(22). https://doi.org/10.1001/jama.1905.52510220021001g.

Larner, A. J., Janssen, J. C., Cipolotti, L., & Rossor, M. N. (1999). Cognitive profile in dementia associated with vitamin B12 deficiency due to pernicious anaemia [3]. *Journal of Neurology, 246*(4). https://doi.org/10.1007/s004150050355.

Leishear, K., Boudreau, R. M., Studenski, S. A., Ferrucci, L., Rosano, C., Rekeneire, N., et al. (2012). Relationship between vitamin B12 and sensory and motor peripheral nerve function in older adults. *Journal of the American Geriatrics Society, 60*(6). https://doi.org/10.1111/j.1532-5415.2012.03998.x.

Lerner, V., & Kanevsky, M. (2002). Acute dementia with delirium due to vitamin B12 deficiency: A case report. *International Journal of Psychiatry in Medicine, 32*(2). https://doi.org/10.2190/C2MU-AHVC-3092-X2TB.

Lindenbaum, J., Healton, E. B., Savage, D. G., Brust, J. C., Garrett, T. J., Podell, E. R., et al. (1995). Neuropsychiatric disorders caused by cobalamin deficiency in the absence of anemia or macrocytosis. 1988. *Nutrition (Burbank, Los Angeles County, Calif.), 11*(2).

Lopes Rocha, F., Flavia de Morao, M., & Barbosa, M. T. (1995). Vitamin B12 deficiency and dementia. *International Psychogeriatrics,* 7(1). https://doi.org/10.1017/S1041610295001876.

Lysne, V., Strand, E., Svingen, G. F., Bjørndal, B., Pedersen, E. R., Midttun, Ø., et al. (2016). Peroxisome proliferator-activated receptor activation is associated with altered plasma one-carbon metabolites and B-vitamin status in rats. *Nutrients, 8*(1), 26. https://doi.org/10.3390/nu8010026. PMID: 26742069; PMCID: PMC4728640.

Markun, S., Gravestock, I., Jäger, L., Rosemann, T., Pichierri, G., & Burgstaller, J. M. (2021). Effects of vitamin B12 supplementation on cognitive function, depressive symptoms, and fatigue: A systematic review, meta-analysis, and meta-regression. *Nutrients, 13*(3). https://doi.org/10.3390/nu13030923.

Marques, S., & Outeiro, T. F. (2013). *Epigenetics in Parkinson's and Alzheimer's diseases* (pp. 507–525). Epigenetics: Development and Disease.

Mccleery, J., Abraham, R. P., Denton, D. A., Rutjes, A. W. S., Chong, L. Y., Al-Assaf, A. S., et al. (2018). Vitamin and mineral supplementation for preventing dementia or delaying cognitive decline in people with mild cognitive impairment. *Cochrane Database of Systematic Reviews, 2018*(11). https://doi.org/10.1002/14651858.CD011905.pub2.

Mielech, A., Puścion-Jakubik, A., Markiewicz-żukowska, R., & Socha, K. (2020). Vitamins in alzheimer's disease—Review of the latest reports. In. *Nutrients, 12*(11). https://doi.org/10.3390/nu12113458.

National Institute for Health and Care Excellence. (2019). B12 and Folate deficiency (Clinical Knowledge Summaries). Available at https://cks.nice.org.uk/anaemia-b12-and-folate-deficiency#!scenarioRecommendation.

Ozdemir, P., Gur, T., Cokluk, E., Isik, M., & Tapan, S. (2020). Vitamin B12, folate levels and somatoform dissociation in conversion disorder. *Journal of the Pakistan Medical Association, 0.* https://doi.org/10.5455/JPMA.31213.

Roffman, J. L., Lamberti, J. S., Achtyes, E., Macklin, E. A., Galendez, G. C., Raeke, L. H., et al. (2013). Randomized multicenter investigation of folate plus vitamin B12 supplementation in schizophrenia. *JAMA Psychiatry, 70*(5). https://doi.org/10.1001/jamapsychiatry.2013.900.

Phillips, S. L., & Kahaner, K. P. (1988). An unusual presentation of vitamin B12 deficiency. *American Journal of Psychiatry, 145*(4). https://doi.org/10.1176/ajp.145.4.529a.

Sahoo, M. K., Avasthi, A., & Singh, P. (2011). Negative symptoms presenting as neuropsychiatric manifestation of vitamin B12 deficiency. *Indian Journal of Psychiatry, 53*(4). https://doi.org/10.4103/0019-5545.91914.

Samson, D. C., Swisher, S. N., Christian, R. M., & Engel, G. L. (1952). Cerebral metabolic disturbance and delirium in pernicious anemia: Clinical and electroencephalographic studies. *A.M.A. Archives of Internal Medicine, 90*(1). https://doi.org/10.1001/archinte.1952.00240070010002.

Saraswathy, K. N., Ansari, S. N., Kaur, G., Joshi, P. C., & Chandel, S. (2019). Association of vitamin B12 mediated hyperhomocysteinemia with depression and anxiety disorder: A cross-sectional study among Bhil indigenous population of India. *Clinical Nutrition ESPEN, 30.* https://doi.org/10.1016/j.clnesp.2019.01.009.

Schedvin, G., Jones, I., Hultdin, J., & Nilsson, T. K. (2005). A laboratory algorithm with homocysteine as the primary parameter reduces the cost of investigation of folate and cobalamin deficiency. *Clinical Chemistry and Laboratory Medicine, 43*(10), 1065–1068. https://doi.org/10.1515/CCLM.2005.186 (PMID: 16197299).

Shulman, R. (1967). Psychiatric aspects of pernicious Anaemia: A prospective controlled investigation. *British Medical Journal, 3*(5560). https://doi.org/10.1136/bmj.3.5560.266.

Silva, D., Albers, U., Santana, I., Vicente, M., Martins, I. P., Verdelho, A., et al. (2013). Do MCI patients with vitamin B12 deficiency have distinctive cognitive deficits? *BMC Research Notes, 6*(1). https://doi.org/10.1186/1756-0500-6-357.

Snow, C. F. (1999). Laboratory diagnosis of vitamin B12 and folate deficiency: A guide for the primary care physician. In. *Archives of Internal Medicine, 159*(12). https://doi.org/10.1001/archinte.159.12.1289.

Spies, T. D., Cogswell, R. C., & Vilter, C. (1944). Detection and treatment of severe atypical deficiency disease. *Journal of the American Medical Association, 126*(12). https://doi.org/10.1001/jama.1944.02850470016005.

Sugden, C. (2006). One-carbon metabolism in psychiatric illness. *Nutrition Research Reviews, 19*(1). https://doi.org/10.1079/nrr2006119.

Syed, E. U., Wasay, M., & Awan, S. (2013). Vitamin B12 supplementation in treating major depressive disorder: A randomized controlled trial. *The Open Neurology Journal, 7*-(1). https://doi.org/10.2174/1874205X01307010044.

Thorpe, S. J., Heath, A., Blackmore, S., Lee, A., Hamilton, M., O'Broin, S., et al. (2007). International standard for serum vitamin B12 and serum folate: International collaborative study to evaluate a batch of lyophilised serum for B12 and folate content. *Clinical Chemical Laboratory Medicine, 45*(3). https://doi.org/10.1515/CCLM.2007.072.

Tiemeier, H., Ruud van Tuijl, H., Hofman, A., Meijer, J., Kiliaan, A. J., & Breteler, M. M. B. (2002). Vitamin B12, folate, and homocysteine in depression: The Rotterdam study. *American Journal of Psychiatry, 159*(12). https://doi.org/10.1176/appi.ajp.159.12.2099.

Tufan, A. E., Bilici, R., Usta, G., & Erdoğan, A. (2012). Mood disorder with mixed, psychotic features due to vitamin b12 deficiency in an adolescent: Case report. *Child and Adolescent Psychiatry and Mental Health, 6*. https://doi.org/10.1186/1753-2000-6-25.

Upadhyaya, S., & Sharma, A. (2012). Obsessive compulsive disorder due to B12 deficiency: Justification required. *Indian Journal of Psychological Medicine, 34*(3), 298. https://doi.org/10.4103/0253-7176.106045.

Vakilian, A., Razavi-Nasab, S. M., Ravari, A., Mirzaei, T., Moghadam-Ahmadi, A., Jalali, N., et al. (2017). Vitamin B_{12} in association with antipsychotic drugs can modulate the expression of pro −/anti-inflammatory cytokines in Alzheimer disease patients. *Neuroimmunomodulation, 24*(6). https://doi.org/10.1159/000486597.

Vidal-Alaball, J., Butler, C., Cannings-John, R., Goringe, A., Hood, K., McCaddon, A., et al. (2005). Oral vitamin B12 versus intramuscular vitamin B12 for vitamin B12 deficiency. *Cochrane Database of Systematic Reviews*. https://doi.org/10.1002/14651858.cd004655.pub2.

Whyte, E. M. (2002). Cognitive and behavioral correlates of low vitamin B12 levels in elderly patients with progressive dementia. *American Journal of Geriatric Psychiatry, 10*(3). https://doi.org/10.1176/appi.ajgp.10.3.321.

Young, L. M., Pipingas, A., White, D. J., Gauci, S., & Scholey, A. (2019). A systematic review and meta-analysis of b vitamin supplementation on depressive symptoms, anxiety, and stress: Effects on healthy and 'at-risk' individuals. In. *Nutrients, 11*(9). https://doi.org/10.3390/nu11092232.

Ziegler, O., Sirveaux, M. A., Brunaud, L., Reibel, N., & Quilliot, D. (2009). Medical follow up after bariatric surgery: Nutritional and drug issues. General recommendations for the prevention and treatment of nutritional deficiencies. *Diabetes & Metabolism, 35*(6 Pt 2), 544–557. https://doi.org/10.1016/S1262-3636(09)73464-0. 20152742.

CHAPTER EIGHTEEN

Vitamin B_{12} and predatory behavior in nematodes

Wen-Sui Lo and Ralf J. Sommer*
Max Planck Institute for Biology Tübingen, Tübingen, Germany
*Corresponding author: e-mail address: ralf.sommer@tuebingen.mpg.de

Contents

Abstract

The round worms or nematodes are the largest phylum of animals with an estimated species number of more than one million. Nematodes have invaded all ecosystems and are known from all continents including Antarctica. Parasitic species infest plants, animals and humans often with high host-specificity. Free-living species are known from marine, fresh water and soil systems, the latter of which contain many culturable species. This includes *Caenorhabditis elegans*, a species that was developed as one of the most prominent model systems in modern biology since the 1960ies. *Pristionchus pacificus* is a second nematode model organism that can easily be cultured in the laboratory. This species shows a number of complex traits including omnivorous feeding and the capability of predation on other nematodes. Predation depends on the formation of teeth-like denticles in the mouth of *P. pacificus*, structures unknown from *C. elegans* and most other nematodes. Here, we review the current knowledge about the role of vitamin B_{12} for the predatory behavior in *P. pacificus* and correlate its role with that on the physiology and development in *C. elegans*.

Vitamins and Hormones, Volume 119
ISSN 0083-6729
https://doi.org/10.1016/bs.vh.2022.01.006

1. Introduction: The ecological diversity of nematodes

Nematodes (roundworms) are the dominant phylum of animals in terms of diversity and abundance. They account for an estimated 80% of animal species and specimen on the planet and they are found in all ecosystems and on all continents (den Hoogen et al., 2019). As parasites of plants, animals and livestock they are a major threat for agriculture and animal breeding. In humans, there are around 30 parasites, several of which can cause severe health damage and even death. Most parasites have complicated life cycles with free-living and parasitic stages and in some species, different developmental stages are specialized for distinct hosts requiring complex transitions during ontogeny. Given this diversity of ecological strategies, introductory texts in nematology are often lengthy but some authors have tried to provide short introductions for general and student audiences (Weischer & Brown, 2000).

In soil communities, free-living nematodes are part of all trophic levels and thus, a key component of the food web (Lazarova, Coyne, Rodriguez, Peteira, & Ciancio, 2021). Nonetheless, their cylindrical body shape is surprisingly uniform with strong specializations largely restricted to the head region reflecting different feeding structures and feeding strategies (Yeates, Bongers, De Goede, Freckman, & Georgieva, 1993). Food sources for free-living nematodes include bacteria, algae and fungi, but also other nematodes in predatory, or hemolymph, blood and other host tissues in parasitic species (Bongers & Bongers, 1998). In some nematodes, larval stages and adults have different food requirements and again others are omnivores. Unfortunately, the complete details of feeding, digestion, nutrient uptake, and metabolism are not known for any species. Thus, the evolution of physiological and metabolic differentiation in nematodes remains currently unknown. Fortunately, however, the capability to culture certain nematodes indefinitely under laboratory conditions is providing increasing information about the molecular biology, genetics and biochemistry of these organisms. In the last decade, such work has also begun to cover nutrient uptake and detailed investigations of metabolism. Therefore, this contribution only covers the two model organisms *Caenorhabditis elegans* and *Pristionchus pacificus* with their simple life cycle and indefinite propagation in the lab (Brenner, 1974; Sommer, Carta, Kim, & Sternberg, 1996). We will largely focus on *P. pacificus* and its derived predatory feeding behavior that is influenced by vitamin B_{12}.

2. *Caenorhabditis elegans*—A model system approach for modern biology

C. elegans is often considered a "shooting star" among model organisms in modern biology. While the fruit fly *Drosophila melanogaster* or the fungi *Neurospora crassa* and *Saccharomyces cerevisiae* are subject to intense research for more than a century, credit is usually given to Sydney Brenner who started working on *C. elegans* in the 1960ies and published his first major paper in 1974. However, it is important to note that culturing *C. elegans* under laboratory conditions was a prerequisite for Brenner and was established through seminal work by Nigon (1943) and Dougherty, Hansen, Nicholas, Mollett, and Yarwood (1959). Together, these and following studies have provided detailed information of the nutritional requirements of *C. elegans*, which can also be applied to many other, though not all, free-living nematodes. Specifically, worms are grown on agar plates with so-called "Nematode Growth Medium" (NGM), using a monoxenic diet of *Escherichia coli* (Stiernagle, 2006). Usually, the uracil auxotroph and thus, slow growing *E. coli* strain OP50 is used as *C. elegans* food source because it allows easy visualization of worms on the agar plate. Importantly, a sterol source has to be externally provided to continuously grow nematode cultures, which is usually done by adding cholesterol. Chemically defined media for axenic growth are also available (Lu & Goetsch, 1993), but are only used for specific experiments as cultures grow slowly.

Indefinite culture of *C. elegans* with a simple diet and a cheap agar plate set up opened new research avenues in developmental biology, neurobiology and genetics. Thereby, several key features of *C. elegans* made the establishment of a new model organism possible in the first place. (1) The life cycle of *C. elegans* is completed in three days (20 °C) under the culture conditions described above. (2) *C. elegans* propagates as self-fertilizing hermaphrodites. That is, animals are "males" during a short period of post-embryogenesis and form around 200–300 sperm (dependent on the genotype and the exact type of bacterial diet). These sperm are used to fertilize oocytes that are formed in the adult stage. If a standard 6 cm plate is inoculated with a single *C. elegans* larva, two generations of worms can complete their life cycle within one week and thousands of genetically identical individuals are formed that will use up the bacterial food. Transfer of worms to new plates is easy and routinely performed using a dissection scope. (3) *C. elegans* cultures and mutant strains can be frozen in liquid nitrogen and thereby, stored indefinitely.

The aforementioned criteria made *C. elegans* a superb model in basically all areas of modern biology. Most importantly, fast and powerful forward genetic screening and the elucidation of the complete cell lineage provided researchers with unprecedented systemic insight (Sulston & Horvitz, 1977). Reverse genetic tools, first through RNA interference and now through CRISPR/Cas9-based engineering allow the precise manipulation of the *C. elegans* genome (Fire et al., 1998). Notably, *C. elegans* was the first multicellular organism to have its genome sequence known in its entirety (The C. elegans Genome Sequencing Consortium, 1998). We refer the interested reader to the various summary articles that provide an overview about current knowledge of *C. elegans* at WormBook and WormBase.

3. *Pristionchus pacificus*—A model for complex trait formation

P. pacificus was one of the first nematodes to be developed as a "satellite model" to *C. elegans* for detailed evolutionary comparisons (Fig. 1A) (Marx, 2021; Sommer & McGaughran, 2013). First described in 1996, this species shares many technical features with *C. elegans* (Sommer et al., 1996). (1) rapid growth under laboratory conditions using the same agar plate set up (Fig. 1B), (2) hermaphroditic mode of reproduction enabling the easy implementation of forward and reverse genetic tools (Han et al., 2020; Sommer & Sternberg, 1996), and (3) long-term storage in liquid nitrogen. Additionally, following the original sequencing of the *P. pacificus* genome (Dieterich et al., 2008; Rödelsperger et al., 2017), multiple *-omics* platforms have been established, including proteomics (Borchert et al., 2010), transcriptomics (Rödelsperger et al., 2018; Sun, Rödelsperger, & Sommer, 2021), epigenomics (Werner et al., 2018) and phylogenomics (Prabh et al., 2018). Various developmental processes and their molecular specification mechanisms have been investigated in great detail, resulting in the general notion of "developmental systems drift" (Wang & Sommer, 2011). This term describes the observation that homologous developmental processes in different species, such as *C. elegans* and *P. pacificus*, are regulated by different molecular mechanisms. For example, the formation of the vulva, the egg-laying system of nematode females and hermaphrodites, is a homologous structure that is formed by homologous cells in both species. Nonetheless, the associated molecular processes differ fundamentally with EGF/Ras signaling inducing vulva formation in *C. elegans*, whereas Wnt signaling having the corresponding function in *P. pacificus*. Importantly, such

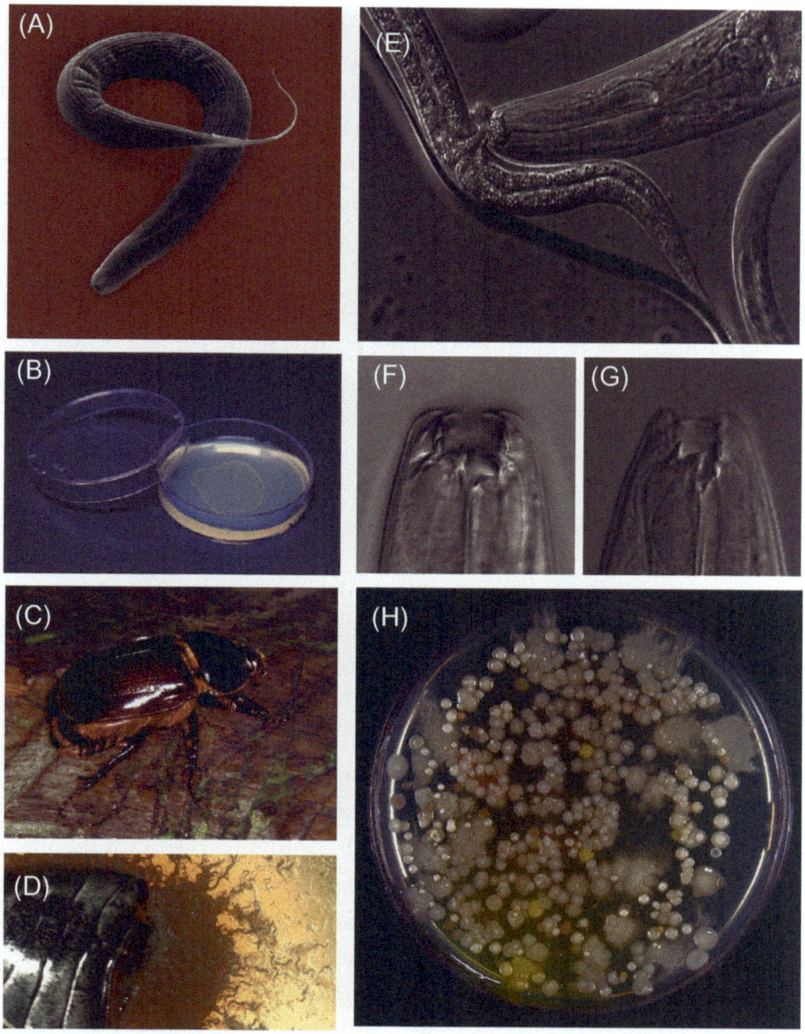

Fig. 1 The ecology of *Pristionchus pacificus*. (A) Scanning electron micrograph of the 1 mm long *P. pacificus*. (B) A 6 cm agar plate with a lawn of *E. coli* as food for nematodes. (C) The rhinoceros beetle *Oryctes borbonicus* from La Réunion Island, on which P. *pacificus* is frequently found. (D) Worms inhabit a beetle cadaver with thousands of animals flourishing one week after the beetle´s death. (E) predation of *P. pacificus* against a *C. elegans* larval prey. (F) The eurystomatous mouth form with a wide buccal cavity and two teeth (only one is clearly visible in this plane of focus). (G) The stenostomatous mouth form with a narrow buccal cavity and a single tooth. (H) Diversity of bacteria cultured from a beetle cadaver.

developmental systems drift occurs in the absence of major genomic alterations because the majority of signaling pathways, *i.e.* EGF/Ras, Wnt or Insulin signaling, are all highly conserved between *P. pacificus* and *C. elegans* with most genes in these pathways being 1:1 orthologous (Pires-daSilva & Sommer, 2003). Several developmental processes have been studied and compared in detail in these two nematodes providing comprehensive insight into the evolution of developmental mechanisms (for review see, Sommer & McGaughran, 2013).

In the last decade, research on *P. pacificus* has shifted towards the formation and evolution of novel and complex traits. This worm is well-suited for such studies because it belongs to a family of nematodes, the Diplogastridae, that shows a number of striking life history traits unknown from *C. elegans* and most other free-living species. *P. pacificus* can be found in soil but is often associated with insects (Fürst von Lieven & Sudhaus, 2000; Susoy, Ragsdale, Kanzaki, & Sommer, 2015). For example, the original isolate of *P. pacificus* PS312 from Pasadena (CA, USA) was indeed found in a soil sample (Sommer et al., 1996). However, the most reliable source for these nematodes are scarab beetles, namely chock chafers, stag beetles and dung beetles (Fig. 1C). By now, thousands of strains of *P. pacificus* are in culture from around the world resulting in this species being thought to have originated in Asia. Indeed, a large diversity of strains has been reported from Japan (Herrmann et al., 2007), but more recent studies also found *P. pacificus* and many more *Pristionchus* species in China (Kanzaki et al., 2021).

The association of *P. pacificus* with scarab beetles is critical for the life history traits to be discussed below. *Pristionchus* nematodes associated with the adult insect are exclusively found in an arrested developmental stage, called dauer larvae (Renahan et al., 2021). Only when the beetle dies, worms leave the dauer stage, become adult and feed on the developing microbes on the beetle carcass (Fig. 1D). However, the natural death of adult scarab beetles is exclusively in the soil, which has prevented an easy access to carcasses. Several recent investigations aim to understand the succession and dynamics of *Pristionchus* nematodes with scarab beetles in soil settings (Meyer et al., 2017). These studies are complicated by the fact that only adult scarab beetles live above ground, whereas the rest of the life cycle, from the egg to the pupal stage, occurs below ground and is hard to follow. However, decapitation of beetles and placement in specific cages in the wild and in the laboratory have recently provided a first overview on *P. pacificus* population density, succession and dispersal. Interestingly, worms show a biphasic boom-and-bust strategy at the beetle carcass associated with a rapid dispersal to find new beetle hosts (Renahan et al., 2021). The dispersal is restricted to the arrested dauer stage, a specialized larval stage that is induced by several stress factors including starvation (Ogawa, Streit, Antebi, & Sommer, 2009).

4. *P. pacificus* predation, cannibalism and phenotypic plasticity

The association with scarab beetles has initiated intense studies of various life history traits. First, several studies investigated the chemo-attraction mechanisms involved in *Pristionchus* finding the appropriate beetle hosts. Surprisingly, chemosensation studies revealed that *Pristionchus* species exhibit strong differences in their chemoattractive profiles (Hong & Sommer, 2006). When the species-specific beetle cues were finally identified they included beetle sex and aggregation pheromones, suggesting that the nematodes used a hitchhiking strategy relying on insect communication. For example, *P. maupasi* was attracted to phenol, the sex pheromone of one of its major hosts, the European cockchafer *Melolontha melolontha*, whereas other *Pristionchus* species were not attracted by phenol (Hong, Svatos, Herrmann, & Sommer, 2008). More recent studies have focused on the mechanisms of beetle pheromone recognition in *P. pacificus* (Cinkornpumin et al., 2014; Renahan & Hong, 2017) and the associated neuronal circuitry (Hong et al., 2019).

The most striking feature in *Pristionchus* nematodes is their predation behavior, an example of a novel and complex trait (Fig. 1E) (Wilecki, Lightfoot, Susoy, & Sommer, 2015). While *P. pacificus* can grow and propagate on bacteria indefinitely, it can supplement its diet by feeding on fungi, yeast and other nematodes. Predatory feeding involves changes in pharyngeal pumping and requires the neurotransmitter serotonin (Ishita, Chihara, & Okumura, 2021; Okumura, Wilecki, & Sommer, 2017; Wilecki et al., 2015). This novel feeding behavior is based on a morphological novelty, the formation of teeth-like denticles in its mouth (Fig. 1F and G) (Bento, Ogawa, & Sommer, 2010). Most importantly, these denticles occur in two discrete forms resulting in alternative phenotypes: worms either develop a narrow mouth with a single tooth (called "stenostomatous," St; Fig. 1G) or alternatively, a wide mouth with two teeth (called "eurystomatous," Eu; Fig. 1F) (Ragsdale, Müller, Roedelsperger, & Sommer, 2013). This developmental decision is executed in late larval development and is irreversible once the worm becomes adult. The ability to form two alternative mouth forms represents an example of phenotypic plasticity and highlights the importance of environmental influences on developmental decision making. Environmental influences can easily be manipulated in the laboratory: if *P. pacificus* is grown on agar plates, most strains are preferentially Eu. In contrast, when grown in liquid culture with the similar *E. coli* diet they are preferentially St (Werner et al., 2017). Thus, mouth-form plasticity represents a trait that can easily be manipulated by genetic and environmental factors allowing detailed insight in associated molecular factors.

Most research in the last decade has focused on the molecular mechanisms and the evolutionary and ecological significance of mouth-form plasticity (Ragsdale et al., 2013; Susoy et al., 2015, 2016). Specifically, genetic and molecular studies revealed developmental switches that sense environmental inputs (Ragsdale et al., 2013; Sieriebriennikov et al., 2018), which control a downstream gene regulatory network (GRN) that guides the execution of one of the alternative forms (Bui & Ragsdale, 2019; Sieriebriennikov et al., 2020). These molecular insights allow detailed investigations of the evolutionary principles associated with the origin of morphological novelties (Casasa, Biddle, Koutsovoulos, & Ragsdale, 2020). The entirety of mouth-form plasticity research in *P. pacificus* cannot be completed summarized here and we refer the interested reader to recent reviews (Sommer, 2020; Sommer et al., 2017).

Finally, the predatory behavior has to be seen in the light of potential cannibalism. Interestingly, although nematodes are the most speciose animal phylum no organismal self-recognition had been described in nematodes. Testing for predation between different *Pristionchus* species and different *P. pacificus* strains revealed substantial aggression and killing. In contrast, each tested strain was able to distinguish self from non-self, indicating the existence of a precise self-recognition system (Lightfoot et al., 2019). The first molecule involved in *P. pacificus* self-recognition was the small peptide SELF-1, an approximately 60 amino acid long peptide with a hypervariable C-terminus. Changing individual amino acids in this C-terminus will eliminate the recognition of self and will cause killing (Lightfoot et al., 2019). Together, mouth-form plasticity, predation and self-recognition represent a set of complex traits in *P. pacificus* that are unknown from *C. elegans* and allow a detailed analysis of the underlying genetic and molecular mechanisms and associated evolutionary and ecological principles. Below, we will summarize current knowledge of the role of vitamin B_{12} on predatory behavior in *P. pacificus*.

5. The role of vitamin B_{12} for the predatory behavior in *P. pacificus*

Akduman and co-workers started an unbiased investigation of the dietary influence on various predatory-related traits in *P. pacificus* (Fig. 1H) (Akduman et al., 2020). Specifically, they tested a collection of bacteria isolated from *P. pacificus*-associated environments, primarily deriving from soil and scarab beetle samples (Akduman, Rödelsperger, & Sommer, 2018). The standard *P. pacificus* wild type strain PS312 was grown on these bacteria and after two generations predation-associated traits were investigated. For that, predation was first analyzed in so-called "corpse assays," in which five predators were offered an access of prey for 2 hours and afterwards, the number of dead corpses was counted under the dissection scope (Fig. 2). For a more

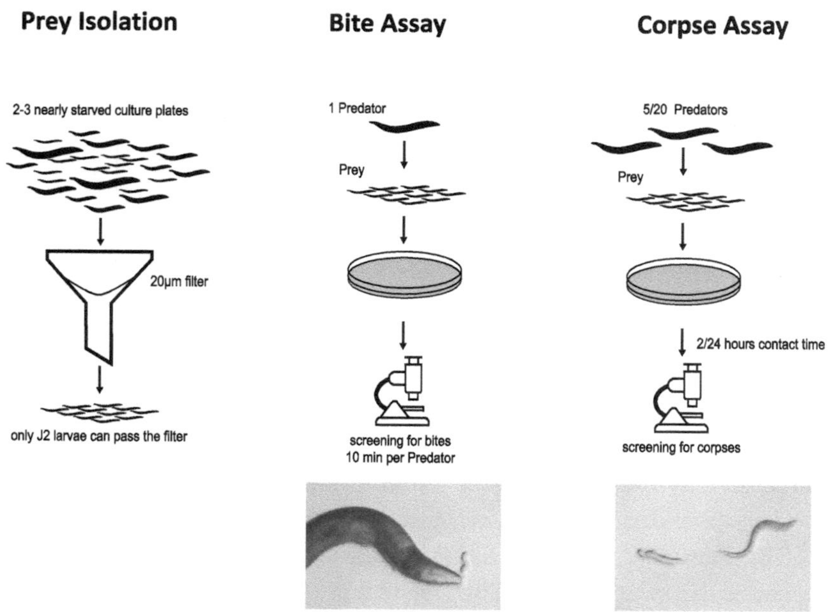

Fig. 2 Nematode predation assay. Left: Prey animals are isolated from several starved culture plates through filtering with only young juveniles passing the filter. These larvae are ideal prey items in predation assays. Middle: Bite assay with one predator being observed for 10 min under a dissecting scope for attacks against an access of prey. Right: Corpse assay with either 5 or 20 predators attacking for 2 or 24 24 h and corpses being counted at the end of the experiment. *Modified from Lightfoot, J. W., Wilecki, M., Roedelsperger, C., Moreno, E., Susoy, V., Witte, H., & Sommer, R. J. (2019). Small peptide mediated nematode self-recognition prevents cannibalism.* Science, 364, *86–89.*

direct observation, "bite assays" were performed by directly observing single predators for 10 minutes (Fig. 2). That allowed predation to be broken down into three distinct events: bites, successful bites and feeding, respectively. It is important to note that the switch from bacterial to predatory feeding is associated with a strong reduction of pharyngeal pumping (Wilecki et al., 2015). In this analysis, "bites" refer to the reduction of pharyngeal pumping rhythms of the predator that are however, not successful to kill a prey victim. In contrast, "successful bites" are those that result in the simultaneous rupture of the prey cuticle. Importantly, not all successful bites result in subsequent feeding. Therefore, feeding has to be independently categorized.

Bacterial diet has a strong influence on killing efficiency resulting in a fourfold difference of corpses (Akduman et al., 2020). For example, *P. pacificus* grown on a standard *E. coli* OP50 diet results in around 80 corpses in standard corpse assays (Fig. 3A). In contrast, worms grown on *Hafnia* LRB96 or *Chrysobacterium* LRB11 only produced around 40 corpses, whereas a *Novosphingobium* diet resulted in up to 160 corpses in a 2h-time period (Fig. 3A). But the story is more complicated than that. Follow-up investigations using bite assays revealed that higher successful biting and killing will not necessarily translate into higher feeding rates. When the predation activity of *P. pacificus* was analyzed over a 10-minute period for *E. coli* or *Novosphingobium*-derived worms, bites and successful bites were strongly increased on the latter (Fig. 3A). In contrast, the amount of active feeding remained unaltered and *Novosphingobium*-derived worms rapidly moved away from the carcass actively searching for new prey (Akduman et al., 2020). This phenomenon is strikingly similar to "surplus killing," a phenomenon that has been described from multiple animals. Surplus killing is known from wolves, bears but also spiders and insects (Kruuk, 2009). However, the causation of surplus killing is subject of active debate for decades.

To identify the mechanisms causing increased killing efficiency of *Novoshingobium*-derived *P. pacificus,* a dietary sensor approach has been used (Fig. 3B). Transcriptomic analysis of *P. pacificus* identified differentially expressed genes upon diet switching and allowed building a *Novosphingobium*-induced and an *E. coli*-induced reporter line. These lines were tested for expression on a *Novosphingobium* library of more than 4,000 mutants. Several *Novosphingobium* mutants were found to abolish the induction of the *Novosphingobium* dietary sensor in *P. pacificus* and several of these mutants affected vitamin B_{12} biosynthesis (Fig. 4) (Akduman et al., 2020). Indeed, *Novosphingobium* mutants defective in vitamin B_{12} biosynthesis did not induce more bites and successfully bites (Fig. 4A and B). However, vitamin B_{12} supplementation to a *Novosphingobium* vitamin

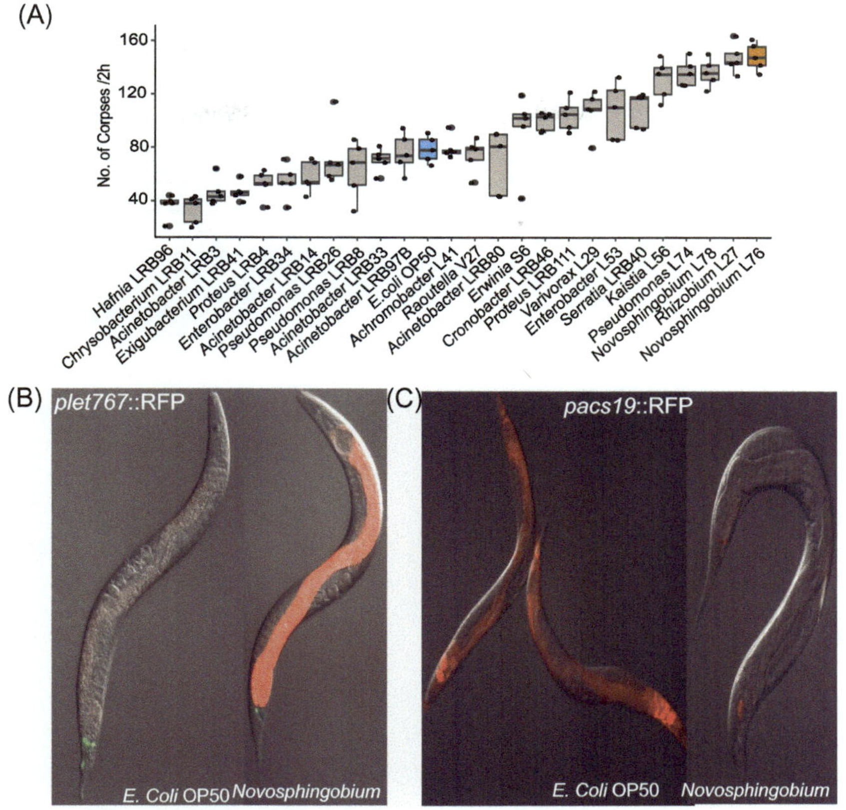

Fig. 3 The influence of diet on predation. (A) Culturing *P. pacificus* on various environment-derived bacteria heavily influences nematode predation as indicated here by corpse assays. (B) An expression construct of the gene *plet767* represents a dietary sensor that is only expressed when worms are cultured on *Novosphingobium* but not on *E. coli*. (C) The gene *pacs19* represents a sensor that is only expressed on an *E. coli* diet. *Modified from Akduman, N., Roedelsperger, C., Lightfoot, J. W., Roeseler, W., Witte, H., Lo, W.-S., & Sommer, R. J. (2020). Bacterial diet-derived vitamin B12 enhances nematode predation and surplus-killing.* The ISME Journal, *10.1038/s41396-020-0626-2.*

B_{12} biosynthesis mutant restored the original killing efficiency. Furthermore, when worms grown on an *E. coli* culture that was supplemented with vitamin B_{12} were tested, they also showed higher killing efficiencies (Fig. 4A and B). Together, these experiments indicate that the higher killing efficiency of *P. pacificus* observed on worms grown on *Novosphingobium* requires vitamin B_{12}. Furthermore, they indicate that vitamin B_{12} is an important micronutrient that when added externally to a culture can affect complex predatory behavior in nematodes.

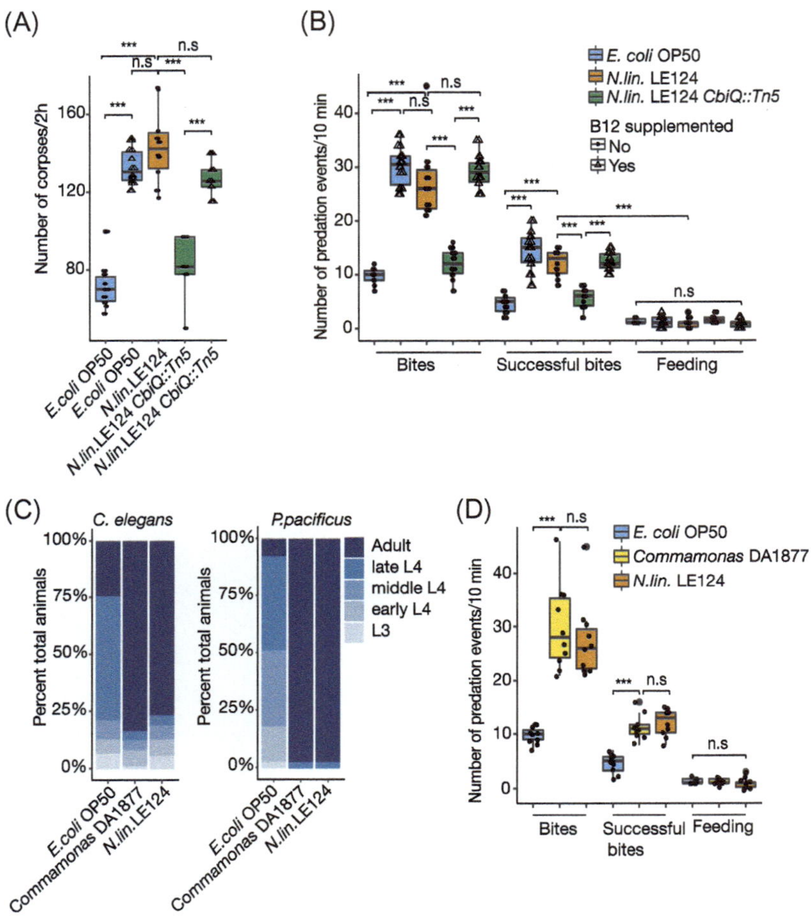

Fig. 4 Vitamin B_{12} influences predation and growth of *P. pacificus*. (A) A *Novosphingobium* diet (orange) induces a higher killing rate than an *E. coli* diet (blue). In a *Novosphingobium* mutant defective in vitamin B_{12} production (green), killing rate is largely reduced but can be enhanced by vitamin B_{12} supplementation to both, the mutant and the *E. coli* control. (B) A more careful analysis in form of bite assays indicates that while bites and successful bites are influenced by vitamin B_{12}, the actual feeding rate is vitamin B*12*-independent. (C) *Novosphingobium* and the vitamin B_{12} producing bacterium *Commamonas* DA1877 enhance the developmental speed and growth of *P. pacificus* and *C. elegans*. (D) A *Novosphingobium* and *Commamonas* diet have a similar effect on predation. *Modified from Akduman, N., Roedelsperger, C., Lightfoot, J. W., Roeseler, W., Witte, H., Lo, W.-S., & Sommer, R. J. (2020). Bacterial diet-derived vitamin B12 enhances nematode predation and surplus-killing.* The ISME Journal, *10.1038/s41396-020-0626-2*.

6. The role of vitamin B_{12} for the development and physiology of *P. pacificus* and *C. elegans*

The identification of vitamin B_{12} as an important micronutrient affecting predatory behavior was originally surprising, but not unprecedented. While new for a behavioral trait, previous work by Marian Walhout and co-workers had identified a strong influence of vitamin B_{12} on the speed of postembryonic development in *C. elegans* (MacNeil, Watson, Arda, Zhu, & Walhout, 2013; Watson et al., 2014). These authors showed that a *Commamonas* diet would accelerate development, have a reduced fecundity, and a shorter lifespan relative to *E. coli*-grown worms. These original studies have helped developing *C. elegans* as a powerful model for studying vitamin B_{12}-dependent processes (Yilmaz & Walhout, 2014). Subsequent investigations shed light on the molecular mechanisms of vitamin B_{12}-dependent processes in *C. elegans*. Like in humans, vitamin B_{12} is a co-factor for two enzymes in *C. elegans* (Watson et al., 2014). First, the methylmalonyl coenzyme A (CoA) mutase (MCM) is involved in the breakdown of the short fatty acid propionic acid. Second, methionine synthase (MS) uses vitamin B_{12} to convert homocysteine to methionine in the cytosolic methionine/*S*-adenosylmethionine (SAM) pathway. Interruptions of both pathways in humans causes devastating diseases and mutations in these pathways in *C. elegans* and *P. pacificus* allowed to elucidate the role of vitamin B_{12}. For example, a mutation of the *Cel-metr-1* in the SAM pathway will make *C. elegans* unresponsive to vitamin B_{12} and affects development and fertility (Watson et al., 2014). Similarly, in *P. pacificus* CRISPR/Cas9-derived mutations in *Ppa-metr-1* of the SAM pathway, but also *Ppa-mce-1* of the propionic acid breakdown pathway failed to respond to vitamin B_{12} supplementation and did not show enhanced killing efficiency indicating that both vitamin B_{12}-dependent pathways are important for the predatory behavior in *P. pacificus* (Akduman et al., 2020).

The two nematodes, like humans, must obtain vitamin B_{12} from their diet. *Commamonas* and *Novosphingobium* have pathways for vitamin B_{12} synthesis, whereas *E. coli* OP50 misses these synthesis capabilities. Indeed, *C. elegans* and *P. pacificus* when grown on a vitamin B_{12}-rich diet, such as *Commamonas* and *Novosphingobium,* develop more rapidly in comparison to *E. coli*-fed worms (Fig. 4C). Therefore, under standard laboratory culture conditions nematodes have to obtain vitamin B_{12} from the culture medium. While worms are not completely vitamin B_{12}-deficient on an *E. coli* OP50 diet, vitamin B_{12} is one of the limiting growth factors. Work by Bito and co-workers grew *C. elegans* in a peptone-free medium to create a complete vitamin B_{12}-deficient situation. Such animals had a reduced brood size, a reduced lifespan, were shorter in length, but had an extended life span (Bito, Matsunaga, Yabuta, Kawano, & Watanabe, 2013). Physiological studies linked these phenotypes to the altered activity of the aforementioned enzymatic pathways. These studies highlight the power of nematode genetics and they provided one of the first detailed analyses of a micronutrient in *C. elegans* and its influence of various life history traits (Yilmaz & Walhout, 2014). The complementary studies in *P. pacificus* show that also complex behavioral traits heavily dependent on the nutritional status (Akduman et al., 2020). Together, these nematode models can shed light on various aspects of micronutrient function that can through its systematic studies help identifying the effect and mechanisms of vitamin B_{12} but also other micronutrients.

7. Vitamins as limiting factors for nematode growth—A genomic perspective (from laboratory to natural habitats)

While vitamin B_{12} is a critical micronutrient that regulates DNA metabolic pathways to maintain the host genomic stability and tissue homeostasis, only a few archaea and bacteria have the ability to synthesize this molecule. The synthesis is a genetic and metabolic burden as the *de novo* synthesis of vitamin B_{12} requires at least 19 enzymatic steps to produce an active compound from uroporphyrinogen III (Martens, Barg, Warren, & Jahn, 2002). Consequently, in eukaryotes, vitamin B_{12} is acquired directly or indirectly from prokaryotes. In humans, the primary source of vitamin B_{12} is through the consumption of animal products, whose vitamin B_{12} is generated by the associated microbiota. Alternatively, vitamin B_{12} can be acquired through symbiotic interaction with bacteria. For example, in cicada genomic studies of its endosymbiont *Hodgkinia cicadicola* indicated that the ability to produce vitamin B_{12} is vital for maintaining the symbiosis (Łukasik et al., 2018; McCutcheon, McDonald, & Moran, 2009). In addition, several studies also indicate that the source of vitamin B_{12} for microalgae is through the direct interaction with bacteria (Croft, Lawrence, Raux-Deery, Warren, & Smith, 2005; Grossman, 2016).

In *C. elegans* and *P. pacificus*, while vitamin B_{12} can be obtained through the consumption of *Comamonas* and *Novosphingobium* in laboratory condition, how nematodes acquire vitamin B_{12}, and how vitamin B_{12} affect the metabolic state and fitness of nematodes in nature is largely unknown. The ecology of *C. elegans* and *P. pacificus* have begun to be discovered in recent years. They are mainly associated with the microbe-rich environments, such as rotting plant matter and decomposing beetles, respectively. The microbiome of the habitats has been characterized based on the 16S rDNA sequences (Berg et al., 2016; Dirksen et al., 2016; Meyer et al., 2017; Renahan et al., 2021; Samuel, Rowedder, Braendle, Félix, & Ruvkun, 2016). Yet, the function of the bacterial community cannot be inferred based on the taxonomic profiling. Nonetheless, several bacteria isolated from the gut of *C. elegans* have been sequenced. The whole genome sequence shows that *Pseudomonas lurida MYb11* and *Ochrobactrum vermis MYb71*, the members of the core gut microbiome, harbor genes for vitamin B_{12} synthesis (Dirksen et al., 2020; Zimmermann et al., 2020). Importantly, however, it is not clear if and how much these enzymes are expressed *in situ*. In addition, it is unknown if vitamin B_{12} is released into the environment, or if *C. elegans* needs to harvest vitamin B_{12} through cell lysis. Importantly, bacterial species who need vitamin B_{12} also depend on surrounding species to produce this nutrient. In human, only 25% of the bacteria in the gut microbiome have vitamin B_{12} synthesis genes, while 80% of bacteria encode B_{12}-dependent enzymes (Degnan, Barry, Mok, Taga, & Goodman, 2014). Therefore, complex host–bacterial interaction for vitamin B_{12} are presumably common in nature.

References

Akduman, N., Rödelsperger, C., & Sommer, R. J. (2018). Culture-based analysis of *Pristionchus*-associated microbiota from beetles and figs for studying nematode-bacterial interactions. *PLoS One*, *13*, e0198018.

Akduman, N., Roedelsperger, C., Lightfoot, J. W., Roeseler, W., Witte, H., Lo, W.-S., et al. (2020). Bacterial diet-derived vitamin B12 enhances nematode predation and surplus-killing. *The ISME Journal*. https://doi.org/10.1038/s41396-020-0626-2.

Bento, G., Ogawa, A., & Sommer, R. J. (2010). Co-option of the endocrine signaling module dafachronic acid-DAF-12 in nematode evolution. *Nature*, *466*, 494–497.

Berg, M., Stenuit, B., Ho, J., Wang, A., Parke, C., Knight, M., et al. (2016). Assembly of the Caenorhabditis elegans gut microbiota from diverse soil microbial environments. *The ISME Journal*, *10*, 1998–2009.

Bito, T., Matsunaga, Y., Yabuta, Y., Kawano, T., & Watanabe, F. (2013). Vitamin B12 deficiency in *Caenorhabditis elegans* results in loss of fertility, extended life cycle, and reduced lifespan. *Federation of European Biochemical Societies*, *3*, 112–117.

Bongers, T., & Bongers, M. (1998). Functional diversity of nematodes. *Applied Soil Ecology*, *10*, 239–251.

Borchert, N., Dieterich, C., Krug, K., Schütz, W., Jung, S., Nordheim, A., et al. (2010). Proteogenomics of *Pristionchus pacificus* reveals distinct proteome structure of nematode models. *Genome Research*, *20*, 837–846.

Brenner, S. (1974). The genetics of *Caenorhabditis elegans*. *Genetics*, *77*, 71–94.
Bui, L. T., & Ragsdale, E. J. (2019). Multiple plasticity regulators reveal targets specifying an induced predatory form in nematodes. *Molecular Biology and Evolution*, *36*, 2387–2399.
Casasa, S., Biddle, J. F., Koutsovoulos, G. D., & Ragsdale, E. J. (2020). Polyphenism of a novel trait integrated rapidly evolving genes into ancestrally plastic networks. *Molecular Biology and Evolution*, *38*, 331–343.
Cinkornpumin, J. K., Wisidagama, D. R., Rapoport, V., Go, J. L., Dieterich, C., Wang, X., et al. (2014). A host beetle pheromone regulates development and behavior in the nematode *Pristionchus pacificus*. *eLife*. https://doi.org/10.7554/eLife.0322.
Croft, M. T., Lawrence, A. D., Raux-Deery, E., Warren, M. J., & Smith, A. G. (2005). Algae acquire vitamin B12 through a symbiotic relationship with bacteria. *Nature*, *438*, 90–93.
Degnan, P. H., Barry, N. A., Mok, K. C., Taga, M. E., & Goodman, A. L. (2014). Human gut microbes use multiple transporters to distinguish vitamin B12 analogs and compete in the Gut. *Cell Host & Microbe*, *15*, 47–57.
den Hoogen, V., et al. (2019). Soil nematode abundance and functional group composition at a global scale. *Nature*, *572*, 174–198.
Dieterich, C., Clifton, S. W., Schuster, L., Chinwalla, A., Delehaunty, K., Dinkelacker, I., et al. (2008). The *Pristionchus pacificus* genome provides a unique perspective on nematode lifestyle and parasitism. *Nature Genetics*, *40*, 1193–1198.
Dirksen, P., Assié, A., Zimmermann, J., Zhang, F., Tietje, A.-M., Marsh, S. A., et al. (2020). CeMbio—The Caenorhabditis elegans microbiome resource. *G3 (Bethesda)*, *10*, 3025–3039.
Dirksen, P., Marsh, S. A., Braker, I., Heitland, N., Wagner, S., Nakad, R., et al. (2016). The native microbiome of the nematode *Caenorhabditis elegans*: Gateway to a new host-microbiome model. *BMC Biology*, *14*, 38.
Dougherty, E. C., Hansen, E. L., Nicholas, W. L., Mollett, J. A., & Yarwood, E. A. (1959). Axenic cultivation of Caenorhabditis briggsae (Nematoda: Rhabditidae) with unsupplemented and supplemented chemically defined media. *Annals of the New York Academy of Sciences*, *411*, 176–217.
Fire, A., Xu, S., Montgomery, M. K., Kostas, S. A., Driver, S. E., & Mello, C. C. (1998). Potent and specific genetic interference by double-stranded RNA in *Caenorhabditis elegans*. *Nature*, *391*, 806–811.
Fürst von Lieven, A., & Sudhaus, W. (2000). Comparative and functional morphology of the buccal cavity of Diplogastrina (Nematoda) and a first outline of the phylogeny of this taxon. *Journal of Zoological Systematics and Evolutionary Research*, *38*(1), 37–63.
Grossman, A. (2016). Nutrient acquisition: The generation of bioactive vitamin b12 by microalgae. *Current Biology*, *26*, R319–R321.
Han, Z., Lo, W.-S., Lightfoot, J. W., Witte, H., Sun, S., & Sommer, R. J. (2020). Improving trangenesis efficiency and CRISPR-associated tools through codon optimization and native intron addition in nematodes. *Genetics*, *216*, 947–956.
Herrmann, M., Mayer, W., Hong, R., Kienle, S., Minasaki, R., & Sommer, R. J. (2007). The nematode *Pristionchus pacificus* (Nematoda: Diplogastridae) is associated with the Oriental beetle *Exomala orientalis* (Coleoptera: Scarabaeidae) in Japan. *Zoological Science*, *24*, 883–889.
Hong, R. L., Riebesell, M., Bumbarger, D., Cook, S. J., Sarpolaki, T., Carstensen, H. R., et al. (2019). Evolution of neuronal anatomy and circuitry in highly divergent nematode species. *eLife*, *8*, e47155.
Hong, R. L., & Sommer, R. J. (2006). Chemoattraction in *Pristionchus* nematodes and implication for insect host recognition. *Current Biology*, *16*, 2359–2365.
Hong, R. L., Svatos, A., Herrmann, M., & Sommer, R. J. (2008). The species-specific recognition of beetle cues by *Pristionchus maupasi*. *Evolution & Development*, *10*, 273–279.

Ishita, Y., Chihara, T., & Okumura, M. (2021). Different combinations of serotonin receptors regulate predatory and bacterial feeding behaviors in the nematode Pristionchus pacificus. *G3 (Bethesda)*, 11. https://doi.org/10.1093/g3journal/jkab011.

Kanzaki, N., Herrmann, M., Weiler, C., Röseler, W., Theska, T., Berger, J., et al. (2021). Nine new *Pristionchus* (Nematoda: Diplogastridae) species from China. *Zootaxa, 4943*.

Kruuk, H. (2009). Surplus killing by carnivores. *Journal of Zoology, 166*, 233–244.

Lazarova, S., Coyne, D., Rodriguez, M. G., Peteira, B., & Ciancio, A. (2021). Functional diversity of soil nematodes in relation to the impact of agriculture. *Diversity, 13*, 64.

Lightfoot, J. W., Wilecki, M., Roedelsperger, C., Moreno, E., Susoy, V., Witte, H., et al. (2019). Small peptide mediated nematode self-recognition prevents cannibalism. *Science, 364*, 86–89.

Lu, N. C., & Goetsch, K. M. (1993). Carbohydrate requirement of *Caenorhabditis elegans* and the final development of a chemically defined medium. *Nematologica, 39*, 303–311.

Łukasik, P., Nazario, K., Van Leuven, J. T., Campbell, M. A., Meyer, M., Michalik, A., et al. (2018). Multiple origins of interdependent endosymbiotic complexes in a genus of cicadas. *Proceedings of the National Academy of Sciences of the United States of America, 115*, E226–E235.

MacNeil, L. T., Watson, E., Arda, H. E., Zhu, L. J., & Walhout, A. J. M. (2013). Diet-induced developmental acceleration independent of TOR and insulin in C. elegans. *Cell, 153*, 240–252.

Martens, J. H., Barg, H., Warren, M. J., & Jahn, D. (2002). Microbial production of vitamin B12. *Applied Microbiology and Biotechnology, 58*, 275–285.

Marx, V. (2021). Model organisms on roads less traveled. *Nature Methods, 18*(3), 235–239. https://doi.org/10.1038/s41592-021-01086-7.

McCutcheon, J. P., McDonald, B. R., & Moran, N. A. (2009). Convergent evolution of metabolic roles in bacterial co-symbionts of insects. *Proceedings of the National Academy of Sciences of the United States of America, 106*, 15394–15399.

Meyer, J. M., Baskaran, P., Quast, C., Susoy, V., Rödelsperger, C., Glöckner, F. O., et al. (2017). Succession and dynamics of *Pristionchus* nematodes and their microbiome during decomposition of *Oryctes borbonicus* on La Réunion Island. *Environmental Microbiology, 19*, 1476–1489.

Nigon, V. (1943). Le déterminisme du sexe chez un nématode libre hermaphrodite (*Rhabditis elegans* Maupas). *Comptes Rendus Societe Biologie, 137*, 40–41.

Ogawa, A., Streit, A., Antebi, A., & Sommer, R. J. (2009). A conserved endocrine mechanism controls the formation of dauer and infective larvae in nematodes. *Current Biology, 19*, 67–71.

Okumura, M., Wilecki, M., & Sommer, R. J. (2017). *Serotonin drives predatory feeding behavior via synchronous feeding rhytms in the nematode Pristionchus pacificus, G3 (Bethesda)*(7), 3745–3755.

Pires-daSilva, A., & Sommer, R. J. (2003). The evolution of signaling pathways in animal development. *Nature Reviews Genetics, 4*, 39–49.

Prabh, N., Roeseler, W., Witte, H., Eberhardt, G., Sommer, R. J., & Rödelsperger, C. (2018). Deep taxon sampling reveals the evolutionary dynamics of novel gene families in the Pristionchus genome. *Genome Research, 28*, 1664–1674.

Ragsdale, E., Müller, M., Roedelsperger, C., & Sommer, R. J. (2013). A developmental switch coupled to the evolution of plasticity acts through a sulfatase. *Cell, 155*, 922–933.

Renahan, T., & Hong, R. L. (2017). A species-specific nematocide that results in terminal embryogenesis. *The Journal of Experimental Biology, 220*, 3238–3247.

Renahan, T., Lo, W.-S., Werner, M. S., Herrmann, M., Rochat, J., & Sommer, R. J. (2021). Nematode biphasic 'boom and bust' dynamics are dependent on host bacterial load while linking dauer and mouth-form polyphenisms. *Environmental Microbiology*. in press.

Rödelsperger, C., Meyer, J. M., Prabh, N., Lanz, C., Bemm, F., & Sommer, R. J. (2017). Single-molecule sequencing reveals the chromosome-scale genomic architecture of the nematode model organism *Pristionchus pacificus*. *Cell Reports*, *21*, 834–844.

Rödelsperger, C., Röseler, W., Prabh, N., Yoshida, K., Weiler, C., Herrmann, M., et al. (2018). Phylotransciptomics of *Pristionchus* nematodes reveals parallel gene loss in six hermaphroditic lineages. *Current Biology*, *28*, 3123–3127.

Samuel, B. S., Rowedder, H., Braendle, C., Félix, M.-A., & Ruvkun, G. (2016). Caenorhabditis elegans responses to bacteria from its natural habitats. *Proceedings of the National Academy of Sciences of the United States of America*, *113*, E3941–E3949.

Sieriebriennikov, B., Prabh, N., Dardiry, M., Witte, H., Rödelsperger, C., Röseler, W., et al. (2018). A developmental switch regulating phenotypic plasticity is a part of a conserved multi-gene locus. *Cell Reports*, *23*, 2835–2843.

Sieriebriennikov, B., Sun, S., Lightfoot, J. W., Witte, H., Moreno, E., Rödelsperger, C., et al. (2020). Conserved hormone-receptors controlling a novel plastic trait target fast-evolving genes expressed in a single cell. *PLoS Genetics*, *16*, e1008687.

Sommer, R. J. (2020). Phenotypic plasticity: from theory and genetics to current and future challenges. *Genetics*, *215*, 1–13.

Sommer, R. J., Carta, L. K., Kim, S.-Y., & Sternberg, P. W. (1996). Morphological, genetic and molecular description of *Pristionchus pacificus* sp. n. (Nematoda, Diplogastridae). *Fundamental and Applied Nematology*, *19*, 511–521.

Sommer, R. J., Dardiry, M., Lenuzzi, M., Namdeo, S., Renahan, T., Sieriebriennikov, B., et al. (2017). The genetics of phenotypic plasticity in nematode feeding structures. *Open Biology*, 7, 160332.

Sommer, R. J., & McGaughran, A. (2013). The nematode *Pristionchus pacificus* as a model system for integrative studies in evolutionary biology. *Molecular Ecology*, *22*, 2380–2393.

Sommer, R. J., & Sternberg, P. W. (1996). Apoptosis limits the size of the vulval equivalence group in *Pristionchus pacificus*: A genetic analysis. *Current Biology*, *6*, 52–59.

Stiernagle, T. (2006). Maintenance of C. elegans WormBook. In *The C. elegans Research Community,* WormBook. https://doi.org/10.1895/wormbook.1.101.1.

Sulston, J. E., & Horvitz, H. R. (1977). Post-embryonic cell lineages of the nematode. *Caenorhabditis elegans*. *Developmental Biology*, *56*, 110–156.

Sun, S., Rödelsperger, C., & Sommer, R. J. (2021). Single worm transcriptomics identifies a developmental core network of oscillating genes with deep conservation across nematodes. *Genome Research*, *31*, 1590–1601.

Susoy, V., Herrmann, M., Kanzaki, N., Kruger, M., Nguyen, C. N., Rödelsperger, C., et al. (2016). Large-scale diversification without genetic isolation in nematode symbionts of figs. *Science Advances*, *2*, e1501031.

Susoy, V., Ragsdale, E. J., Kanzaki, N., & Sommer, R. J. (2015). Rapid diversification associated with a macroevolutionary pulse of developmental plasticity. *eLIFE*. https://doi.org/10.7554/eLife.05463.002.

The C. elegans Sequencing Consortium. (1998). Genome sequence of the nematode *C. elegans*: A platform for investigating biology. *Science*, *282*, 2012–2018.

Wang, X., & Sommer, R. J. (2011). Antagonism of LIN-17/Frizzled and LIN-18/Ryk in nematode vulva induction reveals evolutionary alterations in core developmental pathways. *PLoS Biology*, *9*, e1001110.

Watson, E., Macneil, L. T., Ritter, A. D., Yilmaz, L. S., Rosebrock, A. P., Caudy, A. A., et al. (2014). Interspecies systems biology uncovers metabolites affecting *C. elegans* gene expression and life history traits. *Cell*, *156*, 759–770.

Weischer, B., & Brown, D. J. F. (2000). *An introduction to nematodes*. Sofia: Pensoft.

Werner, M., Sieriebriennikov, B., Loschko, T., Namdeo, S., Lenuzzi, M., Renahan, T., et al. (2017). Environmental influence on *Pristionchus pacificus* mouth-form through different culture methods. *Scientific Reports*, 7, 7207.

Werner, M. S., Sieriebriennikov, B., Prabh, N., Loschko, T., Lanz, C., & Sommer, R. J. (2018). Young genes have distinct gene structure, epigenetic profiles, and transcription start site regulation. *Genome Research*, *28*, 1675–1687.

Wilecki, M., Lightfoot, J., Susoy, V., & Sommer, R. J. (2015). Predatory feeding behavior in *Pristionchus* nematodes is dependent on a phenotypic plasticity and induced by serotonin. *The Journal of Experimental Biology*, *218*, 1306–1313.

Yeates, G. W., Bongers, T., De Goede, R. G. M., Freckman, D. W., & Georgieva, S. S. (1993). Feeding habitats in soil nematode families and genera—An outline for soil ecologists. *Journal of Nematology*, *25*, 315–3331.

Yilmaz, L. S., & Walhout, M. (2014). Worms, bacteria, and micronutrients: An elegans model for our diet. *Trends in Genetics*, *30*, 496–503.

Zimmermann, J., Obeng, N., Yang, W., Pees, B., Petersen, C., Waschina, S., et al. (2020). The functional repertoire contained within the native microbiota of the model nematode Caenorhabditis elegans. *The ISME Journal*, *14*, 26–38.

CHAPTER NINETEEN

Determination of cobalamin and related compounds in foods

F. Watanabe[a,b,*], T. Bito[a,b], and K. Koseki[a]

[a]Division of Applied Bioresource Chemistry, United Graduate School of Agricultural Sciences, Tottori University, Tottori, Japan

[b]Department of Agricultural, Life and Environmental Sciences, Faculty of Agriculture, Tottori University, Tottori, Japan

*Corresponding author: e-mail address: watanabe@muses.tottori-u.ac.jp

Contents

Abstract

The microbiological assay of total cobalamin (vitamin B_{12}) by *Lactobacillus delbrueckii* subsp. *lactis* ATCC7830 is now used worldwide in food analysis because of its high sensitivity, low running cost, and no expensive instruments. It has been recently reported that some foods contain a substantial number of inactive corrinoid compounds, some of which are active in this bacterium. These results indicate that the microbiological method must be replaced with high-performance liquid chromatography or liquid chromatography/electrospray ionization-tandem mass spectrometry as there can specifically determine biologically active cobalamin. Nowadays, powerful tools, such as immunoaffinity columns, purify cobalamin simply and specifically. In this chapter, we

Vitamins and Hormones, Volume 119
ISSN 0083-6729
https://doi.org/10.1016/bs.vh.2022.01.009

summarized the determination methods of cobalamin and related compounds in foods. Various inactive corrinoids found in foods were also characterized.

1. Cobalamin

Vitamin B_{12} or cobalamin is one of the water-soluble vitamins. It has the largest molecular weight (1355.4) and the most complex structure among vitamins (Watanabe & Bito, 2016) and is generally known as a red vitamin. Vitamin B_{12} has a nucleotide including 5,6-dimethylbenzimidazole coordinated with the cobalt atom located at the corrin ring center (Fig. 1). Generally, the scientific use of the term "vitamin B_{12}" is restricted to cyanocobalamin (CN-Cbl); in this chapter, cobalamin (Cbl) refers to all potentially biologically active vitamin B_{12} compounds. CN-Cbl, hydroxocobalamin (OH-Cb), or sulfitocobalamin (SO_3-Cbl) is readily converted into methylcobalamin (CH_3-Cbl), a coenzyme of methionine synthase (EC 2.1.1.13) involved in methionine biosynthesis, and 5′-deoxyadenosylcobalamin (AdoCbl), a coenzyme of methylmalonyl-CoA mutase (EC 5.4.99.2) involved in amino acid and odd-chain fatty acid metabolism in mammalian cells (Chen, Crippen, Gulati, & Banerjee, 1994; Fenton, Hack, Willard, Gertler, & Rosenberg, 1982).

2. Cobalamin sources for humans

Cbl is biosynthesized only in some bacteria and archaea, and there is no Cbl biosynthesis systems and Cbl-dependent enzymes in higher plants (Watanabe & Bito, 2018a). Therefore, Cbl is an unusual vitamin that does not exist in plant-based foods such as vegetables, fruits, and grains. According to USA/Canada and Japan's dietary reference intakes, the recommended dietary allowance (RDA) for adult men and women is determined as 2.4 μg/day, the lowest value among all vitamin's RDA (Shibata et al., 2013). Milk is reportedly an important Cbl source for humans (Fedosov, Nexo, & Heegaard, 2019). Although cattle mainly feed on pasture that does not contain Cbl, various microorganisms, including Cbl-synthesizing bacteria, inhabit their ruminant stomach. Bacteria in the ruminant stomach synthesize Cbl, which is absorbed in the intestine, transferred to the blood, stored in the liver and muscle, and secreted into milk. The Cbl found in meat and dairy products, including cheese and yogurt, is derived from Cbl-synthesizing bacteria in the ruminant stomach

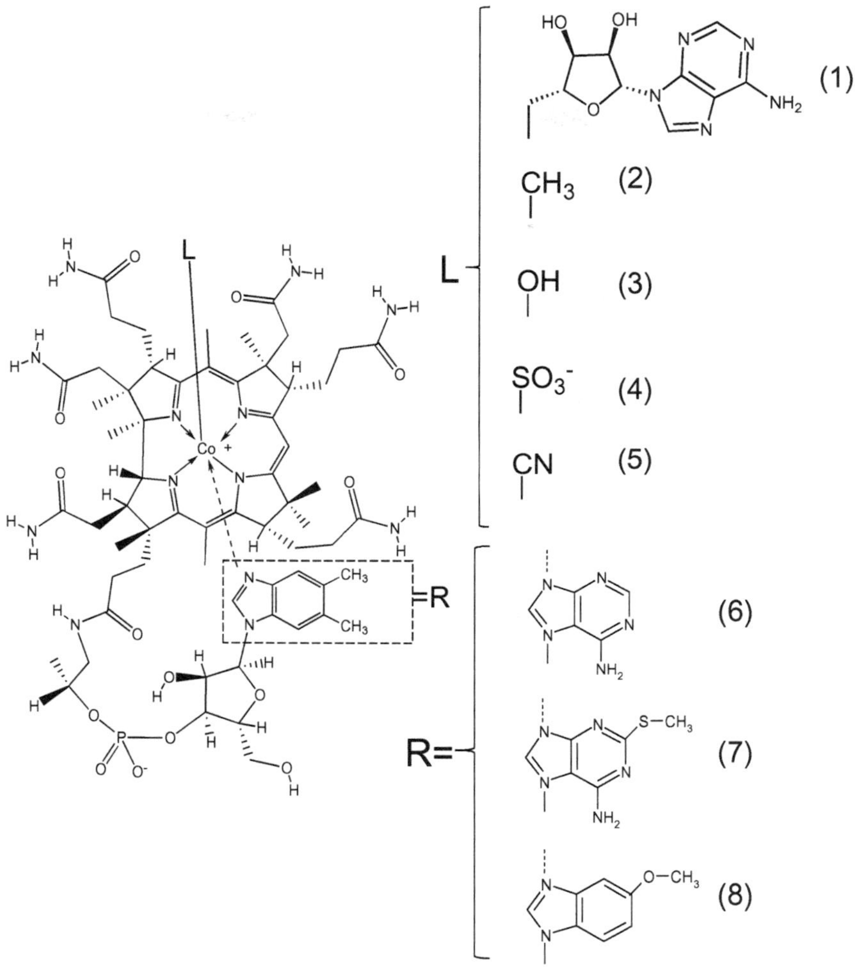

Fig. 1 The structural formula of cobalamins and related compounds found in foods. (1) 5′-Deoxyadenosylcobalamin, (2) Methylcobalamin, (3) Hydroxocobalamin, (4) Sulfitocobalamin, (5) Cyanocobalamin (or vitamin B_{12}), (6) Pseudovitamin B_{12} (or Factor IV), (7) 2-Methylmercaptoadenylcobamide (or Factor S), and (8) 5-Methoxybenzimidazolyl cobamide (or Factor IIIm).

(Kawashima, Henry, Ammerman, & Littell, 1997). Conversely, the Japanese and French obtain most of their Cbl intake from seafood such as fish and shellfish (Yoshino et al., 2005). Moreover, some edible algae contain a substantial amount of Cbl. About half of eukaryotic algae, including seaweed, do not require Cbl for growth but appear to take up Cbl from exogenous environments and utilize it for their cellular metabolism

(Croft, Lawrence, Raux-Deery, Warren, & Smith, 2005). Most Cbl present in the marine environment is reportedly synthesized by Thaumarchaeota inhabiting the deep sea (1000–200 m) (Heal et al., 2017). Thus, certain edible algae are a good source of Cbl for the Japanese (Bito, Teng, & Watanabe, 2020).

3. Determination of cobalamin in foods

A highly sensitive method is required to quantify Cbl in food (Watanabe & Bito, 2018b). According to AOAC Standard Method Performance Requirements for Quantitative Measurement of Vitamin B_{12} in Dietary Supplements and Ingredients (SMPR)® 2016.01747 (van Breemen et al., 2017), 0.001–1,000,000 ppm (100 ng/100 g–100 g/100 g) are required for Cbl analysis ranges. The quantification limit is $\leq$0.0005 ppm ($\leq$50 ng/100 g). In the case of 5 ppm (500 μg/100 g) or less that corresponds to food Cbl content, recovery (75–115%), repeatability relative standard deviation ($\leq$12%), and reproducibility relative standard deviation ($\leq$20%) are required. Our preliminary experiments indicated that a considerable high amount (approximately >20 g) of particular standard samples must be used by the high-performance liquid chromatography (HPLC) method to meet such quantification limit. The dilution factor of the sample to its extract strongly affects the quantification limit.

3.1 Extraction

Cbl exists as a free form that is not bound to proteins in vitamin supplements and drugs, but in foods and biological samples, Cbl is bound to enzymes and transport proteins (Andrès et al., 2005). Relatively high concentrations of free Cbl in pharmaceutical products can be directly quantified by HPLC (Wongyai, 2000), but in foods and biological samples, before quantitative analysis, Cbl must be liberated from the protein-Cbl complex and partially purified by various methods.

Various Cbl compounds, including the coenzyme forms, must be released from proteins and converted into a single stable form of either CN-Cbl or SO_3-Cbl. Boiling with potassium cyanide (KCN) at pH 4.0–4.5 for 30 min is usually used as a food Cbl extraction method (Bito et al., 2016). Treatment with protease or α-amylase (or both) before the boiling extraction to improve Cbl extraction efficiency has also been reported (Guggisberg, Risse, & Hadorn, 2012).

Sodium metabisulfite ($Na_2S_2O_5$) is used in the boiling extraction to avoid the use of toxic KCN. Various Cbl compounds are converted into SO_3-Cbl during $Na_2S_2O_5$ treatment. CN-Cbl is used as a standard Cbl compound in the AOAC microbiological quantification method since SO_3-Cbl has the same activity as CN-Cbl in quantitative bacteria growth (Muhammad, Briggs, & Jones, 1993). However, SO_3-Cbl is not commercially available. The KCN-boiling extraction method has a significant advantage in HPLC or liquid chromatography/electrospray ionization-tandem mass spectrometry (LC–MS/MS) analysis because commercially available CN-Cbl can be used as a standard.

3.2 Pretreatment and purification before cobalamin determination

In the case of the samples containing interference materials, high salt, or viscous substances, phenol extraction (Toohey & Barker, 1961), the column chromatography using Amberlite XAD-2 or -4 (Lamm, Höring, Renz, & Heckmann, 1980; Watanabe et al., 1999), solid-phase extraction (Bito et al., 2016) are used as pretreatment methods in Cbl extraction. Phenol extraction and Amberlite XAD-2 or -4 column chromatography had been used for large-scale purification of Cbl compounds from massive foods and biological samples. However, solid-phase extraction methods are generally used to determine Cbl using HPLC and LC/MS–MS.

Among various solid-phase extraction methods used to determine Cbl using HPLC and LC–MS/MS, Sep-Pak C18 cartridges were widely used. However, as far as we know, recovery of Cbl is considerably low (approximately 43%) because C18 resins were slightly leaked during the treatment with more than 75% (v/v) ethanol concentrations (Fig. 2). When the leaked resins are removed with a membrane filter, the recovery was significantly improved (approximately 93%) (our unpublished data). In advantages of these C18 cartridges, Cbl can be readily concentrated from diluted Cbl extracts and foods with low Cbl concentrations (Santos et al., 2007). Recently Oasis hydrophilic–lipophilic balance (HLB) cartridges (Waters Corp.) are used as a simplified solid phase extraction method before LC–MS/MS analysis (Pérez-Ferández, Gentili, Martinelli, Caretti, & Curini, 2016). Conversely, Cbl immunoaffinity columns [EASI-EXTRACT vitamin B_{12} Immunoaffinity Column (RBRP 80 and 88), R-Biopharm AG, Darmstadt, Germany] were used to purify Cbl and its related compounds specifically from KCN-treated Cbl extracts before HPLC and LC–MS/MS analyses (Tanioka et al., 2014). Recoveries of Cbl and

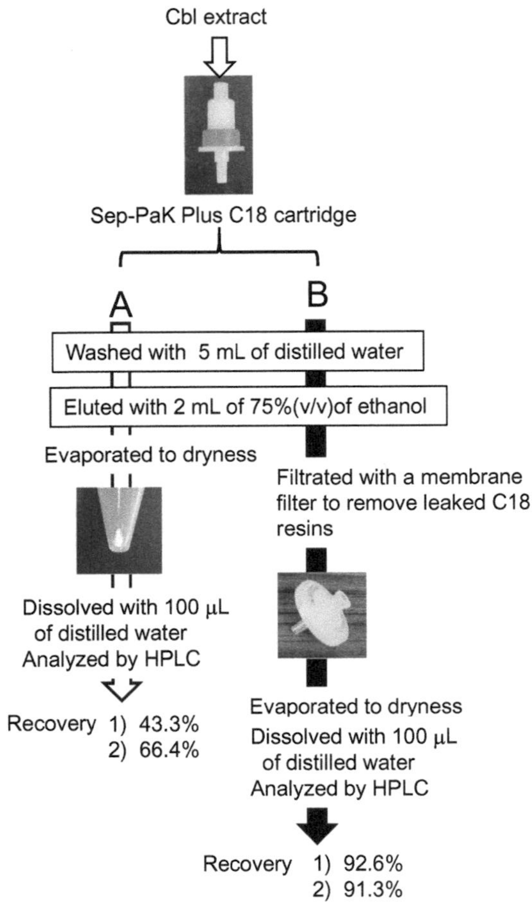

Fig. 2 Recovery of cobalamin by the treatment with Sep-Pak Plus C18 cartridge. Cbl was extracted from a CN-Cbl standard solution or a certified reference material [SRM1869, Infant/Adult Nutritional Formula II (milk/whey/soy-based)] by the KCN-boiling method. Each extract (5 mL) that contains 80 ng of B_{12} is treated with a Sep-Pak Plus C18 cartridge (Water Corp., Milford, MA, USA) that had been activated with 5 mL of ethanol and equilibrated with 5 mL of distilled water. The C18 cartridge was washed with 5 mL of distilled water, and the Cbl compounds were eluted with 2 mL of 75% (v/v) ethanol. (A) Eluate was evaporated to dryness under reduced pressure, dissolved in 100 μL of distilled water, and then determined by a C18 reversed phase HPLC. (B) Eluate was filtrated with a disposable membrane filter unit DISMIC®-25AS (Toyo Roshi Kaisha, Ltd., Tokyo, Japan), washed with 1 mL of 75% (v/v) ethanol, evaporated to dryness under reduced pressure, dissolved in 100 μL of distilled water and then determined by a C18 reversed-phase HPLC. (1) CN-Cbl standard and (2) SRM1869.

pseudovitamin B_{12} (adenyl cyanocobamide or factor IV, AdeCba) replaced with adenine as the base of the lower ligand (Fig. 1) were reported to be approximately 93% and 90%, respectively, during the immunoaffinity column chromatography.

3.3 Microbiological assay

The microbiological quantification of Cbl by *Lactobacillus delbrueckii* subsp. *lactis* (formerly *Lactobacillus leichmannii*) ATCC7830 is widely used in food Cbl analysis because good calibration curves are given in the range of about 10–80 pg (Bito et al., 2016). However, deoxyriboside or deoxynucleotide (known as an alkali resistance factor) found in foods has a growth-promoting effect equivalent to Cbl to this bacterium (Cook & Ellis, 1968). Therefore, Cbl values may be overestimated in food containing rich deoxyribosides or deoxynucleotides because this bacterium substitutes these compounds for Cbl (Watanabe et al., 2012). The AOAC microbiological assay method (AOAC 960.461) does not adopt the correction for the alkali-resistant factor (Angyal, 1996).

Also, AdeCba does not show physiological activity in humans, but it has Cbl activity in this bacterium. The microbiological quantification method cannot separately quantify Cbl and AdeCba.

3.4 High-performance liquid chromatography (HPLC)

Among the HPLC methods published as food analysis methods, AOAC official method 2014.02 (AOAC Official Method 2014.02, 2014) meets various parameters of AOAC SMPR® 2016.017 (van Breemen et al., 2017). According to this method, Cbl is specifically purified from food extracts using the Cbl immunoaffinity column (RBRP 88) described above and then determined by a C18 reversed-phase HPLC or ultra HPLC. Cbl was detected at 361 nm, which is commonly used to detect Cbl.

Contrastingly, the AOAC2011.10/ISO 20634 method (AOAC Official Method 2011.10, 2011) that does not use the immunoaffinity column for sample purification is reported. Cbl is partially purified by gel filtration HPLC column and then quantified by a C18 reverse-phase HPLC. In this method, the detection wavelength of Cbl used is 550 nm. The effect of interference compounds on the detection of Cbl during HPLC is lower at 550 nm than at 361 nm, but the sensitivity of Cbl is lower at 550 nm (molecular extinction coefficient: $8730\,cm^{-1}\,M^{-1}$) than at 361 nm (molecular extinction coefficient: $27499\,cm^{-1}\,M^{-1}$).

In the method that we have used for Cbl assay, Cbl compounds, after they have been purified using both Sep-Pak C18 Plus cartridge and Cbl immunoaffinity column (RBRP 80), are determined by a C18 reverse-phase HPLC (Okamoto et al., 2020). Although Cbl compounds are isocratically eluted with a 20% (v/v) methanol solution containing 1% (v/v) acetic acid in this method, the linear-gradient elution of Cbl is adopted in the AOAC 2014.02 and AOAC 2011.10/ISO 20634 methods. Isocratically elution of Cbl compounds by HPLC showed that Cbl and its related compounds such as AdeCba could be completely separated and determined.

3.5 Liquid chromatography/electrospray ionization-tandem mass spectrometry (LC-MS/MS)

LC–MS/MS or LC/MS is developed to determine biologically active Cbl in milk (Zironi et al., 2013), cheese (Zironi et al., 2013), and fermented cereal (Chamlagain, Edelmann, Kariluoto, Ollilainen, & Piironen, 2015). In these methods, Oasis® HLB (Lee et al., 2015) or the immunoaffinity column (Chamlagain et al., 2015) were used to purify Cbl compounds from food extracts before analysis. Moreover, various inactive corrinoid compounds for humans were identified in foods using LC–MS/MS (Bito et al., 2014).

3.6 Others

Although other methods such as capillary electrophoresis (Chen & Jiang, 2008), atomic absorption spectrometry (Vinas, Campillo, Lopez-Garcia, & Hernandez-Cordoba, 1996), chemiluminescence method (Lweno et al., 2020), immunoassays (Hampel et al., 2014), and a biosensor based on surface plasmon resonance (Gao et al., 2008) are reported in the determination of Cbl in foods, these methods are not widely used in food Cbl analysis.

4. Inactive cobalamins and related compounds in foods

Humans have a complex intestinal absorption system of dietary Cbl (Green, 2017). A gastric Cbl-transport protein, the intrinsic factor (IF), plays a vital role in the intestinal absorption of Cbl because the IF has the most strict-binding specificity of Cbl molecule (Fedosov, Fedosova, Kräutler, Nexø, & Petersen, 2007). The IF–Cbl complex formed is internalized into the distal ileum's mucosal cells via receptor-mediated endocytosis (Green, 2017). Thus, humans can select biologically active Cbl among various corrinoids found in food using the IF-mediated intestinal absorption system (Alpers & Russell-Jones, 2013). Therefore, corrinoid compounds that show no or very low affinity to human IF would become inactive corrinoid compounds. From our studies so far, inactive corrinoid compounds found in food would be classified into two categories; (1) naturally occurring cobamides and (2) Cbl compounds with chemically modified side chains of the corrin ring.

4.1 Naturally occurring cobamides with different base moiety in the lower ligand

AdeCba is the dominant corrinoid compound synthesized by certain bacteria (Hoffmann et al., 2000) (Fig. 1) and found in human feces (Allen & Stabler, 2008). AdeCba was identified as a major corrinoid in *Spirulina* tablets used as human dietary supplements (Watanabe et al., 1999). Other edible cyanobacteria also contain substantial amounts of AdeCba (Teng, Bito, Takenaka, Takenaka, et al., 2014; Teng, Bito, Takenaka, Yabuta, & Watanabe, 2014). (Cyanobacteria can synthesize AdeCba *de novo* (Tanioka et al., 2010)). Shellfish is a primary dietary source of Cbl in humans (Yoshino et al., 2005). However, certain types of edible shellfish (herbivorous sea snails such as abalone and turban shell) contain substantial amounts of AdeCba, specifically accumulated in their viscera (Teng, Tanioka, Hamaguchi, et al., 2015). Edible insects are considered to have the potential to contribute to global food security, but only a limited number of edible insect products are commercially available (Nowak, Persijn, Rittenschober, & Charrondiere, 2016). Commercially available edible insect products contained a substantial amount of AdeCba (Okamoto et al., 2021).

2-Methylmercaptoadenylcobamide, named Factor S, was reportedly found in human feces (Allen & Stabler, 2008). Factor S was detected in commercially available edible insect products (Okamoto et al., 2021) and apple snail (escargot) products (Teng, Tanioka, Bito, et al., 2015). No other study reported this type of cobamide in food.

5-Methoxybenzimidazolylcobamide is known as Factor IIIm. A considerable amount of Factor IIIm was found in escargot products (Teng, Tanioka, Bito, et al., 2015). Some kinds, not all, of *Chlorella* products used as human supplements contained a substantial amount of this type of cobamide (Bito et al., 2016). However, the occurrence of Factor IIIm has not been reported in other foods. These results indicated that these naturally occurring cobamides did not exist in primary dietary B_{12} sources such as meat, milk, and fish.

AdeCba hardly binds to IF because of very low affinity (Stupperich & Nexo, 1991). Although there is no information available on Factor S binding affinities to human IF, 2-methyladeninylcobamide (Factor A), having a similar structure as Factor S, reportedly showed very low affinity to IF (Fedosov et al., 2007). The result suggested that Factor S hardly binds to IF. However, IF reportedly showed a similar affinity to B_{12} for 5-hydroxybenzimidasolylcobamide (Factor III) (Stupperich & Nexo, 1991), which has a similar structure as Factor IIIm. The result suggested that Factor IIIm can bind to IF, suggesting that Factor IIIm might not be an inactive corrinoid compound judging from the binding ability to IF.

4.2 Cobalamins with chemically modified peripheral side chains of the corrin ring

An unnatural corrinoid compound, Cbl [*c*-lactone], was found in certain dried mushroom fruiting bodies but not all (Teng, Bito, Takenaka, Yabuta, et al., 2014). Cbl-*d*-monocarboxylic acid (25%) and Cbl dicarboxylic acids (41%) were found in shrimp innards and their products (Okamoto et al., 2020). These B_{12} compounds with modified side chains of the corrin ring are not reported in other foods. These chemically modified Cbl compounds reportedly showed very low affinities to IF (Stabler, Brass, Marcell, & Allen, 1991).

5. Conclusions and future directions

This chapter has discussed the microbiological quantification of total Cbl by *L. delbrueckii* subsp. *lactis* ATCC7830, noting its advantages and limitations, and using more robust techniques like HPLC or LC–MS/MS, which can separately determine Cbl and various inactive corrinoid compounds. We propose that the microbiological method must be replaced with HPLC or LC–MS/MS assays. Knowing the exact amounts of biologically active Cbl in foods is essential to intake an appropriate amount of Cbl for preventing Cbl deficiency. Moreover, it is important to identify foods containing inactive corrinoids because the intake of large amount of inactive corrinoids appears to induce Cbl deficiency (Bito, Bito, et al., 2020; Bito, Teng, & Watanabe, 2020).

Acknowledgments

All authors contributed equally to the preparation of this manuscript and have approved the final version.

The authors declare that there are no competing financial interests.

References

Allen, R. H., & Stabler, S. P. (2008). Identification and quantitation of cobalamin and cobalamin analogues in human feces. *The American Journal of Clinical Nutrition*, *87*(5), 1324–1335. https://doi.org/10.1093/ajcn/87.5.1324.

Alpers, D. H., & Russell-Jones, G. (2013). Gastric intrinsic factor: The gastric and small intestinal stages of cobalamin absorption. A personal journey. *Biochimie*, *95*(5), 989–994. https://doi.org/10.1016/j.biochi.2012.12.006.

Andrès, E., Afenberger, S., Vinzio, S., Kurtz, J. E., Noel, E., Kaltenbach, G., et al. (2005). Food-cobalamin malabsorption in elderly patients: Clinical manifestations and treatment. *The American Journal of Medicine*, *118*(10), 1154–1159. https://doi.org/10.1016/j.amjmed.2005.02.026.

Angyal, G. (1996). *Methods for the microbiological analysis of selected nutrients* (pp. 63–65). Massachusetts: AOAC International.

AOAC Official Method 2011.10. (2011). *Vitamin B_{12} in infant formula and adult nutritionals*.

AOAC Official Method 2014.02. (2014). *Vitamin B_{12} in infant formula and adult/pediatric formulas*.

Bito, T., Bito, M., Asai, Y., Takenaka, S., Yabuta, Y., Tago, K., et al. (2016). Characterization and quantitation of vitamin B_{12} compounds in various *Chlorella* supplements. *Journal of Agricultural and Food Chemistry*, *64*(45), 8516–8524. https://doi.org/10.1021/acs.jafc.6b03550.

Bito, T., Bito, M., Hirooka, T., Okamoto, N., Harada, N., Yamaji, R., Nakano, Y., Inui, H., & Watanabe, F. (2020) Biological activity of pseudovitamin B_{12} on cobalamin-dependent methylmalonyl-CoA mutase and methionine synthase in mammalian cultured COS-7 cells. Molecules, 25(14), 3268. https://doi.org/https://doi.org/10.3390/molecules25143268.

Bito, T., Teng, F., Ohishi, N., Takenaka, S., Miyamoto, E., Sakuno, E., et al. (2014). Characterization of vitamin B_{12} compounds in the fruiting bodies of shiitake mushroom (*Lentinula edodes*) and bed logs after fruiting of the mushroom. *Mycoscience*, *55*(6), 464–468. https://doi.org/10.1016/j.myc.2014.01.008.

Bito, T., Teng, F., & Watanabe, F. (2020). Biologically active vitamin B_{12} from edible seaweeds. In G. A. Ravishankar, & A. R. Rao (Eds.), *Vol. 1. Handbook of algal technologies and phytochemicals* (pp. 105–110). New York: CRC Press.

Chamlagain, B., Edelmann, M., Kariluoto, S., Ollilainen, V., & Piironen, V. (2015). Ultra-high performance liquid chromatographic and mass spectrometric analysis of active vitamin B_{12} in cells of *Propionibacterium* and fermented cereal matrices. *Food Chemistry*, *166*, 630–638. https://doi.org/10.1016/j.foodchem.2014.06.068.

Chen, Z., Crippen, K., Gulati, S., & Banerjee, A. (1994). Purification and kinetic mechanism of a mammalian methionine synthase from pig liver. *Journal of Biological Chemistry*, *269*(44), 27193–27197. https://doi.org/10.1016/S0021-9258(18)46968-2.

Chen, J. H., & Jiang, S. J. (2008). Determination of cobalamin in nutritive supplements and Chlorella foods by capillary electrophoresis-Inductively coupled plasma mass spectrometry. *Journal of Agricultural and Food Chemistry*, *56*(4), 1210–1215. https://doi.org/10.1021/jf073213h.

Cook, E. A., & Ellis, L. N. (1968). Variations in the growth response of four different vitamin B_{12} assay microorganisms to the same tissue and standard preparations. *Applied and Environmental Microbiology*, *16*(12), 1831–1840. 4881959. PMC547780.

Croft, M. T., Lawrence, A. D., Raux-Deery, E., Warren, M. J., & Smith, A. G. (2005). Algae acquire vitamin B_{12} through a symbiotic relationship with bacteria. *Nature*, *438*(7064), 90–93. https://doi.org/10.1038/nature04056.

Fedosov, S. N., Fedosova, N. U., Kräutler, B., Nexø, E., & Petersen, T. E. (2007). Mechanisms of discrimination between cobalamins and their natural analogues during their binding to the specific B_{12}-transporting proteins. *Biochemistry*, *46*(21), 6446–6458. https://doi.org/10.1021/bi062063l.

Fedosov, S. N., Nexo, E., & Heegaard, C. W. (2019). Vitamin B_{12} and its binding proteins in milk from cow and buffalo in relation to bioavailability of B_{12}. *Journal of Dairy Science, 102*(6), 4891–4905. https://doi.org/10.3168/jds.2018-15016.

Fenton, W. A., Hack, A. M., Willard, H. F., Gertler, A., & Rosenberg, L. E. (1982). Purification and properties of methylmalonyl coenzyme A mutase from human liver. *Archives of Biochemistry and Biophysics, 214*(2), 815–823. https://doi.org/10.1016/0003-9861(82)90088-1.

Gao, Y., Guo, F., Gokavi, S., Chow, A., Sheng, Q., & Guo, M. (2008). Quantification of water-soluble vitamins in milk based infant formula using biosensor-based assays. *Food Chemistry, 110*(3), 769–777. https://doi.org/10.1016/j.foodchem.2008.03.007.

Green, R. (2017). Vitamin B_{12} deficiency from the perspective of a practicing hematologist. *Blood, 129*(19), 2603–2611. https://doi.org/10.1182/blood-2016-10-569186.

Guggisberg, D., Risse, M. C., & Hadorn, R. (2012). Determination of vitamin B_{12} in meat products by RP-HPLC after enrichment and purification on an immunoaffinity column. *Meat Science, 90*(2), 279–283. https://doi.org/10.1016/j.meatsci.2011.07.009.

Hampel, D., Shahab-Ferdows, S., Domek, J. M., Siddiqua, T., Raqib, R., & Allen, L. H. (2014). Competitive chemiluminescent enzyme immunoassay for vitamin B_{12} analysis in human milk. *Food Chemistry, 153*, 60–65. https://doi.org/10.1016/j.foodchem.2013.12.033.

Heal, K. R., Qin, W., Ribalet, F., Bertagnolli, A. D., Coyote-Maestas, W., Hmelo, L. R., et al. (2017). Two distinct pools of B_{12} analogs reveal community interdependencies in the ocean. *Proceedings of the National Academy of Sciences of the United States of America, 114*(2), 364–369. https://doi.org/10.1073/pnas.1608462114.

Hoffmann, B., Oberhuber, M., Stupperich, E., Bothe, H., Buckel, W., Konrat, R., et al. (2000). Native corrinoids from *Clostridium cochlearium* are adeninylcobamides: Spectroscopic analysis and identification of pseudovitamin B_{12} and factor A. *Journal of Bacteriology, 182*(17). https://doi.org/10.1128/JB.182.17.4773-4782.2000.

Kawashima, T., Henry, P. R., Ammerman, C. B., & Littell, R. C. (1997). Bioavailability of cobalt sources for ruminants. 2. Estimation of the relative value of reagent grade and feed grade cobalt sources from tissue cobalt accumulation and vitamin B_{12} concentrations. *Nutrition Research, 17*(6), 957–974. https://doi.org/10.1016/S0271-5317(97)00061-4.

Lamm, L., Höring, J. A., Renz, P., & Heckmann, G. (1980). Biosynthesis of vitamin B_{12}. Experiments with the anaerobe *Eubacterium limosum* and some labelled substrates. *European Journal of Biochemistry, 109*(1), 115–118. https://doi.org/10.1111/j.1432-1033.1980.tb04775.x.

Lee, J. H., Shin, J. H., Park, J. M., Kim, H. J., Ahn, J. H., Kwak, B. M., et al. (2015). Analytical determination of vitamin B_{12} content in infant and toddler milk formulas by liquid chromatography tandem mass spectrometry (LC-MS/MS). *Korean Journal for Food Science of Animal Resources, 35*(6), 765–771. https://doi.org/10.5851/kosfa.2015.35.6.765.

Lweno, O. N., Sudfeld, C. R., Hertzmark, E., Manji, K. P., Aboud, S., Noor, R. A., et al. (2020). Vitamin B_{12} is low in milk of early postpartum women in Urban Tanzania, and was not significantly increased by high dose supplementation. *Nutrients, 12*(4), 963. https://doi.org/10.3390/nu12040963.

Muhammad, K., Briggs, D., & Jones, G. (1993). The appropriateness of using cyanocobalamin as calibration standards in *Lactobacillus leichmannii* A.T.C.C. 7830 assay of vitamin B_{12}. *Food Chemistry, 48*(4), 427–429. https://doi.org/10.1016/0308-8146(93)90330-I.

Nowak, V., Persijn, D., Rittenschober, D., & Charrondiere, U. R. (2016). Review of food composition data for edible insects. *Food Chemistry, 193*, 39–46. https://doi.org/10.1016/j.foodchem.2014.10.114.

Okamoto, N., Hamaguchi, H., Umebayashi, Y., Takenaka, S., Bito, T., & Watanabe, F. (2020). Determination and characterization of vitamin B_{12} in the muscles and head innards of edible shrimp. *Fisheries Science, 86*(2), 395–406. https://doi.org/10.1007/s12562-019-01397-x.

Okamoto, N., Nagao, F., Umebayashi, Y., Bito, T., Prangthip, P., & Watanabe, F. (2021). Pseudovitamin B_{12} and factor S are the predominant corrinoid compounds in edible cricket products. *Food Chemistry, 347*, 129048. https://doi.org/10.1016/j.foodchem.2021.129048.

Pérez-Ferάndez, V., Gentili, A., Martinelli, A., Caretti, F., & Curini, R. (2016). Evaluation of oxidized buckypaper as material for the solid phase extraction of cobalamins from milk: Its efficacy as individual and support sorbent of a hydrophilic-lipophilic balance copolymer. *Journal of Chromatography A, 1428*, 255–266. https://doi.org/10.1016/j.chroma.2015.07.109.

Santos, F., Vera, J. L., Lamosa, P., de Valdez, G. F., de Vos, W. M., Santos, H., et al. (2007). Pseudovitamin B_{12} is the corrinoid produced by *Lactobacillus reuteri* CRL1098 under anaerobic conditions. *FEBS Letters, 581*(25), 4865–4870. https://doi.org/10.1016/j.febslet.2007.09.012.

Shibata, K., Fukuwatari, T., Imai, E., Hayakawa, T., Watanabe, F., Takimoto, H., et al. (2013). Dietary reference intakes for Japanese 2010: Water-soluble vitamins. *Journal of Nutritional Science and Vitaminology, 59*, 67–82. https://doi.org/10.3177/jnsv.59.S67.

Stabler, S. P., Brass, E. P., Marcell, P. D., & Allen, R. H. (1991). Inhibition of cobalamin-dependent enzymes by cobalamin analogues in rats. *The Journal of Clinical Investigation, 87*(4), 1422–1430. https://doi.org/10.1172/JCI115148.

Stupperich, E., & Nexo, E. (1991). Effect of the cobalt-N coordination on the cobamide recognition by the human vitamin B_{12} binding proteins intrinsic factor, transcobalamin and haptocorrin. *European Journal of Biochemistry, 199*(2), 299–303. https://doi.org/10.1111/j.1432-1033.1991.tb16124.x.

Tanioka, Y., Miyamoto, E., Yabuta, Y., Ohnishi, K., Fujita, T., Yamaji, R., et al. (2010). Methyladeninylcobamide functions as the cofactor of methionine synthase in a Cyanobacterium, *Spirulina platensis* NIES-39. *FEBS Letters, 584*(14), 3223–3226. https://doi.org/10.1016/j.febslet.2010.06.013.

Tanioka, Y., Takenaka, S., Furusho, T., Yabuta, Y., Nakano, Y., & Watanabe, F. (2014). Identification of vitamin B_{12} and pseudovitamin B_{12} from various edible shellfish using liquid chromatography-electrospray ionization/tandem mass spectrometry. *Fisheries Science, 80*(5), 1065–1071. https://doi.org/10.1007/s12562-014-0787-x.

Teng, F., Bito, T., Takenaka, S., Takenaka, H., Yamaguchi, Y., Yabuta, Y., et al. (2014). Characterization of corrinoid compounds in the edible cyanobacterium *Nostoc flagelliforme* the hair vegetable. *Food and Nutrition Sciences, 5*(4), 334–340. https://doi.org/10.4236/fns.2014.54040.

Teng, F., Bito, T., Takenaka, S., Yabuta, Y., & Watanabe, F. (2014). Vitamin B_{12}[c-lactone], a biologically inactive corrinoid compound, occurs in cultured and dried lion's mane mushroom (*Hericium erinaceus*) fruiting bodies. *Journal of Agricultural and Food Chemistry, 62*(7), 1726–1732. https://doi.org/10.1021/jf404463v.

Teng, F., Tanioka, Y., Bito, T., Takenaka, S., Yabuta, Y., & Watanabe, F. (2015). Occurrence of biologically inactive corrinoid compounds in canned edible apple snails (Escargots). *Food and Nutrition Sciences, 6*(12), 1071–1077. https://doi.org/10.4236/fns.2015.612111.

Teng, F., Tanioka, Y., Hamaguchi, N., Bito, T., Takenaka, S., Yabuta, Y., et al. (2015). Determination and characterization of vitamin B_{12} compounds in edible sea snails, ivory shell *Babylonia japonica* and turban shell *Turdo Batillus cornutus*. *Fisheries Science, 81*(6), 1105–1111. https://doi.org/10.1007/s12562-015-0920-5.

Toohey, J. I., & Barker, H. A. (1961). Isolation of coenzyme B_{12} from liver. *Journal of Biological Chemistry, 236*(2), 560–563. https://doi.org/10.1016/S0021-9258(18)64403-5.

van Breemen, R., Betz, J., Evans, L., Hall, A., Jennens, M., Kreider, R., et al. (2017). AOAC SMPR 2016.017. Standard method performance requirements (SMPRs) for quantitative measurement of vitamin B_{12} in dietary supplements and ingredients. *Journal of AOAC International, 100*(1), 294–296. https://doi.org/10.5740/jaoacint.SMPR2016.017.

Vinas, P., Campillo, N., Lopez-Garcia, I., & Hernandez-Cordoba, M. (1996). Identification of vitamin B_{12} analogues by liquid chromatography with electrothermal atomic absorption detection. *Chromatographia, 42*(9–10), 566–570.

Watanabe, F., & Bito, T. (2016). Corrinoids in food and biological samples. In F. R. S. Atta-ur-Rahman (Ed.), *Frontiers in natural product chemistry* (pp. 229–244). Sharjah: Bentham Science Publishers.

Watanabe, F., & Bito, T. (2018a). Vitamin B_{12} sources and microbial interaction. *Experimental Biology and Medicine, 243*(2), 148–158. https://doi.org/10.1177/1535370217746612.

Watanabe, F., & Bito, T. (2018b). Determination of cobalamin and related compounds in foods. *Journal of AOAC International, 101*(5), 148–158. https://doi.org/10.5740/jaoacint.18-0045.

Watanabe, F., Katsura, H., Takenaka, S., Fujita, T., Abe, K., Tamura, Y., et al. (1999). Pseudovitamin B_{12} is the predominant cobamide of an algal health food, *Spirulina* tablets. *Journal of Agricultural and Food Chemistry, 47*(11), 4736–4741. https://doi.org/10.1021/jf990541b.

Watanabe, F., Schwartz, J., Takenaka, S., Miyamto, E., Ohishi, N., Nelle, E., et al. (2012). Characterization of vitamin B_{12} compounds in the wild edible mushrooms black trumpet (*Graterellus cornucopioides*) and golden chanterelle (*Gantharellus cibarius*). *Journal of Nutritional Science and Vitaminology, 58*(6), 438–441. https://doi.org/10.3177/jnsv.58.438.

Wongyai, S. (2000). Determination of vitamin B_{12} in multivitamin tablets by multimode high-performance liquid chromatography. *Journal of Chromatography A, 870*(1–2), 217–220. https://doi.org/10.1016/S0021-9673(99)01200-5.

Yoshino, K., Inagawa, M., Oshima, M., Yokota, K., Umesawa, M., Endo, M., et al. (2005). Trends in dietary intake of folate, vitamin B_6 and vitamin B_{12} among Japanese adults in two rural communities from 1971 through 2001. *Journal of Epidemiology, 15*(2), 29–37. https://doi.org/10.2188/jea.15.29.

Zironi, E., Gazzotti, T., Barbarossa, A., Devicenti, C., Scardilli, M., & Pagliuca, G. (2013). Technical note: Development and validation of a method using ultra performance liquid chromatography coupled with tandem mass spectrometry for determination of vitamin B_{12} concentrations in milk and dairy products. *Journal of Dairy Science, 96*(5), 2832–2836. https://doi.org/10.3168/jds.2012-6451.

FSC
www.fsc.org
MIX
Papier | Fördert gute Waldnutzung
FSC® C083411

Kolibri 360 GmbH
Ferdinand-Jühlke-Straße 7
99095 Erfurt, Deutschland
produktsicherheit@kolibri360.de